the Metaphysicians' Desk Reference

including the revised Formal System of Metaphysics

By

Jon Gee

ISBN: 1-4107-1908-1 (e-book)
ISBN: 1-4107-1909-X (Paperback)

Library of Congress Control Number: 2003091068

This book is printed on acid free paper.

Printed in the United States of America
Bloomington, IN

1stBooks - rev. 10/31/03

Here is wisdom (or folly)

Let you who have knowledge of geometry enter here

Let you who have understanding calculate

Table of Contents:

on metaphysics and being a metaphysician

Despite many qualified scholars claims to the contrary there is no contextual history for metaphysics. While they casually claim that any denial of their world view is a corruption, they willfully overlook the fact that their world view, no matter how popular, is equally only a distortion of existing doctrinal evidence by dint of interpretation.

The definition given by Webster's new world dictionary for the word metaphysics conveys the crux of this dilemma most concisely: 1. the branch of philosophy that deals with first principles and seeks to explain the nature of being or reality (ontology) and of the origin and structure of the universe (cosmology); it is closely associated with the study of the nature of knowledge (epistemology) 2. speculative philosophy in general 3. esoteric, often mystical or theosophical, lore 4. the theory of principles (or some branch of knowledge) 5. popularly, any very subtle or difficult reasoning.

Thus, it is taken gratis that metaphysics is philosophy, and, moreover, speculative. Therefore the set of research to which metaphysics may be given access for reference is confined dually to the studies of philosophy, a non-science, and speculation, a non-methodology. In particular, the most broad definition (1) has little room for philosophical studies other than those given (ontology — the study of being; cosmology — theoretical conceptions of universal origins; and epistemology — the study of letters and learning) though it does offer a broader, more liberal definition of these than would be found to describe them elsewhere (explaining ontology as inclusive of the study of reality, cosmology as the continuing structure of the universe, and epistemology, most liberally of all, as the study of knowledge in itself). Definition 4 is the most vague, while definitions 2 and 5 the most well known and accepted. Definition 3 is probably the most accurate, but still a slander by affiliation with theosophy (a non-religion).

Therefore, while what I have addressed in this work does, almost entirely accidentally, fall well within this broad and vague definition, so would anything else that is considered scientifically marginal or scholastically disagreeable.

The reason for this is quite simple. For as long as metaphysics has existed, anyone who specializes in it has been dubbed a metaphysician. There are no duties or goals in being a metaphysician in the standard sense, so, something like a glacier picking up and moving mountains along its way, it has accumulated the detritus of all the schools of thought it has brushed upon in passing. Most of these are philosophical, dealing with sophia, the study of wisdom. Some are religious, though, as they are not religious in the sense that they are accepted by orthodoxy, they are shunned as cults such as theosophy, dealing with the wisdom of God. Now this is an altogether bogus sham, simply because, according to these schools and cults, there is no subject of which they study; there is no such thing as wisdom.

From its earliest conception in its current existing form, wisdom has been shunned and frowned upon as an act of proof that they posses it by those very same teachers who claim to be or who are accused of dabbling in it. Socrates' dictum, that true wisdom is knowing you know nothing, translates equally well into the modern terminology of a computer programmer trying to program artificial intelligence into a computer and give it a consciousness similar to our own when s/he inputs conundrums as axioms such as "the following statement is true; the preceding statement is false." This is known, roughly around the edges, as the problem of wisdom.

In QBLH, the Hebrew people have dealt as thoroughly with wisdom as did the Buddha in the Dhammapada. Both, as would Socrates and subsequent schools of philosophy, separate wisdom from knowledge. They accept that there can be such a thing as knowledge, but that it cannot, or at least, does not, exist in the same realm as such a thing as wisdom. The explanation usually given for this is that wisdom is an ideal thing, and knowledge a real thing, and that, rhetorically, the ideal sits atop the real like oil on water, never the two shall mix. But this is merely allegory. What is worse, whenever anybody since these ancient beliefs has spoken of wisdom, or spoken in wisdom, whatever that means has been lost on the ears of those who do not posses understanding of such allegories. All discussion on the matter has become clothed and cloaked in allegory, to the point of absurdity, where it would seem that true wisdom means ultimate ignorance, as everything about it is possessed by its exact opposite meaning.

This is a criminal shame. Long has it been bemoaned that people are stupid, petty creatures, easily distracted and dismayed; that they fear the new, and that all of their actions of hatred are based on the recoil of this fear; that they fear change, and would lie down on the tracks of a train before getting on board. None of this is true. In the name of wisdom, what is true, is that people pretend to be ignorant. They hide what little they know to feel secretly dominant and superior, waiting for their loved ones to prove their codependent need for them by guessing what is on their minds. This is what is called wisdom: this sabotage of the wheels of progress for the purpose of passive aggressive, in the closet, pedantry. Wisdom has come to mean nothing more than that quote variously attributed to Voltaire and Twain, that "it is better to keep one's mouth shut and appear the fool than to open one's mouth and remove all shadow of a doubt." This is wisdom, then: simply knowledge playing opossum. In the name of this false truth has followed all the battles between what ought to be cooperative schools of scholasticism, and the literary necessity of never saying anything that no one has ever heard before unless it is a quote by someone everybody knows and loves: the abnegation of original thought as being imaginative creation in favor of clinical redundancy of data for fear of an inquisition.

So if this is the mysterium tremendum of philosophy, and wisdom the acidic anticoagulants of that womb from which metaphysics was born, then what can we expect of metaphysics; what are the guidelines for duty of a metaphysician? As defined above, true metaphysics left the home of philosophy long ago, and has passed through various schools of thought and religious cults, gaining in illumination from each. At one point in the late middle ages and early Renaissance a school called the Rosicrucians set down one rule they had observed was in practical usage among metaphysicians since the time of the gnostic Therapeutae: "hele." And to this I would add but one phrase more, before one goes out to set the world aright: "metaphysician, hele thyself." From this all else, in good course, will follow.

introduction to the first edition

This book you are holding is a Great Work. It is so not just in the sense that it is the completion of an exertion of effort in toil, for such as is for the lowly as well as for the lofty, and this book has little to offer by way of the lesser recourse. This is a work in the alchemical sense, and this has a very specific meaning. Since the days of Flammel the exact meaning of alchemy has become lost, and it is commonly misinterpreted in its modern senses. It is held by current scholars of alchemy that it is psychological in its basis, though this is sacrosanct beyond the fact that alchemy preceded the study of psychology as such. So we may say, by this deformed route, that this book is an alchemical action upon the substance of the psyche, and that, thus, the accomplishment of its Great Work is the subject of Self Transformation. This is and is not altogether true. For, by way of this circuity, such as self transformation may very well be carried out through the right understanding of this book. That is to say that, the unenlightened reader who passes through these pages and gains from doing so an understanding of their contents will emerge upon their end illuminated as to their implications, not only for their fields of reference, but for the individual's life itself.

But this Great Work is alchemical in more than merely a psychological sense, since, with as much good intention, the same as can be said regarding the process of reading any other book, from the Cat in the Hat to the phone book, and of the act of reading as much as the act of consuming, since the mind's eye serves much the same function towards information as does the mouth to food. So, in exactly what sense is it meant to say that this book is a Great Work of alchemy?

Let us look further back, to earlier theories about the nature of the alchemical art. Originally it was a part and party to the craft of magic. However, as much difference as there is between Flammel and Jung there is between Merlin and Houdini, because, as the craft of magic has become more practically oriented, so has the art of alchemy become more speculative. Originally these two schools were quite the other way around, and were, in fact, the very inversion of what they are today. Whereas, during the European dark ages, it was alchemy that was involved with practical matters such as dealt with metallurgy, and magic with the esoteric and arcane studies such as astrology, now it is magic that is held to be the science of picking locks, and alchemy associated seemingly rhetorically with psychology.

Originally, it was alchemy that was the craft, and magic the art. So a Great Work of alchemy, which meant, then, the refinement of the philosopher's stone, was considered to be the practical application of the esoteric philosophies of magic. Thus, the Great Work of "self transformation" entailed little more than the right proper study and application of certain aspects of the arcane wisdom of scientific mysteries to produce some sort of concrete sediment for the furtherance of meditation. And it is, most definitely, in this sense that this book is a Great Work.

For it is the completion of such a process, and in itself its own form of philosopher's stone, offering a form of eternal life in the infinite depth of its contemplation.

By way of introduction to this Great Work's most flawed first edition, it would cushion the impact of its reception most to offer up sections regarding its authorship, literary style, scientific methodology, and finally, in brief, its findings and substantive material. This should allow the reader trepidatious of such heady gravitas as is generally implied by the terminology of its overtly grandiose title to accept on general and self assuring grounds the down to earth and lightheartedness of the matters herein discussed. While some nowadays enter too easily into self transformation, the vast majority swing pendulously opposed to it, and feel threatened in their traditional mindsets by the very thought of change. Since it is not my intention to incur or pander to either of these mindsets, but simply to put forward a set of data for the review of a class of peers to be determined entirely by those drawn to an interest in the subject matter, I am going to set forward only so much by way of introduction as will seem fit as I am writing it, and that only to address such pitfalls and shortcomings of this first edition of which seem necessary to forewarn the cautious. In all cases I will endeavor to be, after my own fashion, brief and direct.

introduction to the author

Since one of the first and foremost pitfalls of any author writing their own introduction would be vanity, I should endeavor as regards myself to separate that aspect of my personality from my pride in the accomplishment of this written word. I do not wish for any effect of this writing to befall me, whether it be praise and adoration or shame and mockery; rather I am content remaining altogether autonomous from all the questions raised by any and every aspect of this book. Let me be the first to admit that for every piece of probable fact herein raised there is a distinct lack of empirical evidence, and that for every seeming solution herein presented arise a wealth and welfare of further inquests and inquiries that would glut even the most technically proficient and/or imaginative minds. I do not have all the answers, either. It was only my goal, in writing this, to advance the debate of such theories as are herein addressed, opening the possible channels of communication between otherwise closed minded and competitive schools and branches of science and study.

Personally I do not fall into the fields of any of the areas of investigation herein addressed by way of profession, or even, in all truth, remote interest. On the contrary I felt compelled out of inability to comport myself as would do justice to my standing and my greater society without being driven to set down some of my casual observations regarding the fields on which this data touch. In truth, and not either to make jest or mockery, it has been my personal life experience that any and all who would seek for higher answers than those given as axioms, postulates and theorems will be highly frowned upon and generally shunned for wont of a more faithful humility than befits the lay person, or even the average specialist.

This motive of specialisation has gone largely overlooked in the scholastic society. Due to specialisation, however, no scholar may contradict another without first having already proved their case after the same manner as the initial scholar. Imagine if the forensic pathologist could not offer new evidence to the detective without first having walked a beat as a street officer. Yet this is exactly the expectation of those in the lofty offices of scholasticism, and it goes, insofar as it is ignored, unrecognised or unaddressed, wholly without saying whether it is right or wrong. The result is a stultification of ideologies such that, rather than recognizing all studies branch out from the same root, differing branches are made to compete, and someone without the same resources as their fellow comrade with and by which to test their theory along the guidelines of the sacred, set in stone scientific method is essentially made to feel it necessary to nonetheless struggle among the ranks for the fresh air of research results, lab time and statistical survey data. In the end, since this is not proper science, but Darwinian/Malthusian opportunistic cutthroat capitalism, whoever is richest will end up getting credit for the discovery and refinement of a proof, no matter how vague and far fetched their actual formulae are, while all those who offered assistance along the way by means of speculating, sequencing, code cracking and transformations will be marginalized out of text books and forgotten by history.

All this goes to show that the days of Einstein are over, and that, even then, perhaps, it was only necessary to select for public spectacle a semite whose theories, while radically new at the time, are, in retrospect, essentially flawed and for the great part unworthy of blind acceptance, to counterbalance the overwhelming horror against humanity that was the Holocaust. It stands to reason, then, that the first and most essential application of his theories was to the building of a thermonuclear bomb to weigh the atrocity against his own people by the Nazis with the equally horrific genocide of the Japanese by the nation who adopted him. Before you adjudicate against me on this matter due to the well funded popularity of Einstein take into consideration that his rise to international prominence is the exception and not the rule in the field of science. It was only as an absurdity of that ultimate absurdity, war, that raised him out of the German ghetto and set him, by opposing the crimes against humanity of his own country, at the pinnacle of international celebrity for providing a simple cosmic formula for a further war crime against another country altogether. As a final recrimination notice the fact that the anthropic argument that — without his contribution I would lack the freedom I take for granted by challenging and opposing the morality of his contribution's application — is at once specious and tautological. One is ultimately forced to wonder, if they were in the position of Einstein, but knew what would be the ultimate outcome of their knowledge, whether one would go on ahead unimpeded by guilt to seek fame and celebrity, to boot personal immortality by dint of reputation, or whether one would simply scribble the results down in some safe hiding place and let themselves be taken happily off to the concentration camps?

It is because of such moral tautologies and ethical quagmires that I am publishing this book not for the reason of seeking attention to be drawn to myself. I would like to state for the record that by no means should any of the information contained in this volume be warped and distorted by being applied to anybody's personal cause. True justice does not follow from crusades made by those whose faith is unquestioning. The statue of justice is blind because she holds the scales, which one can feel if they are tipped to either side, while Lady Liberty's eyes are open, for it is she who holds the book of days to be read by the Eternal Light of the Torch.

introduction to the style

As to the style of writing little need be said. It will not appeal to everyone and, I am sure, large segments of the text will fly completely over many people's heads. Other portions of it an educated minority will undoubtedly understand just well enough to be violently offended by or to find horrifyingly disagreeable. Since I am relatively disinterested in much of this subject matter myself, I have paid little attention to most of what is said in this work during the process of writing and editing it. If some of it, most of it, or even all of it does not go down easily with you, remember that I am on your side in this regard, and that I agree with very little of it myself. That part of it which is scientific would be better said in more exact, scientific terminology, such as formulae and axioms containing constants and variables. The parts of it that touch on religion would be better expressed by a wizened elder of whatever faith from which the terms are borrowed, such that, rather than overlapping in a pastiche hotchpot, each could be illustratively enumerated upon in patient depth and detail by a well versed and qualified critic. However, since metaphysics is situated precisely in between science and religion, the subject matter requires that I dwell at times on either one or both, and since there are no existent colleges offering certified public degrees in this field, it is impossible to find a qualified metaphysician to do the job I have herewith endeavored myself alone to accomplish. Now, this task being done, I set it down in the utmost disgust, having performed the miracle of birthing this book I am exhausted by the effort and the product, being altogether infantile, is quite malformed and disgusting to me in my fatigued delirium. Perhaps if, like a newborn, it could offer some tactile reaction to my presence, as its creator, this feeling of distaste would cease and be abated, however not until the work has met with the critical standard of rigorous peer review, a paradox I have already described, can I retain any sort of response to the task whatever. Only then will I be able to determine if the style floats, and therefore must be burned as witch, or falls flat from the surface, sinking like the proverbial philosopher's stone falling out of water into air, and is therefore to be thought of as too innocent and naive to be of any real further interest. In either case, by the very act of publishing this alone, I am effectively throwing the baby out into the world along with the bath water. After all, its buoyancy isn't based on its style — that is the flow of the stream of consciousness. Its gravity is determined by its content. Hence.

introduction to the methodology

The methodology utilized herein will no doubt meet with the most question and doubt by the tired and rigorous speculators in that new religion, science, and that old science, religion. It is the same methodology that has been utilized throughout all of history by those calling themselves, or being accused thereof by others, as are philosophers. The reason for this is that, when metaphysics first began, with Aristotle, he had the bad taste to be following in the footsteps of his teacher Plato, who is most well known for his transcribed teachings of Socrates, who had, in turn, followed those earliest of all speculators in the sciences, the Greek elementalists who were known, in Plato's time, as philosophers. Now metaphysics is no more, in fact, the domain of philosophy than is science. But because, until Pythagoras and Euclid, no one was known in all prior existence to have speculated in pure number and geometry, which is the cornerstone of science, all thoughts that were considered to be transmundane were called "ideal" and philosophy arose as the early study of the ideal. Thus, though nowadays this would be like following a book about the calculus used by Nasa to propel shuttles into space with a book about trial law, Aristotle, acting in the most extreme bad taste, had the gaul to write inclusively about metaphysics and ethics, and class them both under the auspices of philosophy.

Now metaphysics, like algebra, calculus, trigonometry, geometry, astrophysics and quantum mechanics, is a serious science, with no room for the type of abstracts and ambiguities of arguments over right and wrong. According to the later scientific method, what is right in science is what can be proven to occur in nature, and what is wrong is what cannot. However, by putting his metaphysics right alongside his ethics, Aristotle was, either deliberately out of a sense of moral obligation, or, more likely, out of laziness, degrading the status of this earliest of sciences in much the same way that alchemy, the earliest form of chemistry, was degraded by being classed along with magic and conjuring. In short, while the ideals of metaphysics, like any modern science, are strictly black and white, by calling it a philosophy, Aristotle did it the criminal injustice of making its ideals seem vague and gradiated, a wrong from which metaphysics has never in full recovered.

Thus, because metaphysics was, at its birth, considered philosophy, it has been approached only by mystics, seers and skeptics, but never taken seriously as a necessary part, and in fact, fundamental trunk of the tree, of which all the branches of modern scholastic science are merely recent offshoots. In this regard philosophy is altogether unapologetic. The methodology of philosophy, and not that of pure science, remains the key to unlocking metaphysics.

The methodology of philosophy, for its part, is quite clear, and quite in opposition to the pure reason of scientific method. While science progresses by heuristics, with a clinically hermetic seal of approval by mainstream society, philosophy, and therefore metaphysics as well, has always followed the road less traveled, beating around the bush about it, and remained therefore esoteric and marginalized. From the ergot of Classical Greece and the rebirth of meditation for its own sake by DesCartes, through the syphilitic ravings of Nietzsche, Kant's critique of pure reason, Husserl's epoché and Sartre's nausea, philosophy has always been based on the earliest practises of mystery cults — prophecy through dreams and inebriation, and this remains the practise today. Wittgenstein, Baudrillard, Barthes, Benjamin, Batailles and Bey, drunk off their own words, spew forth gibberish and nonsense, while Foucault, obsessed by the recriminations of insanity, and Chomsky, equally obsessed with those of free speech, spout ridiculous rhetoric at once as true to others and as meaningless to themselves as the most well funded political candidate campaigning for constituency. It is, because of that first bastardization by Aristotle, to this vast swamp of hollow words and haunted psyches that the truly lofty and exalted metaphysics has been cast in shame for its entire existence thus far. Nonetheless it remains at the center thereof, and the subject of all debate therein. It was, after all, the ethics of metaphysics Aristotle called ideal. It was for the purpose of studying metaphysics that DesCartes embarked upon his self abnegating meditations. It was metaphysics Nietzsche was reaching out to and crying out for in Zarathustra. It was metaphysics that fueled the desperate befuddlement of Kant and Heidegger. It was metaphysics Husserl was looking for and literally trying to see with his own eyes when he discovered the epoché. It was metaphysics that Sartre tried to distill from being en soi, being por soi, and from nothingness. In the hermeneutics of Wittgenstein, Baudrillard, Barthes, Benjamin, Batailles and Bey there is the longing for a greater metaphysic than they can apprehend in existing culture. Foucault and Chomsky, too, are seeking for the new ideal that will rescue them from deportation into their nightmare dystopias, and both obey the rigorous methodology of pure science, to which metaphysics was destined to aspire.

introduction to the material

The material herein contained is culled from three separate sources. In the chronological order in which these were written, first is a work copyrighted and published in 2000 called "negative zero" from which the essay "on the tables of the eight and the ninth" is culled. Second is the Formal System of Metaphysics, copyrighted 2001, being published here for the first time in its revised form. The revision to this material is the exclusion of most of the accompanying essays, except for "dimensional molecules," "instantaneous manifestation," "on Samsara" and "a trend;" also a couple of paragraphs have been added under the "hexagrams" section to describe a formula for making magic number squares given by Clifford A. Pickover in "The Zen of Magic Squares, Circles and Stars," and the relationship between the 64 hexagrams of the I Ching and the 231 gates of Kabbalah. The remainder of the material comprises the rest of the MPDR proper and, along with the revised formal system of metaphysics, this work is copyright 2002. It is this material that warrants the most revision in terms of references, end notes and citations. In all cases where the specific work of another person or organization is given it is so alongside their name.

As to the contents of "Historia Singularitatis," the descriptions of subspace, hyperdimension, pure dimension, the correspondence of Platonic Solids, elements and forces, the descriptions of microwave gravity tachyons as torus shaped, the description of black holes as containing wormholes, the dual sided surface of spacetime/timespace, the description of the n-dimensional multiverse as being outside of and containing the local universe, and phi/pi as being decomposable from the multiplication tables of eight and nine by the process herein described, all of these are unique ideas, though, I believe, to a great degree, ubiquitously innate. I am not saying that any of these are my copyrighted inventions, nor am I claiming to be the first person to discover or final person to prove the existence of any of them. It is simply my goal to introduce these topics into discussion by a group of educated peers for the purpose of furthering scientific theories and stimulating scientific speculation.

My theories about God are irrelevant. Whether they are true or not is as impossible to prove as it is whether I even actually believe them or not.

I. Metaphysics

All of recorded history is the documentation of humanity's search for truth. In every word we have ever spoken, in every gesture, in every jot we have ever made to express our feelings are all borne this same end: that we should know truth as we know ourselves — and this has been the eternal, unquenchable spirit of our species, and the driving force behind the evolution of our consciousness. We cannot question our urge to question. Without the desire for meaning, there would be no meaning for our desire. This pursuit of knowledge is found archetypically in all our eldest mythologies, and is incorporated as a fundamental cause in the progress of our technology. It can be called by all words, and no word will do it complete justice. Many call this inherent longing of the human condition for its own transcendence metaphysics.

This name is apt. It states what we have: that is, knowledge of what is, and it states what we desire: knowledge which surpasses this. It provides a center and the momentum for a potentially perpetual expansion into the unknown alike what we have already discovered is most probably the condition of our very universe itself. But this is metaphysics on a divine level, elevated to a state wherein it describes itself alone. Can it be brought down to the human level, where we, in our humility before the yawning expanse of time, have only just this moment grasped the use of tools? Can metaphysics be utilized and still contribute its full scope of description?

Two affirmative answers to this derive from different theories. The first states that it is possible due to dimensionality: the microcosm and the macrocosm are the same fractal. The second deals with assignation, and realizes that equivalent terms apply to all scales. These two theories are the two hemispheres of the human mind — one intuitive and structural, the other formal and precise. Their agreement is the mind, a concentrated focal point of consciousness, with which and of which we ask our question. By this medium then, shall we attempt herein to construct a formal system of metaphysics that renders exact insights and completes the gap of inquiry.

Before we can do this however, we must first compress metaphysics into a meaningful context. As the bridge between the individual and the eternal, it has always taken the form of speculation. In accordance with natural selection this speculation has been tested and, when proven, made into a law. We have the laws that govern our physical nature, the laws that govern our behavioral ethics, and the laws that govern our beliefs in the divine. I will show that all of these stem from the eternal quest to permute question into answer, and then I will show you how this may be done universally for all cases.

A. Metaphysics as Law

Let us examine this question, for it is pleasing to the memory of youth: "Can God create a rock too heavy for Him to lift?" This question is a paradox of so fine a nature that, were it to be looked upon as a flaw in the crystal of reasoning, it would be one so precise, it would be more exact than the most accurate cut intended to render the crystal more perfect. The blemish in complexion it shears astray is that of logic itself, for here we have the logic of divinity, of will, and of proportion, all lumped together into a single conundrum. Nowhere else may we so concisely consider the ultimate arbitrariness of these assumptions. Without these three angles the crystal would fail to glint with an implicative twinkle, not catch our eyes at all, and more than likely drift from our lives as though no more than a passing fancy. But by offering explanation of these things, it appeals to our instinct. By providing them, if and when it can, it is the very essence of beauty. Allow me to clarify.

As a living organism humans must depend on three perspectives. Up and down, side to side, and at non-right angles. These compose our entire scope of perception. Were one of these to fail we would feel an innate uncertainty as to what might be approaching us from that quadrant. We would, most likely, find increase in the stimulus of the other perspectives to serve as a form of compensation for our disability.

Now these relate to metaphysics in this way: as we are guided to look down, our mind moves inward on itself, and we search ourselves very deeply. As we look up we search the heavens for peace, hoping perhaps to find some similarities between our internal composure and the forces of nature. This is the divine aspiration, and it legislates religiously. Similarly we look about for balance, for stimulation and for sustenance. We approach what we desire, and retreat from what would upset our equilibrium with the environment. This involves both time, therefore, and other beings. This is the logic of will, and it legislates ethically. Finally we may look at objects that are before us and wonder about their existence. How are they created, how do they change, how do they apply to us? This is the perception of proportion, and it legislates physically. And all of these together are the essence of metaphysics: the triangle that defines a circle; the three dimensions that define a cube; the birth, life and death cycle of time. They are metaphysics by shape and letter of definition — the spiritual is that which is beyond physics; the ethical is that which is our relationship to that beyond ourselves; and the proportional is expansion of physics.

So, in the plea of our youthful curiosity, we discover already the underpinnings of all known manners of reasoning. The answer is therefore impossible, because it is a paradox of these three logics. Thus it can only be subdivided into further specifications of the request to determine at exactly what juncture the paradox occurs. We already know it is in the realm of divine jurisdiction, so the only real response that remains is "how?" Why this question cannot be answered by anything but another question is further described later, when I deal with the Platonic paradox of questions in general.

Throughout all of history such questions as this have arisen in the mind of humanity. All that we have as the basis for our understanding today is the practice of creating and maintaining data in this way. Despite all our accomplishments, we self-aware beings remain like a child, still trying to get our bearings and discern order in what is.

1. the Spoken Tradition

When we were what we call today primitive, when all of humanity was tribal, and much of it still nomadic, when there were as many people on the entire Earth as there are now in some single city, we had only just received the most sacred of the inheritance left us by our last evolutionary leap, that had made us upright-walking, land-dwelling hominids several hundred thousand years before. Communication was very limited, and the art of speech was still being cultivated simultaneously by disparate peoples. We had achieved a stage of such complex cerebellar functioning that we could stimulate the nerve centers in our brain at will, generating from the well of experience both recollection and innovation. New explanations became apparent, new solutions to age old problems. Survival was becoming easier.

The earliest distinction that can be considered indicative of civilization is between the chief of the tribe and the medicine man. Here is the one who preserves the ethical tradition, and here is the one who preserves the physical and spiritual tradition, both passing on their legacy as instruction from parent to child. Here we find the birth of myth, where the characteristic patterns of survival behavior began to be specified into replicable routines, and preserved in the form of archetypes. It is interesting to note, as well, that the first medicine "men" were probably women, who discovered the medicinal properties of certain herbs by gathering while the men were hunting. From these times only art remains to express the minds of our ancestors. This was likely also the place of the medicine man, which suggests that they were responsible as well for the creation of the written form of language, discovering the possibility of impressing clay slabs or making marks on unfurled river reeds concurrent with the introduction of agriculture.

To some of these tribes the acceptance of writing never arrived, and they have remained throughout the millennia, up to this very moment, of the belief that only the oral tradition is divine, and that all other modes of information transferal are a form of blasphemy. A great amount of the history cultivated and nurtured by society remains in the dark as a result of this.

2. Early Writings

The earliest forms of mythology are of this patriarchal system and its rebirth mechanism. We are given the archetype of the wizened elder and the wandering fool, the chief and the shaman, and told of heroes that went from tribe to tribe throughout the world spreading the practice of writing, ascending outside the limitations of space and time, and inscribing the very essence of consciousness on the scroll of the heavens. But the event is much more important than this. It represents the time in the lives of our own ancient ancestors when the mind first became completely awake, and its evolutionary rate broke from that which defines all other life into the flourishing burgeoning of awareness to which we are accustomed today, although it is a blossom unique in all the universe.

The metaphysical breakthrough that occurred was the distinguishing of the physical from the divine. It allowed an entirely new class of people to come into being in the already well established ethical structure of the tribal hierarchy. The scribes were medicine men who offered the economical opportunity of specialization, and, unlike the often mystery enshrouded individual medicine man, were obligated for their financial survival to serve the ethical dictator. Thus, the civilizations that flourished during the era of the old written tradition were often empires, as we see is the case in Egypt, the Mediterranean, the Middle East, Meso-America, China, and, most recently, Europe.

3. Later Writings

In each case the expanded tribe could only grow as vast as its internal politics would allow, and soon each would realize that reorganization and almost inevitably recollapse were imminent. The middle class began to take on more of the burdens of legislation. In most cases this was ethical, and the bureaucracy gradually replaced the empire. In others it was physical, and there was an upsurgance in the arts as the public's surroundings became progressively demystified. In a few this took the form of a revolution in divinity, and the oligarchy of an established religion became increasingly secularized.

Eventually, every society so far has had to undergo several internal schisms during the stage of the written tradition, by which we measure all of recorded history. Such changes predominantly occur in the form of social movements, and these are not recognized until their task in the social restructuring process is complete. At that point they become obsolete as an activity, and are given over wholly to the domain of history. As such, history is not to be confused with society. The story of society may appear to be history in the present tense, but we must learn from the mistakes that we make, and it is a rare and wise individual indeed who knows a mistake before it can happen.

During the period of bureaucracy, naturalism and secularism, several distinctions mark the passage of time. The foremost of these is that the written word becomes the property of the common man. The wisdom of the ages, once held exclusively by the proprietors of legislation, is now disseminated in such a way that the responsibility for the generation and maintenance of legislation may be seen to follow it. This time period is an excellent opportunity for intertribal trade to occur, for it allows avenues for further expansion internally. But this also raises the issue of conflicting ideas, and the curse of cognitive dissonance. A mind holding two opposing viewpoints in itself will grow infirm, and therefore legislation must remain active when dealing with all aspects of metaphysics.

4. Wars

It has proven to be the exception to the rule that cognitive dissonance is dealt with in isolation. Most frequently it manifests itself as two or more parties who have come to cherish the apparently incompatible values at hand. Nature dictates that a stability between them must exist, and therefore a resolution to this issue is indicated.

Wars, like paradoxes, can occur in all sectors of the legislation of metaphysics. In fact, as I will prove shortly, they must. Without imbalance there would be no movement, and without movement there would be no becoming. This is a simple fact of the action of time, and has been legislated physically as the conservation of energy. But this sounds simple.

In reality, wars can be felt many years before they occur, and always change history indelibly. They provide simultaneously a massive release of kinetic energy and mental energy, as well as a clearly designated winner and loser. In this situation, the loser is whoever's energy is expelled faster, and the winner becomes the loser's parental figure, using the remainder of their energy exporting their version of the data to the loser. This is what is meant when it is said that "history is written by the victors," but we need neither believe nor disbelieve this concept. After all, the archetype of the villain is as necessary to mythology as is that of the hero, and the man who spoke the quote above was indeed one of the most infamous characters to have ever walked this planet, and yet himself a loser. In fact, a direct relationship has come to exist in the later written tradition between the outstandingness of information and its frequency of consideration. This has relevance to fact and fiction, and will be discussed in due course, but we may say that in the very recent past, particularly the modern and postmodern periods, the amount of energy expended has far surpassed in scale the quality of the generated solutions. This is taken by many as the dark side of humanity, or perhaps as a curse of the realization of individual legislation of metaphysics. But these ideas are obviously mere superstition, retroactive reactions remnant from the days of the nomadic, spoken tradition, when less property had to reference equivalent sums of memory storage and metaphysical legislation, and therefore everything was taken with a regard that has become irrelevant today. It is merely the luxury of surplus data, which can be viewed as icing on the cake, so long as you are not one of the ones that goes starving.

5. Modern Philosophies

The result of war is such that the wealth of information exchanged in its wake is always greater than the amount over which the initial disagreement occurred, and, if we are to believe that the universe is ultimately governed by a self-correcting set of rules, whether they are comprehensible by the human mind or not, we must come to believe that this is also greater than if the conflict were to have been resolved with the opposite result. The same is true of the effects generated by internal readjustments in the form of social movements and the passage of time with which they wrestle. All this means is that metaphysics always raises more questions than it answers, and in this way grows exponentially.

This has led to one of the most profound legislations of metaphysics' divine aspect, that, as man's population grows according to the same equation, thus are we alike the image of God. But surely, as with all of religion's opaque dictates, this means more than it says; which itself only goes to demonstrate how not only does the image of man correspond to the spirit of the divine, but there is hope for correspondence even in letter. In fact, this is what has been striven for all along, for what purpose does writing really serve?

First it marks down to discount potential subsequent redundancy. Also, by its implied authority, it alleviates the option of replication with error, obligating instead repetition with modification. This is an efficient force, but we are still no closer to understanding what it is expected to achieve.

One of the practices of modern philosophy is the discernment of this inarguably very fine point. As a result of excess detail due to specialization, no short supply of theories exists. We may feel much alike the stone mason who, so absorbed in his work, only too late for rescue comes to realize that he has walled himself in.

6. Technologies

Long before the beginning of history, before we sought to domesticate metaphysics, we were very little different from our cousins in the ape family. They enjoy life as they are, understanding without communication complexities of our universe comfortable comprehension of which we may well have sacrificed to engage in our own sort of intellectual discourse over. As more and more people are eclipsed by awe at the luxury of data surplus in our possession, we begin to wonder: what would be the result if, like the library of Alexandria, it were suddenly to be gone? Is the satisfaction with pure existence elevated towards divine ecstasy by our consciousness so innate that we would succumb to it despite there being a dissatisfactory conclusion to our reason d'etré, or would there be such a vacuum caused by its absence that our savage emotions, tamed only in its presence, would rise up and cry out? Perhaps the solution is both, for one can easily imagine early mankind, at a certain, now long forgotten and unrecorded moment, making the evolutionary leap to communication, only to channel all their anguish at sacrificing eternal silence into a desperate desire for its recapitulation, or at very least clarification of its cause, some purpose for its loss. I know how upset I become when I only momentarily misplace my car keys.

This manifests an unspoken, though widely evidenced, movement toward the creation of artifacts signifying or dependent upon our current degree and particular form of metaphysical legislations. Let us again consider the ape family. The primary way in which they differ from us, for even they utilize a minimum of almost completely non-repetitive and systemless communication, is in the use of tools. There is proof that monkeys can be taught to use tools, even those based on complex systems, and that, using these tools, they can freely communicate complex ideas. There is also proof that technologies such as poured metal latches were developed identically in the monolithic masonry of disparate peoples without contact between them, that wooly mammoths discovered flash frozen in Siberia had recently fed on tropical vegetation, and that tachyon microwaves move opposite the standard arrow of time. It is not the proof we are concerned with, but the implication. If our genetic cousins are capable of the same feats as we, then at some moment, or for some reason, be it by choice or symptom, they must have declined the opportunity to utilize them which we accepted, and that this one simple matter has made all the difference between us.

We first discovered ourselves to be physical beings when we began utilizing external objects as components or practical tools, as weapons or destructive tools, and as toys or creative tools. It is perhaps this event which began the formal systemization that generated organizational communication, metaphysical legislation, writing, wars and even my car keys.

B. the Formal System

The formal system of metaphysics should not be confused with the traditionally accepted definition of this term. The genealogy of the word metaphysics traces its roots back to Greek words meaning beyond the phenomenal, sensational or logical realm of experience, deduction or insight. The primary difference from this standardized meaning of the methodology of the system described here is applicability to specified details of inquiry, yielding verifiable results. What metaphysics means in the broadest sense is the quest for meaning, as opposed to the hierarchical procedure offered by this substructure. We may expect, using this method of metaphysics, to yield results to questions even of a deeply philosophical nature through a series of logically navigable equivalencies, but there will always be just a little more that remains to be revealed. Such is the nature of metaphysics.

1. Memory Castles

With this important distinction dealt with let us focus on what this system is. To do this it is perhaps best to speak initially of memory castles. The use of memory castles for information storage and retrieval is an ancient practice, of little value in the world of today, but when people were more nomadic, and there were less possessions to serve as referentials, it was a far more common technique of mystics. The memory castle itself is merely a large, projected space before and around a person, where they concentrate their consciousness. Different points within this space, like the different rooms, hallways and libraries of a castle, are then imagined to access different sorts of data. This is much the way a modern computer works, and is really little more if anything than a realization of the mental macrocosm.

Now, once one has imagined for themselves a memory castle, a system of metaphysics is a natural next step. Imagine that the entire apparatus compresses into a compact form. First compress it to the size of a doorway, then to a window, then to a fruit or a sports ball, and finally, if you like, to the size of one of your own eyes — floating just above the center of your forehead. Each of these sizes has a myriad of applications in the realm of information storage within the universal unconscious, but for our concern we will be dealing with the model at the size of a fruit or a sports ball. Briefly, I will go into the applications of this size in more detail, but at this time an explanation of the substance of the memory castle or metaphysics model is best.

Use of memory castles and most other forms of mysticism and metaphysics is the domain of a select proportion of the world's population, and even by them is treated with such regard that it is rarely made a public spectacle. Most of what we now know of mysticism and metaphysics has been passed down amidst and between these peoples, predominantly from originally an oral folk heritage. Many similarities mark their descriptions of the substance of externally projected consciousness, which is often called no-mind, nothingness or the void, the abyss, ether, trance, light, the all, the higher self, and by any number of environmentally all-encompassing anthropomorphications for things that go bump in the night. All agree upon an essentially holographic tabula rasa that is subsequently muted or illuminated and fractalized by the act of perception. It is also believed by most that this state precedes being, and that the process of spiritual seeing is equivalent on another dimensional level to that of physical becoming. More recent metaphysicians have forwarded this realm as a tyling board for their own temple building, and see all externalizations as necessary in function. Others share the belief that manifestation is natural, but seek to find commonalities of character or form to unify the universe. Ancient and modern mystics alike caution the aspirant regarding this sacred space, however. It is best to remember that when you look long into an abyss, the abyss looks long into you.

2. the Changing Mind

Based on these assertions of the substance's potency we may now enter into consideration of the model itself. To do this, quickly construct a minimal memory castle, consisting of a single room of right angle walls connecting parallel ceiling and floor, such that the area of each wall is equal to the area of the ceiling, and that the area of the ceiling is also equal to the area of the floor. This should be a perfectly cubical room, with no elaboration as of yet.

Next, step backwards out of this room. This is merely a technique to demonstrate a fact which should become obvious at this time, that one may be aware of the existence of the wall behind one and the corners of the ceiling and floor that define it, but one cannot see all six sides of a cubic room if they are inside. Technically it is impossible to "step outside of" your memory castle, since you are not in any way leaving behind any portion of your memory by doing so.

What is really being accomplished by this exercise is the creation of the memory castle in a higher dimensional level, transforming the cube into a hypercube. What seems to be the further distant wall of the cube is really also the external side of a second cube, and if you were to walk to one side or the other around what you believed to be the simple cubic room in which you were standing, you will find that the shape of the closer wall expands to become its own cube, the shape of the further wall collapses into its own cube, until finally, as you pass around the wall that joins them, that was originally the wall directly facing you in your castle, the larger cube will begin to be compressed and the smaller cube will begin to grow. By the time you have traveled around so that you are facing in the opposite direction you were to start with, as if you had stepped forward from your memory cube and turned around, it will be the closer wall that is large, and the more distant wall which is small, but of course, by now, you will understand that these are not really walls, but cubes. Another way of looking at this would be to imagine that the cube casts a shadow opposite you that is itself also a cube. As you pass by the corners of the closer cube, the two cubes will seem to join into a rectangle, and as you look at the the further cube through one of the walls of the closer cube it will seem to be contained within it. This is how the storage of memory occurs in the mind, and the way in which detailed past experiences are easily accessed from single perceptions. A further expansion of the hypercube into a hyper cross is possible by shrinking the cube down to the size of a fruit or a sports ball and rotating it about in your hand so that you see every wall casts its own cube, and that the central cube, which was the foundation of your memory castle, is only the central cube within them.

It is this central cube which we shall continue to consider in its compact form, but remember, it remains a fourth dimensional object, and each relationship that we can discover pertaining to its surface is also a description of its storage potential in terms of depth. For example, think about a sphere enclosed in the volume of the cube. This sphere represents your mind, as it occupies the space where you stood when you were inside the cube in its form as a memory castle. Now imagine that these two, the sphere and the cube, relate to one another gyroscopically. Turn the cube in your hands and consider how the sphere shifts about before coming to settle again into its original orientation. Different parts of the sphere are now in contact with the surface of the sides of the cube. These places represent juncture points of idea. This is where information is exchanged between the consciousness and the unconsciousness of your memory storage unit. As this rotates around in any direction these points will change, but there will always be exactly six points of contact, because your mind perfectly fills your central consciousness within the memory storage unit, no matter what. When you change your mind, it is really the relationship of these six points that is changing. You cannot change one point without changing the other five as well at the same time.

Now two things can cause the mind to change. One is decision, and the other is symptom. In the first case it is the sphere of your mind that moves first, and therefore effects motion in the cube which it is touching. In the latter case it is the cube that moves first, due to some change of one of these points in the environment, and this causes the sphere of your mind to have to adjust slightly to new coordinates.

To arrive at the system of metaphysics which is going to be described in this text it is necessary to imagine a second cube inside the sphere. Six of the corners of the cube will align with the six points where the surface of the sphere contacts the surface of the memory storage cube, and two of the corners of the second cube will lie on the surface of the sphere where it arcs down beneath or within two corners of the memory storage cube, such that they are opposite one another. To construct this cube it is possible to imagine looking at the memory castle cube directly above one of its faces so that the corners of the further wall of it touch the inside of the circle defined by the center of the sphere. Another way is to imagine that you are once again standing inside the cubicle memory castle, that the sphere of your mind is the area that you take up, and then construct an image in your own mind of another cube, thus placing it inside the circumference of your own consciousness. However you might want to go about doing it, the same effect will be accomplished. A third way to get to the metaphysics model is to expand the sphere of your mind until it touches the midpoints of all the cubes on the hypercross that can be extended out from your central memory cube, until it is the corners of your original memory cube that touch the surface of the sphere from inside. All of these techniques, although seemingly different, will yield the same results.

This is because of the nature of the hypercube. Remember when you went walking around it the first time, and saw how it seemed to expand and collapse; but at one point, when you were standing even with the center wall that joined the two cubes, they would have appeared to be the same size, and you would have been looking at a rectangle. If you were to step back into the cube at this point you would find that it would immediately snap back down into a cube around you, and you would be back at the center of your original memory castle, although probably having no idea which way you were facing. This point is called the antipode. The existence of a point such as this in fourth dimensional space means that two objects, which from one angel appear separate (or adjacent), will appear from another angle appear to be exactly overlapped (or singular), and in all cases have exactly the same area. The fact that you can construct the metaphysics cube within the memory cube means that they have the same area, and that the surface of the sphere acts as their antipode. In the walking around your memory castle exercise, if you were to turn and walk away from the memory castle in a perfectly straight line instead of walking around it, it would appear the same, but decrease in size as you increased your distance from it. There would be no antipode. This would certainly not mean that you had lost your mind, though! It would demonstrate the difference in the right angle construction of the memory castle and metaphysics cube as opposed to the spherical nature of your mind. Without the mind to establish their relationship the two cubes would only be two, plain, three dimensional cubes, but in the presence of your mind, they become strongly relative in the fourth dimension.

In short, for the purposes of our disposition the memory castle cube you created first and the metaphysics cube which we shall consider in greater detail will be synonymous, but the real interaction of them will be left for your mind to sort out, just as in our model.

3. Parts

There are a total of ten individual nodes that we will examine here, in conjunction with how they are situated in position to one another. Six of them are coplanar with the memory cube, and these six plus two more are co-circumferential to the sphere of the mind. The final two are internal to the model, and determine its relationship with the sphere of the mind and the memory cube by acting much the way the balance central to a gyroscope acts to steady it while it spins.

First let us look at the six coplanar nodes, which I will refer to as the six fundamentals. In their purely ideal form they were discerned by the ancient Greeks and incorporated into the earliest systems of logic and reasoning that were applied in daily life and philosophy since the Golden Age. We know them as the fundamentals of higher reasoning, but they are subtly more significant than this. It is demonstrable that apes and other forms of animal life can utilize the projective consciousness technique in establishing relationships with objects and other beings that they then depend on by memory to come to necessary conclusions about their environment. As such the most basic component of metaphysics is grounded deeply in the survival instinct. But the identification of these six fundamentals represents a considerable leap forward in the application of this force as an organizing order. It is difficult to say exactly how this was accomplished, but it is known that it was done so at a time in the history of the world similar to the Renaissance, when the study of the natural sciences was practiced freely and there was an open air for the dissemination of new ideas. Studies in logic corresponded to studies in geometry and astronomy, the development of materialistic cosmologies and ideologies. It is best to put our six fundamentals in this context to show that, although their moment of conception remains mysterious, their reflection in varying fields is unquestionable.

The six fundamentals of reasoning, and the first six nodes of our metaphysics model, are How, When, Where, What, Who and Why. These are questions that are ingrained in human thought to such an extent that for millennia they have served as the foundation for all logical reasoning, and remain so today, without fail, and without a jot of change or addition to them. They have, however, been inculcated into no particular hierarchy, as is the case with much of the rest of the arcana and esoterica that was developed during the Greek golden age. This is because they can survive autonomously from all other patterns of organization, proving to be a key that unlocks them all, rather than a part that operates within each devoid of others. Much of geometry and astronomy was forced into an underground condition during the European dark ages, when the practitioners of these schools were divided by their faith. But throughout, these six fundamentals of reasoning were applied to all areas of thought and exploration, proving to be just as fundamental to the study of scripture as to that of space or the stars. In a way, they may have suffered as a result of this, since both astronomy and geometry emerged from their submersion in the occult strengthened and renewed, but there was absolutely no change to the six fundamentals. They are taken so much for granted, in fact, that their importance has been all but forgotten.

To study them formally it is convenient to construct a model upon which they may be arranged. To do this take the cube that you have assembled in your mind and turn it so that you are looking at it from above one of its eight corners. It should now appear, if you were to draw a two dimensional representation of it, as a hexagon, with six points each of equal distance from its two nearest neighbors. Now start at the top point and, moving clockwise around the object, assign the six fundamentals to the six points. The uppermost will be how, the upper right will be when, the lower right will be where, the bottom point will be what, the lower left will be who, the upper left will be why, bringing us back to where we started.

This assignation is not casual. It is the result of recognition of the relationships between these points that will be subsequently discussed. It is possible to arrange 46,656 different models using the six fundamentals and the hexagon, but the relationships described by them will only test accurate in special cases where external factors must be set or attuned so that the results adhere to prediction. The arrangement described above is intended to yield results, all other things being equal, such that, regardless of the environment, the answers to the questions can be discerned from consideration of the questions alone. This is a unique model.

Now consider the corner of the cube above which you have placed your eye. It remains unassigned, as does the corner opposite it, underneath your cube. These will be referred to as the points of had and not respectively, although their assignation is arbitrary and only pertains to formal conclusions to be reached in specific cases.

Lastly, in the very center of the cube, imagine another point. This point is the binary juncture of fact and fiction, and again only pertains to the solution of specific cases. The nodes of had and not and that of fact and fiction are not fundamental, and therefore will be excluded from discussion of the fundamental relationships and discussed as separate entities.

These are all the relevant parts of your metaphysics model. With this in hand you can answer any question in the world. It is entirely a matter of determining simple relationships between the questions. Allow me to demonstrate the simplest method of this.

Holding the metaphysics cube in your hand you can easily comprehend that the angle between any one of the corners and its nearest neighbor along an edge is ninety degrees. But if you drop a dimension, as I have already suggested is a helpful exercise, then you can see that the angle described by each corner is one hundred and twenty degrees. This means that the metaphysics cube is describing the full rotation of one circle twice, or the circumference in degrees of two full circles. This is in keeping with what we have already discerned for the hypercube, since the sphere of the mind functions fourth dimensionally as well (or there would be no continuity of character from one moment to the next) and therefore is itself a hypersphere, depicted as a torus, which is like a doughnut, or a sphere within a sphere, and possesses an antipode, just as does its cubic equivalent.

This does not mean that you have two personalities, although this explanation would nicely account for such psychological assertions as Freud's superego and id, with the ego as the antipode, and Jung's anima and animus with equal volume. The explanation with which we will concern ourselves here is far simpler, suggesting no new names for the same old psychological behaviors, and its deduction more elegant mathematically. What we are dealing with is a sphere inscribing the hexagon, and another inscribed by the hexagram defined within the hexagon. This will open up a field of prediction hitherto unexplored in the study of mental process, which will later go a long way to explaining the abilities of the unconscious.

A hexagram is two equilateral triangles conjoined such that one overlaps the other facing in opposite directions so that they share the same center point. In three dimensions this translates to the conjoining of two tetrahedrons to form a steloctahedron. Imagine a sphere surrounding the metaphysics cube you are now holding in your hand such that its diameter is equal to the cube's diagonal. Now imagine a sphere inside that cube such that its diameter is equal to the length of one side of the cube; it will touch each face of the cube in its center. This is the fourth dimensional mind, derived dimensionally from the metaphysics model. It was there all along, but by studying the most simple aspects of the points on the cube it becomes realized. In this way the solutions to all humanity's questions are hidden in plain view, so obvious that we tend to overlook them.

4. Relationships

There are three different types of relationships between these six fundamental nodes. The first is around the edges of the cube or the sides of the hexagon. This is the shortest relationship, and if you imagine your metaphysics model as a unit cube, then the dimension of each of these will equal one unit. The next is between the corners of the conjoined triangles of the hexagon or tetrahedrons of the steloctahedron. This relationship is between diagonals, and the diagonals transecting the faces of a unit cube are equal to the square root of two. The last relationship is between nodes on opposite corners of the cube, and involves the diagonal of the cube itself. These relationships connect each point to every other in the metaphysics model.

In addition to these there are measurements that can be made for had and not, which are the corners of the cube that conceal one another as one looks directly down at it from above them. Each of these has a one unit measurement from the three points closest to it, a square root of two measurement for those slightly further away, and a cubic diagonal measurement between them. As I have said before, their attribution is arbitrary, and therefore I will not go into specifics about which of them lies closer to which other corners. Suffice it to say that to whichever is assigned the closer corner of the cube: when, what and why will be one unit off; and how, where and who will be a square root of two measure away. For the further corner the inverse of these assignations applies; and in both cases had and not will be a cubic diagonal apart.

Lastly there is the central node of fact and fiction. It is one half the cubic diagonal from all the other positions. This means that it has one relationship for all of them, irregardless of the nature of their inquiry.

I want these relationships to be as easy to use as possible, and therefore will dispense with the pleasantries of disposition on their origins and nature. It is enough to accept that they exist, so that we may see with usage how they work. I promise that, by following this method, we will soon discover that they are as possessed of potential as they are elegant.

5. Questions and Answers

The six fundamentals of reasoning are simultaneously questions and answers. This is a quaint axiom of logic, a conundrum that occurs as a one-step impossible feedback loop in many formal systems, but it has a highly meaningful use here. The Platonic system of ration, one of the first known to man outside legal practice, consisted of answering a question with a question. The point here was to lead the interlocutor around a series of points to arrive at their initial inquiry from the other side, thus proving to them that they had known the answer themselves all along, and had only been prevented from seeing it at first as a factor of their perspective. This practice has since fallen out of favor among philosophers, whose subsequent endeavors have consisted of trying to unlock ontology by semantics. It is easy to see how this system is not far removed from Plato's original idea, although it has been adhered to in letter rather than spirit. A more appropriate interpretation of the method might be to combine the practice of logical syllogism with the art form of geometry, and plot the points passed along the reasoning person's journey on a map or model of some sort. If these points were standardized, then, all entries into the realm of reasoning would share a single, common reference system. And this is the model of metaphysics. In other words, the answer to the question "what is the model of metaphysics?" lies in a simple rearrangement of the terms of the request — "the model of metaphysics is: 'what is the model of metaphysics?'"

To the interlocutor in the majority of Plato's tales the first time around this was a horrible pain to sort out. At first glance it seems to be begging the question, and little more than a verbal redundancy. This is one of the primary reasons why his avatar Socrates was known as the Gadfly and eventually condemned to execution by poison. There appears little more irony in this ruling than in Socrates's own arguments, afterall, and perhaps that is why he accepted his fate with such passivity. Socrates became a martyr for philosophy, and, like Isaac Newton, was eventually proven to have contributed as much to history as to have perturbed the people of his own day. In this way, subsequent, deepening comprehension of the Platonic technique of rhetoric justifies its continued study, and eventually dawn comes to those who delve into this method of reasoning.

It will be instructional, for our purposes, to postulate that the effectiveness of the method stems from its self-referentiality. That is, it is a closed circuit. Regardless of the path that is taken, if the procedure is executed systematically, the solution will inevitably be deduced. This applies to all questions and questions of all nature. One will translate one answer per rotation, and eventually solve them all.

Starting from a question at any point on the model, you may follow through the rest of the questions, assembling data at each point, until the first answer is obvious. If you continue in this fashion, even if you follow a different path through the hierarchy, you will compile satisfactory conclusions for all of the involved realms. When these are observed in conjunction with one another, the solutions will distill further and further, until they apply to the principal of Ockham's razor — that the simplest conclusion will most probably prove true. Thus, the more rotations undergone, the more questions will begin to turn up answers, until you have them all. If this had only taken so little time during the days of Socrates, perhaps his peers would not have adjudicated his teachings terminally tedious. But alas it has remained to this day for there to have accumulated sufficient experiences in the memory cube for us to have developed rapid processing techniques such as the metaphysics model under consideration here represents.

In the metaphysics model itself there are two means to achieve completion. They are determined by fact and fiction. The first is derivation, and is a practical inquiry towards extrapolating information from the memory hypercube, integrating it into the metaphysics system, and yielding results. The second is the aim of this type of searching. Each of these will be slightly adjusted for the balancing of fact and fiction, which occurs naturally as the system undergoes its rotational functioning.

For the case of derivation what we are examining is the path taken through the system, equivalent to the data stream in a computer terminal, which includes certain subroutines and naturally excludes, to preserve the expulsion of energy by, all others. Since we may follow any systematic path between the nodes, having established the distance relationships between all of them, then each path taken between two, and the order in which it occurs, lends additional determinance to both the working out of the computation, and to the eventual set of answers. Although one need never necessarily pass through the central point of the system in determining the solution to a particular problem, it always serves to impact on the conclusion all the equations reach. Thus, any particular path may be taken, and yet the same path may have a very different value, in fact often the exact opposite, for fact and for fiction. In many cases the two will overlap slightly, and in very rare cases share the same space completely. Derivation as an action is usually considered a more factually oriented function of the system than is aim, for the simple fact that the procedure of derivation is predominantly not predetermined, making it very difficult to fake.

Aim is the very last relationship possible, and pertains to the external motion of the system overall. If you imagine the system as a planet within the gravity of your consciousness, then derivation is equal to rotation, and aim is equivalent to orbit. After derivation is completed, and the data stream passes through the final link at the system's center, then aim becomes the relevant issue. To what event are these answers going to be applied? It is possible to have no aim, only to pass some time, in which case the model before you sits still and slowly spins. If there is an aim, then the system will have two indicative behaviors before ultimate case termination and automatic memory hypercube solution storage. First, before it has reached its final link, you will notice the system begin to speed up in finding solutions. This is the result both of narrowing down the margins of inquest, and of being pulled in the direction of the intended aim. Second, when the final link is achieved, a strong amount of energy will have built up in the metaphysics cube, and it will break towards its aim very suddenly and with an enormous release of power. The former function of aim is the result of a pi spiral, and the latter of a phi spiral, although both constitute the action of symmetry breaking.

This is why it has been important to picture the metaphysics model at the size of a fruit or a sports ball — because it can be either consumed or projected. In the case of fiction, the answers are consumed into a larger plotline, probably involving a complex set of other, related equations. In the event of fact, the metaphysics model can be the determining factor in any dispute. Whoever reaches a more logical conclusion first has the upper hand in debate. This was recognized very well in the era of Socrates, and again, this time in the from of politics, was one of the possible interpretations of his philosophy that would come back to haunt him. Do not forget that the metaphysics model is a structure of projective consciousness comprised of mental energy, and as such there is no limit to its potential for power.

Similar direction of biophysical energy fields are referred to in the martial arts schools of the Orient, particularly Kundalini in India, Tai Chi in China, and Akido in Japan. According to these traditions the conscious directing of mental energy force can serve to massively illuminate an aspirant of mysticism, support the girding of the physique so that it is in harmony with the ambient energy of the environment, or devastate an opponent in the intellectual realm of combat. Having considered this point it is no longer necessary to confine your comprehensive imaging to the size of a fruit or a sports ball. The relationships we are about to describe hold true for the metaphysics model projected at any size.

II. Parts and Relationships

The formal system of metaphysics which we have forged of the methodology of humanity's quest for meaning has been demonstrated to be of both great power and complete simplicity of usage. In this section we will go into further detail regarding the specific aspects of this system so that a more skilled usage of it may, with practice, come to result in a healthy degree of comfort and expertise. I should once again caution you that this is revealed knowledge, and as such, though it is revealed by and composed of the light of mental concentration, itself not to be taken lightly. By the end of this section you will have achieved a degree of experience in this art unequaled by most of human reasoning that has come before, but you must remember the nature of metaphysics, that there is always more to learn, and even as thorough as I can make it, this basic introduction is really only the first and the smallest of steps in a much greater and much more significant journey.

We will also look a little bit more deeply into some cloudier aspects of the model, where geometrical paradoxes and digital transcendentals become a deciding factor of associative logical functions. Although the implications of these are vast, my explanations will be kept as minimal and unobtrusive as possible, for the sake of leaving them to be worked out by you in your exercising of the system. There remains a good deal to be said of these in and of themselves, but for our introductory purposes here it will prove sufficient to discuss them only in their proper place, and not devote too much time to them. They should prove to be a formidable challenge to you in seeking to add to the tally of precise relationships with which I will start you.

Lastly, by way of an example solution I will be working with the question proposed earlier, pertaining to God and the stone. By this example I do not mean to offend anyone, or to imply that certain legislations are proper in place of others, for by doing so I would be tainting my integrity as an impartial scribe. I make use of it merely by way of a standard, paradoxical inquiry, both as proof of the Socratic-Platonic method of reasoning itself, and my formalization thereof. Also I should think that it is rather cute. It does have a certain appeal to the curious and innocent desire for wisdom which is an underpinning and the essence of metaphysics.

All this being put forth, let us proceed to our exposition on the parts and relationships of the formal system of metaphysics.

A. Shorts

The first of the relationships we will look into is that which lies between the corners of the metaphysics cube, along its edges. In two dimensions the projection of this outline is easy to navigate: a simple hexagon with each leg equal to one unit, such that if it were inscribed in a unit circle each of the six points of fundamental reasoning would lie on the circumference. Each turn between points, as has already been demonstrated, is one hundred and twenty degrees, and the total number of degrees for the hexagram is seven hundred and twenty. In three dimensions the path navigated between the six points is a little more complex, as at each point there is a ninety degree turn that steers us from the edge of one face to another, such that the total number of degrees of the cubical path is five hundred and forty. In four dimensions the course we plot thickens even further, as we see that we may move clockwise or counterclockwise through the pattern. All of these features will be discussed in more detail in due time. For now it is best simply to keep in mind that our model is functioning simultaneously in the second, third, and fourth dimensions, and that, as a construct of mental energy, is itself a singularity of the first dimension.

We will, for our purposes, follow the path around second dimensionally and clockwise, as this will give us the quickest and most succinct tour, and establish a firm basis for the requisite understanding that each node functions as part of the system as a whole, and depends entirely on connection to the others for insight. Thus, the course which we will follow will lead us first from how to when, then from when to where, from where to what, then from what to who, from who to why, and from why back to how. Imagine that we are at twelve o'clock on the unit circle, where the uppermost point of the fundamental hexagon joins the circumference.

1. How —> When

Let us ask how and let us ask when as if they were one question. Any event that occurs at a specific time must occur in a specific way, and any event that occurs in a specific way must occur at a specific time. If the event is unspecific, then it occupies a period of time, which can be scale compressed to a specific moment, at which point the event will be bound specifically as well. Likewise any specific event can be seen to take up, no matter how instantaneous, a relative period of time, and therefore yield more complicated details.

For any event to occur it is bound to specific necessities of function. These particulars reveal results in all three branches of metaphysical legislation, which are case specific dependent upon the event. Consideration of How begins before the event and ends after, incorporating initiatory function, proper function, and terminatory function of the event which are all internal to the mechanism of the moment.

In the example of God and the stone we are dealing with a specific function over an indeterminate period of time. We can conclude this based on what we know of the function of lifting and the duration of God. We see already that the most important points of this particular question are really Who and What, but we will come to this relationship soon enough, and focus on its relevance to the example inquiry at that time. It is enough at this point to use the solutions provided for What and Who that are inherent to the question.

2. When —> Where

It is a self evident fact of the existent universe that for any event there is a direct relationship between moment and location. Prior to all other contingencies we may conclude any event's almost completely full nature simply from this relationship. It implies scale, as these two are precisely identical as description of time and space. You cannot describe one without the other being described as well. They invert easily from question to answer as from idea to form by prevalence of closed systems.

This relationship is also exclusively factual, as opposed to philosophical, and as so directly opposite and parallel to its philosophical equivalent, as removed and opposed to it as possible. Due to its position in the hexagon it is therefore also one of the two most independent relationships from the others, and completely interdependent upon and amongst itself. This means that an answer to one implies immediately and unequivocally a solution to the other, and that therefore it is a relationship which has a potential solution in and of itself.

In our example these are both fields rather than wells due, once again, to the nature of the participant. For a man lifting a stone it would be equally easy to compress the exertion of physical energy into a specific event as it would be to expand it to describe an entire range of duration and place, but for the case of God it is impossible to compress the information of this relationship down to a concise mathematically expressible form, as God is the nature of an all expansive field.

3. Where —> What

The relationship between Where and What is less direct than between When and Where, but about as direct, if not just slightly more so, than between How and When. We can say then that the node of How is more philosophical, and that the node of What is more practical, but this is only by a very discrete and ultimately insignificant margin. Scale correspondence, or the first dimensional functioning of the cube, can make this difference appear magnificent, but as both are necessary, the point is ultimately moot.

Also, the node of What pertains to being, and there is a quantifiable correlation between being in itself and doing, or being for others, such that What can refer to either an entity or an event. Again, due to the predominance of closed systems in these two categories, What and When can often depend on one another for composition or designation. What also determines quantity, which corresponds to the space occupied.

In our example What is the consideration of both the stone and the acts of creation and lifting of it. This is basically a semantic differentiation, but underscores the fact of its ontological application. As such it doubles its implicative potential, and renders itself more useful in the derivation of solutions.

4. What —> Who

These are virtually already the same question, the only significant difference between them being that of objective versus subjective respectively — the same ontological division inherent in the What node, but liberated from semantics into a more ideal philosophy. We may specify the difference here as being one of ability to distinguish between themselves, wherein the lower node is incapable of this perception, and the higher, later one defined by it.

In the ethical realm this distinction remains the least accepted, as often one is spoken for by actions more so than conditions. We may take this as a form of ethical laziness allowed by the absolute indistinction upon this point by divine belief. It is often the case that when there remains dispute or contention in one field of metaphysical legislation for the realm of one node or relationship, it effectively weakens efficacy in the other fields as well.

For our example question we find these treacherous logical refractions profoundly underscored as neither aspect of this relationship is capable of exact definition — What referring both to noun and verb semantically, and God referring both to Who and What ontologically.

5. Who —> Why

This relationship is parallel to that of When —> Where, and is its philosophical equivalent. It is possessed of exactly as much ration as its sibling, in fact is composed of exactly the same reasoning, but you may think of it as the next higher dimensional expression thereof, distinguished from it only by the distinction of subject from object, or right angle from non-right angle metaphysically. This is not to say that When —> Where is exclusively hexagonal or that Who —> Why is exclusively cubical, for, as you have already experienced, each is possibly both. What this interrelationship of relationships should be taken to imply is that, just as are When and Where for one another, so Who and Why are synonymous.

It is impossible for either Who or Why to subside in isolation from each other. They are two ways of saying the same thing. For any event in space-time where a subjective entity exists, there will be a field of motive surrounding them that led them to that event. Similarly no event in space-time can occur without a motivating force, which will imply a pattern of implication equivalent in potential to the capacity for explanation unique to the subjective. In the realm of divine legislation this often leads to anthropomorphication that is contrary to purely materialistic observations, but is none the less substantiable following this train of thought.

Our example question makes all this quite clear, as the answer to the more philosophical aspect of it is a factor of its ontology — essentially expressible so: "Why does God do anything?" In the case of this subject, the being under consideration is defined in such a way so as to rule out conditional symptomatics, leaving only choice or free will to hold the account. It should be noted that, though Why is answered as easily for subject as object, physics has proven to be the least adequate branch in rendering solutions pertinent to the prediction of behaviors for variables bound by open systems and all related subjects such as entities.

6. Why —> How

The relationship here is interesting, as it is the only one that seems to imply direction, and the direction it implies is counterclockwise. It is rather as though the How node were charged in such a way that it were universally repulsive. Let us look at this rather oblate generalization specifically. For any event the function described is more intrinsically dependent on how it will operate, as this governs the material aspect, than on why, which is often a field covering a much broader period of time, both before and after the material functioning itself. In other words, How is like a particular and specific manifestation of Why, comprising all the esoteric components in physical form, but not obviously rendering them all except in effect.

Thus, if this relationship alone were to be considered, only by allowing the event to occur would one be able to deduce Why from How. Deducing How from Why can occur without the enaction of the correspondent event, but almost inevitably necessitates consideration of other relationships to define the cause and effect of the unrealized event.

In our example this is the pivotal trait of its paradoxical feature. How and why God would create a stone too heavy for Him to lift are both dependent upon the occurrence of the event itself. As Why cannot be determined from the open set of free will to which it is in this case referred, the only question that remains insoluble in terms of referentiality from the original proposition is, as was postulated initially: How?

7. Clockwise and Counterclockwise

If you are not understanding the nodes so far, the relationships between them may seem very much like a mad tea party, but don't worry. As Kafka would have said if he were a door mouse, "there's all the treacle in the world. But not for us." I assure you that, as we delve into further relationships, you will see the nature of the nodes more clearly revealed, and I encourage patience. You may well be wondering to yourself, on the basis of our example question, what good does the formal system do in transmuting questions from one form into another if it cannot answer them immediately. I hope that you realize the foolishness of this. We have widdled down a very complex paradox in our example to a single formal inquiry, and this process has been as easy as watching the hands turn on a clock as the time rolls past.

But let us pause briefly and consider this point, as we have gotten ahead of the schedule implied by the sloth of the remainder of intellectual pursuit and may stop for a moment to partake of the smell of some roses that have grown up along our way. The determination of the potential for the cyclical nature of certain events in time, and therefore the measurement of its passage by the establishment of regular intervals, was probably first discovered by observation of the movement of shadows cast by the sun while during its passage in the sky throughout the day. We can easily observe this in a sundial, and the very concept of right hand rotating circular measurement of time derived indeed from the placement in the Northern hemisphere of a sundial's orientation along the axis lines of the Earth's magnetic field. So we have the right-handed clockwise and left-handed counterclockwise.

How do these pertain to the formal system of metaphysics, as surely it bears no direct relation to the passage of time in the material realm, as it is an ideal, and therefore possesses no relevance to the sun, the external, material source of light by which this rate is reckoned. And yet we are told by philosophers and metaphysicians throughout all of history to factor in the idea of time, as well as to bear in mind the doubling of all material objects, especially the luminous, in the realm of the divine. Do we have another paradox, a mere axiom of practice, a holdover from superstition? Or is there something to this, by which to gain further insight into the model under consideration?

Let us again consider the concept of God, only this time not with the air of naivety implied by the wanderlust of the child, and let us also consider how such a distinction as this, which surely occurs in the realm of ideas, even though in reflection of natural change, yet still is an amaterial aspect of Time. Imagine once again your memory cube, enclosing the sphere of your mind, and encompassing the cube of the formal system of metaphysics model. Now imagine the hypercross of your memory cube, or, if you previously imagined the sphere of your mind as conjoining the midpoints of the cubes in your memory hypercross, as has its purpose, then imagine a cube surrounding the sphere in this state, and imagine this cube as the center of a still larger hypercross. This or similar extensions can be continued on ad infinitum, and so, for brevity, skip ahead to the end and imagine a hypercube of infinite proportion and a corresponding sphere, also infinite in boundary. The cube in this capacity represents Time, and the sphere, God. This is what was meant when it was said earlier that God is the nature of an infinite field. From this, following the last seemingly immeasurable leap, it is tempting to go a step further, and ponder over the insight of where exactly is God, then. But here we come to an impasse.

Take the nature of the electron. We know of its existence, yet cannot prove it in both time and space at once. When we measure its position, its velocity changes, and vice versa. We recognize this as a human shortcoming, and it is generally accepted that the electron itself is not God. But can we not, in the utmost of our humility tempered curiosity, propose that there may be a similarity in idea? If Time were taken to be a probability field, similar to that in which the electron exists, but expanded infinitely, would not God then be alike the elusive particle therein, which can be sensed but not detected, has effect, but not situation? So where would this leave our meager system of metaphysics?

We can imagine God as alike the sun, circling above us in the sky of pure Time, and shining down in the mental realm on the model of our conjoined memory, mind and metaphysic. Therefore its parts would cast a shadow, and this shadow, with the changing of God's mind, would change, alike a sundial. And therefore we would find, at first removed as we penetrated the ideal, and now given back to us by the divine, as much a miracle as resurrection, the concept of direction within a closed system of inherent motion. Clockwise and counterclockwise, wherein both are equally wise.

So what conclusions can we reach from this? Only that both our relationships can occur in either direction. But this doubles the number and nature of our relationships. We can see, already, how arbitrary it is to assign dimensional nature to our nodes, because, as soon as they begin to relate, that dimensional nature will begin to oscillate. The hexagonal becomes the cubical, and the cubical becomes the hypercubical. And all of this occurs even if we ourselves do nothing. We have reached a level where we see that such exercises as walking around our hypercube, though amusing, can no longer strengthen us. Even when we sit perfectly still, the hypercube revolves about before us. Instead let us turn our attention to effects of this meditative art, and discover what further insights await from contemplation.

8. Paradox of the Diagonals

One remarkable feature of the formal system of metaphysics model lies in its dimensionality. As we have already seen, this is intrinsically malleable, but what relationship does it present besides the hexagon —> cube —> hypercube function? To study this let us start with the hexagon and then permute it into the cube, selecting a particular measurement by which to assess what is actually changing when we say that dimension has been changed. For our purposes let us take the unit measure one, which should be applied to both the segments outlining the hexagon and the edges of the unit cube without dispute. Next let us study the outcome of this measure as the metaphysics model alters dimensionality.

In the second dimensional hexagon we can trace an arc with a center at any one point from either of the two points nearest it through the center of the hexagon. Thus we may say that the radius of the hexagon is equal to one of its sides, or one. This naturally implies also that its diameter is two.

In the third dimensional cube we may attempt to find this same line, but it will not be the diameter measure of the cube, which would connect the midpoint of a face with the midpoint of the face parallel to it. The value of the line we are searching for has now become the value of a diagonal, joining one corner of the cube to the corner opposite it. But this line will still divide in half at the center of the object, just as if it were a radial measure. Therefore imagine the line that connects the uppermost and lowermost corners of the cube. We should expect this measure to be, as we have already constructed, two.

To find this measure it is first necessary to construct a right triangle whose hypotenuse will be the diagonal of the cube we are hoping to find and the right angle of which will be one of the corners of the cube. This appears quite easy, as we already have the length of the edge; it is one. Therefore, the triangle being necessarily equilateral according to the structure of the hexagon, the length of the two legs of the right triangle should both be one. But they are not. One of them is the diagonal of a square. So we are back to two dimensions.

Imagine connecting the uppermost and lowermost corners of the cube. The diagonal links them directly, passing through the center of the cube. Half way between them on the surface of the cube is a corner with the distance of one unit from the corner below it (or above it, depending on from which side you look). The other measure will be across one of the faces of the cube, connecting the corner that lies directly aligned with the center to the uppermost or lowermost corner. We will discuss all these distinctions per rotation in a later section.

Now, according to the Pythagorean triangle, we know that the measure of the hypotenuse of a right triangle is always equal to the sum squared of its two legs' squared sum. So we can deduce the value of the diagonal of the square by constructing a right triangle that bisects the cube's face, using two of the edges as its legs. This yields a triangle with two legs equal to one unit each and an unknown value for the hypotenuse. To figure for the value of the hypotenuse we first square the value for the legs, and then add. The square of one is one, so for both legs the value remains one. Then we add these, totaling two. The value of the hypotenuse is then the square root of two, or, more simply, whatever value multiplied by itself once will give us a total value of two. The problem is that the square root of two is an irrational number, meaning it can be taken out to the nth decimal place, where n approaches infinity. Therefore, not only is the value of the diagonal of a square not a clean real number, it is approximately greater than one.

This means that the value of the triangle we calculate for the diagonal of the cube must be constructed using a leg which is an edge of the cube and has a value of one and a leg which is the diagonal of a square and has an irrational value greater than one. This remains a simple procedure, however, due to the nature of the functions we have to perform. As before we are dealing in the Pythagorean formula, so all we need do is consider the square of both legs. As we found before, the square of one is one. Likewise, the square of the square root of two is two. So the values of the legs total three. Therefore the value of the diagonal hypotenuse equals the square root of three. Concise, but still not what we predicted from the hexagon.

So what we have found is that simply by noticing the alignments of the cube's parts as it rotates in our mental projective space we may discover conundrums of number logic. We could as easily conclude that this lies with some fault in the formal system as deduce that as dimensionality increases the values of the same amount of space decline, but neither of these is an accurate picture of the truth.

These and the further esoteric applications we shall discuss shortly are merely impossible feedback loops in the logic functioning of the formal system. Without these it would not generate a field in and of itself, and could have little bearing outside curiosity for the purpose of reasoning. With them it acts rather like a dynamo or a computer, wherein data is entered in one state and transformed into another. This is the entire crux of the formal system's functioning. If you have understood this consideration, then you may find you can discover other similar structures in general metaphysics for yourself. Even if you haven't, this system will still serve you loyally as a fully operational tool.

B. Middles

The significance of our inspection regarding the cubic diagonal will now become readily apparent as we utilize it as the base unit in the next sequence of relationships. The subject of this section is the second dimensional hexagram, the stellated or star version of the hexagon, wherein the sides have been dropped to the points at which lines connecting two nonconsecutive corners intersect. Another method of expression which we will be considering here is the steloctahedron, which is the star version of an octahedron, which itself is produced by truncating a cube. Another way of looking at the steloctahedron, and the way in which I prefer, is as a pair of conjoined tetrahedrons which share the same midpoint. This view offers a clearer mental picture of the similarity between the steloctahedron and the hexagram, which is itself a pair of conjoined triangles sharing the same midpoint. The relationship between these two points is such that the hexagram is the shadow of the steloctahedron, in the same way a square is the shadow of a cube, or a cube is the shadow of a hypercube. So you might think of the steloctahedron in these terms.

The hexagram perfectly describes in its angle sum the measure of a unit circle in degree of circumference. Therefore it is one of the most ideal representations of the sphere of the mind. It has been recognized as such in the art and geometry of countless varying peoples for thousands of years. It also represents the minimal number of groupings possible for same size objects around a center. The most common example of this is a circle of coins surrounding a coin, all of which are of equal size. The steloctahedron is merely an extension of this, as all of its faces are made up of equilateral triangles, such that there are eight points emanating from a central form completely obscured by the intersections of their bases. The total measurement of angles on all of its exposed faces is four thousand three hundred and twenty, or the circumference in degrees of twelve circles. This is obviously an expression of the same device as the hexagram, but elevated by one dimension.

We will continue moving clockwise around the formal system, following the nodes as they connect along these diagonals. After doing so some traits unique to the model in this capacity will become apparent and we may wish to consider them.

1. How —> Where

This is the practical aspect of the upward pointing triangle or tetrahedron. It pertains to necessity, the resources of location that are available for function, and also ingenuity, or the process of creation of complex forms from base states. Without this pattern situation would fail to arise, and situations, along with free will, are the primary mechanisms of legislation.

There is a strong relationship between this line and the node opposite it. The degree of angle native to the node of Who defines the length of this leg, and we see this manifest in experience as well. A large factor of who one is is determined situationally, by condition, and vice versa, by free will. I will discuss this more thoroughly as I go through the remaining relationships that pertain to this form.

For our example question this relationship is also pivotal. Where God would create the stone raises the differentiation of internal and external, as the universe is all we know, and yet God is thought to be outside of it and, in a way, containing it. So the stone could only be created in the physical universe, which we already know is of a manageable proportion to God, or within God Himself and outside the material universe, which would render it impossible for the mind of man to comprehend, according to the majority of divine legislation on the matter. This means there can be two right answers to How: either physical or divine.

2. Where —> Who

This relationship is the base of the upward pointing triangle or tetrahedron. Here we find the beginning of intermingling between the objective and subjective which defines the essence of philosophy. It is the permutation of evidence into proof, and as well by way of this that question gives forth answer. Thus this is not only the base of this form of the formal system, but also the foundation of its implication on the contemplation of ideal forms in general.

For any location, as has already been observed, there will be a motive field equivalent in potential to the open set of an individual. This leads to strong associations that transcend the physical but that only border on the divine. Ethically this relationship can only imply the angle opposite it, that of the node of How, as this answers the formative details that define the presence of a being within situation, their purpose and its extent.

In our example this question pertains to the location of the idea of God in relationship to the form of that idea. It is only because of the ontological definition of God that He sustains the position that He does in any and all logical hierarchies of metaphysical legislation. If we were dealing with a different individual, the location would be fixed physically, and this would imply a set of laws governing the angle opposite. But instead we are dealing with the almighty, which implies only infinity.

3. Who —> How
Of the two philosophical relationships in the upward directed form, this is the more applicable. This relationship embodies the expression and remanifestation of energy, from the personal to the practical, and, in terms of explanation, associates them back to another alike signature. There is a certain standard that exists in a craft guild of what constitutes fine work, and those capable of surpassing this level become known for it to such an extent that they may be identified even in their absence by their style or method. Similarly one may identify a repeating pattern of expression in a friend that goes unrealized by them, which would serve to change their entire perspective.

In this the relationship pertains to Where as well. Who a person is is very often identified with how they do certain things, but both of these are tempered by the conditions of their environment. This relationship goes a long way to embodying the entire concept of free will, but only as the measure of an angle defined by place. This is the last component to be considered in the initial triangle/tetrahedron.

With the case of our example in mind let us ponder over the unique creation of a stone, possible only through a process spanning a duration so much greater than a human life-span that it obviously rules out all but the most advanced natural or some supernatural force. When dealing with the Creator of All we must remember also that the creation can occur outside these rules, but may yet be manifest within them. Thus the answer to how can the stone be created pertains to divine legislation, but the answer to how it may be lifted remains potentially physical.

4. What —> When
This relationship pertains to the natural flux of outward or inward flowing of energy, again best evidenced by the shell jumping, light emitting or absorbing properties of electrons. At all times matter is being converted into energy, and vice versa. The direction of this conversion is entirely a function of time. The same process is evidenced in the activity of mental energy, in the form of creation or consumption of information. We are, of course, on some level, doing both at all times, but depending on the nature of the moment, our activities will be predominantly one or the other.

As such one cannot occur without the other, and function is inherently determined by timing, as is timing a process governed by the progress of function. As we have already seen the activation, function and termination aspects of What take up a field of time, even if this field is reducible to an instant. Similarly, time is marked by the passage of intervals, which are themselves dependent entirely upon the functioning of the components by which they are enacted. Often, we even associate the patterns of mental energy with their physical detritus, and this pertains greatly to ethical legislation.

For our example the nature of God again supposes that When is less important than What, although we may use the relationship given by the relativity of time and space in physics to underline the fact that the event can occur outside of this closed system. The creation of something, such as the stone, may be thought of as possibly a mental activity, such as the creation of a thought, or even the creation of our formal system of metaphysics. But the act of lifting implies a necessary time correlation in the domain of a closed system.

5. When —> Why

Here we have the relationship rendering the formula for duration as a function of stamina, or perseverance. We generally know, according to the legislation of physics, and to some extent this can be proven to occur in other realms of metaphysics as well, that an event will continue on, all other things being equal, until acted upon by another event. This appears to be similar to the process in the first form of the triangular or tetrahedral relationships of condition, but one is primarily an ethical consideration, while the other is an equation more of the domain of physics. In this most closed set of legislation, all other things are almost never equal, and it is an inherent event that determines change in pattern.

It is also generally temporal, as the rate of matter-energy exchange is only a constant for a period, which can also be used to render interval. At the beginning and end of this period there is usually an alteration of internal composition, which is rendered in proportion to an open set of potential, frequently expressed in physics by use of an irrational number at the moment of symmetry breaking.

In our example a large portion of motive pertains to liberation from the closed physical-temporal system. Only a divine being could conceive of accomplishing a task which is obviously paradoxical to a being confined within a lower set of logic.

6. Why —> What

This relationship governs the formation of material as one moves between an open set and a closed set, and is generally substantive of the sum that occurs in equations at the moment of conversion. These figures tend to transcend ration in such a way that they are more than the sum of open and closed sets.

It sounds rather esoteric, especially as it pertains back to so many other relationships already covered as well, but is, in reality, really quite simple. Imagine the ejection or absorption of a photon during energy shell shifting by an electron. The energy in itself is not effected, as the charge of the electron remains the same. Neither is the mass, as the combination of the charge of the electron and the mass of the nucleus remains constant. But the event of luminescence or darkening occurs.

Let us ponder our example question, to clarify this further. We have already demonstrated the impossibility of comprehending divine free will, and specified that the two actions involved in the single event occur in systems differently bonded — the act of creation of the stone in the mental domain of idea, and the act of lifting in the fixed physical time-space coordinate system. So this relationship deals with the event itself, as divine free will lowers itself to interaction with the material realm, to create the node of What in noun, the stone, to be lifted.

7. Male and Female

We have dealt now with the two essential forms of middle measurement, which themselves may be considered second or third dimensionally as you wish. We have seen that their relationships are a little more complicated than those of the shorter measurements, though their dimensional binding remains unchanged. We have seen the distinction of several aspects of the formal system, such as between semantic and ontological, between practical and philosophical, between open and closed systems, between free will and condition, and between matter and energy. It is imperative to understand that all of these distinctions are really only between two expressions of one idea, as the two hemispheres of the brain originally contributed formation and legislation in the mental projection by which we created the formal system of metaphysics model. Not only do they inherently agree with their apparent supplements, and not only in substance, in end as well, but they are all complimentary to each other as a whole. First I will clear up these distinctions in the order they have cropped up, and then I will go into the implications of the distinctive nature inherent to the formal system.

When we talk about philosophy as rhetoric, little can be said that does not break down to the distinction of semantics and ontology. This was true in the time of Plato and it has been a traditional trait of scholars to this very moment. It is difficult to speculate on the nature of the mind without considering the nature of formal thought, especially in the textual and contextual studies of semantics, and the nature of formal being, as evidenced in the countless tomes devoted to understanding our existence. But by what right, besides this tradition, do we even keep these two schools close? What can really be said by one if it cannot be said apart from the other? This, though, is there greatest strength. We have, in our attention to convenience of categorization, inadvertently assumed that a relationship exists. That the systematic data collection of the mind and the formation of the substantiable are indeed similar. What an astonishing feat for humanity! The mind and matter, essence and existence, are two aspects of the same nature, two sides of one coin. All philosophy has proven, despite the muddlings of its inquests, is done so not by distinction, but by its very being: what we call philosophy has metaphysical counterpart, provably.

This brings us to the distinction between the philosophical and the practical, which, though considered most often as ethical, is little more than an extension of the preceding distinction outside the realm of pure research and debate. When we discuss the philosophical we are usually referring to the realm of the ideal, first described by Plato, or to its pursuit as engaged in by all the subsequent adherents to his notion. This sets itself somehow against the practical, as the latter defines itself by pertaining exclusively to the "real" world, and all related concerns we deem needful to survival, such as economy, commerce, politics and stimuli. But already metaphysics defines itself as the bridge between these aspirations, proving itself to be a functional, socially useful tool that, even yet, is composed in philosophical ideal. This is the case both with general metaphysics as the quest for meaning to human endeavor and natural organization of information, but also to our formal system.

What is really being debated here is the distinction between open and closed systems. It is thought that all things whose measurement is ultimately subject to change form open, or all-inclusive, fields, and that all things that behave in repetitive and analyzable ways over great periods of time fall into closed, or fixed, wells. Thus, for example, the behavior of the mind is open, while that of the body is closed. This distinction is expressed intrinsically in potential, which is the substance of time. So it has, again through no intention of its own, given us the stuff of the ideal realm from which philosophy is made and semantics and ontology are unified. This is achieved by moving between open and closed systems, by doing which we find that their combination is equal to more than their sum alone. Results, even in the strenuously skeptical practice of physics, tend to concur with this prediction, as the digits generated at moments of symmetry breaking, when one pattern gives way to another of a higher order, are often irrational. More on this will be said shortly.

The most classical distinction between open and closed systems occurs within the ideal world of philosophy, and has been the issue of much debate by ethical legislators for several hundred years, and that is that between free will and condition. Free will's position dictates that all events are ultimately the outcome of forces internal to their mechanism, such that only choice determines result. In opposition to this determinism states that conditions in the continuum external to an event have equal influence on its enaction, and that symptomatic conditions delimit potential. To this day the issue has remained unresolved, because it has not, as is often the case with ethical legislation, been systematically formalized. It is, instead, left to be recorded, case by case, in hopes that eventually accumulation of evidence for one or the other will outweigh its contestant and resolve our quagmire. This again, is purely laziness, and itself good evidence of the least common denominator of all possible decisions being the one settled upon in spite of favorable conditions.

The final particular distinction is between matter and energy, and this is the comedy of the cosmos. We know that our universe is made up of some substance, which we shall call quantity x. It seems that this substance can be either matter or energy, but only in such a way that there is never more of them when combined than quantity x, and in such a way that it really doesn't matter if there happens to be more matter or more energy at any given moment. Because of the convenience of the evidence we have at hand, thus, we say with the utmost confidence that the universe is made of matter and energy, but as readily admit that we don't know of what higher order substance these really represent but facets. If only we had a tool of some sort, perhaps we could extend our meager reach! We may learn much from observing our simian fellows attempting to reach a dangling banana while given the opportunity to utilize stackable boxes or even a stick, for there is really very little remarkable difference in the long run between us.

This brings us back to our formal system, a little more enlightened about the fields relevant within its creation. Seeing how easily these disparities resolve themselves, then, how could you have initially found it so impossible to imagine answers being evolved only from the substance indicative of question alone? But there is more. To understand the relevance of distinction within our formal system it should prove beneficial to examine the relationship between the forms in the middle measurement model. We see in the hexagram the conjoining of two triangles and in the steloctahedron two tetrahedrons such that they face opposite one another and share a common midpoint. This is the geometrical expression of the idea of distinction. We observe diversion in their intent, and unification in their core.

So let us compare these, to make one final distinction, to male and female. This has been the practice of metaphysicians throughout the ages, and therefore should neither be looked at now as redundancy nor obfuscation. This is, afterall, a revealed art, and we must always bear in mind that we are merely learning from those who have gone before, contributing of our own only what is seemingly absolutely necessary and seemingly otherwise unavoidable, remembering all the while that the information is not ours any more than it was theirs, and is first and last of reality, posessable by no one, and eternal. Just as, according to the Hermetic tradition, are the ideas of masculine and feminine. In traditional metaphysics the upward pointing triangle is usually taken to represent the masculine aspect, and the female is the triangle angling downwards. This it is hypothesized reflects genital orientation.

One of the most appreciable similarities of man and woman is their ability to see the differences between them. This beauty is so ingrained upon our consciousness because it is a factor of propagation — creation by unification of opposites. It is of us. So is this distinction between the two triangles, or tetrahedrons, of the formal system. It propagates motion by the interaction of distinctions. Without it, there would be no movement, and without movement it would not generate an attractive field, which we will discuss when we look into its center. All the relationships are dependent on this movement, and all serve to stimulate it, so that, as it occurs, so does permutation. By it question is converted to answer.

Let us look at these forms specifically: first the masculine and then the feminine, in order according to the occurrence of their points in a clockwise progression beginning from the uppermost node. The male triangle is composed of How —> Where —> Who, and the female triangle is composed of When —> What —> Why. Both of these become tetrahedrons when related to the points of How and Not, but for our purposes here we will only be considering them second dimensionally, because either triangle can be related to either of the two other nodes to render it a tetrahedron, and therefore there are really four possible structures that can occur at that level of reasoning, two male and two female. This becomes a factor when reproduction of idea occurs between dimension, which we will observe eventually.

The male triangle deals with philosophy, in particular ethical issues of free will versus determinism. This is because the nature of its relationships is inverse to the nature of its nodes. We have seen already how each relationship in itself is governed by the angle of the node opposite it, and this function applies holistically. The "angle" of the female triangle, or the overall essence of its nodes, is philosophical, and so it is the male triangle that must deal with philosophy. The parts with which it serves to do this are more practical in their nature. How is the most practical, followed by Where and then Who, exactly in clockwise order. These also progress from most closed to most open sets in solution, corresponding to degree of ideal.

The female triangle deals with practicality. It must confront and rationally translate the functioning of the masculine nodes into a reasoning relationship. But it does this with nodes better served themselves to philosophical logic. The progress of them from philosophical to practical is neither as clearly defined by the rotation of the hands on a clock as is the masculine: Why being the most philosophical, followed by When, and finally What. But this counterclockwise sequence does correspond exactly with progress from an open set of solutions to a closed set, representing use.

So these are the natures of the primary forms after the hexagon, hexagram and steloctahedron as components of the middle measure. But within these we find further relationships occurring between them. Each individual relationship in one has a parallel in the other. We will look at these, as well as relationships between the nodes opposite them, in clockwise order from the uppermost point, but again let me caution that such an ordering for examination is entirely arbitrary, and as much may be discovered via alternate considerations. These three relationships and the dyad of triangles in which they occur comprise a pentad. This generates the symbol of the rose, whose petals emanate in bunches of five per layer. The rose is also taken as a symbol of the genitals.

The first such relationship we will consider is between How —> Where, with opposite node Who, and Why —> What, with opposite node When. The first thing we can tell from each relationship is whether it is philosophical or practical by applying the inverse relationship to the angle of the node opposite it. This tells us that both of these are practical relationships, with the male more so than the female. How —> Where renders necessity in situation, pertaining to the individual, and What —> Why renders material formation between open and closed sets, pertaining to a moment. So what we are seeing here is a correspondence in matter-energy exchange and alteration to environment; in other words, a change of doing that creates a change of being. At the moment an individual is involved in an event, we may see that there will be surrounding them a specific environment that is created by the interaction of data. They may or may not see all the opportunities that surround them, but these opportunities will exist nonetheless. When one takes an action it increases the probability of a certain outcome and conserves the probability of all the outcomes implied by the other possibilities present, and they will begin to become new possibilities, but still only expressions of the same probability. This relationship represents the physical practice of legislation.

The next relationship is between What —> When, with opposite node Why, and Who —> How, with opposite node Where. In this case the female relationship is practical while the male relationship is philosophical. Thus we are not dealing with metaphysics as hyper-reality, but as an interface between coordinate and ideal. What —> When renders creation or consumption, pertaining to field of potential, and Who —> How renders style of expression, pertaining to an event well in time-space. We may see from this that, wherever the event of exchange in form or state within potential occurs, it does so in a pattern identifiable with an individual open set. When something is created or destroyed, it fulfills or leaves behind its geometrical and relative essence in the realm of idea, and this has a direct binding to a preexistent set of ideas inherent to an individual or group of like-minded individuals. In short, nothing can come into being which is not symbolic of something else, but in order for the connection between these two modes to be translated, the presence of an entity is required. This relationship represents the ethical practice of legislation.

Lastly we come to the relationship between Where —> Who, with opposite node How, and When —> Why, with opposite node What. These concepts are all related along philosophical lines, the male and female performing this in harmonious association. Where —> Who renders the transition between objective and subjective in a motive field, pertaining to function. When —> Why renders inherent stamina, pertaining to event. We may see this as the containment of the fixed set of stamina within the open set of motive, where the exertion of the former arouses interaction between subjective and objective concerns in the latter, in the performance of any deed. This relationship seems best to describe physics, so, looking for the node opposite it, we assign to it the divine practice of legislation. I will return again to the idea of the opposite node defining the relationship, especially as it pertains to the three legislative branches, very soon.

8. The Presence of Irrationals

As we have beheld, the entirety of these relationships occur along lines defined by the diagonal of a square, the measure of which we have found to be the irrational quotient of the square root of two. This figure is approximately 1.41421, but continues on post-decimally, infinitely. It is one of the select number of sums that, in their indefinite capacity, represent pure potential.

Would you be surprised, then, to learn that, in the pattern of the hexagram or steloctahedron, another irrational occurs in the context of the measure of its height? This second irrational number is phi, calculated by dividing the square root of five by two. Its solution is roughly 1.61803, but, just as does the square root of two, continues on without end. This represents the diagonal of the cube, subtracting the pinnacle of the small tetrahedral stellation of the steloctahedron, or the distance from one node to the midpoint of the relationship its angle describes.

In other words, both the relationships of the nodes in the hexagram, and the relationship between them and the nodes opposite them, are hung upon transcendence. These are points, within the formal system, associated with moments of symmetry breaking. What we are looking at is the exact instant of transformation from question to answer. In the case of the formal system we are moving between an open set and a closed set, but these numbers mark the same effect regardless of which direction we travel.

They occur frequently in quantum mechanics in the determination of spin, the inherent angular momentum of subatomic particles, which occurs in isolation from all external influence, but can only be accurately observed in vector space when the particle is acted upon in such a way that it changes form or state, or in other words, when a break in the symmetry of its established pattern occurs, and it becomes governed by a new pattern.

They also occur in another field of physics, and a most unexpected one: that of botany, as the two primary branching patterns of all plants. When plotted graphically these numbers render ascending spirals marked at intervals by symmetry breaking. At the points along the spiral where symmetry breaking occurs, branches form on trunks, stems form on branches, and leaves form on stems. The entire geometry of the fractal growth pattern of a tree can be summed up in either one of two single irrational numbers. And this is the case for all of these which occur in nature. Whenever you have looked at any thing from the smallest flower to the tallest tree your entire life, you have been looking at transcendental math.

The implication of these numbers' occurrence within the formal system is one of reproductivity. They serve as the male and female aspects that, between them, synthesize solutions. We have just seen how all distinctions inherent to the system are self correcting, and now we are looking at the mechanism by which this occurs.

The best way of understanding this is to correlate it to the ideological distinction between image and symbol, such that each relationship generated between two second-consecutive nodes presents a picture, and the imposition of the node opposite gives it meaning. We will continue to consider this as we look at the next set of relationships, which are between opposite nodes. It is interesting to note that the square root of two divides the diagonal of the cube in such a way that phi remains, but if we do not immediately exclude the additional area, then we are left with another familiar symbol, that of the cross. Just as the symbolism of male and female was handed down to us from the Hermetic tradition, there is a similar school of metaphysics that deals particularly with the image of the rose which was rendered by the union of the sexes in combination with the symbol of the cross. The adherents to Rosicrucianism traditionally consider the six square cross, which is the unfolded cube, and second dimensional shadow of the hypercross we have already constructed. In the next section it will become necessary to deal directly with this symbol, as we consider the relationship between the nodes occupying opposite corners in the folded six square cross which is the cube of our formal system of metaphysics.

C. Longs

When we considered the relationship of the node opposite each internal triangular or tetrahedral relationship, was it as right to call it definitive as if we had been looking at the relationship of the relationship to the node between? Afterall, it is the node that occurs between two second consecutive nodes that actually measures in angle the length of that particular relationship. In the case of the node opposite the relationship, what we are in fact measuring is that relationship which occurs parallel to the one we have agreed it defines. This is because each relationship between triangular or tetrahedral leg and the node opposite it is the inverse relationship. The node between two nodes positively defines that relationship, and the node opposite positively defines its parallel.

Now the relationship of the node between to the node opposite is always identical — the two nodes are in agreement in degree of legislation in terms of set system. Each composes a different branch of legislation, and governs parts of relationships in the lower levels of measurement. We will again be dealing with them in a clockwise procession that allows simultaneously a metaphysical progression in the subject of legislation from the base to the ascendent. As we go through them you will see how they function within the formal system, and come to perceive a much more profound wisdom of reasoning. By moving through them you may see how they are determinant of any and every event, and begin to make applications from them to both the greater implications of the formal system to the quest for meaning and the further relationships within the system that define formation of logical conclusions.

To illustrate the relationships between opposite nodes we will again refer back to our example question, now greatly transformed from the bemusing ponderance of naivety to such an extent that it appears rather monstrous. I will attempt to deflate the haze surrounding this question during consideration of these relationships, and hopefully it will begin to compress down into a more stable form as we go on.

1. How <—> What

The first of these nodes is the peak of the smaller triangular projection or tetrahedral stellation of the relationship between Why —> Who, and the latter is the peak of the smaller triangular projection or tetrahedral stellation of the relationship between Where —> When. Each of these is the node that occurs between the nodes in relationship. These parts represent a square defined by the isosceles triangle of How —> When —> Why joined with that of Where —> What —> Who; another expression of this same combination is of their three dimensional tetrahedral equivalents, which form a miniature cube nested within the cubic formal system, creating its hypercube. We will see in these relationships that this hypercube is defined in three states by the combination of the stellations peaked by each of the opposite nodes. This process is the very creation of a well from the parts of a larger field, or in short, answer in question.

Another way of seeing this event is the separation of symbol from image. We begin with a relationship of irrationals, and end with a unit Platonic. That for How <—> What determines the natural structure of the event, arranging all the aspects of environment into a pure image, altered only within situation by the transmigration of relationship. This is, again, the legislation of physics, the most closed system.

The best way to understand the relationship of the nodes opposite is in looking at the conservation of potential in their conjoining, where two corners of the joined triangles overlap, and are absorbed into one another. The triangles in the second dimensional shadow of the center cube of the metaphysics hypercube are those of the half faces How —> When —> Why and Where —> What —> When, to form the square How <—> When/Where <—> What <—> Who/Why, in which the overlap in nodes occurs between the two practicals and the two philosophicals. In short, the square is formed of three practical nodes, two of which are singular and one of which is combined, and one philosophical, which is a combined node. It is therefore a highly practical form, defined in inverse to the outstanding philosophical node it contains.

Let us look at this in terms of our example question. We have already seen that the creation of the rock can occur outside the realm of physics in the ideal realm of thought, but that the act of lifting it is bound to the smaller, closed set, so we can assume that this relationship pertains to that terminatory feature of the question, despite the fact that we are considering it before we look at the legislation that covers its actual creation, but knowing while we do this that it is at once according to the natural application of a formal system and partially an attribute of the divinity of the creator. In other words, the act of lifting it occurs temporally before the act of creating the stone from the perspective of somebody bound within the closed system of physics, but we will shortly see that this is irrelevant from the perspective of divinity. We may say it is a necessary function, moreover, because only by determining how much can be lifted can the idea of something greater than that quantity begin to be conceived. But again, we must remember this explanation is incomplete, and actually inverted, from the truth that pertains to its creator.

2. When <—> Who

The first of these nodes is the peak of the smaller triangular projection or tetrahedral stellation of the relationship between How —> Where, and the latter is the peak of the smaller triangular projection or tetrahedral stellation of the relationship between What —> Why. As before, each of these is the node that occurs between the nodes in relationship. The square formed by these relationships joins the isosceles triangles of How —> When —> Where and What —> Who —> Why in such a way that the nodes of Why and How overlap, and the nodes of Where and What overlap, such that it is defined as When <—> Where/What <—> Who <—> Why/How. We see from this that it is a partly practical and partly philosophical relationship, defined by one more philosophical practical single node, one completely practical double node, one philosophical single node and one half philosophical, half practical double node respectively. Thus in total it is evenly balanced between practical and philosophical single nodes, and the nature of its relationship must be determined from examination of its double nodes. One of these is completely practical, while the other is evenly balanced between practical and philosophical, so in majority this is a more practical relationship.

Looking at it in terms of image and symbol we may see that this is the beginning of symbolizing the image. The composition of the inner cube of the metaphysics hypercube has turned from being entirely rational to being more contemplative, with the closed system of time being dependent upon the subjective interactions of its objective parts to render accurate interval, and the entirely open set of the individual put in opposition to this in the form of semantic and ontological entity. This is, in essence, an expression of the anthropic principle, which states that only because things are the way they are do we exist in a way where we can come to understand them as such. In other words, the reality of the environment and the reality of its observer are interdependent in such a way that they represent relative states of the same substance and in various manner co-create one another.

What we are considering here is the legislation of ethics, the partially closed and partially open system by which all of human behavior is sought to be apprehended and predictable. The relationship here dictates that determinism is only a weak function set against the field of free will, as it is itself shaped by the interaction of components within event. Time, however, as well as consciousness, can be idealized in such a way that fate must be one of the considerations at this point. Some decisions are necessary within situation, and it may be that these together comprise the parts of a larger functional system that dwells therein. We will deal more thoroughly with these notions when we consider subsequent nodes.

In our example we are spared all such weighty oppositions of subjective conclusion, as we are dealing with the divine, who we have already established is liberated from determinism as He is from all Time. The very nature of the behavior under consideration has given us evidence of this positioning, and so we are left for this relationship with a universally applicable solution rendered in a completely open set.

3. Where <—> Why

The first of these nodes is the peak of the smaller triangular projection or tetrahedral stellation of the relationship between When —> What, and the latter is the peak of the smaller triangular projection or tetrahedral stellation of the relationship between Who —> How. As before, each of these is the node that occurs between the nodes in relationship. The square formed by these relationships joins the isosceles triangles of When —> Where —> What and Who —> Why —> How in such a way that the nodes of How and When overlap, and the nodes of What and Who overlap, such that it is defined as Where <—> What/Who <—> Why <—> How/When. Just as with the last relationship, the single nodes are opposed to another as one closed set in the practical realm and one open set in the philosophical realm, so we must look more deeply into the doubled nodes to discern the nature of this relationship. As before, there is a predominance of practical considerations, but we may see from considering prior relationships that they are also of the more philosophical nature in their verticals, particularly How and When, which, when related to What and Where, are seen to be the more open of the pairs. So even though it is highly practical, we may see this relationship is by majority of interaction philosophical.

As such it is the completion of individuation of symbol from image. The nested cube of the metaphysics hypercube has gone completely from being purely self defined and functionless except in intra-relation with itself to being purely defined externally and functional exclusively in reaction or response to input, from being in itself to being for others, and what we are dealing with now is purely consciousness as it exists in its form of an open field determining all motivation. The irony here, and the real miracle of the hypercube, is that consciousness is simultaneously inside and outside of the environment. Let us return again to the sphere of the mind: it is situated in relationship within the hypercube in such a way that the corners of the internal cube touch its circumference and the facial midpoints of the larger cube touch its circumference, in which the inner cube represents event and the outer cube represents the metaphysics model. But the sphere of the mind is at the same time a hypersphere, and thus the corners of the metaphysics cube also thought upon the circumference of another, larger sphere, which is also circumferential to the midpoints of the faces on the still larger cubic memory castle.

A square is the shadow of a cube, and a cube is the shadow of a hypercube, and this is the basis of dimensional geometry with which we have worked so far. So of what then, is the hypercube a shadow, as we are seeing it now to be contained within the triply nested form of the memory castle? This can only be a shape of the fifth dimension, just as the hypercube is of the fourth spatial dimension. So we are really running the gamut of all the dimensions in our model. Let us review them as they collapse: the memory castle is fifth, the sphere of the mind is fourth, the formal system of metaphysics model is third, the map of event is second, and, as will be explained shortly, the internal node of Fact and Fiction is first.

With the completion of the symbolizing of image, which renders meaning utterly relative to consciousness, we come to the legislation of divine belief, the most open system. With divinity comes the recognition of the interaction of dimensions, and the production of meaning as a process thereof. For any event, in other words, space-time coordinate and the field of potential particular to the duration of function are synonymous.

What this means in our example pertains to the location of the stone. It is governed by the completely open field of God's free will, and as such occurs where the external aspect of the mental hypersphere contacts the surface of the memory cube, implying that it is not created from nothing, per say, but from the immaterial purity of data collected in the series of lift actions determining potential limit. The fact that it cannot be brought down and contained exclusively within the closed set of event is equivalent to the concept of its being too heavy to lift. When it is reduced, therefore, it does not translate into a classically defined three dimensional form, but remains predominantly a mental projective ideal. In other words, it becomes, when passed through the sphere of mind toward event, the formal system of metaphysics. By the functioning of the formal system in creating event, however, we see that the stone is transformed as well into the action of lifting, and is therefore no longer impossible. So we see that such a stone can exist, in the mind of God, but that, as it is brought down into reality, it becomes increasingly possible to lift. Its potential is thus inverted, and this proves the presence of the perpetual functioning of the formal system, as it was promised, by transforming an insoluble equation into an elegant solution. In short, by learning how to manipulate our mind to fluctuate between dimensions using the formal system, we have also learned how to conceive of and accomplish feats outside the pre-established domain of our reasoning, and thus to expand it, in the manner of metaphysics, as is done perpetually by the free will of the divine.

4. Attractive Poles

In all the relationships we have seen thus far there has been proven to be attraction between nodes. In most cases this occurs because of distinctions, which naturally bond to one another as they are dual expressions of a singular idea, and tend, therefore, to be self nullifying. We see this in the short relationships most clearly, in the middle measurements as tempered by the node between and defined by the node opposite, and in the long relationships as exemplary of a hypercubic form. This process is absent only in the three vertical relationships as one looks at the formal system above a corner such that the node of How is seen to be at either the exact top or the exact bottom, but their uniqueness holds true regardless of the positioning of the cubic model. The most practical and most philosophical short relationships, as well as the exclusively practical long relationship, are defined by a homogenous attraction.

This plays an important role in the configuration of the model. We have stated before that each node, in imitation of the central node of Fact/Fiction, is one dimensional, and that consciousness occurs as the result of perceiving multidimensionally. In other words, to be self aware of any part or relationship within the formal system we have created within our mental projective space, we must ourselves be in a higher dimensional level than what we are considering: for a node we must be in at least the second dimension, for a relationship we must be in the third dimension, and for the formal system itself we must be in the fourth dimension. These all correspond with the components we find occupying these dimensions, as without being within a relationship we cannot perceive a node, without being in the metaphysics model we cannot perceive a relationship, and the metaphysics model itself occurs within our mind. But when we get to the perception of our own mind, things begin to get a little bit hairy. To perceive the fourth dimensional mind we must employ our fifth dimensional memory. This implies the relationship in the metaphysics model between When and Who, just as all the dimensions before this point, and from a certain perspective this one as well, have been answered for by the relationship of How and What. To perceive our memory, then, we must employ the relationship of Where <—> Why. But what does this tell us about the sixth dimension?

Let us go back to the first dimension again. We have seen that it explains the form of the nodes themselves. But we have also seen that some of the nodes deal with exclusively open sets while some deal with exclusively closed sets, and most with sets that are somewhere in between. How can this be the case of a purely first dimensional form? Obviously we have overlooked something here.

The way to come to an understanding of the dimensionality of your complex image of the formal system is to use it once to create an event. You will see as this happens that as they are considered, each of the opposite nodes defining stellations lowers itself down into the center of the cube until it is parallel to the short relationships between the two nodes to either side of it. By doing this, a hypercube is formed, and it is defined three times, once physically, once ethically, and once divinely. After this is completed the formal system ceases to be confined by the third dimension only, and takes on the function of a hypercubic fourth dimensional shape. This brings about a series of changes to the other dimensions in the model.

Around the event there is a sphere which is formed by the concentration of the mind. The sphere that surrounds the formal system and the sphere that surrounds the memory castle do not go away. They remain, and the sphere of the mind is now triply nested. It has become a fifth dimensional shape. It is a second dimensional circle within a three dimensional sphere within a fourth dimensional hypersphere. These layerings may be legislated according to physics, ethics and belief as well, in an ascending order proceeding outward from the center.

At the border of the mental sphere's most external circumference is the memory cube, which we have already demonstrated is really a hypercross, with six information accessing data bases surrounding a central cubic port. In other words, each of the functions of the nesting of the mind are doubled in memory, for short term and long term recognition capacity. This is the memory cube in its fifth dimensional state. But as event occurs, the inner cube of the central hypercube becomes a hypercube itself, and this entire form slides down a dimension to relate to the formal system model. In other words, memory becomes contained within the mind, and this is the event of the bringing forth of consciousness from the unconscious. The aspect that retains the position previously held by the hypercross model of memory is now held by the unconscious, which is essentially similar in structure, although much less is known of it in detail. It is comprised of a hypercross of hypercubes, each representing different perspectives on event available to, if needed, but outside of, consciousness as it relates to the event itself. Here we find the formation and maintenance of archetypes as established patterns of behavior, a sort of instinct unique and universal to a species fueled by the experiences of our genetic predecessors. You can go one step further than this and experience the interaction of these archetypes as a hyperoctahedron linking the inner cubes of the unconscious hypercross contained within the linking of their outer cubes as mythos. An equal amount can be said for the theory that these hypercross arranged hypercubes represent alternate states of consciousness. But much of this data has been tainted by the need for ethical and divinatory frameworks for the unconscious by past researchers.

At this time I should forward a more metaphysical proposition to explain these unconscious hypercubes. These may be seen as probability wells, which determine possibilities surrounding an event, but are often obscured by the details they involve. Consider their relationship to the rest of the model: they are external to the sphere of the mind in such a way that their external but not internal aspects are in contact with its circumference; they function through memory, which has become a hypercross within the sphere of mind, upon the metaphysics model, and effect the later relationships of its nodes after the formation of event, thereby acting as part of the proper function of event, but not as part of the initiatory or terminatory functions thereof. It is possible, therefore, that only when event has been directed towards one of these unconscious hypercubes, can it formally conclude due to internal mechanisms which will be considered shortly.

The third dimensional event itself is surrounded by the halo of the second dimensional mind, the fourth dimensional formal system encompassed by the sphere of the third dimensional mind and, outside of this, the fifth dimensional memory castle of consciousness, which is itself housed in the fourth dimensional torus of the mind, and this is resting in the the sixth dimensional unconscious hypercross of hypercubes. All of this is concurrent with the formation of event. It seems that we have only added another dimension, that of the unconscious, and compensated by not referring to the specific dimension held by the sphere of the mind. As event progresses this process corrects itself. The mind stabilizes its fluctuations and becomes the fifth dimensional form. Memory is the sixth dimensional structure, the hypercubic hypercross, its centers of short and long term legislation becoming active as hypercubes in themselves. The unconscious becomes seventh dimensional, incorporating simultaneously the aforementioned hypercross of hypercubes and hyperoctahedron to be simultaneously practical and philosophical, with ethics arising as the interrelation between them. So event has created two new dimensions, but we are still a ways from done.

Inside the now hypercubic formal system there is the three dimensional event, the two dimensional relationships, and the one dimensional nodes. But the relationships and the nodes are supernal to the event, as they are contained within the formal system, and this is external to and definitive of the structure and function of the event. From the conscious memory hypercross of hypercubes we recall the information that the cube of event is really constructed of three square sets of relationships. This means that it is intrinsically three dimensional, and that the relationships themselves are intrinsically two dimensional. So what remains to be changed? Again we are brought back to the one dimensionality of the nodes.

As I first said before beginning to explain the changes in the mental projective structure surrounding the formal system of metaphysics that occur in conjunction with an event, the nodes cannot be understood as potential fields one dimensionally. So something must act on them in order to cause them to open up and become second dimensional, or we would not have been able to consider any of the relationships between them we have thus far with as much insight as has been readily available. In fact, worse than this; if the node of Who were really one dimensional, then we would be unable, anthropically, to consider all we have about the formal system of metaphysics at all, for we ourselves would not exist! It seems we are in dire straights just as things are getting interesting.

To answer our question about the nodes, all we need do is observe what occurs when the dimensions interact. This should at least prove to us that we are not one dimensionally bound entities, and it will also go a long way toward solving the mystery of some of the parts and features of our formal system, such as the paradox of the diagonals, the presence of irrationals, and homogenous attraction. To do this we had best consider the last two external nodes in the formal system, and see what their implications are on the others. Then we can get to the heart of the matter, the conclusion of formal event by interaction with the internal and final node.

D. Not

The node of Not cannot be considered in relationship with any particular other node, because it has its own unique relationship to each of them. It negates them particularly, but by doing so does not negate their potential for solution, which is inherent and cannot be distilled by relationship; it only negates their particular implications, or what they mean in relationship to any other node. Thus we stand by placidly while witnessing the destruction by this single node of all we have so far accomplished, and, so greatly unquestionable is its power, so unapproachable is it by logic, that we are entirely without the ability to prevent it from taking its natural course in so doing. This may come as a shock to you — an unjustifiable disappointment. Or it may come as a relief, like a weary traveler feels when setting down their burdens. In either event the function of the node is uneffected, and merely performs its task no differently than the wind blows or the sun shines.

It does not destroy the formal system. Even if you take the formal system and apply it to itself, it will only be an unstable hypercube of nested like within like, which, when it reaches this node, will return to a stable three dimensional state. It should be specified however, that there is the real third dimension, that of objective physics, and then the mental projective third dimension, in which the formal system exists as an ideal. One can argue that no ideal can be destroyed, but this is an overstatement, as we have seen many different practical ideals expressed in the formal system's relationships that are easily canceled out by this one node. It is necessary to change the letter of the law, but impossible to change its spirit.

The node of Not defines by absence. If it is related to another node or to a relationship, it implies the absence of that relationship at that moment in the sequence of the equation. Thus, for every one of the formal nodes that is present within part of the functioning of the formal system, the node of Not relates at that time to the other five. For every relationship between the formal nodes exclusively that is expressed per juncture in the working, there are fourteen connections that occur linked to the node of Not. Therefore we see that at the connection of each node, and with every relationship they form, the rest are implied as well, by absence. This means that no event can come to occur without the interaction of all the nodes and relationships of the formal system functioning simultaneously in some form. As each of the relationships particular to a consideration is carved out into the unique pattern governing its transformation into solution, so at the same time is there a much more complex pattern evolved comprised of all the nodes not used, in order, and the relationships unemployed, in order. This entire codex is contained within the single node of Not, and determines the shape of the patterns formed between the conscious memory hypercubic hypercross, contained within the cube of time, and the unconscious memory hypercubic hypercross, comprised of cubes of time, through the medium of the mental sphere's perpetual expansion and contraction through dimension. It is by these patterns that event begins to be motivated. The opposite of all of these relationships, as we shall see momentarily, is true for the opposite node of Had, and these two set event in motion.

This occurs in the situation determined by the hypercubic formal system of metaphysics model by the elimination of excess determinism by free will. In the moment of decision the possibilities representative of surplus probability begin to change into new forms, but, as I have repeatedly asserted, this occurs in accordance with the conservation of potential such that no potential is either created or destroyed and its overall presence remains entirely isotropic. What the alteration of possibilities creates, however, is a rip tide in the continuum conjoining the substances of matter and idea, and you can feel this effect as event begins to move.

What is the impact on the event of our example? As the act of creation of the stone takes place in the mental realm, what we are dealing with is a spherical open system of ideal event formed from the lowering of the mental sphere into the domain of the formal system. Therefore the stone itself is the third dimensional cubical event, but exists only within the circular halo of mental projective space.

1. The Woods
A follower of Transcendentalism, a philosophical search for solutions through naturalistic isolation, once observed, "I went into the woods because I wanted to live deliberately." What greater archetype for the unconscious memory exists than

the woods? What demon of myth cannot be dreamed up and confronted therein as innocently as children playing, as innocently as a curious question? But herein is the entirety of the unknown. The full implication of all the open sets combined, the dreaded road less traveled; the missed opportunity; the doubt; Not. How easily it lays waste to all forms: as innocently as children playing; as innocent as awe.

Now does it not feel as though we have wandered a little too far? That we have bitten off more than we can chew, or seen too much to know? Here we find the remainder of all the data we have already begun to factor out still silently lurking about, sorrowful and spurned, and full of power. Realized by the discovery of a single, seemingly harmless node, this beast of incompletion looms all around us now.

We have reached a level in our contemplations when it is easier to see more distant relationships than those immediately before us. How can we defend ourselves against what we cannot even rightly expect? Have we seen the forest for the trees? Another great philosopher once said, "it's a common fact that, as one approaches a potential conclusion, the answers become so obvious they are more easily overlooked." I will deal with the nature of these two philosophers presently.

We see ourselves surrounded by strange forms and myriad relationships. How lost we might feel, if it were not for a single shining fact: the metaphysics model cannot destroy itself. If the node of Not is applied to all the other aspects discussed, the formal system will remain alone, a single isolated three dimensional cube defined externally by six formal and two informal nodes placed on its corners.

We may feel that we are in the woods, but it is only illusion. To penetrate the woods is to leave nothing behind, to confront nothing. It is merely the sudden reflex of dread as event is set in motion, as it begins to be acted upon by the informal nodes. But many philosophers cannot surpass this point. They see their desires lie more with cataloging the ultimately irrelevant implications more rightly discarded than in accepting the necessity of the activation of event. If their were an empire that bound the free will of the mind, it would be this empire of nothingness. And yet, knowing it to be a trap, we cannot but walk right into it. It is temptation itself, and might yet lead to what we desire, despite the fact that it inherently expands itself instead of collapsing down the variables into the closed set of a solution.

The reason for this is energy. Just as an electron traveling the circuit of a shell more distant from the nucleus will constitute a higher amount of energy expenditure as it completes its revolution in the same amount of time as one of lower energy in a tighter orbit, so the outer cube of a hyper cube must express a higher level of energy as well, and so on throughout all the dimensions, as greater and more complex forms are seen to perform their functions in an amount of time determined by the most compact component of them all, event. So what we are naturally doing is following the ebbing tide of energy as it is ejected outward with the reshaping of possibilities until it reaches the outermost dimension in the functioning model, at which point it begins to double back upon itself and we must confront it.

But in so doing we find the entire issue of right and wrong. This arises as a simple factor of exclusion: right is what works, and wrong is whatever is left over. But to determine this solution we must have ourselves anchored in a sense of self. If we have left no desire to return to event, then we will truly be lost in the woods of the seventh dimension, at which point no amount of reasoning with the implications we have described and which, by ejection from event have become irrelevant, will get us turned around, we will have sacrificed our sense of self, and lost all bearings of right and wrong. At a certain point in the operations of the formal system, sacrifice of sense of self too becomes necessary, if one wishes to again return to the infinite measurements of the divine realm which is the ultimate, transdimensional extension of the system. But at this point we have only just discovered what is meant by sense of self, and it would be meaningless to sacrifice it while it remains an image. To make it meaningful we must consider it in the context of the node opposite that of Not, which we will do shortly.

2. The Way

Rather than applying Not to the formal system itself, which has allowed us to see the incision of superfluous implications and possibilities irrelevant to event, we can get back to familiar territory quite easily by applying Not to itself. This is implied by the nature of homogenous relationships, but also implies self-referentiality amongst each node which allows us to see them again Platonically, as simultaneously a question and an answer. What Not defines as question is what it rules out as answer.

So we have gotten back at least as far as the formal system, which we have stated emphatically all along is nothing more than a systemization of Platonic rhetoric. But let us now return even into event, and see what transformations can take place having dismissed such a great quantity of matter. Afterall what remains now to designate the structure of event is only the minutest of fractions compared to the amount of data already considered.

The form of event remains governed by the intersection of the three squares generated from the three legislative branches. Not has begun to motivate it, such that it is now in dynamic interaction with the relationships of the formal system, and through them the short and long term aspects of the three branches in the hypercubic hypercross of conscious memory. Beyond this the excess relationships that are created by the functioning of the hypercube which do not pertain to the particular event created are stored, when the event becomes motivated, in the six hypercubes arranged in a hypercross of unconscious memory. All of this is transacted by the dimensional fluctuations of the sphere of the mind.

When event is set in motion it becomes fourth dimensional. Its velocity is equivalent to traversing of distance, and the traversing of distance per component is equivalent to intervals of time. This does not effect any of the other dimensions above it, all of which we have numbered, purely for the sake of convenience in classification, and which do not change their form as a result of event, and therefore do not change dimension, as dimension is based on more than just order of occurrence, but on complexity of structure. The motion and therefore duration of event does however open up the dimensions lower than it. When event was created, we saw how three two dimensional relationship planes came together. Once event begins to move, these squares are cubed. Although it doesn't change the way it looks per se, the formal system of metaphysics begins to serve as a fifth dimensional extrapolation, storing information pertinent to the shape and duration of event in the cubes of each of its faces, like a hypercross. The reason the dimensional form of the formal system is not changed by this function is that it is canceled out by the application of Not, which simultaneously serves to take back one dimension to stabilize the formal system. Therefore the formal system neither becomes three dimensional by application to it of the node of Not, nor becomes a hypercross relative to the information storage of the relationships definitive of the event. A hypercross would be storing all the information relative to the central cube, whereas the formal system at this time is only storing the information left over after the node of Not.

So even though the formal system itself does not change dimension, the nature of its relationships does. They become three dimensional, and so the nature of each node becomes two dimensional. So we see, as we have been doing all along, that the only way to know the nature of the nodes is in relationship, and the only way to know the nature of the relationships is by example of event. This is necessary to understand what is implied by the concept of self-referentiality intrinsic to each node. Imagine the mental projective space occupied by each node as a square plane, with the node as a point at the center of any one of its edges. Its relationship within itself traces a circle around inside that plane until it comes back to meet its origin. Another way of looking at this would be to place the node at a corner of the plane. It then must follow through four transformations of directed intent before coming back to itself in the form of an answer.

Although this is useful, we shall see shortly that there is another node which is two dimensional, and all the individual nodes are really still one dimensional. By its interaction they have relationships which determine their inherent nature as self-reflexive and therefore two dimensional, but they remain within the context of the formal system only one dimensional expressions. Because of this second dimensional feature in the presence of event, however, we can see that it is possible for a node to represent an open set, in the form of a circle, or a closed set, in the form of a square.

E. Had

It would certainly seem that the node opposite Not should be Which, since it is by which relationships remain in effect of the node of Not that event is defined. But if this were the case then a certain aspect of these relationships would be undefined: that of their necessity. In any event, the possibilities that differentiate it from all other events are the polar opposites of arbitrary. Things must occur in the way they do for an event to be known. If even one thing were different, then all things in the event would be different. Granted this change would be, by definition, imperceptible, and it can be argued that two events differing only very slightly, such as by the time at which they occur, can yield results so similar that they may be applied as identical. But this takes for granted that all other things in the events under consideration must be exactly equal, and it is usually only through conscious tinkering with them that this can be made to serve. In nature, where we have already observed things to never be equal, but in perpetual flux, not even the differentiation of time or location will produce identical results, only results of an open set which we can classify by comparison of composition, according to their particularity. The tendency is towards creation of the unique, and generally similarities only occur due to the influence of very large sets contained within the node of Had.

The presence of this node is the evidence of the conservation of dimension between the negating function of Not and the expanding function of event. Like all the other nodes it possesses second dimensional self-referentiality, imbuing it with the potential to express either an open or closed set, but is localized in the first dimension. Where this node and the node before differ from the six formal nodes is that they are relative to the existence of event in function. The six formals can be expressed as compactly in a second dimensional hexagon as in a third dimensional cube, but, taken in unity with them, the two informal nodes can only be expressed in the third dimensional construct, rendering depth. The relationship between them is that of the diagonal of a cube, and this divided by the relationships of any of the nodes surrounding it separates the two nodes to form an inner cube of event. In other words, event is composed of the three branches of legislation as they are divided from the formal steloctahedron, but is itself an embodiment of the combinations of the two remaining stellations of Not and Had.

The meaning of this is that the relationship of Had and Not serves as the combination of the three branches of metaphysical legislation. The square root of three diagonal between them serves as the axis of rotation for the cube of event. Just as Not served to initiate this momentum, Had determines its termination. The internal composition of the node is comprised relative to the event, and implies in field form the parts of event's function, such that event serves to amend possibilities, and then is transformed into another event when these possibilities have been changed. At this point the functioning of the formal system begins anew from its simplest state, and progresses through dimensional expansion until another conclusion is reached. The potential present is the same for every event, but assumes new forms according to the relationships that occur in their order as necessary for the event.

Because it cannot be stored in the formal system as a hypercross, the data structure of the node of Had is stored in the conscious memory hypercubic hypercross. For any event that occurs there is a change in the patterns formed between the hypercubes of conscious memory that remains permanently accessible after the event has concluded. Consciousness, if not the consciousness of one individual, then the consciousness of another, can extrapolate the pattern determinative of any event from the hypercubic hypercross of the memory castle. All that is required is the sphere of the mind to interact with the formal system.

For our example question, this represents the completion of the formation of the rock, and the absence of the action of lifting. It can be created and uncreated within the open set of the mind, but it cannot be created as a closed set, because this would imply duration of existence, and duration of existence is irrelevant in the light of the sphere of the mind. To recapitulate the entire process, we have seen the nature of the rock defined by the context of the question according to the relationships of the formal system. We have also defined the entirety of the parts involved in its hypothetical creation and seen how they differ from their infinite extensions in the realm of the divine, compared to which, if the divine were thought of as fourth dimensional, the entirety of the formal system and memory castle would be no more than a first dimensional dot at the center. We have speculated the necessary functions as divine will lowered itself through this system to the event of the rock's creation, and finally we have come to the conclusion that it cannot be conceived of as purely physical, due to the influence of the divine will, and cannot be lifted, due to the function of the formal system upon the event itself.

To understand why the rock cannot exist in a way that we could record its measurements and remain impossible for God to lift at the same time we must look further at the difference between physical and divine legislations. We have seen already that the best place to view these is in the vertical relationships as the formal system is viewed as a hexagon such that How is at the apex, where physical legislation stems from the short relationship between When and Where, divine legislation from the short relationship between Why and Who, and ethics the long branch joining How and What. The depth of the cube — the relationship between Had and Not — is the union of these three legislations, but it is equal in length to the relationship of ethics between How and What. Therefore ethics also represents a combination of legislations, as we have seen by its being partially an open and partially a closed set, but in this case it is the combination of physical and divine legislation. The difference between these diagonals has already been covered in the Paradox of the Diagonals.

1. The Law of Man

The law of man is traditionally governed by determinism and negation. It takes the physical functions of right and wrong, or, known by their psychological terms, do and don't, and expresses them as morality, the difference between "good" and "evil."

Aside from God there is no imaginable ultimate good, and therefore only an asymptote reflected in the concept of evil. Most people are considered by one another to be capable of good and evil in equal part, but always end up acting more frequently on one than the other. Good refers to what is necessary, and evil to what is harmful, but their relationship is more complex than mere duality.

We have dealt with a number of distinctions defined by their supplementarity amongst themselves and their complimentarity to each other, but Good and Evil have traditionally been held as above and removed from this rule, such that Good does not complete Evil, nor does Evil Good, and they cannot be held relative to any other distinctions. They have been taken to be absolute, and therefore used as a form of divinity. But they are not divinity, only a form of short hand legislation for man's behaviors justified by their fictional liberation from the application of relativity.

Aside from being objective and metaphysically non-right angled, Good and Evil are names that humans call themselves in an attempt to reconcile that which is above to that which is below. Therefore their applications are subject to condition. As such they must be limited, and the most convenient method of doing this is by negation.

Hence the oldest forms of ethical legislation remnant of the written tradition are efforts to set down all existing behavioral limitations in closed sets, attributed to divine dictate, and indeed these are the seeds of all subsequent ethical legislations, which have become numerous and exclusive where the originals were few and universal. What we humans consider first now is negation, and most condition. The definition of Evil, by both the closed set of the early, divine laws and the subsequent, partially closed set of states' legalities, can be justified by an attempt to derive the origin of motive from situation according to the specifying of the mental perspective on the event into a more closed set of specific conditions. This does not, however, imply that the event under review is in any way Good.

We deal, in modern, ethical legislation terms, only with negation and the negation of negation: "guilty" and "not guilty." If someone is guilty then they knowingly acted in Evil. If they are not guilty it does not necessarily make the outcome of the event in which they were involved less Evil. So it becomes a tacit assumption that there is more Evil in the world than Good, and this is taken as proof by the rule of relativity for the existence of the divine in opposition, and this, circularly, is used to justify belief in the self-consciousness of the concepts of Good and Evil. So even though it was not intended to do so originally, the movement of ethical legislation towards concentration on the particular has conditioned our belief that morality is above conditions, thus proving that some relationship, such as supplementarity or complimentarily, must exist between Good and Evil.

Many people believe that "what goes around comes around," but it has to change derivation and aim while doing so. If Good were considered as we have considered the nodes of the formal system, it would occupy the midpoint of one side of a square shaped plane, and Evil would lie directly opposite it. If you traced out a circle connecting them, you would see the relationship they share that prefers to remain hidden — that of their relativity to one another.

What made them enforceable initially was the argument that they constituted evidence of divinity, but as we have just proved, they are relative to themselves, and only the law of man can be reflexive and conditionally dependent. But we have also seen that the aim of ethical legislation to render a more closed than open set from an open set of perspectives on a growing number of closed sets of events necessitates consideration of each event as situational, and therefore implies that Good and Evil are relative to other distinctions as well, which go into the determining of an event.

So we have seen the philosophical side of the legislation of ethics as a combination of the legislations of the divine and the physical, but because it functions simultaneously as a closed and an open set, there must be a practical aspect as well. This is where the economy comes in. It reinforces mathematically what the concepts of Good and Evil idealize, and that is the basic approach/retreat mechanism of the bio-survival circuit. The primary behavior covered by ethical legislation is that of business, as this has, since the time of the earliest tribes when the first distinction of role occurred, been the occupation of the minds of the majority of the people. It attempts, identically as morality does, to render the opposite of its own composition by evolution. Morality seeks to generate an open set (divinity) of closed sets (laws) with the foundation of a closed set (condition), and economics seeks to generate a closed set (market) of open sets (investments) originating from an open set (capital). The reason the economy is an open set despite the fact of money being objective is that it occurs within the context of a free market imitative of the free will of man. As much as morality is guided by the open set of situation towards the closed set of determinism, so economy is guided by the closed set of finance towards the open set of free will. So the first complimentarily to morality is economy. Both these occur within the domain of the realm of Had. Had determines by necessity the justifiability of an event morally and the possibility of an event economically.

2. The Law of God

To reign in Hell is the same thing as to serve in Heaven. This is because Hell represents the ultimate, situational, closed set of the individual, and heaven the ultimate, unconditional, open set of the infinite potential dimensions of the universe. By bringing to bear control over the former, one has a self to sacrifice to the latter. If the divine is anything, it is the potential of the field of Time, measured in dimensional interval, and governed by the infinite extension of the mind transcending dimension.

"Do what thou wilt shall be the whole of the Law" of the divine. The universal mind cannot be bound to symptom, as any situation in which an event occurs is localized within the open set of God, and therefore inapplicable to it holistically — except as constituting a sample of the fiber of space-time's existence, and even this need not reflect directly the physiognomy of God. He is thus bound only by free will. None of the legislations familiar to man apply. Physics, ethics, and belief are all irrelevant. "When nothing is true, everything is permitted," where truth pertains to measurability.

These two quotes come to us not from philosophers, but from shamen. It is the purpose of the medicine man's existence to aspire to divine knowledge, but their burden to know that such is definitively impossible, and ultimately irrelevant. The beliefs of man regarding God only give form to ritualistic interface with the divine realm, but can hold no bearing there. Without belief there would be no foundation for science, and without science little hope for survival, for belief presents the illusion of enclosure of concepts that are, in reality, infinite, and science is dependent upon such closure of sets. Where the philosopher represents the way of ethics, oscillating between the physical and the divine, shamen have become confined to the wilderness of, at once, both the physical and the divine, but accursed of the guidance of ethics which, it has proven, is ultimately foreign to these realms anyway.

We have seen the legislations that arise between the formal nodes, and we have seen the legislation that arises in combination of these between the informal nodes. Let us now set forth the proposition that the legislation formed of interaction between the informal nodes is that of geometry, as this is the unification of physical, ethical and divine legislation, and can be seen to occur in each. So when we say that the Law of God supersedes all forms of human attempts at imposition of legislation, we are saying that divinity is beyond geometry. Geometry, it can be equally stated, is of divinity, as much as it is as well of ethics and physics, but only in a natural ratio which the true realm of divinity surpasses. Hence, the structure of sets within sets that defines the law of man is dimensional, and geometry is the measurement of dimensionality; but the Law of God is beyond dimensionality, and therefore immeasurable by geometry.

When applied to geometry in the presence of event, human legislation of the divine manifests three subdivisions corresponding to the construction of geometry from the three legislative branches. These are Light, and Love, and Life, and occur in an interrelated dynamic with the three branches. They are expressions of the raw potential compressed in and implied by event, though they are most commonly experienced only when the event is of a divine or revelatory nature. Therefore we see that geometry is more than legislation, but is not more than divinity. Again, I must assert the fact that the subdivisions of divine legislation are bound within geometry, and therefore pertain only to the law of man, though they are taken to be the substance of the Law of God in such a way that, by providing evidence of their origin from these, the belief in the divine nature and origin of the other three branches of legislation, as they are expressed in combination as geometry, derives entirely from this attribution. What it means to be bound within geometry is that these are conditions determinant of situation, and though representative of the field of potential rather than particular possibility, cannot therefore completely or accurately speak for the absolute free will of God. I leave it up to you to deduce the manner in which all this proves true.

Divinity is usually anthropomorphised, and so its geometry is one of relationships between archetypal entities. Light, Love and Life align with subject, anima and animus in all their various combinations. For example, if the subject of a particular legislation regarding divinity were Light, it would have as supplements anima, which could be either Love or Life, and animus, which would be either of Love or Life its supplement was not, such that together they are all complimentary. This reintroduces us to the pattern of the five petaled rose, as well as to the cross of male and female. It is also associative, before the determination and overlap of subject and assignation, with the hexagram, and these two are the fundamental forms taken by prior formal systems of metaphysics in the West and East respectively. In our last evolution we will consider these concepts in more detail, and see what further applications arise from the study of idealized geometry on our formal system of metaphysics. But this pertains to aim and will be dealt with in due course.

Now we must ask the conclusive question: Is it possible to integrate a flowing network into the continuum of potential in such a way that they are interactive and record? The answer to this is provided in the central node of the formal system model, that of fact and fiction.

F. Fact and Fiction

Although events in reality are defined by complex sets of variables all of which have influence upon one another individually, in groups, and upon the event as a whole, each variable itself can be seen in simple binary, either/or form. It is the nature of the open set of potential that possibilities are never truly mutually exclusive, although we often perceive them that way, and the best solution is frequently both/and. But to study the transformations of our formal system in the final stages of its definition of reality, let us consider for now a binary example. In the situation of writing this paper I may say to myself, "did I leave the door open?" Now there is obviously only one right answer to this. Either I did or I didn't. But in my mind both possibilities exist, and therefore both effect the situation. For example, the fact that both possibilities exist allows the question itself to arise, which otherwise it would not, and I would probably be writing about something different.

To compensate for this obvious distinction between the closed set of the physical and the relatively open set of idea, we say that one solution is fact and the other solution is fiction. The terms "real" and "imagined" are often applied, although inappropriately, because both outcomes exist in my imagination, and my imagination is just as significant to my situation as what would be objectively described as "real" by an observer. Whereas in the former case there is a specific binary relationship, in the latter neither term is exactly exclusive.

The terms of fact and fiction in themselves are somewhat unimportant. It is possible to construct an elaborate relationship between them as complex as the entire system and all its extrapolations thus far, although they are a single node, because they are its core and, serving as a kind of seed, are the center of each of its parts individually, in groups, and as a whole. But to do so would be a waste of time here, and is best left for further study on your part. Suffice it to say that, once the distinction between the concepts of fact and fiction has been grasped, the rest of the formal system of metaphysics will be a permanent part of your faculties, and moreover no more challenging than child's play.

What is really important to go into here is the effect that this last node has on the rest of the formal system. I shall refer to this effect as manifestation, which is at least an oversimplification. We have seen already how the action of the mental sphere serves as a carrier for the patterns of potential information both inward toward event, and outward toward the memory castle. We have seen how each nested aspect is equivalent to a dimension, but we have not yet addressed the nature of the sphere of the mind itself, nor seen how information being transferred from event is stored. To do this we must focus on the node of fact and fiction, as it is the point of origin for the inward and outward oscillations and for the stream of data to be retained.

1. The Fluctuating Field — Introduction

Compared to the sphere of the mind itself, which passes through each of them sequentially as possibility, and maintains dominion in all as potential, the dimensions themselves are relatively stable. Relationships are constantly forming, flowing and evaporating between them and within them, but always along fixed, geometrical patterns. In this way redundancy balances entropy, and the formal system operates the gamut of its functions remaining stable throughout. But the sphere of the mind is quite different. To understand it we must examine it dimensionally, and then, once we have grasped a full understanding of its shell, probe the depths of it to comprehend its motions.

Clear the formal system from your mind, and imagine a simple circle. But you already know that, even in the stillness of the space surrounding that circle, the afterimage of the formal system lingers. This is, and cannot be otherwise, so to resist it would be to no avail. Instead accept it, and observe its effect on the circle at the center of your attention. The circle now displays incomprehensible patterns, reflected from the formal system. They pass around it like clouds before the sun. The patterns are greater in number around the edges of the circle, and larger in the center.

Make the circle into a sphere, and the patterns will become clear, ordered. You see that the patterns are the outlines of shapes, and that these shapes are regular in size, and that they appear different only because they completely surround the sphere. When you look at the shape upon the surface of the sphere at the point closest to you it is directly aligned with your vision, but as you look away from the center towards the rim of the sphere, the alignments of reflection appear more angled, allowing you to see that a great number of patterns surround the sphere, although their memory is already fading, and all that remains is the shapes on the sphere. These shapes are your thoughts. The one that is closest to you is the one occupying your focused, concentrated consciousness. Adjacent to this are thoughts which pertain to it, are related. Around the very edge of the sphere are the myriad thoughts of your unconscious, ever present, only waiting for the sphere to turn that they may be brought, in turn, to your attention. But we know that each thought is the same size, and that they are, like the points on a sphere, potentially infinite.

Now focus intently upon the sphere, making the thought which is at the center of the sphere as clear as possible. You will see that, the more you focus on the sphere, the more the thought that occupies your consciousness will resemble the sphere, until, after forgetting all the other thoughts on the surface of the sphere, you will see that it is an identical sphere, within the sphere of your mind. It reflects on its surface images of all the thoughts that were on the surface of the sphere of your mind, but, by focusing on only one thought and one alone, you have cast them out of your mind, and now they exist only as parts of that thought; as afterthought. I will return to this more in a moment.

You should at this time recognize that you have created a hypersphere, and are now free to impose the working of the antipode upon it. Imagine that you are walking around the sphere of your mind, or that you are turning it about with an invisible hand, all the while keeping perfectly focused on the thought closest to you. You will see that the thought upon which you are focusing will grow larger, until it surpasses the size of your mental sphere, and then your mental sphere will begin to shrink down until it is the size your thought was after one half rotation. If you wish to you can keep doing this until you become comfortable with their interchangeability.

There is another way to look at the two spheres as well. Imagine flying up above them, or holding them under your eye, and you will see them assume a new form. The larger sphere will cave in on itself, until the thought at the center of the smaller sphere has become its own sphere, and the large sphere appears only to be a round ring with reflections outside of and inside it. Now we are looking at your mind as a torus. It reflects thoughts that are external to it, and it reflects the thoughts which are reflected on the thought at the center of your consciousness.

Finally picture this: the thought at the center of your consciousness is the circle that you first imagined. It is upon the surface of the sphere of the mind. The torus that surrounds this can be cross-sectioned to form a sphere. There is another sphere outside the torus, and it constitutes the realm of your perception, or your mental imaging space. So lastly, where are "you?"

This is the process undergone by the mind as the formal system of metaphysics unfolds itself. This is why the sphere of the mind cannot be said to be bound by a single dimension, as can its cube based counterpart, the formal system, because even before it begins it is already all the dimensions in one. We cannot argue which came first when we are focusing on the thought of the formal system itself, because this is a paradox. By applying the formal system we could deduce an answer, but this answer might not satisfy us. It is that "some suggestions come to us from the future, while other suggestions come to us from the past." These two, like "reality" and "imagination," are really one. To understand this of fact and fiction is to understand the entirety of geometry.

It is fact and fiction which motivate the entire formal system, as well as the sphere of the mind. All the relationships that we have studied thus far, between the nodes of the formal system, between the relationships between these nodes, between the formal system and event, between conscious memory and the formal system, between unconscious and conscious memory, and within the nodes themselves arise from and conclude at the node of fact and fiction.

Fact and fiction cause the metaphysics model, including the aspect of the memory castle, to fluctuate as does the sphere of the mind. This occurs as fact and fiction interact with both the naturally fluctuating sphere of the mind and the naturally stationary or inert formal system of metaphysics, as the central point of all involved. So it is really the sphere of the mind that initiates the fluctuation of the formal system of metaphysics, through the node of fact and fiction. What is implied here is that the node of fact and fiction is a microcosm of the process of dimensional fluctuation and that the motivation of the sphere of the mind and the formal system of metaphysics arises out of its intra-action. Thus, each is older than the other, and both age simultaneously at the same rate.

This motivating field and the sphere of the mind are one, when the sphere of someone's mind is present, for the waveform of probability that we are describing here has been proven to collapse only under observation, although one supposes as well that it is present even when no mind is around. The infinite thoughts which can occur on the sphere of the mind are thus comprised of the closed sets of information which are translated between experiential possibility and mental recollection through the formal system, and can therefore be measured. Because the wave of probability is objective, or non-right angled, we can determine fixed location for it more easily. To do this we may make use of the hitherto mysterious diagonals, irrationals, and attractive poles. Let us study a specific function of the formal system, and see how these components contribute.

2. The Fluctuating Field — Application

One of the necessary tasks of the formal system is to transmit the answer to a question provided by the creation of a real or hypothetical event into the vast associative storage space provided in the memory castle. We have already observed that the exact moment of the answer's creation is determined by both the carrying outward in the fluctuating field of a wave of superfluous data from the node of Not, followed by the returning of that tide through the filter of the node of Had. So we will pick up there and follow the answer as it is transported from the now dissolving event outward into the also collapsing dimensions of the memory castle.

Now we may deal with the node of fact and fiction as a microcosm for dimensional fluctuation. It, in itself, bridges the gap between the sphere of the mind and the hypercubic formal system of metaphysics memory castle. No other node does this; fact and fiction alone holds responsibility for this position. As we shall see shortly, there is a strong binary aspect involved in the final stages of functioning of the formal system, particularly in the interaction of the nodes of Had and Not, but in field form through fact and fiction as well. It can be said that these nodes are practical and philosophical equivalents. They determine what is real.

Had and Not accomplish this determination for the event, but as informal components of the formal system they are ultimately referential to the transmutation of information. Fact and fiction pertain not only to the event, but to all the other structures reflexive in event. That is, by examining Had and Not, what is true for an event may be strictly determined; only by examining fact and fiction, though, can proof be provided even for the existence of all the very structures involved in that determination. This is from where reasoning derives, and is the significance of the centralization of the double-node. It can be said that the ultimate definition of an event's reality isn't what's true *of* its composition, but what is true *for* its composition — the context of reason in which event occurs, that is, fact or fiction.

To fully study the function of fact and fiction, and to understand its inherent dimensional fluctuation, it is relevant to consider its function in each of its dimensional levels, or each of the parts of the fully functioning formal system of metaphysics model. We shall go through these expanding outward from the node, just as we have with the sphere of the mind's expansion from its lowest dimensional level. Following this, and in conclusion to this section, we shall see the final dimensional level to be included in the aspects of the sphere of the mind. As we do this, remember, this can be done with any of the other nodes as well.

The first of these levels, that we have already touched on briefly, is the dimension of event. Event, in terms of physical legislation, has definition as a singularity of space and time, that is to say, it is internally unique unto itself and relative to others like it by categorizing of external similarities. With some certainty we can assert that event has existed as long as time, as the lowest common denominator of physical interval, but it will be added shortly that this does not mean it is indeed the lowest level of measurement possible from a dimensional perspective.

Fact and Fiction pertain to event in aim, assembling its unique composition and its relativity according to binary terms equivalent to Had and Not, but with open, inclusive, rather than closed, exclusive sets. If the event is fact it will have greater bearing on the practical, physical aspect of reality. If it is fiction it will have a more direct impact on the philosophical, ideal aspect of reality. Both of these realms being equal, as we have seen in the legislative branches of the formal system, we may see that an event will represent the transformation of a fixed amount of energy towards either realm of possibility. It is also common for an event to function as a translation between these realms, with input in one realm and output in the other. These translations can be roughly rendered as distraction and inspiration, and we shall return shortly to the exact application of these terms.

This level of attribution is equivalent to the application of the measure of the line connecting the uppermost and middle nodes on the hexagonal shadow of a cube as one to the full component measure of the dimensional potential. This is accurate according to the measure of a compass tracing down from the distance between two of the formal nodes external on the hexagon, with one of the nodes as an origin, to connect the other external point to the center of the hexagon. The reason for doing this is simple: this is the measure of event: $1 = 1$, a simple equation in a simple plot.

The second form of which fact and fiction are the center is the formal system of metaphysics itself. When it is said that the effect of an event occurs, it is meant that there is an outward dimensional expansion in the form of a change to the situation embodied by the metaphysics model in the exact information structures of its constituent relationships carried by the fluctuation of the sphere of the mind in the form of raw potential. When this occurs it marks, temporally, the end of one event and the beginning of its transformation through translation of potential, from a closed to an open to a closed set of possibilities, into another event. This is the origin of the concept of event as the simplest measure of temporal interval. But the change in the information patterns in the formal system of metaphysics implies an even smaller measure, pertaining to potential rather than the exclusively physical. We shall revisit this again shortly.

The transformation of information in the formal system of metaphysics that is initiated by and determinate of event pertains to fact and fiction once again in aim.

What we are seeing now is an outward wave of potential emanating from event. If this is fluctuating at the same frequency as the full spectrum of the physical cosmos, or fact, it will double back upon itself with half intensity to generate a new event. The remaining half energy of the potential wave will continue onward to the memory castle, where it will be stored. This is evidenced by the very fact of cause and effect itself: not only does one event beget another, but it is remembered as well, accessible when the sphere of the mind is present.

If the sphere of the mind is absent, the wave of potential doubles back with full capacity upon event. For fiction, the wave of potential passes through the metaphysics model, and will only turn back once it has reached the memory castle. This is because of the philosophical nature of fiction, and is evidenced by the necessity of it being passed through the sphere of the mind before redirecting its potential for generating another event. As we shall see shortly, the sphere of the mind, though present at every level, is the medium for movement of potential in the form of an outward ripple of information only when it cannot be otherwise.

The second level of attribution to geometric function is that of the formal system of metaphysics to the measure of the same line as before, that following from an arc connecting the center node to one of two consecutive external nodes on the hexagonal representation of the formal system, the other point being the origin, only this time measured as the diagonal of a unit square, which totals the square root of two. The reason for this is that, not only is the formal system the second consecutive representation following the first dimensional level as represented by the measure of the leg as one, but it is expressed as a hypercube nesting event, and therefore expressible as a square root, or the diagonal of a cube nested within a cube.

Now we proceed along our wave of effect to the lower form of memory, that of conscious memory. Here we begin to see more interesting results. As the memory castle expands through the functioning of the formal system of metaphysics, it becomes a hypercubic hypercross, with six new formal system structures surrounding a central seventh. This means that the node of fact and fiction is also multiplied, and therefore what is seen at this stage is fact and fiction coming into contact with themselves.

The intellectual action described here is as complex as it may sound. This is one of the most immense implications of the node of fact and fiction because, as it will soon be explained, this process is itself inherent to the individual node. To see every implication in its most basic, or at least initial, form, we must examine each hypercube of the hypercross of conscious memory, of which there are the long term and short term cubes for each of the three legislative branches — physical, ethical, and religious — and the central cube, with its triple nesting of memory cube, formal system and event, from which the potential wave emanates.

There are a total of twenty-four outcomes from the interaction of the central node with its satellites, including fact to fact, fiction to fiction, fact to fiction and fiction to fact, multiplied by the number of orbital formal systems. Each of these occurs within one of the short or long term legislative branch memory storage cubes, and it is not necessary that any of them relate with each other during this process. After each has been determined autonomously, some of the potential will be reflected back down towards the metaphysics model, to impact upon event if the sphere of the mind is present, and the remainder of the wave continues out toward the realm of unconscious memory. Though this is true for both fact and fiction, these terms have both become relative after this point by interaction with themselves.

Conscious memory can be measured by the third geometrical mystery, that of the rendering for the cut defined as equivalent on a compass to that connecting one external point on a hexagon to the next one, and joining one of these external nodes to the middle, and called half the diagonal of the unit cube, which is measured as one half the square root of three, or the square root of 1.5. The significance of this is that this is only half of all of consciousness, which includes the other half of subconsciousness and the entirety of unconsciousness as well. At the same time, this is the measure of the radius of the structure as it occurs for the first time in geometrical form in three dimensions, indicating that the sphere of the mind's radius is equivalent to half of the sum of its reflected thoughts, or half of infinity. We shall consider this again later.

Finally we come to the hypercubic hypercross of the unconscious memory, a hypercross of seven hypercubes, six on all sides of the central seventh, wherein each of the hypercubes is itself a hypercross of conscious memory, so that there are seven of these, six arranged around a central seventh, and each of these is composed of seven hypercubes of the formal system, six arranged around a central seventh, and in the center of the formal system in the center conscious memory hypercross in the center cube of the unconscious memory hypercross is the event, from which the wave of potential emanates. We can see already the parallels to the thoughts on the sphere of the mind as a reflection of these reflections.

There exist in unconscious memory six shadow events generated by the replication of conscious memory, and these represent simultaneously different potential perspectives on the initial event, or potential memories, when the sphere of the mind is present, and possible events that are accessed by effect, whether the sphere of the mind is present or absent.

Each of the events contained within the unconscious memory contains a node of fact and fiction, and these are realized as archetypes. Each of the thirty-six formal systems surrounding these also contains a node of fact and fiction, and these are realized as myth. These are both realized by manifestation, which we will speak of shortly. Now the nodes of fact and fiction for the archetypes and the nodes of fact and fiction for the myths interact to determine internal from external, and this is how aim is set. An archetype can be fact or fiction, and a myth can be fact or fiction. Because the myths themselves are comprised of multiple fact and fiction nodes it is even possible for them to be both fact and fiction equally, and because the six archetypes interact with one another to influence the overall probability of an outcome realizing itself as consecutive event, it is possible for these to be balanced equally between fact and fiction also. Here the determining node is that which was central all along, and is always irregardless of the presence of the sphere of the mind. So when the wave of potential reaches this point, because at this point it contacts the sphere of the mind, it turns around on itself. It can be either fact or fiction according to the total tally of fact and fiction nodes along the way.

There is admittedly no known equivalent to the castle of unconscious memory, but there is the mystery of attractive opposite nodes. This pertains to the basic cubic structure by piecemeal, but it also changes over time, and therefore references the fourth dimension as well. Here we find not a measure of spatial distance, but a measure of energy expenditure duration. Thus, what we are dealing with is a balancing between opposing sets of binary charges. What and How are both positive; Where and When are both positive. Why and Who are both negative; Had and Not are both negative. Fact and fiction are dually positive and negative. This evenly balances the positive and negative charges.

3. The Fluctuating Field — Conclusion

Though our model ends here, the fluctuation between dimensions does not. We have seen that event serves loosely as the smallest certain interval by which to measure physical time, but that potential implies one smaller; we have seen that the presence or the absence of the sphere of the mind pertains to the redirection of the wave of potential after it has passed through the autonomously functional formal system of metaphysics model; we have seen that event begets event by manifestation. Let us deal with each of these three points in terms of the three branches of legislation, assessing the first to the physical, next to the ethical, and last to the religious. Thus we are looking at physical interval, measured in potential; ethical attendance of the mental sphere; and religious manifestation.

First we find the case of physical interval. Event is a singularity. What can potential be then, but a measure of dimension? We tend to think of potential in the binary terms of probability, zero being null, one being certain. But potential can be expressed dimensionally: in terms of ascent through a series of probabilities, each additional level expanding the limit of certainty. Thus, potential is itself as good a measure of interval as event, but pertinent to a lower level of dimension.

Now on to the ethical angle on the influence of the mental sphere. We can use the tree falling in the woods metaphor, only substituting meaning for physical substance; otherwise, matter for matter. We can say that it is all difficult logistically from here on out, but worth the journey. "Work out your salvation with diligence."

Lastly, looking at the religious implications of manifestation, we can deal with the point I was hoping to make with this essay all along: that isolations of disease outbreak and metaphysics breakthrough can be correlated, perceived through the aspect of the sphere/cube, physical/spiritual dichotomy, and are utilizable through the fact and fiction node. Examples of this include: free will over aethyr and the fever of mind over matter, astrological Cancer and spiritual AIDS.

One interesting side-effect of manifestation is the breaking of concentration. When working for a long time on any given project, it is occasional for the focus of a participant to waver, and possibly confront the subject of the work smiling, as though it were for the first time, and yet like some distant memory. This is to say: be careful what you wish for, it just might happen.

III. The Platonic Solids and the Hexagram

We may say that it is easy to be distracted, or to get what you want. The problem is maintaining this, or deserving it. We can do what we like, and others can do what they like, but when these are combined, a new situation arises. This is one of both inspiration and confusion, and over all of love. Thoughts on the surface of the sphere of the mind are only perceived as thoughts from inside this sphere. From outside they are as real as the kinetic energy of phi, Freud's name for the transception of physical potential. Therefore what we perceive in others is true for ourselves in a way that makes our remembering this as difficult for us to perceive as an illusion. This provides concise metaphors but also has consequences. That is, we may perceive as fact or fiction, and be confronted by reality as the opposite.

All these things come naturally, and can be seen to be defined only separately as the will and geometry. We will discuss the implications of these under their forms as the Platonic Solids and the Hexagram. Yet, dimensional leveling remains the best way to explain them, however this accomplishes itself temporally and amounts to the use of words or images, and the selection of specific impressions thereof. It is through these that we see our truth unlocked, and through which to you I am conveying all of this.

A. The Platonic Solids

These are all that we have. From all the events in all the world in all of recorded history, these are the only regular solid polyhedra; and this is all there is in all the world, subsisting in all natural forms. For every argument there has ever been there is only this set of five to serve as proof, and all else is merely replicative elaborately. In the ancient world these were used as meditative tools to obliterate consciousness, that is, to perceive an object in active consciousness and then, through removal, perceive this reflected in subconsciousness and unconsciousness.

Understanding them is like understanding the lyric of the poet. They are like a foreign language, and yet they are in the same tongue spoken by you. They are like the thoughts of a child: profound, yet blind. They are a new revelation, excavated in time, that we are awake and together. Together we discover whether we are doing right or wrong, just as the five solids are together, and prove reality.

These are the tetrahedron, the octahedron, and the isocahedron comprised of triangles, the cube composed of squares, and the dodecahedron crafted of pentagons. These are arranged according to their digital value, as being products of their sides and vertices that complete a system of pairing. This system is the lowest common denominator (edges per face) times the greatest common factor (faces), and shows us that the icosahedron (20 sides of 3) and the dodecahedron (12 sides of 5) both total 60; that the cube (6 sides of 4) and the octahedron (8 sides of 3) both total 24; and that the tetrahedron alone is 12. Thus the sum of all of these is 180, and therefore they may be rendered as measures of degree on a a 360 degree circle.

This means they can also be assembled as a measure of the color spectrum. When positioned on a half circle with infrared to ultraviolet as the base line, the visible spectrum presents itself as angles within this limit, forming the cross-section of a rainbow. The color spectrum provides us, in the case of light, the same thing as do the Platonic Solids, in the case of form, that is, a knowledge of the presence of pattern in event.

This pattern is evident in the composition of event, and therefore implied in the conception of event. We shall therefore examine them in this order. But we must remember: all of these occur as one at once, in a single measure of potential. Therefore their order is ultimately irrelevant, because all are simultaneously relevant. It seems when we extend ourselves into the realm of effect, that we are incapsulating ourselves in terms of time, but this is only an illusion. The information is, and always has been, accessible. There is no temporal distortion necessary.

A brief note should be included about the difference in attribution presented here from the classical or traditional attributions. To do this we must say something about the history of mysticism — the study of the implications of metaphysics. In its origin in classical Greece it was initially philosophical and geometric. The former school, started by Socrates, and the latter school, begun by Pythagoras, were merged by one man; Plato — who wrote extensively on both schools, is responsible for naming the five natural solid polygons as we know them this day. After him, it became common practice to make use of them in studying all aspects of nature, from the mental through the phenomenal. In accordance with his observances, subsequent schools and scribes attributed these in a particular manner; that is, the tetrahedron was associated with the fire element, the octahedron with that of water, the icosahedron with air's, the cube with earth, and the dodecahedron with cosmos.

When I first discovered this arrangement, and compared it to that of producing faces by edges, I had thought that only the tetrahedron and dodecahedron would be inverted. My argument was that, in studying the contributions of the various schools in organization of their information, it is an often occurrence that there is such an intentional inversion of terms, implying an attempt to obfuscate impact of certain discoveries. This was done mostly when one school maintained dominance as a centrally utilized legislative system — such as during times of utilitarianism — but this was *not* the case in classical Greece, when direct Democracy was at its peak. Yet consider the sayings of Socrates and you will see why he was poisoned. The very idea of picturing an idealized state other than the existent one is, admittedly, not patriotism. In short, it is not outside the realm of possibility that, in incorporating the two schools, a tentative amount of disinformation was considered wise. This conclusion is demonstrable using the formal system of metaphysics, by deduction of aims. Usually, when one school is dominant, it forces the other into being seen as fiction.

After reasoning thus, I realized that this was not the only difference between the two systems. I therefore assembled a two column tally of all the attributions from both systems such that the rows were aligned by solid shape. I then noted the differences in placement between the two of elemental attribution. In terms of the "spaces" occupied only by each of the solids and its correspondent attribution, their differences assume a specific pattern that ascends in numerical value. The assembly of this pattern dictates that the difference in "spaces" of Earth is zero, water is one, fire is two, air is three, and cosmos is four. I encourage you to assemble such a table for yourself, to confirm or disprove my results.

Therefore the manner in which we shall go through the solids is in accordance with the numbering system of differences between the traditional and the digital systems of attribution, and the attributions we will be using derive from the digital system. The reason for doing this is that it allows access to a greater referential realm of readily explicable information than otherwise.

Now the breakthrough comes. It is slow and preposterous, as much a sentencing as experimenting. But the information orders itself. We will have dealt with a logistically jumbled mess of facts and concepts, but how can we make it work for us? The fact that it arranges itself naturally implies it has a goal.

We shall deal with the information summarily in the following form, which, in accordance with cosmology, is chronological. First we will look at the numerical change between the initial and digitally translated systems. Then we will look at the color attributions as they occur on our new digitally mapped arrangement, assigning each number a color. Next we will look at the elder elemental attribution and the equivalent attribution of an elemental force. Finally we shall assess their function in the formal system, and consider the implications on their geometries of dimensions, particularly dealing with space and time.

1. Cube

Fundamental is the cube, for it can be trusted, despite being known. What is known is dreamed of, for it is merely whatever is believed.

Having said so we are prepared for our introduction to the cube as it was known by those from whom the six fundamentals of reasoning and the five Platonic solids originated. It was known by the element Earth, due to its solidity, but little more than this can be said. Earth was possessed of certain seeming similarities, but these were irregardless of the cube in their own ways.

Outside of this not much was known. It is probable that the reason for this is that the Greek thinkers were attempting to describe observational data of which they had some working knowledge and experience, that was subsequently lost or obfuscated, and which would await as an inherent, natural fact for rediscovery by modern physicists.

The cube itself was used commonly as a remembered unit, as in the formation of certain minerals and gems, and in the making of bricks, stones, or blocks. But it seems to have failed to gain such familiarity that its more cosmic implications were known even covertly until such breakthroughs as Newton's second law of motion were made aware to the entire population of the world simultaneously.

Still, this was not until a long time after the geometric properties of the cube were discovered, and a while yet after they had been demystified, as they had become in the interim. So, aside from the addition of the knowledge given us by the hypercube, the cube has not changed much over time, and all the hypercube implies really is that the cube doesn't change much over time. But with this discovery, more becomes revealed regarding possible attributions to the cube.

In this, and the following sections, we will examine the attributions of an elemental force and a color, both according to the digital placement of the shapes when they are arranged according to edges by faces on a half circle, forming a spectrum that has already been discussed. For the cube the arrangement yields elementally the weak nuclear force as equivalent to Earth, and the color attribution of blue, due to the doppler effect on their potential spin.

The weak nuclear force can be thought of as the most attractive of all the elemental forces. It determines that the spin one force carrying particle the gluon holds together quarks in the proton and neutron, and the proton and neutron in the nucleus of an atom.

According to confinement these particles are always bound in such a way that they create self-nullifying, stable particles. When this fails to occur unstable particles, such as mesons (a quark with an antiquark) and glueballs (several gluons), may be produced. As the energy level of these particles is increased they display asymptotic freedom, meaning that they behave more like observable free particles.

In other words, only in the mortar and not in the bricks is there even the possibility of anything other than the darkest of blue shifts being displayed. The cube's attribution to force and color being reasonably simple, we can only wonder at the infinitude of its implication. Let us next discuss the implication of aim following event, then discuss the implication of space, and in conclusion discuss the implication of time, which we have already touched on briefly.

When aim follows event it takes the form of an outward flowing wave of probability in the medium of potential which effects disturbance in units of information. So let us examine where the relationship between the weak nuclear force and the color blue comes into this model.

In each single unit of information there is a similarity with the particles of the gluon and quark variety. They are compressed and held together in confinement from the shells of probability that surround them. As it is impacted from outside, the probability shell gives in at one point, and refers the force through to the core, where it changes the composition of the information unit, or quantum. This might not be in a directly physical sense, but rather in the realm of spin. For information this has no bearing, the two are as one and the same thing. We will return to this model as well for each subsequent corresponding section. Here we find the similarity is in cohesion.

Finally let us discuss the implication of space-time. Just as in the formal system of metaphysics model, a unit of information and the event from which its effect emanates are separated only by the wave of probability, which itself is carried on the dimensional expansion and collapse of the sphere of the mind, so does the sphere of the mind, at its most approximately infinite extension, separate space and time.

What we have seen up until now of the formal system of metaphysics has amounted to little more than this observation of reality: there exists a strobing between moments, which constitutes the duration it takes for the universe to process the information being applied at all times. The Buddhists recognize this, and I will go into what I mean by this as briefly as possible. To do this we must examine the transition from the formal system to its dimensionally full extension as space-time.

The reason the cube was chosen as the model for the formal system is unclear. There are six triggers of reason, yet eight corners to a cube. Why not attribute these to the six sides? What a jolly children's toy, then! A die fit for God.... But in attributing them to the corners, and specifically those that outline the shape when it is looked at above one corner, which itself is not included, seems to be aspecific in its purpose.

Of course, we have found many attributes of the model in this configuration, but they are no less arbitrary than would assigning meanings to the nodes as faces have been. The only thing it necessarily takes for granted is the addition of Had and Not to the six fundamentals of reasoning. As we can easily see, these depend on awareness of the strobing, which can only be comprehended with a knowledge that transcends time, or the functioning of formal metaphysics.

What we can deduce from this is that what is implied by the cube is more than merely a single attribution to a natural science, but the specific attribution of motion to the formal system of metaphysics. As we have already seen, this is not motion in terms of trajectory, but as projection. It occurs as a wave of probability emanating from the formal system either inwards or outwards.

Therefore it is not only a cube that is implied in the formal system's functioning, but a hypercube. Furthermore, the hypercube is a fourth dimensional shape, and the fourth dimension is the dimension of time. So not only is time a factor due to the strobing functioning of the formal system, but effects its shape to match it dimensionally. In short, what we can deduce is that time is, in essence, a cube.

Time is the measure dividing the inner from the outer cube of a hypercube. We know that these are one and the same cube, so we can say that time is the measure of a cube from itself. This seems quite natural, for if you have a cube standing on your desk, for example, it will not change its own measure, but it will be changed by time. Its atoms will be gradually sifted around as though between two walls that are all around them, the planes of orbit in electron shells — the walls of a cube between the third dimension and the fourth dimension. This is of the probability field surrounding atomic nuclei, the probability field in which its electron shells are stored. It constitutes the medium in which our universal strobing occurs. We can also say that this strobing is time, and that its medium is potential, its substance probability, and its essence information. If we let it, this is what makes us.

2. Octahedron
There is a change in numeric sequencing of the solids according to elemental attribution for the first time in our investigation of the solids in the case of the octahedron. Thus the number assigned to it according to the amount of displacement to it caused by translation is one unit.

The implications in the difference between this and the attribution of zero to the cube are various. It might be interesting to make note of the idea that the shape of the numbers zero and one have certain similarities to their surroundings on other dimensional levels — the cube would be enclosed in the sphere of the mind and the octahedron the essence of directed intention — when, later in the text, you consider the meaning of motion, and might ask yourself to distinguish between forces and substances.

Whereas the cube's function was essentially formal, that of the octahedron is experiential. This is so because its elemental force is permanently and all permeatingly attractive. This accounts for the color attribution of indigo, being a more minimal blue in the doppler shift of the light wave carrying its particles' data.

This pertains to veiled messages, or doubly encoded information, because of the diminishment of spin on the particles.

Consider the waves in the ocean, as our ancient ancestors once did, and see the transverse shape of them. Their base is pulled backwards, in the opposite direction of the wave itself, forming a plateau with a spiral foundation, where the water curls under. It is being drawn toward a wall of upward flowing water, which is the onslaught of a higher plateau, and these two are connected by an upright, forward twisting spiral, known as the wave front, or crest.

The two spirals are the same in measure, and differ only between horizontal and vertical. Also, the horizontal one, flowing towards the wave-front, leads the vertical one, flowing upwards. This is how the particles of $H2O$ within that wave are pulled around in the disruption of the water. The ones at the tips of the spiral spin the least, and this is where foam floats.

The particles begin to roll around to their original position before they are acted upon by propulsion, and this creates a sort of reverse force, that carries no energy. This is known as reverse spin, and it is proven to exist by the laws that govern classical physics. When a force, such as a wave, acts to generate spin in an object, such as a particle of $H2O$ for example, or a graviton, as we shall see shortly, counterspin is simultaneously created, although it usually occurs in a higher dimensional medium, and so is dispersed without generating measurable effect.

In the case of $H20$ it is swept up and absorbed in the momentum of the wave in the medium of molecular collisions. In the case of a graviton, it is generated only in the realm of probability, where motion is measured in interval and observed as physical, just as time is seen as at once a force and an object. We will come back to what this means for information when we look at that interval measure's impact on the formal system. For now let us examine also the graviton, which, as its name implies, is the force carrying particle for the elemental force of gravity.

Gravity is no different from a wave. It exists in static form as a field, and this divides itself into two dimensions when it is disturbed by the presence of mass. At these events it emanates forth from the disruption like waves in water from a distortion to the standing wave state or base energy level. The force of gravity is perpetually attractive. It is a one-way wave, pre-reflection. Hence its blue shift, and hence the Golden Division relationships in a side-slice of its wave front.

Now we can discuss the implications on the formal system of the concept of doubling information. The art of this is utilizing the storage space of reverse spin as well as classic spin. Both create a wavelength, and these counterbalance one another.

When they are equal there is no spin at all. When one is more probable than the other, the lesser is a measure relevant within that of the larger; however, since it operates at a greater frequency, it is thought to pertain to the greater information embodied in the energies of higher dimensions as opposed to storage space available, since the smaller is essentially identical to the larger, aside from size, and therefore like a fractal of its partner. So, in the formal system, we can see how the raw units of dimensional information, also known as quanta, transform, and what they can contain when they do so.

Let us here limit our attentions to the implications of gravity upon space-time. Space is little more than a field of gravity, and time the movement of that medium by its own effect. In short, space-time is very malleable according to the presence of gravity. It flows like water, yet is infinite. It passes through us like moods, emotions. Knowing that the octahedron is the very shape of motion, according to its simultaneous spin and reverse spin applications, still, can we rely on these mere implications as necessarily indicative?

As we study the shifting in shape of shadow that occurs for each Platonic solid as it is passed through dimensions, we can add the factor of an oscillating motion in each lower dimensional shadow, such that, when one is in a stage of conjoinment, at the point of apogee, the other will be in a state of full alignment. Thus, since it is not necessary to assign space or time to either the shadow, or the shadow's shadow, per se, there is no use quibbling about it, as they all amount to the Qliphoth, or forms of motion, a topic fully elaborated upon elsewhere.

All that all of this means is that the motion introduced in the incremental measure of the octahedron between the cube and hypercube will carry through all of the subsequent solids on both the level of space and time. This is the first shape to connote such motion, and is the method of splitting a cube and hypercube, which is the first motion of the expansion and contraction of the sphere of the mind.

On the smallest level of pertinent effect, that of one event leading to the next, we can see the presence of the octahedron as well — necessary to the transformation of dimension, such as between the solid third and the medium fourth: Gravity causes change in time. This means spin and reverse spin are present in the form of motion of information units, or quanta, in the formal system, electron clouds and real particles in the third dimension, probability fields in general for the fourth dimension, potential in general for the fifth dimension, and the sphere of the mind, for reasons discussed elsewhere, for the sixth dimension.

What this conveys to the formal system model is that, with knowledge of the one-way wave, we can separate expectation of the pre-reflexive from the reflected. Each is one way in itself, but due to confinement, the second wave will overlap the first. The distinction between them is between effect as verb and as noun. There is no aim to the pre-reflexive mode, yet there is aim in the reflected wave — its crest will have a measure closest to the center of the pre-reflexive wave. This creates similarities to the movement of particles in standing wave form, but in this case the distortion to the preexisting field is the motion of the metaphysics model.

Aim is the wave of potential that ripples outward from basic fact and fiction in a field of one or the other as an alteration to it by its opposite. Either fact or fiction that is not the field or medium amounts to the motivating energy within it.

From the perspective of water this is the H2O molecules of a wave, or potential from the perspective of gravitons. According to symmetry breaking, which we will come to shortly, each of these is only a higher or lower energy density form of the others — gravitons being one energy wavelength different from Hydrogen and Oxygen atoms. Potential acts the part of medium for aim, as it does the part of energy for a graviton field. It merely occurs as a substance separating the manifestations, the particle of pure dimension.

That aim is a reflected wave of potential is pertinent because of the presence or the absence of the sphere of the mind. When we watched the wave of potential started by the ejection of excess definitive information from event, at its conception, by the node of Not, we saw that this occurs without aim. It is an unconscious function of the formal system, connecting nodes of the formal system with their mirrors as far removed as the unconscious memory castle, for the first time, but the wave of potential started by that same event once it is determined fact or fiction does posses aim, and is therefore a product of awareness.

Remember that the single node of fact and fiction serves as a microcosm of all the sphere of the mind in one, including its surface reflections as thoughts (even those of the node of Not) and its resident formal systems up to and including the unconscious memory castle. So what we are looking at is not really another wave at all, just the same outward emanating wave acting as a reflected wave passing through the inversion of dimension at the bridge between the sphere of the mind and the node of fact and fiction. This happens perpetually as each event follows from the last, and this creates the appearance of temporal strobing.

3. Isocahedron

The Isocahedron is the second shape to undergo numeric transformation in its translation from traditional to digitized attribution. But the number described by the units it moves in a chart of elemental alignments is double that of the last one, and therefore represents only duplication rather than creation from nothing.

The similarity in color assignation and the attribution of elemental force is again based on the distortion of the emission spectrum by the doppler effect on the transverse wavelengths of photons. This is an elemental force that is recognized by rapid retreat, hence a red shift to their appearance.

Next, we can examine the isocahedron as indicative of fire, or the weak nuclear force, both representative of action in the potential effect of one event on another.

In this mode there is only activity, although there remains effect and medium. The activity is such that, rather than the directed change seen in rippling water, there is constant, direct change, by the effect of the medium.

This manifests depletion — what is, in weak nuclear reactions, radioactive decay. This leads to the understanding of time as a destructive force, although this is a limited perspective on the true nature of what occurs. In fire, we see the transformation of fuel into ash, usually carbon. It is interesting to note that humans, animals, plants, and most other solid matter in the solar system are made up of burned and fused remains of the gas that feeds our sun. Life, as we know it, is carbon-based. Yet still, as such, even this material is prone to gradual deterioration.

We know from examination of many other minerals that make up our world, that one of the factors of aging is the transformation of matter into energy. Such is the case for most other of the solid materials in our solar system than carbon, and, accordingly it is supposed, so such is the case in the rest of the universe as well. So what man is of fire itself, most else is of radioactive decay.

The particles of radioactive decay are three massive vector bosons which display symmetry breaking with photons at energies greater than 100 GeV. They also have the same spin as photons in all states. It is by radioactive decay that the half life of all the periodic elements is determined. It should also be noted that this force causes the nuclear reactions of the sun.

The resemblances of this force to fire are implicitly apparent. Both are of the vertical spiral of the transverse wave. The proof that this is the case for radioactive decay lies in the spin of its particles, which lacks reverse spin except as pure, informationally inaccessible potential. These, like photons, have a spin of exactly one. What spin means is that it takes a certain amount of rotation before a point comes back to its original position. When reverse spin is present it causes longer or shorter paths to be taken by the point as it travels back to its original position, and these paths are called vectors, from which the massive vector boson takes its name.

Thus, forward propulsion is at a maximum, but reverse spin is completely compensated, as it is in the upper parts of the vertical spiral front of a transverse wave. Fire is alike this shape because of its composition as it arcs into its fuel supply, the air. It has a broad base, but the heat within it is swirling. This causes only half of the base to rise, and then half of the risen flame to curl, etc. creating an upward spiral. This spiral is a phi spiral. We have already discussed the deduction of phi, or the Golden Division, from the pentagram, and in this solid do we see that shape for the first time, though it is not flat, and is formed out of triangles.

What follows aim in the expansion of the formal system is action. Here we see the outward ripple in potential constituting effect transform itself from activity in a passive medium to pure activity. The waves consume the water, the entire pattern now consisting of the spiral shape of the wave seen from its side.

This is spontaneous symmetry breaking. It constitutes a fundamental change in the realm of unit interval time measurement, or patterns that occur or repeat over time. This effects the formal system, space, and time.

Just as when the cube changed from the cube to the hypercube and back over time, and the octahedron represented the interval stage of the transformation of one to the other, so does the isocahedron come into play as the measuring of that interference as entropy. It represents the stage at which the one becomes the other, or the transformative action itself.

Here no further reliable intervals remain, as they are all consumed. But remember that they were only ever expressed as potential to begin with, so it is only by seeing them each revealed individually at the time of their destruction that we can even prove that they ever existed potentially as well.

It should also be noted that at this point it becomes irrelevant whether we are looking at the formal system from the side, or down along the series of actions separating one event from the next, so we will describe them as though the two were identical in implication as well.

The exhibition in outer space for this aspects elemental forms have already been discussed, but the meaning of the substance of space itself was only connoted. Here again we should examine the application of phi.

Phi occurs as a potential geometric measurement within the form of the isocahedron, and plays a very important role in the stimulation of motion in the physics of the universe. It represents the azimuthal angle that any elliptical or spirally orbiting object will use to break its course and move to a new orbital position when acted upon by another object or force, and is determined as a spiral between the original and the new orbits by the effect of the object's mass and its degree or amount of internal spin. This holds true for the interactions of all particles, and for objects on all scales, all other things being equal.

So really, by the presence of phi, what we are seeing is a break from the pi, or round, spiraling pattern of time, into a diversity of component patterns which are being destroyed, and which all together assume the shape of a single phi spiral, contained in the shape of the isocahedron.

It is correct to point out now that it is with the isocahedron we see a progression through the spectrum not counterclockwise, as has been the case up until now, but clockwise, begin.

The best example relating the shape of the phi spiral, which is as familiar as the shape of the eye itself, to fire, is that of the flame of the single wick candle. The shape of this flame is an exact phi spiral when measured from any direction. If one could see the waves of heat that emanate from it, though, one would see that they are insubstantial to the air molecules they disturb, and assume no pattern as does the central area of the flame, that part of it that is, no less, luminous.

What this means for space, time and the formal system is that entropy is omnidirectional. It pertains first to the original wave of potential emanating from the node of fact and fiction to create the sphere of the mind, to the implosion of this to create the formal system, and then to the reflection of this to create their interaction. Thus, it flows simultaneously inward and outward on two levels and meets in a third. These levels are potential, probability and information, and move from philosophical towards physical as they pass through each dimensional inversion, offering accordance with the three branches of legislation, patterned in reverse order.

However, the fact that the activity of the medium constitutes entropy, or the increase of disorder and unpredictable chaos, means two things. First, as time is usually measured by entropy, we see that time is represented here as being omnidirectional. This pertains greatly to the physical aspects of the seventh dimension, which is discussed elsewhere. Second, that before passing through pure dimensions, as it must do repeatedly with each of the three states of its manifestation, the wave of probability must become asymptotically disordered. This occurs not only between crossing pure dimensions, but also before the wave undergoes dimensional inversion between maximum and minimum extension of interval, and these two occur in opposition to one another directionally. When one wave is emerging from dimensional inversion, another part of it will be being reflected from the next higher dimension, and therefore entropy overlaps itself, and is omnidirectional.

4. Dodecahedron

The digital translation of the dodecahedron changes three places on a chart of aligned elemental attributions. This is the first prime integer — multipliable only by one to total itself, having no other factors.

This number is of extreme significance to the concept of dimensional leveling. Without it we would not know reality as we do. Because it is prime it means it is the progeny of both two, by addition, and one, by multiplication. The number Three has been referenced many times by symbology, but most of it comes back to dimensional leveling. That is how innate it is in our consciousness.

Its color attribution makes it similar to water in placement on the spectrum, or on the chart of angles, above the larger wedge of fire just as indigo is poised above the cube. The color attributed in this case however is yellow. What we will see with this is that the particles pertaining to the assigned elemental force will posses a lesser red shift to the doppler effected wavelengths of reflected photons; but we will address this presently.

The dodecahedron applies itself to the element of air, especially, although traditionally to the concept of the cosmos. The Greeks, particularly Plato, preferred to liken their idea of the realm combining the ideal and the real to this solid because of their strong ties to the Pythagorean Brotherhood, whose symbol was the pentagram, and this is the only solid comprised of pentagons.

The Pythagorean symbol of the pentagram itself derived most of its meaning from phi, the Golden Division. The Classical Greeks used this measure as their entire standard of beauty, and applied it to everything from sculpture to architecture so that it would perpetually surround and inspire them in their everyday lives.

This may have borrowed from, or was at least perpetuated by, Judaic mysticism, where the ascendent number of the sephira Tifereth in the Kabbalah is five, and the meaning of Tifereth, beauty. But, though our cosmos may be beautiful, it is not beauty alone that is our cosmos.

So when we translate our digital attributions, it is air rather than cosmos that is assigned to the dodecahedron. Humorously this puts me in mind of something Aristophanes said once: "when Zeus is toppled, chaos succeeds him, and whirlwind reigns."

The implications of assigning air to the dodecahedron are simple, once the elemental force produced of it from digital translation is taken into consideration. The field of electromagnetic radiation is about as equivalent in the modern mind with the air as, to the ancient mind, was it with the cosmos.

It is the medium in which information is physically transmitted, though we know information to exist only as probability. Thus, before the advent of technology specifically designed for its utilization, electromagnetism was probably perceived as some sort of higher power.

The electromagnetic force interacts only with charged particles such as electrons and quarks. It determines the exchange of spin one photons emitted when an electron changes energy level shells in its orbit of an atomic nucleus. It also governs the exchange of photons between an electron changing energy shell levels and another electron, which is caused, by absorption of the emitted photon, to fall to a lesser energy level. This force controls most of the events that occur on the subatomic scale, as it effects quantum particles in much the same way gravity does the planets.

At the core of the electromagnetic force is collision. It is by particle displacement that we know of the existence of and the properties of the various universal particles, and by stimulation along electrical or magnetic rays that this is accomplished.

Electrical and magnetic rays are fields of force that occur as wells of potential align along a standing waveform. For electricity this occurs parallel to the transmission of charge, creating a wave of fractally infinite hyperspheres, but for magnets these lines connect one pole to another, repelling their own like, creating a wave of fractally infinite tori, or hyperspheres at antipode. In the combination of these two forces, charged particles oscillate rapidly, forming fields that are centered, yet can be projected — in other words, generating excitation over subject areas.

The impact the dodecahedron has on the formal system is in the form of the release of message.

Just as our atmosphere recycles itself through organisms, so quantum information is liberated from form and exposed in its essence as potential energy before it can cross the gap of pure dimension and reform in a translated version. Thus, in the simultaneous expansion and collapse of the formal system, it occurs as between the square representing the relationship and point or node of fact and fiction, and between the hypercross and the hypercross of hypercubes.

As you can see, after these points the outward emanating, reflected wave of aimed potential carried by the sphere of the mind undergoes dimensional inversion, and becomes the next event in the formal system of metaphysics model. The pattern in the field of potential constituting their interaction goes from being a single phi spiral into an orderly chaos of static and wind.

At this point the field of potential breaks down into infinite pi spirals, and it is thought, therefore, that the dodecahedron is a valuable tool for meditating upon the surface of the sphere of the mind. The excitation it generates acts like a background radiation scan, determining the amount of energy effected through potential by informational distortion.

More than this, sources and strength of interferences can be localized and neutralized, although doing so robs flavor from the message. The proof and implications of this are in space-time, so let us turn our attentions there.

In space it is by the amount of electromagnetic background radiation that we can fix the age of the universe. This includes all light, visible and invisible, all sound, and, insofar as the particle/wave exchange of matter-energy goes, all forms of emission, absorption and reflection by all mass everywhere.

Because space and time are synonymous, the universe's size and age are synonymous, and because time passes, we suppose there to be a change in the size of the universe as well. What we find is an increase of information being carried on the wavelengths that constitute our measurement method.

The electromagnetic spectrum is host to input from an increasing number of sources, creating greater overall interactions. Thus, even if we could find no physical evidence to support the non-static universe hypothesis, such as the homogenous red-shifting of galaxies implying their motion away from us and from one another, there would still exist evidence supporting the same data based on a dimensional hypothesis, where information, through probability, yields event. So not only is the electromagnetic spectrum the medium for the universe's age, but it may display the causation thereof, and therefore imply a solution.

Time is thus what information becomes immediately before crossing the threshold into pure dimension. This is to say merely that the essence of information is potential; that is — that it is its motivating energy, and not its substance. The substance of potential is only present at the level of specifically the fourth pure dimension, though potential acts as an energy between every dimension. It is in this form that we encounter it in the electromagnetic spectrum, and in this way that it acts as the factor for aging in the universe.

5. Tetrahedron

The numerical change of the tetrahedron is the greatest for all of the solids on a chart equating traditional to digitally translated elements. It moves four places, between its original attribution to fire and its digital attribution to cosmos.

Four is the number of relationship because, as we have already demonstrated, relationships occur as planes. These planes insert themselves into our perception within the strobing of event creation. According to them we speak and act with various voices and for various times, and accordingly they are ordered such that we can perhaps, at times, be influenced by them. In short, relationships are all self-referential, and no less real.

The tetrahedron is attributed to green according to the arranging of the visible colors on a half-circle intersected by angles conceived by the numeration of the forms of the Platonic Solids themselves. In this arc, green is central and surmounts the composition, and the angle attributed to it is that defined by the tetrahedron. This is little more than a coincidence of information organization between two finite sets, but it serves to support the attribution to the tetrahedron of cosmos by its pivotal position.

As with the dodecahedron we must discuss traditional and digital attributions of element. Traditionally the tetrahedron was associated with fire, and this, it is thought, was due to its shape.

As such, it embodied the Golden Division in the form of two legs of the triangle bisecting one of the faces of the tetrahedron. The leg that is half of the unit leg of one of the four tetrahedronal faces is in a 1:2 relationship with the leg opposite it across any of the three sixty degree angles of the face. Thus phi is produced and thus, in its spiral form, fire appears. Though phi had meaning, for Heraclitus, as fire, and for Classical Greece as cosmos, the phi/pi spiral, which can also be derived from the tetrahedron, is unique to cosmos alone.

Now the stage is set for consideration of the tetrahedron as cosmos. Geometrically its origins depend on the simplification of other solids. Consider the cube of the formal system of metaphysics model. Now imagine the diagonals connecting every corner, cutting the cube and slicing the faces, so to speak. Remove the shape defined by these diagonals, discarding all excess components of the cube.

The form you now grasp is the steloctahedron — the conjoining of two opposing tetrahedrons. You will notice that it assumes all sizes, from event through fraction of the unconscious memory castle. It seems to have more significance at the size of the formal system of metaphysics model, as each corner will still have functional meanings (and its entire set of relationships be effected but slightly); though considering it at its lowest dimensional level, that of event, you can better see that it is nothing but the intersection of fact and fiction embodied.

Even more than this, each event is possessed of simultaneous influence by both fact and fiction twice, as through the occurrence of cosmos also at the dimensional intersection between the unconscious memory castle and the sphere of the mind, in the form of short term and long term implications. The deciding factor is the central attribution of the tetrahedron, which will be impacted upon by either fact or fiction and adjusted accordingly.

Herein lie great unexplored mysteries of fact and fiction. It is enough now to say that it is the interaction of these two simultaneous midpoints in the conjoined tetrahedron that cause the observed temporal warping of the hypercube.

Physically, the steloctahedron could be equated to either the heart, the gonads or the brain by attribution of relative branch of metaphysics — ethics, physics or religion respectively according to tradition.

Ethically, the double tetrahedron also has a tripartite position as representative of the church, the state and the people — yet the relevance of these forms to one another changes over time, not only socially, but within each individual. This leads to belief, or the gathering of histories for the comparison of experience, both within and with others, which generally lead toward the categorizing of experience by natural/supernatural and good/bad.

The relevances we are discussing now for the three positions of ascent in terms of the legislative branches of metaphysics are the same as the distinctions already described for the elemental solids in terms of space, time, and the formal system. These expressions are all merely names for the shadows cast by the apogee, antipode and perigee of the expanding and contracting solid fourth dimension.

It should also be noted that the shadow cast by the double tetrahedron at its apogee is the hexagram, a form of particular use, as we shall see shortly, in divination — or in the forecasting of future events. This represents one of the spiritual implications of the fact opposing fiction tetrahedrons we now see intrinsic to the metaphysics model. The fact that the divinatory school has been shunned down by its opposition is evidenced in their use of this form only representationally, in the lower dimensional expression.

We should talk briefly of the single tetrahedron, which we can see can be of either fact or fiction, and which composes the concept of cosmos. It is most notable that cosmos, essentially the world-view of a person or party, is determined and conveyable only by quantum informational units of fact or fiction. We have already demonstrated that both of these are equally *real*.

Therefore what forms your cosmos of reality is a single tetrahedron of smaller tetrahedra that are composed of smaller tetrahedra, etc. each more fact than fiction or vice versa. The tetrahedronal form that cosmos takes is determined by conservation of potential: the most complex concept is contained within the most concise form.

There are no direct equivalencies between the tetrahedron as cosmos and the realm of time and space, but there are grand unification theories of physics (g.u.t.s) that relate all the four elements in one through symmetry breaking, and combine the theories of quantum mechanics and general relativity elegantly, and these can counterbalance cosmos. Cosmos, however, is also equal to the philosophical ideal.

The tetrahedron, being thus twice extractable from the cube, is the shape taken by a unit of information. We have seen these to be made of probability, suspended in potential, with an essence of potential energy, and pushed into various patterns by the distortion of their medium in the form of a series of waves, where these waves have many attributions, one of them being the Platonic Solids. The fact that they are tetrahedrons is borne out by their binary directional nature.

It should be noted that the order in which the four elemental attributions occur is precisely the order in which they occurred in the foundation of our world. First there was Earth, the molten magma stage of terrestrial development, followed by Water — the condensation of atoms in the airborne volcanic debris causing inundation and for 80% of our planet's surface to be covered up by the seas and ice; then came Fire in the form of lightning, and the electrification of the created molecular medium for the atmosphere, followed by Air — the final fixed and cyclical processing system for the self-sustenance of the surface of our planet, for it contains — at the tips of its branches — the carbon-dioxide inputting, oxygen outputting plants, and the carbon-dioxide outputting, oxygen inputting us. Finally there is Cosmos: the dictate "As Above, So Below," to describe that, just as in that order was the earth formed, so were we evolved, do we mature, do fruits ripen, do harvests pass, do governments grow, do communities thrive, do our cells function, do our planets and sun form, are our molecules and atoms and quanta made and moved, do the walls and voids stir, are the Platonic Solids, as are the elemental forces, so Time *is*.

We have seen that time can be compressed as a box, or, more exactly, as the space between a thing and itself. In the space of this fissure occur all the rest of the Platonic Solids as representative of forces with various attribution, constituting a measure of this nothing. It should be noted at this time the progression through these states is not fixed — that, just as all colors of the visible spectrum exist simultaneously, so do the Platonic Solids, and therefore there is disagreement between varying cosmologies regarding the organization of the elements, which are usually arranged temporally.

This may appear confusing when it is first encountered, but the existent evidence imputes that these are all equally important, in every order, and this position is reaffirmed by the metaphysics model itself. Though changes seem to be occurring as one reflects on first one, then another, of these thought forms, there are several similarities which ought to remain present to consideration of the solids.

There is always a medium in which effect moves, and all movement therein is effect. It is not, ultimately, the relationship of these modes to one another that fundamentally determines their structure, but their relationship to the sphere of the mind. Finally, just as the dimensional levels of the formal system and the sphere of the mind, all are only as one stage in the metaphysics process, and this stage, just as the dimensions of the formal system and the sphere of the mind, are always present as a measure of potential.

6. Sphere

We have just seen the transformation of one event to the next, at least up until the moment of the strobing of pure dimension, where everything breaks down into something else. This substance is at once pure potential and luminous. It is observable in both the strobing of event and the primary clear light of ylem recognized by the Buddhists.

It is not necessary to study this substance in too great a depth here, so we need only address its effect. We can see that it acts as the interval of event, begetting the future as the translation of the past in the present. We can see that it involves all the elemental forces and their force-carrying particles, which motivate all real particles. We can see that it is the culmination of a very specific procedure that occurs in the context of the formal system. So let us continue to study this process, and see what the next immediate result is in its unfoldment.

At the point where information, having already parted with pattern of motion, now parts with its tetrahedronal unit form as well, we see a release of potential energy into a static field of pure dimensional potential, thus supersaturating it. This moment, you might want to take note, is the final translation between matter and energy, and information, we can now see, occurs as though an intermediary between them. It is likely this universal release of invisible energy between each instant in the field of potential that causes the strobed illumination.

At the moment after this we see the entire system collapse into a single universal, fractal event, going through all the forms and forces on all scales at the same time. In the formal system this occurs between the plane of relationships and the node of fact and fiction, which can be seen as the smallest unit either of the cubic formal system or the sphere of the mind, and between the unconscious memory castle and the mental torus, except that, in reality, the mental torus is the node of fact and fiction, and the one dimensional point is extraneous to the unconscious memory castle.

So, as the strobing occurs, energy is released and a certain amount of potential must be conserved into matter to compensate, and from this the next event emerges.

It emerges as the wave of aimed probability emanating outward from the one dimensional point to the plane of two dimensional relationships, and, simultaneously, as the reflected wave impressing itself from the mental torus to the unconscious memory castle. Really the information contained in the moment being measured by the translation of the probability into information rendered by the planes of relationship is equivalent in quantity to the full amount stored in the unconscious memory castle, since there is only one wave.

During this moment it is entirely possible for you to review the full understanding of the relationships that will define the event about to be fully formed and make decisions regarding your strategies to evoke certain and perturb other paths of repetition of potential patterns by alteration to the arrangement of their constituent information units.

Next the wave moves into the conscious memory castle and the actual three dimensional event. It is an exact copy of the preceding event, aside from the particular pieces of information that were changed by the translation of the wave of probability from possibilities into pattern, into particle, and finally into potential energy, where the change occurs in the realm of spin.

It should also be noted that the planar aspect of event formation immediately preceding three dimensional realization is the same as a shadow cast by each of the hypersolids overlaid, each one storing the change pertinent to their elemental force, and of particular note is the shadow cast by the tetrahedron at perigee, for it pertains to prophecy. It is entirely possible to utilize relationship planes to hypothesize outcome, which I go into later.

All of this only appears to pertain to the sphere cursorily, as a function of the wave of probability being at all times carried by the sphere of the mind. But let us pause for a moment and consider why the sphere was even chosen as the shape representative of the mind.

We have observed its reflexive nature, and stated it as in opposition to the cube, which is representative in its pattern of order of the brain. We have also observed that these evolve congruently, such that there is no answer to which came first, since, in their conception as the one dimensional point of the node of fact or fiction, they are both the same.

Finally we can state that the only difference between the expansion of the wave of probability that, by passing through gaps of pure dimension, casts shadows in the mind of the shapes of the formal system and in reality of the elemental quanta, and the expansion of the wave of probability that creates all extensions of the sphere of the mind, is that the formal system was created with no aim — that is, to cause no specific event — and that the Platonic Solids were created with aim — again from this purely physical view, to cause the sensual aspect of event — and that, from any view, the sphere of the mind is the source of the aim.

Thus, though all the waves evolve from the same root source, they will each be functioning between different dimensions at the same time; the realm of the Platonic Solids is between the shadows of the shapes of the metaphysics model, in the interference pattern created by the crossing of the impressed and expressed ripples of probability. So really, just as this distinguishes the Platonic Solids, the only thing that distinguishes the sphere is its association with aim.

Just as much can be learned by examining the solids as shapes by which to measure the element or force to which they are attributed, so it is probable that the origin of the implication of aim pertains to the ability to predict reflected trajectory, or bounce.

This would seem easier to calculate for the solids, as they all posses certain, very accurate measures which, when acted upon by other forms or geometries, would have exactly and easily measurable alteration according to these calculations, except for spin, which is predictable only for the sphere, which has the most possible number of sides — asymptotically infinite — and yet averages out to have the most rightly calculable reflected trajectory. Thus, as in the formal system of metaphysics it makes no difference whether Aim refers to expansion/contraction or to reflected trajectory, consciousness is a measure of interference.

There is yet one last attribution to the sphere we can discuss: the attribution of the sphere to the sixth dimension. Elsewhere I go into copiously the ramifications of this on Judaic mysticism and psychology, but here let me just say this briefly — that this attribution derives from observation of the first probability wave's progress through dimensions; it generates the point, the plane, the cube, the hypercube, and the hypercross, in order, as it extends outward from a central distortion.

This distortion can be anything, subject or object, answer or question, thought itself or merely Time; though for the purpose of studying our hexagonal formal system of reasoning it was convenient to use first a question, and then, by application of the measures relating a series of nodes, to see how it can be transformed into an answer. This is only one of a multitude of possible examples for distortion to potential, and I stress again that it was used for the sole purpose of decorating it with familiar ideas to introduce you to the formal system.

Once the wave of probability gets to the end of its interaction with the hypercross it is prepared to meet the sixth dimension, and here it finds the sphere of the mind; granted, at its smallest form — the first dimension, but to understand how this is indeed equivalent to the sixth dimension we must study the second emanation.

Now, as I have already described, these dimensions are not encountered in uniform order by all of the three simultaneous forms, expressed as potential, of the single probability wave. For the first projection of the wave the dimensions are encountered in the way I have just described. Then the wave of probability contacts the sphere of the mind as a point, and this changes the nature of the wave so that now it possesses aim.

This contact also changes the structure of the model: it places the base state, due to the ingrained processing procedures of the brain (the post-mind formal system, expressed as a cube, the microcosmic equivalent of the macrocosm Time), and therefore changes the numeration of the lesser dimensions (the square to five and the point to six); but it also divides the conscious memory hypercross into a conscious memory hypercubic hypercross and an unconscious memory hypercubic hypercross of hypercubic hypercrosses, which are the seventh and eight dimensional forms, and which describe certain factors in the processing of time as the passage of three dimensional event intervals, and which I describe elsewhere.

After this the double-wave at once continues outward and is reflected back such that, within, it meets the sixth dimension, and without the sphere of the mind — once again in its smallest form, the one dimensional dot. This is because the torus, the sphere of the mind's equivalent to the hypercross, can also be expanded to accompany more dimensions: it possesses a central dot, a circular plane, a simple sphere, a hypersphere, and a torus; but with the torus the central sphere is the same as the outer tube surrounding it at perigee, the radius of which is the diameter of another identical sphere. Though this is innate to the torus, the non-central, or complex, sphere does have a circular circumference, and this circle does have a central dot — and so here are the seventh and eighth dimensions for the mind.

There is one further emanation for the wave of probability to make before it has restored equilibrium to potential after the initial distortion, and it will do this now also possessed of reflection. As we have just seen, the wave now passes through each of the five Platonic Solids as it approaches gaps of pure dimension, where the constructs of the formal system lie as pure illuminated potential.

Furthermore we should recognize that this process has been happening all along, but was not identified until it became useful to demonstrate changes in the medium through which one event begets another. In the end we are again confronted by the sphere of the mind which, being the sixth dimension, encompasses all those before, including, through the compression of the formal system by the sphere of the mind, the seventh and eight dimensions as well.

7. Dimensions

We cannot say that we are passing through dimensions. We can only say this of an objective measuring model, even one we just imagine. While we are thinking thus, it is true that dimensions are passing through us, for it is dimensions that move, while we are stationary.

What can we say about ourselves and our mental projective space that will be true of dimensions? For this circle of awareness around us constitutes an eye through which we observe dimensions.

We have already noticed how the five Platonic Solids can be used as a measure between any dimension and another. So what is implied is that there is a centralized feature and an intermediate feature — much the same as the focus of our self and the surrounding uncertainty of our mental projective space.

In other words, there is a probability surrounded by potential. This is true for atoms, units of information, dimensions, the sphere of the mind, and the parts and relationships of the full formal system — which, in its base state, is the probability field surrounding event.

When this is seen to apply to the tetrahedron, the hexagram is rendered at apogee, and to this we will devote the next section. This will serve to unlock the nature of dimensions.

We can use the five Platonic Solids therefore as a rule for uncertainty, but we have already seen these are most distinctly arranged according to chronology, and I have shown you as well that potential is the virtual particle produced within the medium of the fourth dimension, time. So here we find a direct reference to the medium in which the patterns indicated by the solids manifest and the implication of a certain order to addressing their occurrence.

Because each of the component forms of the formal system of metaphysics model, like the sphere of the mind, contains those of the lower dimensions, we can see the presence of expansion, indicating there is increasing space between forms as the model ascends through dimensions. The fact that this implies greater gaps between dimensions is measured in the amount of influence the attributions of the Platonic Solids have on the varying structures. This is infinite in potential — or media of expression — but finite in possibilities pertaining to a specific event, therefore causing waves of probability to occur as patterns.

As the gaps between dimensions expand, we can see the patterns delimited by the attributions of the five solids deepen. This means that, as transported by the aimed and reflected waves, there is an increase in combinations of possibilities that will effect event. The forms taken on by the solids and the manifestations of their attributions account, indeed, for all diversity, even opposites, such as particle and wave in physics.

This at once paints large shadows on the wall of our unconscious memory castle, and intensifies the details in the foreground. And all of this occurs in proportions that are determined by the five solids themselves.

When the wave of probability passes through the solids, in the gap between any two adjacent pure dimensions, it is like two hands with their fingers folded together. Sometimes the index finger of one hand will be above, and sometimes it will be the other. When the probability wave passes through, regardless of which direction — because, as far as the potential itself is concerned, the effect of one on the medium would be the same as the effect of both at once — it will be like these fingers becoming confused, and seeming to be joined at all points, the fingers feeling like they are perpetually changing places with one another.

As I have cautioned before, of those who attain it, some never surpass this perception, and thus reincarnate from one moment to the next in high sensory awareness, and may yet remain confused causally. Combining these two observations we can propose that the "fingers" of the five solids aimed and reflected metaphor switch more in the higher dimensions, where, though they themselves are the same size, so to speak, they cast a larger shadow in potential. This metaphor in particular pertains more to Judaic mysticism.

Having seen the location in which we shall measure, applied the Platonic solids, and seen how their interval distorts in size with distance, we can now look for exactly where the solids occur between the pure dimensions.

We have earlier examined the structure of the probability wave as it passed through the solid attributes and assembled its activity as patterns. The solids then manifest their influence as within or upon this wave of probability, in the form of undertow. As the incoming wave of probability approaches from whichever direction, there is a motion in the wells of potential towards it.

During this time they take on the greatest clarity, and are the most implicitly detailed. Here is the entire pattern of effect expressed by the attributions of any of the solids, and this is merely picked up and delivered towards the next stage by the probability wave. For the sake of simplicity I will from now on refer only to the probability wave, and not differentiate between the first, aimed or reflected waves, which, it should lastly be noted, have similarity to the three legislative branches.

So, while this tells us much about the nature of the fields between the pure dimensions, we must still derive data regarding the dimensions themselves. They are, essentially, the opposite of the spaces between them, and this accounts for the strobing appearance of the passage of events. Between each event and the next, reality is passing through any number of dimensions, and each possibility is passing through the entire formal system and back before being transformed into its new form for the next event.

When I say that the dimensions are the opposite of what lies between them, I mean to refer to a distinction only passingly similar to the distinction between light and dark, however. For in the difference between the five solids with attributions and the potential extensions of essential geometry expressed in pure dimension there is as much variety of view as anywhere on earth when seen by either day or night.

In this case, the dimensions are day and the gaps are as night. By using this metaphor we have to concede that the scenery beheld is the same for either day or night, but their difference is really much more intrinsic than even this: the reason we do not see the same colors reflected in diminished light as in full light is because they are not there — the wave lengths do not exist, and the reflected photons are propelled at a different frequency, of enshadowed hue.

So even though a landscape may look the same by day or night to one who would envision poetry in nature, one who looks upon the same scope with a scientific lens sees them to be altered from one another in every way at the most minute of levels. This is all the same for the difference between the dimensions and the solids, except that it is potential rather than photons that set the stage.

Between the dimensions potential exists as a field, in which there are wells of probability. In the pure dimensions there are no probability wells, only infinite potential.

While the former is true for photons among others, the latter answers as to what becomes of the wave of probability carrying the units of information we had just witnessed transforming themselves into potential energy during the influence of the tetrahedron. Infinite potential undulates, transferring the data as pure momentum across the divide. This undulation is the only thing that differs between the dimensions.

Just as in between dimensions the wave of probability is filtered through all the various mediums attributed to the solids, in pure dimension there are vibrational differences in the consistency of potential, and when the wave of probability passes through these, it assumes different patterns of effect as it is translated into motions within a medium whose differences are geometric structures. I discuss the attribution of geometry to the pure dimensions, as opposed to mathematics for the subspaces, in great detail elsewhere, however for now it can be taken for granted that I will rely on these meanings for these terms.

It is accurate only to say that what is reckoned as spin in the movement of a particle is equivalent to the force that forms the geometries of pure dimension. Because this force is present both in the lower energy level, subspace particle and the higher energy level, pure dimensional wave, it can be thought of as their broken symmetry aspect. I will call this force-spin, as opposed to particle-spin, and say that this is the virtual aspect of the fifth dimension as revealed by the release of the potential energy of information from its tetrahedronal form, which we can only now see as though from the position of pure dimension, the stage taken by the wave of probability after it has passed through the fifth solid manifestation phase.

Force-spin manifests geometries in the infinite potential of pure dimensions, and will in the spin of the probability wells between them. This is to say that math has will and geometry has illumination, but the two are merely descriptions of the same vista, as I have already shown using the contrast between light and dark.

Lastly, following the wave of probability as it passes through pure dimension, we can observe the form it will take when it exits. Until now we have been looking at the wave of probability as essentially unified, but here we must stop to examine the order in which it progresses through the Platonic Solids as it emanates either inward toward event or outward toward the unconscious memory castle.

We have seen that there is a specific order in which the wave manifests the solids, but if we were looking at two overlapping waves passing between dimensions, we have only the Judaic, clumsy, interlocking fingers metaphor to aid us. The true nature of this interference pattern is such that all the solids are present in all orders at all times, but that, between any one information unit probability well and another, there will be a certain transference of effect that is in accordance with the attributive elemental solid as which the preceding well functioned.

It is always an interference pattern since the disruption of one wave is as good as the disruption of two to the stimulation of potential, and the probability wave is a constant factor in the formal system's functioning. It is, afterall, the substance of the sphere of the mind, in which the formal system is eternally no more than a thought.

In other words, a wave is carried along within a suspension of probabilities that are free and adrift within a medium of potential from a possibility in the environment of event so that it will specifically follow the same pattern of repetitive manifestation, despite being constantly bombarded by and surrounded by other waves being transmitted from another possibility to or amongst information unit probability wells.

Thus, the form taken by the motive energy transposed in the infinite potential of pure dimension will immediately assume the form of the cube upon remanifestation, regardless of its direction of emanation. This is because of the shape of event, as it is comprised of the double tetrahedron of conjoined fact and fiction, and the precedence of three dimensional interpretation in the form of the cube's excess components.

The equivalent of this in physics is the short life span of antiparticles, which tend to combine with their real particle counterparts and annihilate, releasing energy, which would merely be the momentum of the wave of probability in the formal system of metaphysics model. The shadow cast by this is that of the conjoined tetrahedrons, symbolized by the hexagram. The hexagram itself, as we shall now see, embodies all of the Platonic Solids.

B. The Hexagram

The hexagram pertains to karma, which is the belief that events are predictable according to the ratio of all that is seen to all that is done. In this model an event will more mirror the preceding if it has a high similarity regarding all that is seen during the strobing of pure dimensions, and all that is changed by the translation of probability during its passage through the subspaces between them.

This belief is carried out by observation of the workings of the formal system, as we have seen, in that the hexagram is the shadow cast by event as it enters metaphysical (that is, more-so physical, in this case) reality after emerging from the realm of pure dimension.

The hexagram is alike all things, and potential makes all things possible.

In many cases in the practicing of divination, the hexagram is a symbol that is used. In the form of a six-barred tetrahedronal tripod supporting a crystal ball, the hexagram was used by the oracles of ancient Greece, and has not been out of use since then among gypsies and fortune tellers. The system we will be attributing our attention to here is that of the I Ching, which uses the hexagram in the form of six parallel lines — the deconstruction of the model of two equal, intersecting triangles.

According to all schools of divination, what is seen can be a prediction of what will come to pass, should the present visible and the present experiential differ, as in the case of random eye movement sleep. According to general society, when it does not choose to accept and embrace its older sister prophecy, these are merely the ghosts of the past that are being observed.

These two beliefs seem at first to contradict each other. While we will examine this duality, as well as the substance shared by all hitherto distinctions, in more detail in the next section, nonetheless, it, like most of the other binary conjoinings of fact and fiction information units, forms its own completion, and thus proves it to have a reconciliation. One of the two units representing the intersection of the two schools of thought on prophecy will assume the role of fact and the other of fiction, and this is a necessity of it being carried out in the beliefs of interactive people.

But this only determines dominance, not accuracy. From the perspective of potential both are equally present, and therefore neither more or less true than the other. In fact, the visions of the past may evolve into not only the physical present, but also the visions of the mental projective present.

It is also equally true according to potential that the events of the future, as they occur as thoughts reflected on the surface of the sphere of the mind, may evolve backwards to become the phenomena and perception of the past. Also these may occur at the same time.

Lastly, we might benefit from perceiving that these might not occur in terms of evolution; there might exist permanent imprints in the world of archetypes and myths that simply appear at physically regular intervals. It is also possible that this and evolution occur either in isolation, in unison, or both at once. All of these possibilities are born out in the structuring of the formal system model.

It is beneficial at this time to discuss the three ways of becoming a mystic, which pertain to the three legislative branches of metaphysics. The first way is to be born a mystic, and this is of the religious branch, for many archetypal holy men and women were born more endowed for research and experience in the realm of the ideal. It is also possible to wake up one day and find that you are mystic. This implies that — either you always were so dormantly, or that something has changed within you or your environment to awaken new insights — and this change can be measured physically. The final method of becoming a mystic, and the one with which the formal system concerns itself, is ethical, and involves training and study to hone the inherent potential for mystical wisdom within you. In this case, a vast quantity of that which is connoted by mysticism refers us to divination, and all that it entails.

1. Yin Yang

This is usually best recognized by its circular signature symbol, which can, itself, be thought of as an artistic rendition of a conjoined information unit — where one tetrahedron represents fact, the other fiction, and they form the core of a cube — just as Yin represents true and Yang represents false, and they form a complete circle.

There are many implications to Yin Yang, but I would like to take this opportunity to discuss one of the reasons I chose the terms "Fact" and "Fiction" in particular. There is a certain quantity of literary criticism inherent in these terms. This is entirely intentional, as it is important for the formal system to demonstrate self-referentiality to whatever is its medium, which, in this case, is the written word.

Thus, the formal system can be applied to its own representation, and determined according to those terms. But if the formal system were pictured, the terms impression and expression would be equally accurate. For example, the binary terms of Yin Yang might be used to test for lies in recoding protocol.

Allow me now to diverge briefly to comment also on impression and expression, as it will be useful to refer back to this distinction as well while rendering our considerations of Yin Yang. Although these are perhaps best recognized culturally as art movements, they derive from methods of perception innate to the sphere of the mind.

As it expands the sphere of the mind experiences impression, and as it contracts it undergoes expression. In the functioning of the formal system impression precedes expression — the first wave of probability carried out being that ejected by the node of Not, irregardless of the presence of a reflexive active consciousness.

Thus, to the perception of a mystic, there is a moment during which all that can be known, is known — though this occurs in a vacuum of meaning. Only when the wave of probability is reflected can attributions for Had be assigned and event fully coalesce. In other words, to a mystic, an event can be known before it occurs, though only indefinitely, and similarly its implications — or possibilities of subsequent events — can be reached, but without certainty of apprehension. This is also sometimes expressed under the dynamic of passivity and activity.

Just as the hexagram stands for the similarity in the forms of the forces governing dimensional subspace and the geometry that arises within pure dimensions, so does Yin Yang cover the duality of the individual information units of conjoining fact and fiction and the raw energy they embody which is released in pure dimension.

While the hexagram marks the underlying structure that unifies all the dimensions, Yin Yang is alike the signature of this structure, occurring along the plane of every subspace that divides the measures of the pure dimensions as they occur in the expansion and contraction of the sphere of the mind into exponential intervals.

Thus, without realizing it, we have been relying on Yin Yang all along — it has been the core of every Platonic Solid — the medium of the spaces between dimensions; it has been the veil through which the wave of probability has passed at every stage in the formal system's functioning, and it is the difference between the formal system and the sphere of the mind.

Yet in addition to having all of these duties relative to the formal system and implications for the sphere of consciousness including the dimensions of mental projective space, Yin Yang has a set of meanings that are entirely its own, and from which we can extrapolate still more understanding of these things.

Much could be learned by studying the symbol of Yin Yang alone, but we will also examine this symbol as its own functional model, equivalent in complexity to the formal system, but of much too great a depth, because it is so much older than the formal system, to be fully delved into now.

The symbol which is chosen for Yin Yang is a circle, divided in two by a circular spiral seen from the side; each half containing a smaller circle. The two halves represent Yin and Yang, the male and female, passive and active aspects of metaphysics, and of everything we can observe in nature. The smaller circles within the halves are of the opposite attribution of their domain, and symbolize the dynamic symbiosis of the cyclical system.

In its traditional applications, nothing falls outside the realm of this single logo's resolve. In the context of the formal system we have already seen the evidence supporting this practice. Much of what is taught in traditional schools amounts to the same assertions as we have made here, only expressed in different terms.

What this symbol embodies that is conveyed less eloquently by detailed inspection of the formal system's components is the motivated interaction of dualities. We see them at once working for and against one another, moving towards, away from, and around one another, both causing and created by each other.

So much of this requires the researcher to experience what it is describing in order to have full awareness of their relationship, leaving little to be said without fanfare. We know from observations of the formal system that, at all levels and scales, there is intense interaction between the dualities described by Yin Yang, and it is further possible to infer that these occur in the manner described by this symbol. It remains only to logically elucidate the extent to which this is carried out.

What is described here is a type of permanent imprint, what recently has been proposed as set and setting — what is tentatively hypothesized as inspiration and distraction. It allows for our realism, as well as great philosophy, or at least — great questions.

To what degree are we programmable? To what degree do we exist? Is it as though every thought is real in its own realm, and accessible, constantly? What then is the difference between sensory and brain born experiences? Awake they are the same, that is, dependably similar — like the major and minor keys. They are as different states of comfort, which we go through over time. We look for them like beacons. It is as though ideas have fallen into objects through time.

Yin Yang is in everything. Thus Yin is in everything and Yang is in everything, and this is like the mother and father, revealed in their progeny.

First let us look at Yin, the active, female aspect.

How many of all the thoughts of all the people throughout time have been great things people have said? The very difference between verbal and visual is a great joke on too small a scale. From a certain angle all things can be filtered through Yin — it is like the mouth too small to express all good thoughts, or the pen too slow; it is the freedom of a caged bird.

In the west this is the path of the Light, in essence, of morality. It leads us around in an ever expanding circle, dizzying and confusing us, teasing, then giving chase to us. It is like a great, immense, and savage animal, tamed at the sound of a few ethereal notes on a primitive, folk instrument. There is deep, deep satisfaction in Yin — so much so that it slumbers innocently even while around it and within it, blood runs hot and slow, and fevers rage, and ages of sickness come and go.

Now let us look at Yang, the passive, masculine trait.

The beauty of everything is in its simplicity and clarity. Yang is as functional as every part of the formal system, and the very metaphysics of reality. It is, in itself, just another way of seeing those metaphysics of reality than the formal system. Both are ways of describing the world of the mystic.

In the west this is known as the Great Work, or the perfection of the mind and soul as a temple housing the Philosopher's Stone, the hewn shard of reasoning consciousness that acts as the lens connecting the telescopic and microscopic worlds for the induction and transduction of the universal energy that is the body of God. Because this amounts to man playing the role of God in the context of the children's riddle referenced earlier, it has been associated with the dark path of Yang, and an existential malaise of supernatural inadequacy has settled about the postmodern man's heart, and led to cognitive-dissonance regarding their ancestors' occasionally chemically induced transcendences.

Yin is shy and aversive, yet amazingly accurate. Yang talks itself into corners. Both can be translated according to the constellation of the bio-survival circuit, and both tallied in histories.

There is no specific conclusion to be reached from all of this. We are in light and we are in shadow. In the orient and the occident these have opposite connotations. In the west, white is worn to weddings, and black to funerals. In the east, black is worn to weddings, and white to funerals.

Yin Yang offers a method of accessing the same information contained in the formal system that is as much poetically phenomenological as the formal system is utilitarian and objective. Both systems, by addressing the presence of patterns in probability, offer scenarios for the afterlife, for example. They are equally self-referential — and still their difference in data is like night and day.

Everything observed to be true for physical sensation remains true when examining the formal system in mental projective space, but it is necessary to track their references in another way — either visually or verbally, depending on which way the metaphysical descriptions are sorted while in experience of the system.

Even these are as of Yin and Yang — impression/expression and visual/verbal forming the familiar structure. These references may also be likened to attributions and affiliations — things that are true of the source and things that are true about the source — yet try going to sleep without dreaming. (This refers to the singularity model of consciousness.)

So, it is as though, in potential, we are experiencing every mental thought we have ever had, and every real experience, at the same time, and available through sequence. Some of these are needfully closer and some further away, separated as by space and time. It would seem that comfort is essential, though so is right-action, whose compliment is right-thought. The image they form contains the formal system.

In both of these we see that impression and expression are present simultaneously. Thus, whether in visual impression or verbal expression we must remember that Yin Yang is like a compass, and can assist us in aim. It is as though when the formal system acts as a measuring device, the force it is measuring is Yin Yang, and vice versa. So it is as well with Yin and Yang alone.

Yin Yang can be directed toward many different functionings of the formal system — and these are generally arranged according to the Feng Shui system of directional attributions. Cardinal to this system are the eight trigrams, arranged according to the realms described by the circumference of a compass. This is gone into more thoroughly later.

The essential function of Yin Yang is inversion. Inversion occurs physically as between impression/expression, ethically as fact and fiction, and spiritually as inspiration/distraction. Yet these are only words, beside which we know inversion. They fall into an ascending geometrical pattern that can be represented by three parallel horizontal lines — the trigram. If the line is broken that means it is the experiential impression of male Yin. If the line is unbroken that means it is the objective expression of female yang.

Inversion cannot be taken away from you. Imagine that you are standing in the center cell of your memory castle when it is at its cubic stage. Now invert the inside and the outside of the walls. This can be done on finer and finer and grander and grander scales until nothing exists anymore. In this way you can experience the strobing of reality as events and dimensions invert, and slow this down to experience ylem.

It is possible to will a specific inversion to occur. One can plant a series of suggestions into the flow of universal consciousness. One technique is to describe such a transition from event to event and thus plant the seeds of it in the realm of potential. Therefore it should be likely to produce this same effect in conversation. One way I describe elsewhere is by the splitting of the tongue into old and new, wherein the old acts to create a carrier wave, and the new is learned language.

But these are merely illusions. The real trick in the west is achieving contact with the deceased after death, and in the east of liberation from reincarnation, as though, by finding a way to raise the dead, you can't find a way to make it stop. This too is merely a parlor trick for mystics, but you won't find many around.

2. Trigrams

The trigrams can be arranged, as we have seen, about a compass. According to the tradition of their history, there are two standard attributions by which to do this.

The first is known as the Earlier Heaven arrangement, and is based on the Ho Tu chart of four cardinal and four diagonal directions. In the Earlier Heaven arrangement, clockwise — Heaven is north, Wind to the northeast, the Deep is to the east, Thunder toward the southeast, Earth is to the south, Mountain to the southwest, the Sun is toward the west, and Swamp is to the northwest.

The other is called the Later Heaven arrangement, and is based on the Lo Shu — a fifteen sum magic number square. In this arrangement, clockwise, Water is in the north, Mountain is in the northeast, Thunder is in the east, Wind is in the southeast, Fire is in the south, Earth is in the southwest, Mist in the west, and Heaven in the northwest; with the additional attributes of earth to Mountain, wood to Swamp, and metal to Heaven. The Later Heaven does not show as much deference to the binary nature of Yin Yang as the Earlier Heaven sequence does.

The trigrams are embedded within Yin Yang. They are inextricable. However, they do occur in a regular pattern, and this is the same as one present in the dimensions of the mind described by the formal system.

Traditionally, this has been rendered according to either a solar, a lunar, or a combined calendar — as it is only visible from a fourth dimensional minimum perspective. This pertains to the regulation of inversion, as a function of the interaction of the trigrams, Yin and Yang, and Tao. Each of the trigrams represents a different alignment of triangulated forces.

There are eight of them because they occur along the diagonals of two intersecting, cross-joined squares; traditionally the astrological sigil; this shape is also the cross section, or shadow, of a hypercube at perigee. They thus produce experience at tilted angles to one another, and yet this does not effect their internal composition.

These are finding a way to happen. They arise as the triangular minimal description of a circle, and therefore a plane, such as the second dimensional relationship, which is what Yin Yang is.

They are as shadows of the tetrahedron, and still, the trigram Earth is composed of three broken pairs of lines, or a total of six short lines — which can be thought of as the outlined shadow of a cube, or our familiar hexagon. This relates it to event, to the formal system, and also to the steloctahedron of conjoined fact and fiction, as Tao, the trigrams and Yin Yang respectively.

As collections of Yin Yang lines they represent arrangements of information units, and one way to read their triple nature is dimensionally; thus they pertain to event. This classifies them only metaphysically however, and offers an ecological approach to seeing the probabilistic assemblage of events similar to Shinto.

In other words, impression and expression can be experienced as environmental, and this seems to diminish the significance of the sphere of the mind — or at least, the thought at the center of active consciousness — in impacting on metaphysical functioning. Even this thought is of Yin Yang, though, and so it still has a role in functioning.

In this capacity the trigrams bear similarity to states of consciousness. This is perhaps best understood as a bridge between second dimensional Yin Yang and the third dimensional events of the hexagrams if seen as the seven events present in the unconscious memory castle, plus the hypercubic formal system, at the moment of maximum extension just before the wave of probability is reflected off the one dimensional sphere of the mind.

In this process the trigrams act only as patterns of raw potential transferred across pure dimension, while the hexagrams stand for the subspace events themselves by being combinations of trigrams, either the same or different, depending on the input on event of the transformations in patterns of information units undergone during the periods governed by the Platonic Solids.

Which of the second dimensional Yin Yang triangular codings maintains dominance in the hexagram is determined by the magnitude of change they engender in possibilities. In the arrangement of a hexagram the top trigram represents dominance and the lower trigram primacy.

The trigrams are the essential ingredients of the hexagrams, which will be the next section we discuss. They are the different components involved in event.

3. Hexagrams

There are sixty-four possible pairs of the eight trigrams.

As they can be arranged hexagonally, they are each the shadow of a cubical event.

The sides of the hexagon on which there are broken Yin lines indicate the functions of the formal system to be excluded, and beyond and between them the twisting of the wave of probability as it penetrates dimensions.

Yang lines indicate that the shadow outlines of certain nodes be related.

According to the I Ching, each hexagram represents one in a series of events. This is mirrored in the formal system by the cube being the first of the forms to emerge from dimension, in a pattern that repeats, just as in the I Ching the hexagrams represent different configurations of the information being exchanged in the formal system's routine functioning.

They answer as patterns when assembled, and these patterns determine the construction of a program for the formal system. One pattern that occurs naturally is that of the eight doubles — hexagrams that look the same right-side-up and upside-down. They serve, among other things, to determine two legs of a triangle function that establishes the dimensional depth of the Yin Yang program to be of the second magnitude.

These eight doubles also represent the five elements, including cosmos — which in this case is depicted graphically as the construction of particle and wave. Yin represents earth, The Deep represents water, The Sun represents fire, and Yang represents air. The remaining four doubles can be arranged in pairs so that either Mountain surrounds Thunder or Marsh surrounds Wind to represent the form of a particle and either Wind surrounds Marsh or Thunder surrounds Mountain to represent the form of a wave, both of which are expressed in terms of probability (more solid) and potential (less solid).

The sixty-four hexagrams are best gone into individually elsewhere. Their function is prophecy. There were two schools on prophecy among the ancient Greeks: that of dreams, and that of inebriation. Both of these systems include as a given the implication of scale. This is due to the entrenchment of the philosophy, "As Above, So Below," in Athenian Democracy, as pertaining to the individual vote's synonymity to the voice of the masses. In other words, when you dwell for a moment in the world of the mystic, you will experience, if you like, of the cycling, inversely related eight doubles that they can render themselves as recognizably repetitive events. This is called Samsara, and it is what we try to escape from. It is thought that different patterns of karma effect such.

The ancient Sumerians are credited with the creation of the game of chess, which was brought to Rome, where it was finalized, through Greece. Some evidence suggests, however, that the game originally had its origins in the far east, where it evolved independently into what is known in modern times as the game of Go. These games are played on flat boards consisting of 64 alternating black and white tiled squares. While Go and chess have different pieces and different rules from one another, the board that is common to them both has remained entirely unchanged for thousands of years. It is a simple square of eight by eight smaller squares. When the board is taken by itself and viewed as the skeletal framework for a magic number square, such as have been used for prophecy and divination since time immemorial, the significance of its 64 component cells as they relate to the oracular I Ching becomes crystal clear instantly. the number of eight by eight magic number squares possible using only, and not re-using any, consecutive integers is 6.34435e+88, a number that increases asymptotically as the number of cells tiled in the square increases arithmetically. For example the number of such possible four by four magic number squares is only 880. These figures are produced by the formula for any order magic number square N, where (N-squared) factorial/8, whether N = 8, 4, or any other number of cells per side. To deduce the magic sum of a number square there is another formula, which reads Sum = N (2A + D (N squared - 1)/2), where A equals the starting integer and D the difference between successive terms.

The significance of the eight square, or 64, base tiling in use as a foresightful method of cyclical prediction becomes even more substantial esoterically when we consider the mathematically magical properties of its numerical value alone. Begin with the 231 gates of the Hebrew aleph bet according to the sefer yetzirah arranged by Alexander Rockeach of Wormes into squares comprised of the twenty two letters and twice the ten sefiroth plus twice Daath, front and back, to arrive at a perfectly rounded 44 squared = 1936. To concretize this sum let us also note that 22 squared is 484, twice which is 968, or half of 1936, and that 968/44 = 22. Then let us correspond this ancient Hebrew mysticism with the number magic of the Orient to derive 30.25 from 1936 as being the 64th part, or the product of the divisor two to the fifth. Thus we find that 1936/2 = 968; that 968/2 = 484; that 484/2 = 242; that 242/2 = 121; that 121/2 = 60.5; and that 60.5/2 = 30.25. The square root of 30.25 is 5.5. 60.5/11 = 5.5, and so does 33/6. 33/22 = 4.4, and so does 48.4/11. 1.1 * 10 = 11, and 11/2 = 5.5. 1.1 * 20 = 22, and 22/5 = 4.4. What we are looking at here is another pattern of gnomonic expansion similar to that of the transcendental numbers phi and pi.

The 231 gates can, and perhaps must, be thought of as a very large magic number square that can be extended into three and even four dimensions. It most resembles a 22nd order magic cube comprised of the gematria letter/number equivalencies of the Hebrew aleph bet, with eight interchangeable inlaid magic cubes comprised of twice the ten sephirot plus twice Daath. The sefirot cubes are nested within the center of the aleph bet cube in the form of a magic number hypercube. This pattern of embedding is closest in three dimensions to the Menger fractal sponge, a solid looking lattice that has an infinite surface area, but zero volume. The pattern of propagation through the dimensions of the magic hypercube begins with the small sums 4.4 and 5.5, and continues on through 22, as well as through 64, and can be thought of as a neighborhood set on a surface mapped sheet within an analytic algebraic manifold that is comprised of the hypercubic lattice the spin matrix of which is listed within the construct as the magic sum gematria. The key to unlocking this rigorous hypercube is that 22 cubed is 10648, which, when divided by 2420, reduces to 4.4, and when divided by 1936, to 5.5.

IV. Conclusion

In conclusion I would like to think it proper to take a final guided tour through the constructs we have discussed in their most integrated form. To accomplish this we must choose a specific path, and since I have led you along this far, I would hope that you would trust me to input a suggestion as to this final trip.

I wish to take you along the worn course of the active consciousness. It is present within the sphere of the mind at all the dimensional levels, and therefore is witness to the entirety of the formal system's functioning, including the Platonic and I Ching metaphysics. In closing I would like to make a small reminder that this has been the path that we have been on all along.

With every level of consciousness, active consciousness is present in the capacity of a lower dimensional level of consciousness contained within the first. At times it is external — when it is at apogee, and at times it is central — when it is at perigee. It fluctuates between these across the dimensional levels in the same way as do all the rest of the Platonic Solids.

When we first encounter active consciousness it is as the first dimensional, binary, central fact/fiction node in the formal system of metaphysics. This is not a comment on consciousness, nor either its origins nor aims. Here it may appear under magnification as the basic component of smaller dimensional geometries than the first, should they be found to exist, and as a steloctahedron at that level. In event this can be manifest as a subject or object, and is germane to both sensory and mental projective space. It is, thus, simple probability in the form of clear possibility.

This is the most fundamentally real form of an information unit, which is a socially shared concept accessible diversely by the conscious memory from the unconscious memory of all things that store memory all around you in event, observed when concentrating on the veil of these information units behind and within the perception of all things as possibilities comprised of potential.

After seeing these information units to be as fundamental a part of the reality of our mental projective space as are atoms and molecules to our physical existence, we can then perceive the roll, or the turn, of active consciousness in the presence of true consciousness as it exists in the realm of the second pure dimension. Here it assumes the form of a ring of perception called thought.

Thought acts as a fractal circle comprising the surface of the sphere of the mind, determined in size by the measure of the radius and circumference. Thoughts are separated up into three categories: first, active consciousness — which is the portion of the sphere of the mind through which an exchange of information units is occurring between the potential of event and the other functioning structures of the formal system of metaphysics model; then subconsciousness — the portion surrounding the active consciousness and slightly stirred by its activities; and finally the unconscious, which is the portion of the sphere of the mind left most in shadow by the glimmer of potential's transportation through active consciousness.

It goes without saying that those thoughts most unconscious relative to a particular conscious thought would be those on the opposite side of the sphere of the mind from it, and thus its inversion. If you are keeping your active consciousness aligned with itself, then the thought that ought to reside in active consciousness is the image of the one dimensional steloctahedron of fact/fiction. This occurs in reality as an abstract measure that we seldom use — yet this is only because a naturally determined set of behavioral characteristics usually mark out the flowing aim of the direction of our attention in accordance with the bio-survival circuit's interaction with our multidimensional environment. Though, since in the circumstance of introspection this is fully under conscious control, it would hardly seem to apply.

We generally encounter the measure of our degree of attention in the three dimensional terminology true to the dimension in which the sensate portions of us dwell, that is — as the sphere of the mind or the aura.

Here it is necessary to process a little religious legislation, because the visual observations of metaphysics pertinent to this particular topic are filed under the auspices of spirituality. In specific we are referring to the forms and functions of the soul and spirit.

We can say with some certainty that the soul refers to the full dimensional extension through the Platonically governed dimensional subspace of our sensory aparati, and the spirit refers to the same measure taken, for the moment of attention, to the level at which active consciousness is making direct interface. One is a measure of perception of subspace, the other, dimension. These are in a way equivalent to the aura and the sphere of the mind, respectively. Both are realms accessible to active consciousness.

It is equally convenient to postulate that the body is the probability cast out by the node of Not; that the ethical soul is the rationing awareness of our sensations thereof, embodied as possibilities in a constant state of flux over the node of Had; and the spirit the pure potential ejected by the node of Fact or Fiction.

From the third dimension, in the shared space of the second dimension, we see information units occurring as nodes in a formal system, and possessed of relationships between them that constitute solid lines formed of patterns in probability equivalent to information units. In the case of introspection the thought which would occur is of a circle within which there is a dot. The circle is the thought of thought. It is traditionally called the abyss. The tiny point or steloctahedron at the circle's center is an implosion of the nothingness of thought. It is concentration. This is a constant arrangement in probability, underlying the construction of the vision of our mind's eye.

Just as for thought there were three states of consciousness pertaining to their relative proximity to the reflexively central thought of active consciousness, so these states of consciousness have oppositely reflexive memory castles comprised of spaces formed of the fluid mechanics of probability guided by the geometry of the third and fourth dimensions. The shapes of this access system take up no space, because they are a measure of thought. These occur in the form of remembered patterns that can be at once impressionistic and expressionistic.

The information units that tend to populate these realms more generally take the form of nervous ennegrams or mathematical abstractions.

One of the highest levels of dimensional consideration is the I Ching, and through a process of comparative contemplation it is possible to evoke knowledge of disparate events separated physically and chronologically.

The realm of which the I Ching is one representation metaphysically, the active consciousness can access dimensionally.

Consciousness extends to the fourth dimension. This is the perspective from which you were aware of the thought that possessed your active consciousness. It is a hypersphere surrounding the central sphere, and the place of active consciousness on it is equivalent to a point directly centered above the penetration point of active consciousness in the three dimensional sphere of the mind. We can consciously align these points and provide ourselves with a view of the sphere of the mind in a fuller form. Looking down from the outermost sphere we see the sphere of the mind, upon which these is the circle of thought, within which is the pinnacle of concentration.

Now, the sphere to which we have found ourselves transported is the fourth dimension, the hyperspheric equivalent to the hypercubic hypercross of the conscious memory castle. What this means is that the path of active consciousness finally has places to be connecting, that is, between the fourth dimensional formal system and the fifth dimensional conscious memory through the realm of the hypersphere of the mind.

This constitutes, according to the I Ching, the third tao line in the pattern of a hexagram, and therefore the completion of the first trigram. The layers of pure dimension occupied by the sphere of the mind mark the active or passive influence of that layer of reasoning on a particular consideration in the pattern of the hexagram. These are concurrently reflected in the relationships of the formal system of metaphysics according to a clockwise circular rotation around their hexagonal shadow's outline beginning from the top, as nearly as I can predict.

Active consciousness at this level will remain reducible to a point upon its surface, but it is now also its own sphere. It lingers one dimension below that which comprises the space occupied by the whole of the mind, and at the maximum extension of this shape we find the hypersphere, whose shadowed outline is a sphere.

Now looking down upon it we could observe that every mind is its own thought. This is the impossible loop at the root of consciousness —the concept of the ego as emergent from the id, the "I am" in the affirmative, the personal, the self-conscious, the self-reflexive, the self-aware; a poetry of creation; a feedback loop; infinity.

Let us proceed to the fifth dimension and see what we will find true for active consciousness there. The fifth dimensional form of the sphere of the mind is a torus, containing one real and one irrational sphere in the form of a hyperspheric shell around it, and the real sphere being comprised itself of the hypersphere of conscious memory and the sphere of the mind, on an increasingly stable gradient of descent.

The difference between a torus and a hypersphere is that a hypersphere can only be represented at apogee, in the form of two nested spheres, whereas the torus is always at perigee, although the two are merely differing expressions of the same thing; and if you understand that then you understand a great deal about the substance of the sphere of the mind.

It represents however only the fourth line of a hexagram based on the subspace measures between pure dimensions. When, as in a hexagrammatical core sample, these dimensions, as they are measured internally by the difference between the sphere of the mind and the formal system of metaphysics structure, are aligned, the result is a viewing of the entire hypersphere of conscious memory as a single thought inside which there is a sphere of concentration, a circle of nothing, and a point of implosion. This structure constitutes the mind's eye nicely.

Then active consciousness performs a miracle. It separates itself from its own central mind and creates a shadow-image reflection of itself outside of itself, filling up the irrational sphere of the torus. Thus it bears its way through the sixth, seventh, eighth and ninth dimensions.

In the sixth dimension it manifests the mathematics of a hypersphere of unconscious memory and conscious memory, a circle of the mind, a point of concentration, all within a torus of implosion, that is contained within a larger hypersphere of active consciousness.

In the seventh dimension we include outside a sphere, which is part of a new set, pertinent to active consciousness, and comprised of focus or awareness.

In the eighth dimension we include an amendment to the subset, a circle of concentration.

Finally, in the ninth dimension, the second set concludes in a point of implosion, which can be thought of as connecting with the point of implosion at the very core of the entire mental mechanism.

The transpositions through these active consciousness dimensional levels occur in the functioning of the formal system of metaphysics only as reflection of the wave of probability from the further boundary of the fifth dimensional torus of implosion (or inversion, alternately) where it reads the subsequent hypersphere of active consciousness and valleying array of connotations only by the glint of the most distant point, that of implosion in the ninth dimension, and then relays its charting of the entire set as a closed cycle, ergo a hexagram, back to the central point directly along a probability wave to test its hypothesis.

Whether or not the subset of dimensions for the extension of active consciousness outside the mind actually contains subspaces of physical manifestation is a matter that continues to be researched. This would theoretically equalize the amount of time it would take for a wave of probability reflected off the torus of inversion to reach the internal locus of implosion and the amount of time it would take for that same wave to carry through the torus, unimpeded, to the external implosion of the ninth dimension, even though the distances described by the second set of pure dimensions is naturally exponentially greater than the former, which is a reflection of their containing that much more information .

What can be said for the projection of active consciousness is that it is absolutely essential to understanding the functioning of the formal system. Its extension is like a physiological echo of mental projective space, where both are expressed as potential energy. Active consciousness is what we are experiencing right now, as boringly strange as it may seem. It surrounds us and interacts with the probabilities generating possibilities in our event. This is true within our aura, but in the lower realm of that field there is also a stalk of physical functionality in the form of the body that we rely upon the sensations of the most. Active consciousness is fundamental to our form of existence. We have chosen to perceive this as a species as though it were more disparate with distance, both in individual cases and in general.

When we say that active consciousness has its own specific occurrence, just as does the sphere of the mind itself, in the realm of pure dimension, we mean that this is a literal reality. It has full grounds in referential experience. Active consciousness is the base state of what we experience as consciousness when it is focused outwards.

This hypersphere containing a torus of unconscious memory surrounding a hypersphere of conscious memory containing the sphere of the mind containing the circle of thought, which is nothing, housing the point of concentration, is what we experience when we first awaken from sleep, for, in sleep, our active consciousness submerges itself in the subconscious and unconscious realms of memory and thought. Thus, upon awakening, it undergoes an implosion of unconscious memory, wherein the focus of concentration, or the flow between the two opposite first dimensions, shifts from an internal to an external locus. This is the sixth dimension. It is comprised of light and motion. It is the fourth line on a hexagram.

Next our active consciousness brings itself into awareness. This is a non-instinctual process whereby we acclimate ourselves through a change in conditions between the mental projective set and the physiological set. The result is the bio-survival circuit. It governs the working of the central nervous system. This occurs as after a meal, when entering or leaving rooms, in the presence of great art, etc. It is common enough to imply necessity, but remains a matter of choice.

It takes the form dimensionally of a sphere that is, essentially, our mental projective space. Though it is always at perigee in the other perception, it is akin to the aura of active consciousness in its substance. This is also, however, the sum of the five senses, which have their equivalence to the five Platonic Solids, and so there may be more weight in the previous statement than was first observed. It is herein that all subjects and objects are comprehended, and here where we have constructed the formal system of metaphysics as if using a mirror from what we can sustain to be true based upon introspection.

This is also the seventh dimension and pertains to the nature of event. Until this point in the functioning of the formal system, that is, the moment of reflection of the wave of probability, we have dealt only with event in well form, that is, strictly three dimensional; now we will deal with the measuring of event, and what realm constitutes the suspension of that measurement. In the seventh dimension we come to deal again with the fourth dimension, as, with the beginning of compression of probability within the many shells of the system, event itself becomes fourth dimensional. This does not indicate the immediate effect of change; it only opens event to the option of transformation.

In this capacity we see that event exists in well form, that is, as a measurably closed set of probability. It assumes the same form as does its larger, hitherto fourth dimensional component the formal system, but when this is brought to light, we see event as well becoming alterable over time. It is essentially equivalent in form to the formal system, that is, it is fractally reducible to the conjoining of the tetrahedrons of fact and fiction. Each event, we now perceive, is itself both fact and fiction.

The experience of this is that, things change; different situations occur around us. The question becomes: does this occur as in a suspension, as though we were still in the womb? To what degree do we manifest this? To what degree can change occur? We can answer only that the seventh dimension is the fifth line on a hexagram.

When we begin to observe this, we are entering into the eighth dimension with our mind. Traditionally this is attributed to the crossing of the veil of the abyss. The insubstantiable sephira that exists in this domain is Death, and we find an equivalent to this transition in the measuring of active consciousness's extension beyond itself in the observation of the formal system. Here we see the spherical realm of concentration begin centralisation into attention. We begin to see under a new enlightenment emanating from the measurement between the ninth and the first dimension. We begin grasping at potential, perceiving probabilities, because these are equivalent in measurement to the extent of our eye of consciousness's view.

This is our circular sense of attention. It is circular because it is really second dimensional. It contains all the other forms of consciousness within it though — the sphere of awareness, the hypersphere of active consciousness, the torus of implosion, the hypersphere of unconscious memory, the sphere of conscious memory, the circle of the mind, and a point of concentration or thought. In short, what I am describing to you is a two dimensional representation, but the fact that I don't have to draw it for its meaning to be neatly transferred is a statement of testimony as to its essence, which is pure information in the form of energy. This dimension is equivalent to the sixth line on a hexagram in the I Ching sequence. It is also applicable to the measurement of event in field form, as the substance of time, or the fourth dimension, on that particular scale. This is lastly referential to the peripheral nervous system.

Then concentration implodes. This is the pinnacle of active consciousness's known projection into the realm of the sensory or the mental projective space. In the ninth dimension there are not even any thus far specified attributions. Little more than the measurement of the distance from concentration to it is known. This distance is measured in cycles or repeating patterns of event.

Traditionally this has been attributed to karma — an interactive system involving input to event in the short term being equivalent to output from event after a longer term; but we find that this is not the only attribution possible, as, at this elevation, we can see similarities between the eastern and western systems. Therefore, the same effects are accomplished by what can be called strategies in the realm of event.

This is all equivalent to what is summed up in the I Ching under the space between the sixth line and the first, where it refers back to itself as a cyclical system. This number is included in the measurement of the first order of difference between any hexagram and another, constituted by the number of tao lines interchanged between yin and yang. It is thought that the attribution to the tenth dimension would be the particle form of the substance being measured in this interval, that being tao.

It is possible, for one example of the ninth dimension before we move on to discussing tao, to invest in lifelines. That is, over periods of time determined by the passage of event, we may choose to maintain certain perspectives, forming world-lines out of our archetypes over time, even over many life times. In this way we propagate some while failing to maintain others. So even archetypes rise and fall with time.

If there is a tenth dimension, it is ylem expressed as tao. This, too, is merely a measurement over time of a construction of pure potential, but we must occasionally rely on such to help guide us in metaphysics. In discussing ylem in the context of tao however, we must do much as the ten commandments do for the ten sephira, and undergo inversion of the entire system. This is due to the substance of the implosion of the ninth dimension. Thus we must invert Nothing, to become Something. And yet all things that we may choose at this point are of the qliphoth, or subspace interdimensional manifestation. Within these however we find such occurrences of the open set as electron clouds and the potential of consciousness, and these perhaps indicate flaws in the substance of ylem, where closed sets exist, such as antimatter and wrinkles of microwave vibrations.

Tao is thus the presence of all patterns. In the I Ching it is all 40,320 possible arrangements of the eight trigrams, of which the King Wen sequence constitutes only one, 64-hexagram fragment. This is the dimension of attainment in meditation; it is the projection out of the third dimensional experience of consciousness through the third and fourteen hundredths dimension, or the difference between the pattern described within the sums of products of the multiplication table for the eights and the beginning of that for the ninth, and this physically describes the location of mental projective space, or at least pinpoints the penetrative glance of consciousness.

Now, within all of this, you have finally seen how consciousness divides itself into introspection and inspection of external evidence, and thus can you understand where you have stood all along the path which we have together only just finished wandering. You see that active consciousness is within the mind, and also outside of and containing it, as though it were overflowing from within, and yet it is entirely insubstantial, and appreciable only as potential. It is yours and yet extends invisibly beyond you to encompass and enhouse the thoughts whose reflections manifest in that medium. It answers questions in a process that can be at once internal and within mental projective space, in the form of the formal system of metaphysics model, for example.

In this capacity we can finally see how some of the answers can be thought of as portions of karmic equations that have occurred in the recent or more distant past, or those which are only just beginning to occur. So let me caution you in the investigation of future possibilities without the use of some guide, for answers have multiple meanings in different contexts, and may have traveled much more than we. They may also be brought to us by disparate messengers, many no more real than the Greek versions of prophecy — dream and intoxication, which is to say no more than those synchronicities which slip in when mixing media, or which may be thought of as "hyper-real," having come from irregularities of archetype in the hyperdimensions between pure dimension and the last form of their manifestation in the subspace of the Platonic Solids, which archetypes can be potentially useful both as guides and as a form of communication.

As we have said they may also be brought to us at different times relative to their references both individually and as histories. The point over which to bear caution is the realization that it is much easier to make fearful or reactive responses to foresight, or unsystematized predictions. All practice of this, particularly the refinement of such practice, is herewith denied by the author of this work.

All this means is that you have been within yourself while contemplating the formal system of metaphysics, and thus, having done so, proven the existence of the formal system of metaphysics from a certain angle in your reasoning mind. Thus doing so we must accept the formal system of metaphysics as real, at least from a metaphysical point of view.

We have stood at the base of the model, the single node, and concentrated on from what it is really comprised. We have looked over the lip of relationship into the nothingness of pure thought, and stood upon the firm sphere of the mind which is composed of infinite thoughts. We have jumped up one level to the hypersphere of conscious memory whose sky is the torus of unconscious memory, a sky shared by all minds alike. We have penetrated the outermost hypersphere of active consciousness, and gazed upon the externalized sphere of awareness, itself alike a realized thought or an event. We have seen the levels of this as circular in nature and describing concentration. We have seen these circles as reducible to points of implosion. And we say that all of this occurred within the sphere of our mind. During all of this we have brought along the I Ching and constructed a single hexagram. We can centrifugally spin this hexagram, by unzipping it upwards from the lowest tao line into a clockwise hexagon starting from the top, and have a program describing a certain event which can be plugged directly into the formal system and progenate results.

The formal system of metaphysics has applications in several domains. The tour of active consciousness was only one tour of the physical branch of legislation. There are also examples of tours through the ethical realm, such as the application of artificial intelligence (which is a more physically literatured debate) as well as the more philosophical, spiritual applications of it in Enochian communication systems.

The concept of the rationalizing of artificial intelligence is a straw man set up at an illusory horizon of emergent events in and around the near future. It is basically the externalization of the idea represented by the formal system of metaphysics into a shell called a computer. The question, then, is does this constitute artificial intelligence? The answer is that it does, but that it also resolves the need for a great deal of psychic tension throughout the world. Since most objects are sacred as items of memory storage, the consolidation of task chassis basses into fewer processing resources in a highly expendable market is relevant to a consideration of our species's behavior towards tools throughout history, and especially to the concentration of potential constituting predictions of the near future, and of a prodigal nature. It is as convenient to pretend that artificial intelligence already exists as to want for living in a world where it does.

The tour of Enochian systems begins with the fourth dimension being introduced as the domain of the closest experiences with archetypes, where they are seen as hypercubes of events, similar to inserted hexagrams. In the fifth dimension they are accessible in a hypercubic hypercross configuration for relationships, and in the sixth dimension in a hypercubic hypercross of hypercubic hypercrosses for relativity. Their natural home is the sixth dimension. It is thought that in the fourth dimension they have houses, which are the seventh dimension, and that in these houses they have information, which is the eighth dimension. These occur as changes in probability over time, just as in event, and do so as part of the changing of event. Because they are purely dimensional, they are thought to differ from the manifestations of subspace, which are called, complimentarily, the Qliphoth. They represent types of connections, and can be anthropomorphised as the Angelic Host. The Order of the Golden Dawn had much to do with the contacting of these denizens. This entire system seems to be moved toward synchronicity.

In addition to these aims of using the formal system, there exists a puristic, cosmological approach to it, which is more naturalistic in its description. In closing I will present this description. Imagine a sphere of a transparent material durable enough to maintain its shape, but pliable enough to allow the transport of the energy of a ripple of water.

Now, this sphere is like the universe, for it has no boundary, that is — its boundary is defined, but insubstantiable. We are told several things by the general cosmology of the world, and that is that first in this sphere there was light. This was the pure light of consciousness, or ylem. This is an eternal moment. Next darkness was divided from the light; though there are various other disagreements, this is regularly agreed upon by most cosmologists throughout history. We see that darkness winds up in the subordinate position, beneath the light. Lastly beings emerged in the infinite nothingness that separated the dark from the light. We are also told that the darkness was liquid.

So, imagine your orb half full of fluid, and the other half left full of air. Next, understand that there is earth in the water and that there is fire in the air. A fuller, richer meaning of these terms is implied in this text. These four form a square of which man is the center, and this does two things. It creates a cube surrounding the sphere, and made of the same material, and it creates a cube the sphere is surrounding, made of the same material as well. This is because we are now dealing with substance, and with form, even of energy, and these things are all measurable, all quantifiable. Thus, to measure them in terms of depth, we need a beginning marker, the inner cube, and an end marker, the outer cube. What we are measuring is their combined effect.

Fire moves the air so that wind stirs the glass, earth moves within the water and calls up ripples that quake the sphere. When the momentum of the force transmitted within the elemental medium impacts on the outermost wall, it retreats, turning back on itself. All these same effects we have witnessed in the formal system. Now imagine the entire formal system, all which you know about it, all its dimensional levels, as well as all the dimensions of consciousness, the realms of manifestation in which we dwell, the I Ching's ability to predict events, everything you have learned of it, and know, that it may all be transmitted by discussion of this seed form: the sphere nested within the hypercube.

At the center of the four elements is man, who constitutes the combination of the three pillars of metaphysics — Light and Love and Life, shadowed in the three legislative branches — Physics, Ethics and Spirituality. Together with the three legislative branches the metaphysical pillars make up a sextuplet of strings, as on the familiar guitar. This separates into chords defined along an octave much like one encountered within the I Ching. Thus, the three pillars and the three branches represent all that cannot be taken away from us, a hexagram of event, or one turn around the formal system.

In the same way man dwells in between, so dwell demons below and angels above, the way fish dwell in the sea and birds dwell in the sky. And look upon the rest of our environment for evidence of this model, you will find a wake of bubbles and a shore of shells where sea meets land, a fractal wilderness of waylaid pilgrims turned to trees where the land meets the sky, a corona of bending rays where sky meets fire, and throughout all of these can we find consciousness indelibly signed in the field of potential.

introduction

A. *historia singularitatis*

1. dimensional cosmology (the universe since the big bang)
The universe began in the first dimension. There was a microwave vibration that occurred under the influence of uncertainty and this caused everything. A single

amount of the void spun around itself and split off, forming the first particle of karma or quantum information unit. This particle was a singularity compared to nothing, and thus was compelled about itself with the combined weight of the fullness of the abyss, which was a great greatness. It was forced to begin to consume itself by the emptiness, and this it did with such haste that it began to implode with a force greater than that of the darkness, that is, that velocity known as the speed of a photon, and thus to bend the space-time within it, as it had been bent from the null space and zero time of the void when it was conceived. This turned it inside out quickly, and filled it with light so that it shone then in the pitch. But these were not rays of photons, too slow in the darkness and too easily consumed would they be, but the projection of astral light, that is the microwave gravity particle tachyons, and these are projected as an outward rippling orb. It is said then, that the finger of the creator came down and touched the spot on the globe in the heavens from outside the space that was outside our universe, at the moment of the Big Bang, so that it would be swept away with its generis to become proper space-time.

Then it was the time of the second dimension, when the waves of tachyon luminous microwave gravity stirred the void up into action, and caused more reactions that created particles. These are the events when the four forces were set down, and everything had been called into spin. Time began then, as a measurement of the spinning, a speed that could be measured by the velocity of a photon. Space was conception itself. A single point in null space would be drawn out and then turned about itself, creating polarity. A particle would spring into existence as a self-expanding wormhole tachyon torus in the vast expanse of the nether realm and immediately progenate a stream of similar shapes, that would continue on filling in the lightlessness until they were all a solid throng occupying a region, and causing by their continual exchange of motion between them, which asymptotically approached regulation, the oscillation of that great polarizing force we know as time. These tachyons tended to accumulate themselves then in a topical aura, since they were emanating outward from a center, and so their region of most profound discourse was around the edge of their expansion. It is upon the surface of this three dimensional shape, expanding in the fourth spatial dimension, that the story of our universe continues. By this time the four elemental genres of particle had been formed, and this had given the Light a fine quality, invisible to the Darkness, that of all those less intense manifest fluctuations of those particles slower than the speed of light. This was the material universe that was becoming polarized as three dimensional space on the surface of the fourth dimensional inflation of tachyons.

Once the third dimension began to appear out of the pure heat following the Big Bang, it rapidly accumulated masses in space similar to those underlying its own mechanisms of creation. These are, in order ascending outwards from our planet, stars, galaxies, and the walls and voids. The planets and the stars are spheres, the stars emitting light and the planets reflecting it. The orbits of the planets around the stars and the orbits of stars around the centers of galaxies are both planar, that is, purely based on the polarization principle — that is, that elemental and temporal-spatial opposites attract. In the case of stars and planets this means the star is too weak to attract heavier objects then the solidification of the fine layers of the gas cloud that surrounded it before its fire scorched them making them curl up into spheres. In the case of stars in galaxies, that is that those bastions of the lesser light all fall towards and are caught in the wake of a singularity where microwave gravity has torn a hole in space-time leading at the edges into hyperdimension, and in the center to the abyss outside. The walls and voids arrange themselves in random strands and gaps, the extended projection of the first spurts of probability in the infinite field of potential.

The fourth dimension gives us time. This is the surface upon which we measure a beam of light as it is guided. It is homogenous to the very small and the very large, though we recognize these terms to be relative to our perception, and it makes the smaller particles to move faster and the larger sphere to move slower, although we can project our understanding of the relativity of size onto the relativity of temporal durations. Again it is only the measure of the averaged frequency over wavelength for an area given as pi squared, or a factor of the force of the bending of microgravity, the force that causes all points in the universe to expand apart from each other as microgravity is perpetually self-generative and repulsively charged toward matter-energy, being that is on the degree of frequency where it is thought to be so improbable for it to exist in the confines of our universe, in the presence of its finer aspect, the larger solid particles or the longer wavelengths of energy, that the likelihood of it is so infinitesimal that it is considered antimatter, or otherwise, bordering on being opposite possible reality. The formula for time is thus given as phi over pi, that is the formula for a hypercube that is contained within and surrounding a sphere, that it is set to work measuring the difference of that sphere, so that, as the sphere expands, so does the hypercube.

On the day of the fifth dimension let there be Light, for as we are given to know of consciousness and sleep, and of day and night, so too do we know of the nature of these tachyons. In the proper conditions they can be observed in the three dimensional matter-energy universe, where, true to form they can be measured by instruments before the time it would take a photon resultant from the same events from which they derived to arrive. In these cases we see that they are able to utilize the same factor of the uncertainty of existence as a probability in potential to quantum tunnel through solids, moving from one point on the surface of a virtual particle to a point exactly on the opposite side, not by going through the center of the atom, nor by following a curve defined by the orbit of its electron, but by passing into and then out of the electron itself, which can be at all points on its orbital shell at any time, where it does not manifest trajectory spin as a probability like a photon being absorbed or emitted by an electron, but warp spin as it is swallowed up into itself between the two points, consumed in hyperspace where the point it disappeared and the point it reappeared are the same point, and the tachyonic wormhole itself fills the space between them, such that spin is conserved by the tachyon. The realm of hyperdimension, or the hyper-real warping of the fabric of space-time so that it is always consuming and regenerating itself simultaneously, is the surface of a geometry in pure dimension also, and this is the origin of spin-wave mechanics.

In the sixth dimension there is potential Light, that is, the absence of space as a continuum of vortices, and the absence of time as this substance in motion. Here is the dark pit from whence we started. It is the black hole of the larger universe that ours lives in, between which various frequencies of microwave vibration are shared, though it only looks light because the light of spin burning off pure potential that is our universe is so dim compared to the speed and involution of the Greater Light of this field, equivalent to the electromagnetic torus surrounding the singularity of a black hole as we know them, on the inside of which wormholes to alternate universes form. Thus it is truly here, in the quantum foam of spontaneously upsurgant probabilities, that we see the connection points between such wormholes form as a gravitational microwavelength that is the history of a single tachyon, and thus we see how our own universe formed as well.

The seventh dimension is that of potential information, where all pure data is truly relative and thus it is said to be the dimension of dimensions, that is, the one dimension containing the differing geometries of all the others and providing for them a basis for their continuous contiguity. It is for this reason, for example, that we can say there is no division between multiverses in hyperdimension where the geometry governs fine waveforms, for the same reason there are different divisions in the manifest realm of basic matter-energy exchange governed by entropy. So there is subspace, so there is hyperspace in hyperdimension — the hyper-reality of the multiverse, and so there is the pure dimension of the primary clear light called ylem. To this end they say that the creator rested.

a. the universe, and us, now

Because the speed of a photon measures the time it takes a photon to travel a certain distance as well as that distance itself, when we say that the furthest known galaxies from our own are 11 - 15 billion light years away, it means that the light we are receiving from them today also left them 11 - 15 billion years ago. According to 20th century mathematical calculations, the life span of the different types of stars, the most common individual evolution of which transforms one form of star into another along a portion of its existence called the Main Sequence, is only 100 million years. After this the star spends a short while as a red or white dwarf. The red dwarf star burns out, but a white dwarf star becomes a black hole. These black holes become supermassive until there is no longer any surrounding quantum matter for them to feed their gravity well, at which point they expel the additional matter that expanded their event horizon around the central singularity. At this point in the universe all that remains are naked singularities and fluid dynamic background radiation. This has probably already occurred for the furthest regions of the known universe, if stars remain homogeneously predictable according to universally applicable laws of physics, which depend on dimensions, which depend on geometries. Similarly the nearby galaxy of Andromeda in the Virgo cluster may have already crashed into our own Milky Way galaxy, the light from this not having reached us yet. According to our observations today, this galaxy, the only one in the visible universe whose light is blue shifted (meaning that the source is approaching us) is a little more than two million light years away, which means that if it has already collided with the Milky Way, it would have had to have happened less than two million years ago minus the combined duration of pre-collision trajectories of the two galaxies.

Our solar system began as a giant solid planet the size of the Oort cloud. It was covered in temporal wormholes, and so has become remembered as wormwood. The sun was at its core, and when it ignited, the solid surface shattered and crumbled into an enormous gyroscope. This pivoted around ten times, forming a new planet each time the three rings aligned. The orbits of these planets is unstable over the aeons, and some have come closer or moved further away from the sun. While the Oort cloud has largely dispersed into a loose, spherical field of frozen asteroids far around the outer circumference of the solar system, occasionally comets are still drawn down from within it on elliptical orbits that pass through the solar system. There are various anomalies of the planets that may have been caused by the cyclical drawing down of comets, measured by the sunspot cycle. Of these are included the red spot of atmospheric storm on Jupiter, as well as the sideways rotational axis of Neptune. Pluto itself might have at one time been one of these such comets, as well as Vulcan, the very small moon closest to the sun. On all the moons and planets without atmosphere in our solar system there are large craters that can only have been caused by such space debris, and the asteroid belt separating the solid planets from the gas giants is testimony of an unspeakable cataclysm that probably resulted from the complete destruction of a very nearby planet at a time before life began on earth.

2. negentropic genesis (the viral and bacterial hybrids)

When most of the lands of earth were still underwater, about 3.5 - 3 billion years ago (only about a billion years after the planet formed), in the lava ducts in the rifts between the plate tectonic continents, which rose upward like fingers from the ocean floor, jetting massive streams of bubbles, right at the lips of these, in the very hot, boiling, waters around the inside edge of these, there arose the first microbes.

The lands would go on to rise up out of the sea, and the microbes would fill the entirety of the earth's ocean, making it a fully functional ecosystem for abundant microbial life forms. However, aeons are passing while all this is going on. Day and night, night and day. Warmth and cool and light and dark, and always the exact same stars, all rush by in the blink of the sky.

Eventually the microbes became sponges. These sponges became cnidaria — jellyfish which would evolve into starfish, and anemone which would evolve into seaweed and, perhaps, flatworms. Flatworms would give rise to trilobite echinoderms, and these trilobites evolved into shrimp and brine. The shrimp evolved into lobsters and fish, the lobsters evolved into crabs and sand fleas and the brine, plankton and anemone into sea weed. Sea weed and horseshoe crabs emerged onto the land.

a. the first global cataclysm — the cataclysm of the trilobites

Up until this point, when all life teemed in the sea, there had been no border to evolution. The struggle for survival was easy, so adaptation was slow, and diversification of appearance abounded. However now, life evolved from the trench

103

microbes was faced with the difficulty of new necessity. Here is where we probably come to the first global cataclysm.

One possible postulate is that autotrophs and heterotrophs, that is — those things which feed off of other things similar to themselves and those things that feed off of things fundamentally different from themselves, might actually descend from a division between earthly and alien origins. In any event, the weak photosynthesis of sea weed became the strong photosynthesis of precambrian oak while the crab and flea gave way by mutation to all species of dinosaur and insect.

Notice that, rather than diversity in individual shape or appearance, these lifeforms bred new traits for the species, which were then infinitely repeated in each generation, and these templates changed in shape or appearance almost as often as with each generation. The reason for this was twofold, and indicates the probable nature of the global cataclysm.

If an asteroid had struck the earth when these first species drifted up from the primordial soup, it would have probably upset the crust and the atmosphere most. The result would have been the fracturing of the mantle and the beginning of continental drift below, and EM disturbances and the blackening of the sky above. I believe this asteroid to essentially have become the mass continental shelf we today call Antarctica. It is likely the mineral deposits there that have caused it to move toward the opposite magnetically charged pole. If, as I suspect, that asteroid did harbor an alien life form, then that lifeform would have to have been the first virus.

b. life on land

Thus, life in those times was ever changing and hostile. The dinosaurs had highly developed thalami, but little to no development of the cerebral cortex, much like modern day lizards. They grew in size due to the intense radiation, both caused by the tectonic shifting as EM disturbances along fault lines and by the thinning of the atmosphere at high altitudes due to the ash of the last meteor and the constant, subsequent, volcanic activity, leading to global warming similar to, though to a much more extreme extent, what we are now suffering from due to chloroflouridation of the ozone layer.

At the time this would have worked itself out due to the rotation of the earth. Just as the ash content was pulled toward the equator at its uppermost altitudes, so the warm air trapped in the atmosphere followed, until finally there was global cooling and the polar ice caps began to descend. Meanwhile life went on, more or less obliviously, multiplying and diversifying. The dinosaurs sired mammals and birds.

The first of these such mammals was a psylodon, a mammal that walked like a lizard, with its legs off to the sides of its body, and had a full body tail. As to the dinosaurs, many of them, such as velocoraptors and pterodactyls, started to grow feathers and hollow bones to help them maintain less weight during flying.

c. the second global cataclysm — the cataclysm of the dinosaurs

Then, 65 million years ago, there was another global cataclysm. To this day we do not know for certain what happened to the dinosaurs. We only know that some event devastated the surface of the earth, destroying all these majestic creatures, and sparing only the lowliest of serpents to crawl before the face of the titmouse and the mosquito. If there was an asteroid, it would have had to be much smaller then the last one, because it did not destroy the trees and plants. I therefore propose that it struck where the modern Bermuda Triangle is, and that the distortion to compasses there is the result of the asteroid's massive amounts of minerals and ores. This would have been in the space almost directly between the modern day Yucatan and Florida peninsulas, then on the western shore of Gondwanaland. There is evidence of a 300 foot crater in the Yucatan peninsula, which at that time was underwater, as well as accompanying remains in modern Cuba of a 900 foot tall sediment deposit carried in by the resulting tsunami.

By this time the lands of Gondwanaland (which would later become North and South America, Europe and greater Asia) and Laurasia (Africa, India, Australia and eastern Asia) had been parted wide and the sea flowed in between them. Even then, mysterious forces were acting on a global scale. Ice ages came and went. A wooly mammoth recently discovered flash frozen in Siberia had in its stomach undigested tropical vegetation.

The most probable theory is that, if an asteroid did hit the planet at this time, it triggered a rapid ice age, which did not allow dinosaurs the necessary time for adaptation through random mutations and natural selection, and which rapidly grew the mammals, from the tiniest shrews to the largest mastodons and whales.

3. evolution into humanity (hominids to homo sapiens)

Then there was a revolution in Africa. Some monkeys next to a brush fire that burned off a certain weed decided to get down out of the trees and start walking around exclusively on their hind legs. We know our ancestors were Austrolopithecines who lived in southern and eastern Africa 5 to 1 million years ago, Homo Habilis who cohabited these lands 2.5 to 1.6 million years ago, Homo Erectus who crossed the equator in Africa into the North and spread west as far as the Atlantic and east as far as the Pacific and east Indian oceans 1.7 million to 200,000 years ago, the Pre-neanderthals (600,000 to 230,000 years ago) and Neanderthals (230,000 to 35,000 years ago) of Europe, and Cromagnon hunters of Europe and Canada from Africa through Israel beginning 117,000 to 95,000 years ago all walked the earth before our modern homo sapiens. It is likely that the Cromagnons and the Neanderthal were the father and mother species of modern Homo Sapiens. Australopithecines first migrated out of Antarctica and into southernmost Africa 5 million years ago, and Homo Sapiens migrated up from Antarctica to Africa 100,000 years ago, and again from Antarctica up to South America 33,000 years ago.

a. from monkey to man

We presume these species to be descended from interbreeding between species of monkeys such as apes, chimpanzees and gorillas, who were, themselves, originally derived from small mammals such as lemurs who took to the trees at least several hundred thousand years before. Lemurs evolved through cats from weasels, who had evolved from the first mammals: rodent-size furry lizards, with their legs square to the sides and full body tails that first appeared about the time of the end of the larger dinosaurs from smaller lizards and snakes. However we have not publicly found either "missing link" between wombed mammals and their egg laying ancestors or between early hominids and the family of the monkey, which, considering the level of our species' technological development is probably about an equivalent evolution.

As for the absent interspecies leap between egg laying and wombed animals are of note the platypus and marsupials such as the kangaroo. The platypus, which lays eggs, is a genetic link between birds, with its flat, round bill, and mammals, with its coat of soft brown fur. The kangaroo resembles a large, erect rodent like a jack rabbit, with strong back legs on which it hops, and short, underdeveloped front legs like the tyrannosaurus rex. It carries its young, after birthing them from its womb, in a pouch on its undercarriage until they are fully developed. This clinging of the young to the underbelly of the maternal parent is also seen in koala bears, sloths, and certain types of monkeys and primates, and differs markedly from the nurturing behaviors of other animals, such as the pelican and the lion, which carry their young in their mouths. The platypus, kangaroo and koala bear are all native to Australia, the closest island continent to Antarctica, while the primates are native to Africa and South America, and lions and tigers to Africa and Asia respectively.

As to the missing link between early hominids and the apes, monkeys and primates, it is possible that this stage in evolution occurred on the continent of Antarctica at a time when, again because of ash content and volcanic activity saturating the equatorial atmosphere in the wake of the cataclysm that killed the dinosaurs, there was no polar ice cap. It is equally possible that the remains of the first homo sapiens would be found there, flash frozen, before they were forced in their sea faring boats to the tip of South America — where the earliest fossils of the homo sapien population exist in the Americas (a 14,700 year old campsite has been found in modern Chile, and a 13,500 year old skeleton has been found in modern Brazil), the Cromagnons living to the north in modern Canada (paleolithic artifacts of Clovis people 13,000 years old, the 13,000 year old Arlington Springs skeleton and a basket dating back 12,900 years show their presence from one coast to another) — and up to modern Cape Town, South Africa, as well as around the horn of Africa and up the Nile to Ethiopia.

b. the first way of five and the spread of the thumb

Human cultures have grown with human migrational patterns and evolved from the same origins differently in different regions. The oldest state of cultural exchange is the false path of the Right Hand, which seeks to reveal that all Truths are actually merely lies. It is anarchism, and its aesthetic is surrealism. Not only are all interpersonal reactions to be taken as impossible and absurd, but all of physical reality's consistency as matter within a universe of energy is to be constantly held under conscientious suspicion. This is the path which promotes personal power by revealing the fact that others are lying. The fact that others are lying is true, but this path's self-centered focus on the wrongs and sins and falsehoods of others and tactic of maternal nit-picking is meant to paint the world with a disgust that actually only exists within and for the selves of the individuals who subscribe to this path. It is thought that earlier hominid social units than modern man possessed this level of cognition, for the practice of ritual burial associated with it was shared by later Neanderthals and Cromagnons too. The alpha, beta, omega hierarchy of other pack mammals, and the extended families of the apes both contributed to the Chief, Shaman, hunter-gatherer unspoken order humans evolved in their groups. The true path of the Left Hand is the one which promotes the entertainment of others as the ideal goal of lying. This is the imaginativeness of the artist as opposed to that of the ingenuity of the scientist or the ruthless cunning manipulativeness of the politician. This is the co-creation of reality promised of all shamans by the decentralized forests which feed their mojo.

By the time of the last ice age most of the continents were in the positions that we know them today, and so we can trace the route our human ancestors took as they populated the lands of earth even on a modern globe.

Of the homo sapiens, first there were the Negroes of Africa. Next the Australoids of the bush. Then followed from the interbreeding of these the Mongoloids. Then there were the Americans, who were interbred between the elder native homo sapiens of coastal South America and the migrating homo sapien tribes who journeyed over the Beringian land bridge, between modern Siberia and Alaska, about 16,000 years ago. Finally the last tribes migrated into Europe, which had been occupied by cromagnons, and these Europeans then spread through upper Asia, becoming known as the caucasians. It was, however, probably not until the middle of the spreading of the Caucasoids, at the end of the last northern ice age, that any of these different races began to lose their thick mammalian coats of fur which protected them from ice age conditions, and only then that their genetic traits of difference in physical appearance, such as skin color, began to become visible.

By 20,000 to 18,000 years ago mankind had finally settled in all the lands of earth. At this time the negatively magnetically charged pole was in the north, and this caused the electrolysis (or ionization) of the arctic ocean forming one mile thick glaciers of distilled salt water that covered northern Europe and much of North America. These decreased sea levels by as much as three hundred feet, leaving land bridges connecting India and the orient to Australia and eastern Asia to North America. There may also have been land bridges connecting Australia to Antarctica, which would have been only a semi-polar, temperate continent, free of the ice sheet formed by the negatively charged pole. A land bridge also might have connected Antarctica to South America. The strong electromagnetic field served as a filter for cosmic radiation and was the exact source of evolutionary stimulus the cerebrum needed. Therefore, the earliest civilizations humanity constructed were monumental in scale, economically pro-free trade, and politically democratic. They arose as a network of global coastal trade communities between 22,000 and 7,000 years ago.

4. legacy of long lost legend (Atlantis and Lemuria)

The first of these such centers began more than twelve thousand five hundred years ago in Meso America. This was the civilization of Atlantis. They lived in the area of the modern Gulf of Mexico, and were astronomers, speculators, and geomancers. They made settlements as far west as China (where they were probably the red haired, fair skinned mummy makers of whom the vedas were written), Japan (where a 10,000 year old sunken city has been found) and easter island (where they brought the people to erect over 600 monolithic heads), and as far east as Egypt (where they ended cannibalism in the lower nile and became the generations the Old Kingdom would establish as representative of God forms — the king representing the divided migration routes of man, later represented by Adam Kadmon who would become the twelve tribes of Israel) and the Canary Islands (where a mexican style plaza exists to this day). Throughout the world we see the same burial mound and ley line culture evidenced in the pyramids of Carral and Merubecka and the Nazca desert lines and geoglyphs take root. The same people constructed a stone henge in northeastern America (complete with a speaking tabletop for prophecy) and in northern Britain along the clockwise currents of the northern Atlantic as constructed a city, now beneath the waves, off the coast of Spain. There is evidence that a mighty culture comprised of a multitude of people existed in South America as recently as seven thousand years ago, who farmed on fields surrounded by artificial lakes created by clearing vast flood planes around the higher grounds where they lived, and who created an expansive canal system, so technically accurate that it could channel water uphill, that connected the entire continent with waterways. The oldest known pyramids date from no later than this time period, located at Merubecka (Meru — the great mother mountain, of Becka), as well as a sunken city off the coast of Cuba. At this time the Beringian land bridge united Siberia and North America, the islands of Micronesia were a unified land mass connecting India and Indochina to Australia, and the Giza plateau was a lush tropical garden fed by the cool winds blowing off glaciated northern Europe. This was the age of the QBLH of the tree and of the serpent. The ten holy centers are where we now find pyramids in Palenque, Utzmal, Chichinitza; in Bimini and Alta; in Merubecca; Peru; of the Manoans of Brazil; in Guimar, Spain; the Aggahar in the Sahara; in Giza, Egypt; in Xian, China; and in Yonaguni, Japan, and these were all aligned along the Yukon pole ice age equator.

According to the Kings List of ancient Sumer, Anu, the king of the sky, sent his two sons, half-brothers, to Ki, the earth below in search of the rebel Alalu. Ea-Enki, or Ilu Kur-gal, was ruler of Nibiru, the Great Mountain, and El-Elyon Enlil was the first ruler of the earth. According to the Kings List, kingship was first lowered from heaven to Eridu. Here Al-lulim, and then Alagar reigned. Then kingship was carried to Bad-tibira, where En-men-lu-anna reigned, followed by En-men-gal-anna and then Dumu-zi. Then kingship was carried to Larak, where En-sipa-zi-anna reigned. In Sippar En-men-dur-anna reigned, and then, in Shuruppak, Ubar-tutu reigned. The son of Ubar-tutu was Zi-u-sudra, also known as Utnapishtim. He was the last of the heavenly kings to reign before the great deluge. The son of Enki, called Marduk, dispelled the heritage of Enlil by using the weapons of Alalu against Ninurta and Nergal, killing the last of the heavenly kings Ningishzidda, whom the Egyptians called Thoth, and declared himself Ra, ruling in Babili. These were the ten rulers who ruled in the ten places.

The humans of earth enjoyed absolute liberty in the selection of natural drugs and medicines, and using these to enhance their minds, were able to progress rapidly through all the pitfalls of technological development. If you doubt that this was the case, recall the story of the garden of Eden, or even the amazing tactical resources of the addict seeking out satisfaction in the form of a fix. At that time, there was a sufficient amount of lush psychotropical vegetation covering the land masses of the planet for the human population to enjoy their fill of such fruits as they chose.

Modern past life regressions to this time period describe life under such conditions as in accordance with the Law of One. According to the Law of One, the highest revelation is that all is the Clear Light, and that the movement in this Light is Time, and that one can move about through time freely in the Light. On a practical level it applies that what is good for the one will be good for others, and that what is good for others will be good for the one. Everyone could follow their True Will.

At this time there were vast herds of very large beasts, including the remnants of the age of the dinosaurs, as well as very large mammals, such as sloths, mastodons and wooly rhinos. Pterodactyls of the time were trained so that the people could fly them, however many of the animals fed off the same vegetation that the people were cultivating, and so they presented equally an opportunity and difficulty.

The Law of One provided the people the ability to make incredible technological advances even by modern standards, and they learned how to inscribe geometric patterns of crystal into magnetized stones and to use them in many of the ways televisions, telephones and computers are used today, however these had even greater applications for prophesizing probabilistical outcomes of future events and in medical procedures than any technologies known today. While these technologies and their benefits were shared freely among the people under the Law of One, the desire for use of them eventually outgrew their ability to be produced and distributed and at this time their importance was seen to overshadow even the cultivation of the holotropic planet. It was thought that such technology could be used to control the very fabric of reality itself, while the drugs could only reveal it as an illusion.

At this time civilization began to take hold in centers formed around the largest and most immobile stone technologies, such as stonehenge and MeruBecka. Perhaps the grandest of these was the Altiplano, an irrigated grid around concentric island rings. Those who were descended from the developers of the stone technology and those who were their friends and helped to guard it became the ruling bloodlines of these earliest societies. In order to leave these civilizations it was a requirement that, before one could go, they must write down all they knew so that it could be recorded for later generations. Because the bloodlines saw these people as being self-exiles, yet were eager to learn how to better govern the later generations by study of the works of these rejects of their societies, the bloodlines would be remembered as the Sons of Belial, a word which meant to cast out. This led to the story of Lucifer as being cast out of Heaven, as anyone who wished to return to the lifestyle of the placid pacifist naturalist junky humanist was seen as rejecting the collectivist tech-God of socialized community rule.

Very few human remains have been found in most of these locations, though it has been thought by most archaeologists that such stone sites, called shems, served, alike as they believe did the later pyramids in Egypt, as tombs. This might suggest that the homo sapiens of this era did not bury their dead, instead leaving their remains to rot above ground, to be burned on a pyre, or possibly even consumed. However this is unlikely, as we have already seen that the elder neanderthals and cromagnons, both with less brain capacity than homo sapiens, practised elaborate ritual burials, and, according to archaeologists, there remains the evidence of the neolithic and megalithic shems as, possibly, tombs. There remains another possibility regarding to where all the vast populations of the earth from this time period disappeared, taking with them the entire block of history describing the missing evolutionary link between the fur covered early hominids and the bare skinned, ethnically diverse, differing and dispersed modern tribes of homo sapiens. This, as we shall see, is accounted for, tragically retrospectively, in globalized mythology.

a. the third global cataclysm — the cataclysm of man
Then there was some form of cataclysm. Just as with the cataclysm of the dinosaurs, there is still much mystery regarding this great, earth shattering event. According to history this must be seen as the first event to be recorded by what we have since come to call history, since in this event all prior history, that is, all historical record, ceased to exist. However, from what myths do survive, and more of the flood survive around the world than any other type of myth, including that of a messiah, it is recalled that this event was sudden. In this way it might differ from the cataclysm of the dinosaurs, who may have gone extinct over several hundred thousand years. Instead, the timeframe for the cataclysm of man is from 12,500 to 7,000 years ago, with the worst years being from 10,000 to 7,000 years ago, or between 12,000 and 5,000 b.c. 10,000 years ago the last of the giant glaciers were beginning to break up, causing massive flooding. This continued until at least 7,000 years ago, and, scientists say, due to non-ecologically safe solid fossil fuel consumption causing global warming, it continues to happen to this day, most recently with an iceberg some 2,000 miles across breaking off the coast of Antarctica. This still seems to us today like an interminably long cataclysm, and recent enough to be remembered by all mankind, even better understood than the disappearance of the dinosaurs.

Perhaps the best way to understand our history is to look into the stars, since the light that is just now striking the surface of the earth left some of them in the past at the same time these events were occurring, and because the same stars shown overhead then as now, though earth is always in a different place in its orbit relative to the celestial sphere. History is the recording of the similitude of events on earth and in the heavens.

i. description of the earth and the heavens
The earth is a huge orb that turns slowly around itself in one direction. Because of the metallic ores produced in its crust friction, as well as supported by the

holographic force upon the entire earth by the gravity well generated by the earth's mass and stimulated by its rotation, the earth itself is magnetically and electrically charged. The poles of the gravitational rotational axis of the earth and those of the electromagnetic field do not currently coincide. They are offset from one another by about 11 degrees. It is not known if they originally coincided when either or both of them first began. It is postulated that there has been a difference between them for as long as they have existed, however there is no evidence to support such.

The earth could not have had either of these poles in their present condition earlier than when a large asteroid struck the earth, sheering away a portion of its surface into a debris field in tight orbit around the remains of the earth, and in this way creating what have come to be the earth and the moon today. This, of course, only could have occurred at a time earlier than the iron core of the earth had been smelted from the molten magma of the mantle, and this itself happened long before the gasses given off by the cooling crust atop the lava mantle condensed into clouds and formed the thousands of years of rain that created the ocean, where the trench microbes first appeared even much later, and where our story began.

The moon has very little gravity because it is not of a very dense consistency, about equal to that of earth's mantle, however it has no strong electromagnetic polarity because its mass contains few magnetic minerals, and because its sidereal revolution (27.322 days) and its synodic rotation (29.53 days) are so nearly equal (differing by only 2.208 days due to the movement of the earth relative to the sun, which adds to the position of the moon relative to the sun, effectively canceling out the difference over time by averaging), as opposed to the difference of revolutions and rotations of the earth, which makes 365.25 daily rotations on its axis during one yearly solar orbit, giving earth's much greater mass a much greater electromagnetic field (the only averaging of the difference for which with that of the sun occurs relative to galactic core).

(Even though the same side of it is always facing us because of its synchronous rotation and orbit, the face of the moon that we can see has large, evenly rounded impact craters, implying relatively right angled collisions. The source of any such debris large enough or propelled fast enough to leave such scarring on the fine dust surface of the moon could only have been its nearest, overshadowing, sheltering neighbor, us. In particular are the Copernicus and Ptolemeaus craters, the former much deeper and younger than the latter.)

Because there is no electromagnetic polarity, a compass on the moon would not move, the charged iron pointing any direction the compass is held. There are some places on the earth where compasses turn wildly around because they are in a magnetic bubble where there is no polarity, and are detecting the presence of polarity outside the bubble. One such place is the magnetic south pole. Another is the Bermuda triangle.

One thing that could create such a magnetic bubble effect is an impacted asteroid. It would have high mineral and metallic content — thus becoming strongly magnetized, but because it was not necessarily rotating around a single fixed axis before impact, it would have no polarity relative to that of the earth. Nor would the metallic mass assume the greater, or outside, polarity — the stimulated electrons would homogenize to a disordered state equivalent statistically to the same effect as equal possible attraction to either pole. It is also possible to create a magnetic bubble artificially — whenever electricity is used it generates such an autonomously polarized magnetic field. Similarly it is also possible for a magnetic bubble to be left behind in an area even if the initial instrument that created it is removed. The earth's own electromagnetic field is such a bubble formed by its charged iron core.

It is known from the examination of the orientation of layering in the formation of rock containing deposits of iron around the world that the earth's electromagnetic field has changed the directional charge of its polarity at several times since its formation following the collision that formed the moon. However it is unlikely that the collision of a comet or asteroid would account for this.

The sun also has an electromagnetic field but because it is composed of ignited gas, its rotation is not equally distributed. The surface around the poles rotates faster than the surface around the equator. This causes the middle of the magnetic field to be pulled around along with the equatorial rotation, and causes the electromagnetic field to wrap itself up around the sun. The visible results of this are sunspots — where the invisible electromagnetic field itself is crossing from one to another of its bands, prominences — where some of the surface plasma of the sun follows along one of these cross-jumping bands, and flares — where some of the plasma breaks out of the banding and ejects a jet of radiation into space. When the electromagnetic field is coiled as tightly as it can get, the sun's poles reverse, and the field resets itself.

This happens in a cycle determined by the alignment of the ecliptic with the center of the galaxy.

It is possible that this is an effect that is caused by the determination of the obliquity of the sun's ecliptic relative to the center of the galaxy by the difference squared between the sun's mass and the distance to galactic core, whereby whenever any star's equator aligns with a nearby black hole, such as at the center of the milky way galaxy, its magnetic poles reverse. However, when this happens it may effect the electromagnetic fields of all the planets in the solar system as well. In any event there is some reason that the 25,920 year precessional cycle has been traditionally divided up into twelve signs, each lasting about 2,148 years. Because not only the sun at equinox and solstice move through these ecliptic constellations, but the moon as well, the year has also come to be divided up into twelve months. Similarly, for some great amount of history, perhaps even since the invention of the first sundial, the day has been divided up into about twelve hours, which match also onto the night. It is easy to mark the four seasons by the perihelion and aphelion of the earth to the sun. It is possible to match these also onto the sun and galactic core. Just as the moon is always in a different, but predictable, place in its fixed 11 degrees tilted orbit around the earth when the earth is at perihelion (equinox) and aphelion (solstice), so is the earth in a different, but predictable, place in its fixed 23.5 degrees tilted orbit around the sun when the sun is at perigee and apogee to galactic core. Thus, if any of our local planets in their tilted orbits align with the equator of the sun when it reaches its zenith relative to the center of the milky way, there might be events on their electromagnetic field.

So, similarly, the earth's rotational and electromagnetic poles may have been gradually coming closer and closer to being aligned. When this happens the free energy (gravitational) and the charged energy (electromagnetic) can compound one another, and the earth be transformed into a giant dynamo. The end result is that the magnetic poles reverse, and when they do this they are repelled from their position overlapping the rotational poles. This does not cause the electromagnetic poles to move, however, because they are now held in place by the sun. Instead the rotational axis of the earth is moved in the same direction that the electromagnetic pole moved to overlap it, and to a distance determined by the strength of the electromagnetic / gravitational surge caused by their overlap. It is known that the north pole of the rotational axis has occupied at least three different positions over the past 80,000 years — the Yukon (117,250 to 80,000 years ago), the Greenland Sea (80,000 to 50,000 years ago) and Hudson Bay (50,000 to between 17,000 and 12,000 years ago, most likely 11,600 years ago, causing crustal displacement from 15,000 to 10,000 years ago) before moving to its present location in the middle of the Arctic Ocean.

This can cause many types of other changes as well. Volcanic activity, tectonic shifting, continental drift, rapid glaciation and complete crustal displacement are all possibilities, as well as the displacement of the planetary bodies from their proper orbits or the movement of a body in the Kuiper belt or Oort cloud. There is still no explanation, for example, of the volcanic activity on a moon of Jupiter, geysers on a moon of Saturn, and gas jets on a moon of Neptune, since all of these are outside the asteroid belt, and considered too far away from the sun to receive enough radiation for there to be heat enough for such conditions to exist. The effect all depends on the placement of the planets in the ecliptic relative to the alignment of the sun and galactic core. Since this is a cyclically recurring process, it can be understood to account for any form of naturally occurring global scale event one can imagine, however it can only be linked definitively to the 41,000 year cycle of the ice ages.

It is possible that the earth did not acquire the 23.5 degree angle of inclination of its rotational axis and thus the 26,000 year cycle of precession did not begin until this time. Precession moves the earth's vision of the cosmos 1/360th its circumference per 26,298 days (72 years). It moves one seventy-second the full way around every five years. It has precessed the north polar star from Vega to Polaris over the past 13,000 years, and shifted the alignment between the constellation in the ecliptic zodiac and the spring equinox sunrise in the opposite direction as the course of the moon and the sun (seen via the earth) in their orbits along the same path by one of the twelve constellations every 2,166 years eight months, on average. 2000 years ago the sign of the vernal equinox was Aries, whereas now the first yearly spring sunrise occurs between Taurus and Gemini, as the age of Taurus is just ending, and the age of Gemini just beginning. Thus the zodiac changes relative to the seasons. As the sun's electromagnetic field resets itself when the solar system's orbital ecliptic (the zodiac) aligns with galactic core (at the center of the milky way where it coincides with the constellation Sagittarius), the ecliptic may have been divided into twelve signs or houses (now known as the lunar mansions or months in the solar or sidereal year) to mark a 2000 year cycle of alternating sunset and sunrise in Sagittarius relative to the four yearly seasons of the inclined earth that might have a simultaneous effect upon polar climate conditions, due to electrolysis of salinization, to the alternation of the earth's electromagnetic polarity relative to the resetting differential electromagnetic field of the sun. One way to observe the earth's 23.5 degree angle of inclination from perpendicularity to its plane of orbit around the sun, at least in combination with that of the orbital plane of the moon from the sun, is by seeing that the craters on the moon during the span of one night, as the moon seems to move through the sky as the surface of earth turns around as earth rotates on its polar axis, seem to change position relative to earth's true north.

The most probable reason for the division of the zodiac into twelve signs lies in the mathematics of precession itself. The twelve signs each have three dekans, making thirty six. Each of these dekans has day and night aspects, bringing the number to seventy two. The sum of the three dekans with their day and night aspects, five, times seventy two, therefore, is three hundred and sixty, the number of degrees in a circle, or the five and one fourth days fewer than the number of days in a solar year that were holy to the Egyptians. If we combine the two calendars of the 360 degree year and the 365 and 1/4th day year, they synchronize every 1,461 "Sothic" years. During the eighth Sothic synchronization, 116 solar years after 11,688 Sothic years — or 42,369 solar days after 4,269,042 Sothic days, that is, 4,311,411 solar days of 365 and 1/4th day years, or 4,308,460 days of 365 day years — that is, in total after 11,804 years, some global event transpires. This was recorded in the Sothic calendar of the Egyptians, the Mayan Baktun — where 11,804 years was 227 katun of 52 tun, as well as the 384 and one fourth day/night lunar calendar of the Chinese I Ching, all of which claim to be descended from an elder Atlantean calendrical model. These place the most probable date of the Atlantean cataclysm some 11,781 years ago from the year 2000 AD, or 11,804 years before December 21, 2012, on July 27, 9792 BC.

It is also possible that the supercivilization that erected stone megaliths throughout the world earlier than the building of the pyramids, those who founded the first coastal communities during the last ice age, the people we call the Clovis people, or Atlanteans, discovered the remnants of an even earlier culture. Perhaps what they found were dinosaur bones, since this all occurred in the era when a star in the constellation we now know as the Dragon was above the north pole. Although it is possible that they unearthed evidence of another ancient, lost supercivilization. Modern homo sapiens have existed for 4 precessional cycles. This means that polaris is approximately our birth star. That's why this is the star I know as Lucifer. This also gives modern homo sapiens 104,000 years in which to reach the state we're in. (Consider the fact that most of the modern technological luxury we take for granted is the product of only the last 100 years.)

ii. the Sumerian description and evidence

One account has come to be pieced together from the eldest written records of the ancients, where often we find the celestial bodies used as descriptions for the Gods, or vice versa. In early Phoenician accounts we find record of a near collision between the earth and either TIAMAT (a sphere many modern scholars associate with the planet known since Roman times as Venus), MARDUK (alternately possibly Mars, the moon or Mercury), or NIBIRU (the small moon orbiting closest to the sun called Vulcan). It is mathematically possible, using calculated gravitational impacts on orbital trajectories, to predict that a planetoid could have, at a timeframe before them early enough for the ancients to have kept their mythologies about it, entered the solar system from a more oblique angle and careened past our planet before being caught by the sun's gravity well and pulled in to its modern day orbit. Because its five retrograde cycles per revolution perfectly form a pentacle from which the golden division, or divine proportion, may be derived, there is ample evidence linking this planet to the mythos of beauty surrounding Venus, Aphrodite, and Isis, all later generations of TIAMAT. The Greek myth regarding the birth of Aphrodite (Venus) states that she "sprang from the head of Zeus" (Jupiter). If the planet of Venus had come anywhere near Earth, ever, it would have turned the entire surface of the crust on the fluid magma mantle.

There is evidence from nitric acid in ice core samples and iridium in deep sea floor core samples to support that a pole shift and crustal displacement due to near interplanetary collision might have happened at the time of the end of the last ice age, between 12,500 and 7,000 years ago (about 11,600 years ago) since there were seven massive cometary impacts between 9,000 and 7,000 years ago, causing the Beringian land bridge, uncovered 22,000 years ago, as well as a land bridge in the Indian ocean connecting Asia to Australia and unifying the Indonesian islands, to sink as massive ice sheets moved back from the faces of Europe and North America, raising sea levels worldwide by 180 feet.

Maps survive copied from prehistory showing the continent of Antarctica uncovered by an ice sheet, as it is today. These maps were made by the sea-traveling explorers who built the first stone temples and monuments, many of which have now been reclaimed by the sea since the glaciers melted. If the crust of the earth were turned on the liquid mantle by the very near passing of Venus, then Antarctica would have been slid under the south polar ice sheet. There is evidence that around this same time a large ice sheet was ejected by Antarctica into the Southern Indian ocean, opposite the direction the pole was displaced, although this could have occurred even earlier, perhaps 30,000 years ago, when the magnetic polarity of the poles reversed and the rotational axis recentered itself on the Yukon area north pole.

There may have even been displacement of the continental tectonic plates as a result of this, and one theory is that South America was once at a right angle east filling the entire space of the present north Atlantic ocean, connected at its now southern tip to the interior of the horn of Africa. It is possible that the center of Atlantis was approximately where Florida — a mineral-rich silt peninsula, Cuba — a small island, and the mountainous Yucatan peninsula are today, on the last remaining portion above the molten mantle of the asteroid that killed the dinosaurs. The topography of the Atlantic ocean is such that the eastern and western continental shelves fall off 300 feet out to great depths, and there is a vast mountain range building up to the mid-Atlantic ridge in the center. The reason the plates might have been moved around is that they contained minerals that would be drawn along the magnetic distortions created in earth's gravity well. Electromagnetic distortions occur along fault lines even today due to the enormous friction of the earth's crust pushing against itself. These have been known to produce hallucinations of unidentified flying objects and thin, spectral, gray aliens, usually with a medical sensation — perhaps the experience of a newborn in a hospital.

iii. the Egypto-Graecian description and evidence

The original Egypto-Graecian mythology of Atlantis states that there have been many such utopian supercivilizations that have risen and fallen over time. The sites remain spiritually unified — Antarctica (the DNA womb); Australia (the dream land — where genetics shows our early ancestors immigrated first after leaving Africa); the Altiplano (a golden rectangle plain with an irrigation grid fed from a canal leading to the Atlantic, located in the Andes mountains in South America is the site of the capital city of Atlantis described by Plato — though the destruction attributed to it is most likely that of Minoan Crete, more recent and nearby to Plato; the Altiplano is the only known site of Orichalc, a fusion of monatomic gold, copper and tin produced from antimony and platinum in high spin states created by extreme heating, in existence in the world, and it was supposedly this substance from which were made the two pillars shown to Solon by Egyptian initiates according to Plato); America (the ideal democracy superimposed upon the hunting grounds of mystics). One day soon the entire world will be covered in connections to ancient lost civilizations. They are the result of modern and near future time travel, which resulted in the descriptions of "the ones who came down in ships from the sky" as the Nefilim of Sumeria.

iv. the banishment of the peoples

Some of the Atlanteans fled to the North and would become the Anasazi. They lived peacefully for a great many years, becoming one with nature. Another group of Atlanteans fled to the south, and also returned to a neo-primitive tribalism, beginning the oral tradition that would become the popul vuh while watching their great achievements slowly sink into the jungle. A few of these tribes, such as the Nascans of Ica, Peru, preserved the Atlantean tradition of skull lengthening by application of a magnetic crown. In the middle east, where many survivors of Atlantis settled under the red skinned Adam, Ziasudra's heir, known in South America as Pacal Votan or Quetzalcoatl, (whose name, Adama, means red clay blood of the God of Mt. Zion), the Anasazi became known as the Annunaki, and the Nascans as the Nefilim.

5. the beginning of civilization and keeping record of its history

At this time the Nubian Ethiopians (at the source of the nile) and the Sumerians of the Tigris and Euphrates river valley (who built ziggurats to their sun-kings and preserved a superstition of a war in heaven) were only just beginning to learn the art of pyramid crafting for themselves, and were quickly converted into the mass population of the upper and lower Old Kingdom of Egypt 5,000 years ago to build pyramids (4,500 years ago) in alignment with the stars forming their constellation Osiris (our Orion) to secretly declare the passage of Atlantis to the heavens. They discovered the carved Australopithecine head of Giza and built for it the body of the sphinx, aligned to Leo, aligned the pyramids with Osiris (as did the Maya in the city of Teotioaucan), and so 900 years ago would the Buddhists align Angkor Wat in Cambodia with Draco, all constellations that would have appeared on the horizon 12,500 years ago. This would mark a time when Vega was the north star and the last ice age was just beginning to end — the glaciers that had covered north America and Europe receding, restoring sea levels to their original height and wiping out globally all previous coastal civilization.

It is known that trade between Egypt and South America continued, as traces of nicotine and cocoa have been found in the remains of the mummified Pharaohs, and massive carved stone heads of the Olmecs are markedly African. Also it was not until the New Kingdom that the Egyptians began depicting themselves as of a colored complexion, indicating that the rulers of the Old Kingdom, depicted as fair skinned and manneristically thin beings, may have had a different origin than either the semites (who were usually bearded, as opposed to the Egyptians who were clean shaven) or the Nubians (who were dark skinned). Since we know that the genetically transferred chemical melanin is responsible for skin color we must conclude that the gradual darkening of the Egyptian people was due to interbreeding and not simply exposure to solar radiation in desert conditions.

a. the beginning of social philosophy

The false path of the Left hand is the one which promotes personal power through lying. The one which encourages the formation of cults, of religions, of societies, of Leaders and their Followers, of dominants and submissive personalities gauged in terms of magical hypnotic suggestiveness. This is where all orders of authority, all centralized bureaucracies and all hierarchies of influence and interpersonal political power in situation are concocted, cultivated, and culminated. Money is the only drug of these truly evil realists, and fame the only transcendent heaven and pantheon of immortalized gods.

An inversion between the Right Hand Path and the Left Hand Path occurred at the origin of civilization, when the "savages" (whose minds, if we are to believe the empiricist philosophers of 200 years ago, were "blank slates" and whose lives were "brutish, nasty and short") accepted organized rule under what would have been the first "great dictator," or "just devil," in the entire subsequent political "Leviathan." Inversion opposes one effect simultaneously with its reverse, such as spin and counter spin. This is to balance the time stream, backwards and forwards, with a simultaneous ending and beginning of the same effect, thus dispersing the opposition of the effect on a higher dimensional level, just as occurs with spin and counter spin. All early pantheons were myths of the first metaphysical legislators, whose adopted names were words with specific meanings from different lost civilizations. These demigods were the first lawmakers, thus creating the first covenant of slavery in life for remembered meaning after death, and beginning the persecution complex of the public that remains to this day. It is correct for the people to feel this way, since the masses and civilization survive off of one another symbiotically, and it is only equivalent to our sacrifice to the social structure that it maintains and supports and affirms our orders upon it. The process is based upon the parent forcing what they hate upon the child, in retaliation for the child's utterly liberated, though temporary, youth and inevitable, individuating maturity that are seen by the parent as a commodity for them to keep to have power over their children and as an imposition on their personal lives, respectively. It would prove to be society, however, that would allow our exponential population expansion, a growth pattern identical to that of a virus, and so it is that society is the cultural simulacrum of the viral form. The subsequent struggle between active society and passive culture is merely the continuation of the war between the two genders of organism, viral and bacterial life forms.

b. the beginning of religion

The Order Of Daath began from the cult of the Duat, where Imhotep (Thoth) was the name of Khufu (Osiris) in the ritual that made him the first YHVH. All subsequent offshoots of this original cult share one thing: revolution around the death of time. Greek mythology begins with Zeus killing Chronus. Egyptian mythology begins with the resurrection of Osiris by the magic of Thoth. Hebrew religion begins with Abraham adopting YHVH as his elohim. The history described in Hebrew religion, however, antedates this, and may therefore possibly be an account delivered to Abraham from a more ancient source, including the books of Enoch and apocalypse of Adam. It, as well as Babylonian religion, describe the creation; although these accounts differ in most details they share a common flood myth. In terms of history the concept of Time seems to have been dealt with by the first civilizations early on. At the time they would build monuments. Perhaps this was meant to commemorate the manner of natural cosmic manifestations. One possible conclusion that could have been reached is that the universe itself is a time machine.

i. the myth

The Egyptians' myth involved the betrayal of the King God Osiris (Imhotep, builder of the step pyramid of Djoser and architect of the great pyramid at Giza was probably the first to hold the title of YHVH in the Egyptian mystery cult, which made him Thoth, the moon god, or aura, that watched over the body of the Pharaoh, represented by the constellation of Osiris, while he, the king representing the sun, was asleep) by the 72 conspirators of the Water Serpent God Set — representing the annual inundation of the Nile; His death at Set's hand; His resurrection by His wife and sister Isis, the Eastern or Silver Star, using the mummification technique (flying saucers are disembodied manifestations of cnoptic tachyonic microwave gravity singularity tunneling time machines for the transportation of astral bodies) of the mysterious magician God Thoth, associated with Time; and His redemption by the son of that resurrection, Horus, the Hawk-headed Sun God and reincarnation of Osiris.

This is the same as the slightly elder, but essentially contemporary, myth of the Sumerians regarding the Annunaki god Enki saving Ziasudra from the flood and then creating the Adapa bloodline by mixing his own seed with that of the early homo sapiens inside the womb of his sister and wife Ninti. It was this myth which was adopted by Abram of Ur when he left to enter Egypt, and which mixed with the Egyptian mythology describing the same events that inspired Moses, on leaving Egypt, and the Hebrew compilers of the Pentateuch during the Babylonian captivity.

ii. the archetypes

Osiris (Asar, or Enki) was a deity of spring. He brought with him the golden bough from the times before the flood, and eradicated ailments of the people, leading them instead in silent songs of reverie. The golden bough was the acacia branch, representing the bush that burned but was not consumed, and the ways of silent song were ways of wine, the vine, and dances in trance. Because he is archetypally similar to so many other Gods, prophets, leaders, healers and heroes of different world cultures he was a ubiquitous force in the multiple independent births of civilization, truly a God of the growing harvest, a pangenitor. He is credited with the invention of agriculture, the distillation process for beer, and the taming of wild animals to do the herd work of the land. In Africa it is said he cured all the peoples of the land of cannibalism, or the Voudou practise of human sacrifice and reading of the prophecies of Humbaba in the entrails. Like the later Alexander the Great he is said to have left Egypt to bring civilization to all the peoples of the surrounding countryside. His history is similar to that of the later Pacal Votan of the Mayan Yucatan. Supposedly this great man, too, would rise up from humble origin to become a templar and world traveler. At some point during His pre-deified incarnation, Osiris became afflicted with a wound that would not heal, and His spirit became disincarnate and was guided by Gods through the underworld. Modern consciousness research has led in the direction of the twelve worlds of the Duat being associated with dissociative states of consciousness, however those elitist men's clubs and witches' covens that have preserved the inner essence and the proper gestures and annunciations of the mummification ritual remain elusively unavailable for comment on the accuracy of this presupposition.

Isis (also Ninti, later Sophia), associated with the twin star Sirius at the heel of the constellation Osiris, would go on to be fused with the Sumerian Mother Goddess Inanna, or Astarte, associated with the moon, to become Ishtar, priestess of the bedouins associated with the morning star, Venus, known an Shalam. Belief in her traveled as far east as India, where she was known as Shiva who, with Brahma (Osiris) and Vishnu (Horus) made up the three ages, or Yugas, of each cycle, or Kalpa — being themselves the destroyer, creator and maintainer of Manvantara, the manifest universe, to the Hindus. Along with this calendrical system came the description of an immortal soul that reincarnated through a myriad of lifetimes seeking enlightenment and perfection. They describe a blue skinned race called the Aryans, who have long since vanished from the region, from whom they inherited their kingship by divine right, and from whom they inherited their system of laws, both natural and social. This was the system described in the Rig Vedas. Of course, as the world was short on civilization at the time, a large part of attaining "perfection" was associated with civic duty, and thus the same manual also created the class system, where the slave caste served the land owning caste, and the land owning caste served the kings. In exchange for a life of servitude, the elderly at the time of retirement were exiled to live in aescetic meditative purgative contemplation of their value to the world and brace themselves for the impact of their eventual demise. It was these people who, in their aged wisdom, began the original orders of spiritual attainment, and outlined the parameters of much subsequent research. Thus were begun the world's first sects, or religious cults, devoted to the study of pure number, and naturalistic allegorical encryption systems.

Iblis-Shaytan, Satan, or Set — Sargon the Great, Scorpion King of Sumeria and unifier of upper and lower kehm, KMT, khemet, Egypt, was already an old and earthly soul (dating back as far as 6000 years ago) when he tempted our original ancestor, mitochondrial Ethiopian Eve, Isis of Sheeba, wife of king Djoser, heir to the rule of unified Egypt, in the hanging gardens of Babylon outside of Ehdin in Mesopotamia with the apple of a new civilization, giving reason to mankind's evolution into the trees as monkeys and then down onto their feet on the grasslands. Under him was begun the first system of financial record keeping, which necessitated the creation of hieroglyphics, the first form of writing. Under his rule was the epic of Gilgamesh (gilgal meaning 'water' and mesis meaning 'heir,' derived from the God Enlil, Sumerian for the onset of rain) engraved into mud tablets with reed wedges, describing Enkidu (probably Khufu, whose name derives from Enki, the Sumerian sky deity thought to have written the tablet of testimony), and Utnapishtim (probably derived from Imhotep, but whose character is that of Ziasudra — who built a ship shaped like a cube in which to survive a flood). The unification of these character traits of Sargon (Shemyazza) had long ago been Lucifer, and led the rebel angels in their war against the Atlantean cataclysm as directed by God in accordance with His Elohim Holy Assembly. Even before that he had been so close to God that no other angel dared look upon His true countenance, for there shown a light so bright it was too much even for the ajnas of the most holy Watchers. Then, it is thought, His name was Michael. Michael, as an angel, was eternal, and therefore still is.

iii. the lesser archetypes

These things — described by the Sumerians as between Enki and Enlil, and later between Ea and Marduk, by the Phoenicians as between Gilgamesh and Enkidu, by the Egyptians as between Osiris and Horus, by the Greeks and Egyptians as between Thoth and Hermes, by the Greeks as between Chronos and Zeus, by the Persians as between Ahura Mazda and Zoroaster, by the Hebrews as between Adam and Eve and Cain and Abel, by the gnostics as between Sophia and Ialdabaoth, and later by the Romans as Romulus and Rhemus though with the apis, as well as, to some extent, for Moslems as between Allah and Mohammed — which refer to God, most holy most High, and Michael, the first guardian angel, will all still be true if you substitute the names of Satan, the father of darkness, and Lucifer, thief of the first light, to whom the torch is passed. These are synonymous forces, one angelic, one demonic, though both synchronous necessarily. Between them they comprise rational opposition in the mind; without opposition man is swept up into the presence of divinity and overpowered immediately; without ration he is consumed by cognitive dissonance. It is also true, however, that Lucifer is the son of Satan as well as Satan himself, in the same capacity that Jesus was the son of God, and, more importantly it is thought, God Himself. Since the middle ages the title of Satan, as coven master and priest of Pan, has been the future of Lucifer, the free spirit, just as, in ancient times, YHVH was whoever was the Grand Master of the esoteric Egyptian-Hebrew occult, and thus the person of the true Jesus of Nazareth would hold that same covenant. These are extrapolations through reincarnative systems, indicating that both are, indeed, brothers in manifestation. The beneficial aspects are directly relative to the temporal direction through which the force is moving. God is good in a past to future timeline, and Satan is good in the inversion of this. Whereas God grows, Satan brings about regression and eventually swallowing up. Thus, death is associated with evil.

The psychic continuum, the psychic community, psychic culture, society and government, as well as the international conspiracy of psychics are all the dominion of the archangel Michael, who oversees the running of all interpersonal affairs between the souls of us living beings, which souls are themselves the spirits of the suffering Annunaki, cast down from the Heavens, now forever in what was their Hell, where everything is prettier, yet there is something subtly wrong with everything. They are they who do not know that they are only manifestations. Just as, by practising projection it is possible to induce manifestation from the multiverse into the universe, so too can these manifestations change probabilities and make predictions. So it is with the discorporeal djinn of the ancestral departed, the archetypes and the holy guardian angels. The evidence is the visit to Abraham of the two angels who were God, and also of the visiting at Sodom and Gammhora of Lot by an angel whom he identified as his elohim, as well as the apocrypha of Enoch and the later, Christian apocrypha of Hermas the shepherd, although yoga makes astral travel and extra sensory communion with this plane very easy, becoming popular.

According to modern psychology and religion, the only significant difference for modern man between projection, or wishful thinking, and an actual encounter with an angel, is that, in modern times, projections are associated with the free will of the ego, while angels one encounters with the null willed I AM god name. In ancient times it was categorized that willful or disobedient projections only reflected their summoner's (or, in psychological terminology, their "sufferer's") own lack of self discipline or self-control, which was itself classed as the undesirable trait. When people would get lucky enough to have everything go their way, then they came to attribute it to the intervention of angels answering their private prayers. The superstition deepens to preclude that when your prayers go unanswered it can only be because either you are in some way delivering them wrong (most religions have strict guidelines for an evocative meditative or dissociative state), that you are flawed and imperfect (the theory of original sin), and/or that you are praying to a holy guardian angel who has failed you. When a holy guardian angel, what I call the free spirit, fails its summoner it is said to have "fallen." Thus it ceases to be a projection of one's true will and becomes a manifestation with its own lesser will.

The Annunaki, or Watchers, see the 200 fallen angels — rebel Annunaki known as the Archons to the later Greek speaking Hebrews, whom, according to the ancient Ethiopian prophet Enoch (thought to be Thoth, the Metatron), Michael is one of seven angels (Uriel, Raphael, Raguel, Michael, Sarakiel, Remiel and Gabriel) entrusted with guarding — that they may never escape their projected, manifest, being. The Archons are twelve (Athoth, Harmas, Kalila-Dumbri, Yabel, Adonaiou, Kain, Abel, Abrisene, Yobel, Armou-pieel, Melceir-Adonein and Belias) and seven (Athoth, Eloaio, Astaphaious, Yao, Sabaoth, Adonim and Sabbede), less by Sabaoth, who repented; and their leaders of the fives and tens are those fifty names of the Ancient Ones listed in the Necronomicon, primary among whom is Marduk, the rebel son of Enki. Their Hell is being stuck in the aura of the Enochian Communications System, or the summoning triangle of Choronzon ruled over by the infinite singularity of the all seeing primary clear light. In the subsequent monotheisms these discorporeal djinn, at first associated with our original civilizing ancestors, would come to be seen as angelic and demonic servants of the universal god, and therefore as anthropomorphised electromagnetic energy. Later the importance of the role played by the original civilizers themselves was depleted to Enoch's description of them as the Nefilim Giants — the offspring of the rebel Annunaki with the wives of men. As far as we know they were creating tachyonic wormholes and bending spacetime all over the continuum for aeons before bumping into our little cosmic backwater, where they immediately became saviors to the counterintelligence community. The Guardian Angels over the rebel Annunaki, and the slayers of the Giant Nefilim, became known as the Kherubim, the first winged men, who guarded Adam and Eve against returning to Ehdin, and whose chief was Michael, known, along with Samael (Sargon), as Ialdabaoth (Imhotep).

iv. the prehistory of the archetypes

If we truly think of these as opposing tribes conquering new lands, then the allegorical tales of Enoch and Moses become much more interpretable. The Annunaki and the Nefilim could easily be seen as a foreign force that invaded the valley of Ehdin and was representative of a corruptive influence to the indigenous inhabitants. There was stated to have been interbreeding between the indigenous tribe and the invading hordes, to the extent of creating offspring called the Nefilim, of whom the first priest-king, of the bloodline of Enki, but not an immortal God, was Adapa, later known as Adam. Now whether these tribes came from as far away as the Anasazi and the Nascans, from the recently flooded lands of the Persian gulf, or from what had become the island of Sicily, or even that Ehdin be seen as referring exclusively to Catal Huyuk, is ultimately irrelevant. All of this might be considered allegorical description of society before the end of the last ice age. Then, with the kherubim, came the banishment and exile of the indigenous tribe from their original home in the valley, which hitherto remained guarded against their return by men with wings called the kherubim.

Now, according to this interpretation, "men with wings" might be a little allegorical. To understand what might be meant, we must understand the sacred secret of the indigenous people, and why their valley was invaded and conquered. According to the mythology of the figureheads of the era, who included at the time, Thoth (Enoch), Enki (Imhotep), the Egyptian dynasty from Sargon through Djoser to Khufu and Khefren, and later, Moses, and to Egyptian metaphysics in general a human god (Asar) had been killed and resurrected (Adapa). While this might mean the reassemblage of the dynastic lineage of ancient Atlantis, it was also an allegory for a fundamental truth of medicine: people can be taken apart and put back together again. This is always happening inside of us. As our cells regenerate themselves they replicate themselves. When this happens the DNA inside the nucleus splits into two identical copies. When it is doing this a small organism called ribosomal nucleic acid unzips and replicates the DNA strand. This it does much like a virus feeding off a bacteria. The allegorical equivalent of this is that we were created "in the image of God." Our brain, spine and nervous system resemble a virus; our other tissues support this like a bacteria. Hence the body of Osiris became synonymous with Adam Kadmon.

According to Hebrew mythology, Eve, the wife of Adam, was seduced by Set into accepting His offer of an apple from the tree of knowledge over good and evil — probably an allegory for the use of a psychoactive substance to unlock the mysterious potential of the memory of the ages, that which usually lies dormant within the double helix of our DNA coils. In the myths, the first wife of Adam Kadmon is listed as Lilith, and it is also said that Eve was created by the extraction of a rib from Adam. The meaning of this obfuscation is that Kadmon and Lilith were immortal within the Garden, representing Atlantis, but when Lilith was tempted by Samael (Set) to give birth to Luluwa (who would become Cain's wife), then Eve replaced her as the consort of Adam, who became Adam homo, mortal man, upon being banished. The tempting of the incarnate goddess in Paradise in Hebrew mythology is a parallel to the Egyptian myth of Isis healing Osiris with the medicine of Thoth. In Egyptian metaphysics, Osiris, Set and Thoth are all aspects of the same incarnate deity, Adam Kadmon, just as are Isis (Eve) and Nephthys (Lilith) of His wife, Shekina.

Thus, the "men with wings" refers to the same thing as in Egyptian metaphysics is given the name Hadit, the winged orb. Hadit, like the spermatozoa, is the microcosm and Nuit, like the egg, the macrocosm. Hadit represents the singular particle, Nuit the all encompassing continuum. Thus, the Kherubim may be seen as the anthropomorphications of Hadit, in the same way that Adam (Osiris) and Eve (Isis) are seen as the living parents of Cain and Abel (Horus, king above the winds) — representing the energetic resurrection, and again in turn, of Seth (Set) — representing the erect phallus. They may be seen as the seed of the fruit of the tree of knowledge, spilled within the Garden of Paradise, who at once prevent a new race of half deified offspring, and guard against the return of their original incarnate progenitors.

c. the beginning of recorded history

"Elohim" means "my god," and because it is this personal relationship with a deity that is at the heart of the Hebrew religion it is important to note that it implies only one of many possible higher powers, now called guardian angels, at that time. This was at about the time when Azazael or Abram (meaning he who has Ram, or the table of testimony, containing the history of humanity), came out of Ur into Egypt, where he became known as Imhotep — designer of the first pyramids at Saqqara for king Djoser and, later, of the great pyramid for king Khufu. He

123

imported many people from his homeland of Phoenicia, and they would become the Hyksos, "shepherd kings" over the northern portion of Egypt during the middle kingdom. These people studied the arts of manifestation, telekinesis, telepathy, astronomy and astrology, geomancy and skrying, forming an occult mystery school. Egypt would give birth to the Jewish people, whose religion had been only a loose tapestry of disparate mythologies before being rigorously schooled under Egyptian metaphysics.

Internally the first great pyramid would have made excellent living quarters, and it is possible that it served as Khufu's home while he was alive, during the fifth dynasty, much the same as ziggurats had been in Ur. During this time it may have also been used as a temple of religious ceremonies whenever the other royalty of the lands came to visit him. The air shafts are thought to have contained crushed quartz crystal lenses to magnify the stars at which they were aimed. Then, at the time of his death, the air shafts were sealed up and it was enclosed in a clay wall so that water levels could be gradually increased and blocks floated into position to complete the immense structure. That workers left rather than complete a monument to a dead man, or that some workers died while working in difficult conditions are two possible explanations for why there do not appear to be as many pyramidal worker graves on the plain nearby as rationally one would expect to find for a work of this size.

However construction on the first great pyramid utilized a new technique — building up a large mud mound in a circle around the base of the construction to fill in with water on which to float the massive building blocks into place on a large boat (which is still preserved to this day, buried alongside the pyramid). The only problem with this was that there was no rain, so all the water had to be transported up hill from the nearby Nile, and it had to be drained and changed rather frequently, because the mud and silt would have gotten into the unmortered cracks, as well as obscuring the water's clarity. So the water was let to drain back into the Nile at first, but this caused erosion around the nearby sacred carved stone head for which they had built a stone brick cat body. They decided to use the water pressure to drill into the rock of the plateau, and carved channels for it to flow through. As the pyramid grew taller and taller, as well as would have the mud mound around it have had to, more and more water built up greater and greater pressures, and the artificial caverns were bored deeper and deeper. It is possible that the interior of the great pyramid served as a giant pump at this time, with the table of Ram, or Power, acting as a power source or possibly a cutting instrument; it is known that a small clay jar containing a copper cylinder and an iron rod that, when alkaline rich fruit juice is added, creates an electric charge of one volt was found in Mesopotamia. The workers who died before the great monument's completion became known as the first djinn — or disembodied spirits of the sacred living dead, and this displaced a certain trajectory of electromagnetic spin that came to be known as the ka, the ruach, or the aura, and can be thought of as equivalent to the quantity of reflected light.

d. the lesser archetypal, elder masters

When they were first met by men the Watchers began teaching them in the use of fossil fuels. This would serve to deplete the ozone layer, thus allowing in more solar radiation, which would heighten the energy trapped in earth's electromagnetic spectrum and increase traffic on the Enochian Communications System. In this way the fallen spirits were bound to cause a depletion and plague of the astral environment in which they exist and on which they depend, effectively punishing themselves. These fifty Ancient Ones, the leaders of the rebel Annunaki, or children of Anu (the Sumerian celestial sphere of night known to the Egyptians as Nuit, just as the Nefilim derive from Tefnut, the Egyptian goddess of the air, whose brother is Shu, from which comes the words shem and Sumerians), are survivors of the 72 names of the Shemhamforash, which are given as seventy-two Guardian Angels (one of whom is Michael) controlling seventy-two subservient qliphotic manifestations (known in ancient times as demons or, to us today, as forced coincidences). The Shemhamforash itself was the Name of God uttered by Moses, again another name of Imhotep, (from Moshesh, meaning 'saved from water') at the parting of the Red Sea. However it is more than this. It breaks down mathematically into the 36 dekans of day and the 36 dekans of night, who were divided up as three per each of the twelve constellations. This system was more precise than the Egyptian system of 36 dekans, each ruling over ten days and ten nights, with an additional 37th representing the five annual holidays to make up for the lag in the ten day week, three week month calendar from the actual solar revolution of 365 and 1/4th days.

The knowledge of the Watchers, believed to be recorded in the Table of Testimony, which had been being used as a lid to the king's chamber sarcophagus in the great pyramid, was smuggled out by the Hyksos followers of the deposed monotheist Pharaoh Akhenaten, whom they called Mesis — meaning rightful heir. Thus it was used as a spell to defeat an army. In this event the Shemhamforash was a temporal manifestation that marked the beginning of the Hyksos/Habiru/Hebrew nation's wandering in the desert of the Sinai peninsula, which ended when they entered the lands of Canaan and the Gaza strip. Like any manifestation it had its consequences. Moses went up to the top of Mt. Sinai (Zonei or Zion) and meditated upon the table of the Ram, however when he came back down (the effects of radiation exposure on his face) and found the Zadok Cohein Sanhedrin worshipping a golden apis bull, the sign of Taurus in which the grand cross alignment of the planets was at that time occurring, he is said to have smashed the original tablet. Subsequently he created a replacement, perhaps using a form of scalar wave nuclear alchemy, for it was said to have generated a storm, though this time he inscribed them with laws to govern the people, and delivered the Table of the Testimony of the Ten Commandments, but then died before entering the Land of Milk and Honey. Because it was temporal, though, it had massive ramifications in the astral realm.

After this event, 22 of the Host were spared, in accordance with the letter of the Hebrew Law (that is, the Hebrew alphabet itself). The other fifty were shown the Mercy of God, and dwell now about us in the form of karma in our aura, or personal electromagnetic field, also called the spark of life or the soul. This being thus, our spirit is yet free, though theirs' are not. It is proper to talk about the ramifications of the Shemhamforash in this form if only to assure ourselves that our fate, some 4000 years later, concerns it not.

It is the sole condition of the rebel Watchers that they be real — and are to reality bound even now. Thus it was that they fell as the result of the war in heaven. Now the Archons, like the Watchers, are little more tangible than reflected signals through peoples' minds. They are also known as archetypes, though not all archetypes are the Watchers. They are less tangible than our bodies, though they are bound to us to relay through, and must make use of us like puppets in order to communicate. They all come with Michael. It is a package deal. To know of Michael is to know of Lucifer. To know of Michael (Ialdabaoth) is to know the Shemhamforash (of Samael). To know of Lucifer is to know the Fall of the rebel angels into Samsara. Therefore those who would be agents for Him are also agents for the devil, for the two are one and the same.

To understand the twenty-two who remained, you must perceive the paths on the kabbalah. To understand the fifty who fell, you must understand the role of Lucifer in the war of the rebel angels. Remember that Lucifer (Marduk) became called Iblis-Shaytan at this same time, acting as the right hand of Shemyazza (meaning, the name of Azza, or Azrael), coinciding with the writing of the Pentateuch beginning with the creation of Adam, the first man — based upon Egyptian Atum or Aten, from the Sumerian god Anu, the first law making ruler, named Ur-Nammu who we know as Hamurabi, known in early Egypt as Abram (Azriel, from whence is derived the Egyptian and Hebrew saying 'Amen' — 'let it be'), and who was thought to have been Sargon (Shemyazza).

So let us work backwards from then on. Satan, the face of God, with Maloch, His all-seeing eye of Ultimate Judgment, art S.A.M. (the three mother letters of Hebrew Kabbalah), known in gnostic scripture as Samael or Shemyazza. S.A.M. oppose God. Where Shekina (as Sophia) falls before God (as Adam Kadmon), Choronzon — a null void — appears before S.A.M. Where Shekina (as Barbelo through Koi) conjoins with Malkuth (in the form of Sabaoth) in the kabbalah, Lucifer the Antichrist rises up in the qliphoth (most modern trees of death place Belphegor over the Tagaririm — disputers — on the middle pillar, however it is rightly Lucifer, the Antichrist, who belongs in this place, rather than the degradation, Lucifuge Roffacale, ruling over the Satariel — concealers). The Antichrist penetrates Choronzon, and, as Seth, becomes Christ ascended from Daath. All of this occurs within the kaballah, which is a hypercube, or the shape of one fourth dimensional solid. God remains remote.

e. the ruler over the lesser archetypal

Now the name Michael is a common name. It is originally Hebrew, meaning "who is like God" phrased as a question. The power of the name Michael is obvious, examining its roots. In the Shemhamforash, or 3, 72 letter passages from Exodus that align to form angel names, His name appears at random (Mem-Yod-Kaph). Similarly, in the book of Enoch, an ancient scribe, often associated with Thoth, who lived before the world flood described by all surviving mythologies, the name Michael appears as one of the Annunaki angels who guard those rebel angels who "bred with the wives of men." He is torn between the angelic Shemhamforash and qliphotic Goetia, and plagued by the fate of the Enochian hosts, the eternal punishment of the fallen angels. In the New Testament Apocryphon of John he is also listed as one of seven (along with Ouriel, Asmenedas, Saphasatoel, Aarmouriam, Richram and Amiorps) who have control over the parts of the human body itself.

All of this was being written down by the Hebrews during the Egyptian and Babylonian captivities. Thus it is hard for most people to understand its true meaning, since it is so hard for the average person now to identify with being a slave — a condition alienated only slightly by our modern token exchange system. Just as we believe we serve society, so we believe the greater spirit of society will serve us, however the service of us by our society is only beneficial if greater than our own individual capacity — such as in the case of all available social individuals combined, and if it is made to serve us personally. To this end the Covenant of God was created by the Zadok priests (of Sinai) making humans and "god" mutual slaves, thus allowing the majority of blind believers in the Universal God, usually too distracted by what they assumed was its involvement in their daily lives — which constituted the ultimate truth for them irregardless of what in fact is true of the universe and for God — to get their relationship to the concept neatly squared away however they liked, which mostly resulted in the assumption of the civic spirit. The description of being real slaves, for the real Hebrews, would not be uncommon in the context of their history. In fact it would be found, as the myth of a lost continent and the myth of the messiah, recapitulated in the traditions of many other great world religions of the time, such as Moses had personally discovered in the religion of the Egyptians.

Thus it is difficult for the majority of us to identify with the plight of the fallen angels. Angels, because they are eternal, exist outside of time as we know it. They have always been, and always will be, unlike we who are born, mature, wither and die. Anthropomorphically, they comprise the exterior surface of timespace. Yet some of them, to be precise it is given in Revelation as exactly one third of them, are said to have fallen. What does this mean for an angel? It means that they have entered spacetime as we know it, and are forced to stay within it by the consequences of their acts. In order for any angel to appear to a human, in any form, it must, in effect, kneel before him, and in so doing enter spacetime as we know it. This does not mean that it is mortal, for not even the fallen angels are mortal. It means that it is immortal — that it is alive, and yet cannot die, by any means. It cannot be killed, nor struck sick, nor age. The fallen angels have simply become stuck in this capacity. Yet because they are still angels, they remain eternal. How can this seeming paradox be? It is said that it is by their arrogance that they fell, and thus that trait in them which is manifest in their electromagnetic apparitions, or their immortal shells, was present in them even before the fall, with them, outside of time, in the form of their arrogance. In this way the fallen angels are enslaved to serve that force which they thought was beneath them in the beginning, that being the manifestation of space within the confines of entropic time. While the angels serve God by serving man, the fallen angels are forced to serve man, and in this way serve the principle of the demiurge Ialdabaoth within themselves.

Now to say that these lesser archetypes, the fallen angels guarded by the archangels, are collectively ruled by a single archetype needs a little further explanation to be better understood. The concept of the duality between male and female was, at the time of the flood and the replacement of all lunar menstrual Goddess cults with masculine solar deity cults, itself masculinized. The evidence for this is not only Enki and Enlil in the earliest Sumerian religious records, but, even later, in the Tupian mythology of Brazil, with the story of the brothers Tamendonare and Ariconte, whose bickering, like that of Enki and Enlil, and, arguably, that of Cain and Abel, brought the deluge upon mankind. In the Egyptian tradition this duality infused everything, not only between the husband-brother and sister-wife Osiris and Isis, but also between the loyal and beneficent Thoth and the cunning and treacherous Set. After the deluge, when all prior philosophy had been forgotten to the memory of men, the initial metaphysicians speculated about the original duality as being between light and dark, which the Egyptians identified as the Khabs and the Khu. Most religions began with a creation myth in which the darkness of matter was separated from the ideal light, and then man was separated from woman within the realm of matter. Thus, the division between man and woman was marginalized beside the contemplation of the duality of dark and light, expressed in masculinized roles, as the conflict between Good and Evil.

So to Ptah, the original creator in Egyptian theology, as well as to Anu, the progenitor of Enki and Enlil according to the Sumerian records, there was no such thing as Good or Evil, because they preexisted the duality between aeons marked by the flood. However to the lesser deities there was good and evil. These lesser deities were usually amalgamated idols representing elements of other tribes conquered by the early empires, however in order to integrate them with the solar deity cults it was necessary to place them into a hierarchy, and the earliest form of hierarchy in these empires was the blood line of the royal family tree. So it was said that Ptah had produced Amen and Nuit, who had in turn progenated Osiris and Isis, and so, for these lesser deities, there was not only good and evil, but birth and death as well. The metaphysicians preferred to disentangle these concepts from one another as much as possible at first, so it fell to Isis and Osiris to tell the tale of birth and death, and to Thoth and Set to tell that of good and evil. So it was also with Enki and Ninurta and Enki and Enlil. The Jews, who came immediately after the fall of the Sumerians, would borrow heavily from these metaphysical myths in crafting their tales of Adam and Eve and Cain and Abel, based on actual characters, real or not, from the blood line of the Sumerian pantheon, the Annunaki.

So the greater archetypes are those for whom good and evil was relative, such as Osiris and Isis, while the lesser archetypes were those for whom birth and death were irrelevant, such as Thoth and Set. Now the final point to be made here is — what does it mean for birth and death to be irrelevant to the lesser archetypes? Essentially: that they are immortal. This is a simple concept, however abstract. In reality it takes a little explanation. While all archetypes are transtemporal, that is, relatively eternal and only appearing here and there from time to time, the lesser archetypes are living beings, such as those with the name Michael. Those who are archetypes and know it not are in the larval stage of archetypicality, and are called stereotypes. Those who recognize their archetypal nature transcend this, to become right and proper, their souls aligned with their spirits. Such as these are wise men who appear similarly throughout the ages, usually with different names; yet they are alike, and in their similarities the same, for there is only one wisdom, so there is only one wise man, who was once called by the exalted title the Most High. Ibruim (Imhotep) was the same as Shemyazza (Sargon), just as Solomon (whose name was like the visage of God) was the same as Moses (Akhenaten), etc. Yet how can this be? The answer is simple: immortality, and this is a mystery answered by the shew bread, MFKZT, made from monatomic gold, which may contain telomerase, an anti-aging hormone. The same component led to the drinking of menstrual blood around which the Satanic Black Mass is based and which led to the mythology of vampirism.

f. the good works of man with the service of angels and demons

In China, the elders were already well versed in an oracular divination system derived from yarrow rods cut from bamboo shafts which has since come to be called the I Ching. In this system there are 64 separate possible outcomes, each comprised of six yarrow rods, or lines of yin and yang. In this system yin and yang represented the naturally occurring inversion between polar vibrational wavelengths of chi energy, an energy found inside the body as well as in fields surrounding all things. Thus the hexagrams were thought to replicate microcosmically a macrocosmic probability. These derive from the same system of seventy two, based on the doubling of thirty six dekans (of which there are three per each of the twelve signs in the zodiac), but they subtract eight as being represented by double hexagrams, that look the same right side up or upside down. The I Ching was thought to represent a temporal pattern for which the eight doubles were the standard parameter of inverting fluctuation.

At this time in India the prince Siddhartha lived peacefully in palatial luxury. His father kept secret from him all the suffering of the world, and his mother continually recounted the story of his birth, and how it had been holy. When he became a man Siddhartha discovered the suffering of the world, and set out to become a monk. He joined the elder Vedic ascetics, who had worked their entire lives in misery and poverty and were now prepared in the act of spiritual cleansing for their inevitable deaths. One night he meditated beneath the bodhi tree, and summoned up all of his personal inner demons, and one by one slew them all, so they would never distract him again. He began to teach that inner peace and states of mental calm recovered greater revelations of enlightenment. At that time many people living from as far south as the Indus river valley to as far north as the Tibetan steppes were well versed in Vedic philosophy, the Stoic work ethic of the Hindu caste system. They came to the Buddha to debate and challenge his view of passivity and positivity, and most came to understand and accept his insights and results. The later, Chinese representation of the Guattama, or wise one, as a smiling obese, bald, recumbent man known as the Ho Ti may derive its origin from representations of Thoth of Ethiopia more than as a social commentary on the potential for enlightenment of bourgeoisie.

According to the Hebrew tradition, around this time the Temple of Solomon was built as a House for God. Inasmuch as the Exodus may have represented the land bridge of Beringia between Asia and North America, the myth of the Temple of Solomon may have been a metaphor for the building of the great pyramid. It is at least likely that the pillars of Jachin and Boaz mentioned in its context refer back to the same two stones of the Decalogue, which were in the possession of the Egyptians at the time of Solon, according to Plato, and are supposedly from Atlantis. According to the mythology of Solomon, he had attributed to a ring the power to govern the seventy-two. These were merely collections of potential. They could be based on either the sum of all people who potentially understand the knowledge implied by the number 72, or the sum of all people who are enslaved to belief that is not based on knowledge and awareness of true and accurate facts, depending on whichever would prove eventually to win out over the other, and thus were the source of tremendous potential power. The decalogue stones were subsequently carried out of Israel by Menelik, Solomon's son with the Queen of Sheba, and moved to Ethiopia.

The class system of India spread to the west and influenced the minds of even the early Greek philosophers, considered to be the first social politicians, as opposed to laborers or lords, and it was upon the framework of a class hierarchy that the first utopias were speculated. Both Egypt and stone henge contributed to the wisdom and science of the Greek Hellenic age, followed by the Hellenistic age conquests as far east as India and China by Alexander the Great — who took a town and a wife in every land he visited. This inspired both the later expansion of the territories of the Republic of Rome, until they became an empire following the Gualic or Gaelic Wars with the kilted Celts of France — red-haired descendants of the Indian Aryans — as well as the messiah, or living anthropomorphic ideal, of the Jews. In Meso-America Teotioaucan was built by a race of people who had knowledge of the great pyramids at Giza and was arranged along the artificial waterway known as the way of the dead in the exact same angle as the Giza pyramids, that is, in alignment with Orion's belt.

6. the current or common era (dark ages and fractalizing faiths)

By then, Roman astronomers were noticing an increase of sunspots. At that time, everybody in Rome would have qualified as fit to observe this effect, for it would have been sensed in the form of heat by the flesh itself. Rome was considered the karmic manifestation of Plato's Republic, by the Romans anyway — and it is true that, because everyone considered themselves a philosopher, just as nowadays anybody can read Stephen Hawking, it had constituted a similarity to that idealized utopia, at least until Julius Caesar proclaimed himself the posthumously prodigal emperor and his own aura, his inner circle, turned against him.

The true astronomers knew this to represent the grand cross alignment of the planets of our solar system, which had occurred in the ides of March. Then the sun's electromagnetic field went pop and blew the earth's soul — its spark of life, its own electromagnetic aura — out into the thin plasma sheet that trails in earth's shadow like the tail of a comet today. It is to mark this event that none of the surviving popular calendars today recognize the effect of precession on the zodiac. During the time it took Roman astronomers to observe this effect of the electromagnetic field of the planet, that it had suddenly changed, there was a kind of occultish manhunt to find or produce a messiah, or scapegoat, to symbolize why Caesar had done this.

a. the quest for the resurrection

At this time on the bank of the Dead Sea was Qumran, a monastic library. Here lived a community of freethinkers who had been exiled from the surrounding communities who called themselves the Essenes. (The level of the Dead Sea has dropped so low in the subsequent 2000 years that it is now below sea level, and is full of so much salt it is almost possible to walk on the surface of the water. There are also higher than normal levels of radiation in the area. According to mythology this is the site where Solomon disposed of the 72 Goetic influences he had evoked to build the temple.) The Essenes of the Jews were being covertly infiltrated by the opposing sect of the Nazarenes, from Nosrei Ha-Brith, "keepers of the covenant" (from Nosrem, meaning 'poor') and vice versa, and the result of this was outbreaks of religious zealots wanting to worship at times not given sanction by the Roman occupation. We are told it was from these conditions that the recapitulation of Great Caesar's Ghost was made flesh, from the politics of Athens and the breeding program of Sparta by the same method that Thoth taught Isis to resurrect Osiris in Horus (which Shiva did for Kali in Siddhartha). This is the same, finding of a receptive body and then drawing out of a certain soul, used by Buddhists today to preserve the holy luminescence of Guattama Buddha in the bodies of various Dalai Lamas throughout the ages, just as the Catholic Church duly elects Popes. What the Essene Baptists and the poor Nazarenes were arguing juxtapositions of was a Chinese system at least 1,000 years older that predicted cyclical alignments of the solar system's orbital plane and the galactic center of the bulge in the middle of the Milky Way. Rome's territories did extend into the east, and they had rebuilt and rejuvenated many Alexandrias, though it was easier for their occult to merely infiltrate the nearby Nazarenes than send an agent far abroad to a foreign land, amongst an alien people. A certain Essene — we are told his age was attributed to be 33 — attracted the attention of the Pharisees, the collaborators with the Roman occupation. They would go on to resurrect the cult of Zoroaster (Imhotep) in Christ's title, and thus the oldest surviving monotheism (or belief in a discorporeal, universal, conscious continuum) would produce an anthropomorphication of the unified field in the form of a finite unit (the sun's alignment with galactic core, in Sagittarius, as seen from earth) as it passes through a fixed circuit (precessional change of this position relative to the earth's seasons) in a functional system (the ecliptic zodiac). The rest is history, or at least, so we are told.

In truth, the official version of Christianity, as contained in the canonized Gospels of the New Testament, was written by the Piso family of Rome over three generations, with the help of both Pliny the Elder and the Younger. This was in accordance with an agreement made between early Christian Churches — who possessed no existent dogma of their own, but were essentially gnostic (that is, Egypto-Hebraic) and Coptic (Egypto-Christian) initiation cults whose secret was that Hamurabi (Imhotep) was the same person, in flesh or in spirit, as Sargon — and the family of the Augustine emperors of Rome — which began with the grandson of Lucis Calpernius Piso, who lived during the time of Christ — by Saul of Tsarsis, who changed his name to Paul on the road to Damascus, which was the community name used within Qumran. This would replace the original Essene mythology of the Righteous Teacher (known in the bible as Jesus, John the Baptist, James — Christ's brother, and also as Barabas) and the Wicked Priest (Paul — represented in the bible by both Peter and Judas) that described the Maccabbean uprising, at which time the scrolls stored under the original temple of Solomon and under the second temple of Herod — rebuilt after the Babylonian captivity, were transported to Qumran, and the second temple burned to the ground by the Romans. Neither of these is an accurate depiction of the life of Yeshua Ben Padiah, Jesus Ben Pandera, son of an anonymous Roman soldier and a Magdala temple prostitute named Miriam, a najjar, or carpenter of the holy guild of the second temple, who wrote of the Enochian angels in Qumran, moved to France with Mary Magdalene, bore the bloodline which would later become the Merovingians with her, and lived a quiet, uneventful life until he was at least 55, and whose body is buried along with his wife's at Rennes Le Chateau.

The true path of the Right Hand seeks to bring enlightenment to others by exposing to them the inconsistency of everything, including the dubiety of the seeker, the leader, and all of the reality in which they both exist. Thus the man who tells others the Truth for the purpose both of seeking the truth with them and for finding the truth for himself, merely exposes all the lies of others, by being a bad liar himself.

The entire validity for the authority of the early unified Christian church was based on the accounts of three different people, in writing, who lived at the same time as Jesus and who would probably have counted themselves under his influence, tracing his lineage back to the house of King David, whose son was Solomon, the offspring of Moses and heir of Abraham. The fact that these documents are known Roman forgeries should not matter, because this is not a true bloodline, but a list of rightful achievers of a certain fixed level of initiation in the Egyptian mystery cult, and the necessity of reinvigorating the social status of the myth was great at the time (2000 years ago) also because it had been exactly 4000 years since the Epic of Gilgamesh had been written, at the time of Sargon, and 2000 years since the Biblical account of the flood had been written, at the time of Moses.

b. the miracles

The word Messiah derives from the same source as the name Moses, which means "saved from water." Although this belief does in fact predate even Abraham (whose name is an unfolding of the Habiru language name itself) and is contained in the texts of Zoroaster and, independently, ancient south American tribes including the Maya, the belief in a Messiah, or great unifier, peaked much closer to the time of Christ, with the cult of John the Baptist. Now it is known that John the Baptist knew of Michael, for he was a kabbalist. He referred to Jesus as his cousin, which may have had an esoteric meaning referring to their shared membership in a secret brotherhood such as the essenes of the dead sea. Thus we can see that, by John's recognition of Jesus, he also was inferring a common friend in the angel Michael, who ordered the governments of earth.

The water of the Baptism itself signified to those two men the sunspot cycle, which was at its 2000 year peak at that time, and its consequential disruption of the earth's electromagnetic field and distortion to human brain waves. John had chosen water for this, and thus introduced the subsequent tradition of Holy Water into the Christian myth, because he was a pessimistic pacifist. When Jesus bowed His head beneath the water, as though it were stones that were falling upon Him, John realized that this was more than merely his cousin returned from his time in the desert. This was a man who had been heavily transformed by a vision of God, and so Jesus arose the Messiah. It is difficult to say right now whether John the Baptist was merely overwhelmed by the Glory of the Christos, the creation of his very own hands, and therefore failed to foresee the fall of the Roman Empire as the source for all those hallucinatory stones, however he did subsequently end up with his head on a plate, at least according to the gospel. [It should be briefly noted the few hermetic or Therapeutae trends (in terms of medicine and hygiene) that were interjected into the Roman myth at this point, such as the serving of John the Baptist's head on a plate, Pontius Pilate's washing of his hands, and the sacrament of the last supper, where wine replaces blood and bread replaces flesh.]

It is easier to expand upon the parallel between John the Baptist before the Passion and Paul of Tsarsis afterwards. In fact, had Jesus and His little band not bumped into the Baptist cult, it is unlikely that the chapter of Paul's conversion of the goyim (which would eventually lead to the conversion even of Constantine, the Roman Emperor himself, after the fashion of Pilate washing his hands of Jesus) would have either been written or needed to be.

This event happened for them some time after the historical Jesus had left Roman occupied lands, and gone to live with the celts. At this time Paul, who had created the first network of Christian churches to raise funds for a second Maccabbean uprising by zealots against the Romans, contributed the total sum of his collection plates to James, Jesus' brother in the essene cult, and zealot leader. James staged the burning of the second temple, and then he and several thousand Jewish troops and civilians retreated to Massada, a fortress on a butte-like desert plateau to the east. The Romans gathered conscripts from the surrounding bedouin tribes and made camp all around the fortress, waiting to starve the people out. They crucified a local prophet named Niccodemus, and the zealots stole the body from its grave at the base of the plateau through underground tunnels. When the Romans and Ishmaeli bedouins finally stormed the plateau of Massada they were greeted by the site of Niccodemus' rotting corpse crucified over thousands of dead Jewish bodies who had all died of starvation or committed suicide. Not one living rebel remained.

The result of this event, to which, truly, we can attribute Michael's intervention, was that the goyim quickly became the bride of Christianity, and have preserved to this day a deeply clung belief that the Messiah, or the living body of God, has already walked the face of the planet, died, and gone to live in Heaven — all a profoundly meaningless concept to them.

The reason they continue to hold this belief, rather than addressing life's greater mysteries, is their fear of the Messiah's return foretold in the Book of Revelation, penned some 2000 years ago by John of Patmos. In this account the times of the return of God to the earth will be marked by the presence of many terrifying things, in the form not only of monstrous demons clinging to the mouth of the gateway to Hell, but also ghastly events such as the falling of a celestial body to the earth and the decimation of millions upon millions of souls.

c. the archangel's intervention

Anyone who knows anything about the angel Michael will recognize this as a black op. God, unsatisfied with the Hebrew's obsession with the goyim following the conversion of Saul, when he became Paul, sent Michael to foster in the goyim the belief that the Messiah was arisen and would return, thus, effectively ending the faith of His people that the Messiah was yet to be born, by joining the Hebrews and the goyim in the Christian tradition, forever tainting native tribal purity.

Because Michael had unified the Goyim Christians and the Hebraic Jews under the banner of Christendom, and thus tended to the flocks of two fundamentally different faiths, He was known as an archangel. He was called an archangel for as long as he had been written of, though it was only when He presided at the Dionysian wedding of these brother and sister flocks that the power this implied was known.

i. a war against sacrilege

Subsequently He came to Mohammed in the form of a luminous cloud of smoke in a cave while that Holy Man was at meditation. Then, He was not known to Mohammed as Michael, though His vivid descriptions of the wars in Heaven, and of the roles of Iblis-Shaytan and the fallen angels recapitulated in the later Ishmaeli struggles, indicative of the pyrrhic conquest of Massada, as well as details regarding the particular judgments of God, identify Him indeed as the archangel Michael.

As Mohammed went out amongst the people and began to draw their attention more towards religious matters, the Arab bedouin tribes people found his words rang true. It was they who had come to populate the land whereupon the lode stone had been erected around the capstone of the pyramid. (It is not known how the cap stone and the decalogue stones fit together, or what energy would have to be applied to them to make them function in harmony with the rest of the pyramid if they do. Doing this would be considered a very Holy thing only because so much energy has been generated over the matter between the Muslim and the Jewish people.) The lode stone is a perfect representation of a cube in the same way that the outside area of the sarcophagus in the king's chamber of the great pyramid is exactly twice that of the inside area. They would eventually rise, under the reign of Suleyman, to become the Ottoman Empire, and rule as far west as Spain and as far east as India. One cult of Ishmaili believers would go on to produce very great leaders and important officials of many nations by its premise that one could get the smokers of hashish into politics by convincing them they were only "sleeper agents" waiting for the time to kill.

ii. effects in Asia

Meanwhile, the Samurai warriors of China and Japan, who are thought to be inheritors of the originally Indian cultural heritage of Buddhism, represented a totalitarian, or dictatorial, though also, other than poor peasant monks trained to be ninjas, a classless society and insofar as it was cultural, Christianity spread into northern Asia as well, though this caused a split in the church over the doctrine of the trinity in the west and that of the ascended human in the east. The Christian faith fractalized into the Roman Catholic and Greek Orthodox faiths. The Roman Catholics would go on to produce the Prussian Holy Roman Empire, the Roman Catholic church, each with dominican, franciscan and jesuit sects of monks and friars, and, a little while later, Lutheran Protestantism, which would spawn the Anglican Church, Baptism, and countless other denominations and local churches. The Greek Orthodox faith would spread to northern Asia and become the Russian Orthodox church, introducing the cyrillic alphabet to the nomadic tribes people of the Siberian steppes as far as Mongolia.

The Catholic church was particularly unfit for its marriage to the Jews. They waged brutal pogroms upon them erratically, and sanctioned those conducted independently by the governments of nations. Almost every attempt made by Jewish or concerned factions to modify dogma was thwarted by the papal patriarchy. The height of this was the Inquisition, during which Christians burned one another alive for having a literal vision of the scapegoat as opposed to the romanticized passion play according to the Roman gospels. As the result of this these Visigoth and Semitic gypsy factions adopted trade guild unionism in the mediaeval times and adapted it to sectarian secrecy, creating Free Masonry.

iii. Freemasonry

The first enterprise of Free Masonry was in trade, and to this effect they settled to fix prices and establish a stable banking system. When the merchants came together and agreed upon a range of prices for different goods and services they quickly realized they could more easily read the market, and thus set prices as high as they decided. This created the first non-royal bloodline monopolies in all of recorded history.

The Catholic Church and the European monarchies recognized this problem, as well as the problem of the plagues arising from Moslem Spain, and decided to begin the Crusades, drafting the merchant class's children on campaigns into Muslim controlled Biblical lands on the command of the Pope. Their answer to Masonry was the Knights Templar, created by Hugh De Payens, a Mason, under the king of France and funded by the cistercians. The Knights Templar policed the route into the Holy Land for the Christian pilgrims — along with their sibling order the Hospitalers, performed the first archeological excavations in recorded history at sites such as Mt. Sinai (also known as Zonei or Zion) and the site of the second temple (transporting the scroll library that had been brought back to the temple ruins and buried beneath to the caves near the abandoned community of Qumran, just as they were instructed by the scrolls the Essenes had added), and established a very stable banking network to protect the funds of the Princely vassals who lined their pockets from the Crusades — in this way assuaging proto-bourgeoise Machiavellian aspirations for the sole benefit of the patriarchy. The church owed Gothic architecture to the first stone mason guilds and the alignment of Church locations by ancient geodetics to the Templars. Still, despite all of this, on Friday the thirteenth, a date that has become considered cursed even by common, church going citizens, the church and the royals crucified Jacques De Molay, divested the Templars of all their financial holdings and arrested all those affiliated with them throughout all of Europe. They put on trials and there were public executions.

According to Masonic tradition the modern Scotch rite was begun when a few of these Templars escaped to Scotland and England, particularly to the abbey of Rossy in Glastonbury beside stonehenge. Modern Masons have written of a connection to the elder Catholic Order of Zion through the monastery of Rennes Le Chateau, believed to be the burial place of the real Jesus and Mary Magdalene. It may even have been this same Priory of Sion that encouraged the holy wars as an excuse to send in an archaeological team (the Templars) to Solomon's temple. For essentially all these reasons Freemasonry may have inherited or adopted a warped interpretation of the New Testament from the Cathars, who held that Jesus believed YHVH was a manifestation of Satan, just as the Romans inherited the Essene version of Mithra through Saul.

Free Masonry is in league with Michael as well. In modern Masonic lodges, on the altar room, or vault, floor, there is a pentagram painted. This is surrounded in a circle, and a magician stands in the center. The circle separates the aspirant from the spirits he or she will be conjuring. The pentagram separates them from the communion of the minds of all other humans and five phallanged animals that also roam and range in the electromagnetic continuum in which the aspirant will operate. The primary myth of Free Masonry is the betrayal of the chief architect of the temple of Solomon, named Hiram Tyrian, by three apprentice masons. The three masons represent the first three degrees of initiation. In these first three degrees the initiate reenacts the death of Hiram, the discovery of the keystone of the royal arch, the collection of payment for it and, later, the right proper understanding of its use and recognition for the initiate for having discovered it. The name Hiram was also the name of the king of Tyre, the town in Ur (Mesopotamia) of Hiram's origin.

d. the quest for the Holy Grail

The Muslims and the Christians redefined their territory with the Muslims yielding much of eastern Europe as far south as Turkey, just north of ancient Sumer. However the Roman Catholic papacy was unable to delegate authority over the expanded regions enough to prevent a split between the Christian Orthodox Church, unifying the eastern churches from Greece to Russia by creating the cyrillic alphabet from coptic Hebrew cyphers, and the Holy Roman Empire of Prussia, modern Germany. The Muslims retook the lands of old Israel, but it was not long before they were invaded by mongols from Mongolia — on the Siberian steppes in western China, just northwest of Tibet. These were the legions of Genghis Khan, and were ultimately repelled by the Muslims. However tribes of Visigoths and Ostrigoths swept through the Holy Prussian Empire and sacked Rome.

At around the same time the faith of stone henge and Christianity were fusing into the legend of King Arthur and the Holy Grail, in America, while searching for a turquoise trade route, immigrant Quetzalcoatl worshippers from the south discovered peyotl, erected a cylindrical observatory with the same skill as did the hohocumb northern natives a cubical one, and, during the era of the Chocco Great Houses, turned cannibalistically upon their neighbors. They decimated the Anasazi culture and interbred with their prisoners, creating all the subsequent North American tribes. They then migrated south where they gave birth first to the pyramid building Maya and Olmec, then to the bloodthirsty Aztec, who overran their peaceful, sports oriented brothers, and were then, themselves, overrun by Spanish Conquistadors simultaneous to the Spanish Catholic Inquisition, killing Jewish Gothic immigrants.

For its repeated crimes against humanity the Catholic Church was the subject of a list of grievances aired by Martin Luther in a list of demands for reformation nailed to the door of the church in Prague. This led to fierce debates among scholars and theologians of the time, many of which ended in entire states seceding from the domain of the Catholic Church and beginning reformation, Lutheran and Protestant Churches of their own, such as the Anglican Church of England. It was for this reason that Queen Elisabeth of England commissioned the master magician John Dee and his assistant Edward Kelly in the sixteenth century to construct a system of magic for the Anglican Church that would parallel and protect them from the angelic hosts of Catholicism. To accomplish this Dee turned to the wisdom of Enoch, an apocryphal book excluded from the Latin vulgate, and created the system which has since come to be called the Enochian communications system. This system describes 30 ayres in which dwell a host of 24 seniors over the four cardinal directions, 16 angels each of medicine, of precious stones, of transformation, of the four elements, of natural substances, of transportation, of the mechanical arts, of secret discovery, a host of temporal agents, and a collection of heptarchical kings, princes and ministers.

7. the modern era (industry, agencies and LSD)

The final world cataclysm is a revolution of perception, which has begun among a few individuals, and with various results. This is the age of the QBLH of the sword and of the lightning bolt. It has led, almost as though spontaneously, to the industrial age, a time of rapid social evolution bourgeoning forth grand scale science, politics, and economies that seek to unite the world. The practice of magic, relegated to the woods of the collective unconscious everywhere except India (Kundalini Tantra) and China (Tai-Chi and feng shui) since the decline of ancient Egypt, has begun to reemerge.

This can be evidenced by the folk interest in the supernatural that humanized the Renaissance. The boy genius Wolfgang Amadeus Mozart included initiatory elements in his opera 'the Magic Flute.' This sort of vapid optimism prevailed in European society during the Age of Reason and the American colonization, with the formation of dozens of different fraternally based organizations known as Jacobin or Jacobite clubs, prevalent among which were the Free and Associated Masons. One of these clubs, started by Adam Weishaupt and called the Illuminati, was taken under the wing of the Masons to protect it from persecution by radical Jesuits — Weishaupt's religious faith and the progenitor of many of the Illuminati's mysteries. It outlined a plot to dismantle all forms of social structure by revolution. Such revolutions actually occurred in France, against the monarchy, and in America, against England.

a. the birth of a nation

At first, the intelligence community of the American colonies, the men who would come to be called our "founding fathers" — no more than a good-old-boy network of Masonic lodges in 1776 — had been engaged only in a messianic breeding program, inherited from the priory of zion, and constituting little more than a mysteriously veiled version for the goyim of what is, and always has been, the most biologically fundamental habit of the Jews. In Europe the class system had evolved through the dialectic (thesis-antithesis-synthesis), leading to America arising as the synthesis of a capitalist free market and a representative democratic republic. The rule of American society is rule by the law, where justice is only what can be proven, and where it is easy enough for those with money to determine for which version of reality there will be evidence. To this extent, "fortune favors" is printed on the cash.

Then there was an industrial revolution. Steam technology running off coal in iron engines could produce hotter and hotter temperatures, and when trapped, generate greater and greater amounts of energy. This led to faster and faster modes of transportation. The pony express was replaced by the horseless carriage, now called the automobile, the steam boat surpassed by the steel rail riding locomotive, the farm fell by way of the factory, the city began swallowing up the country, skyscrapers replaced small shops and apartment hotels, the airplane and hot air balloon were invented, and immense steam liners opened up the shores of the entire world to the possible redistribution of populations. The entire country of America went to civil war with itself over a machine replacing cotton picking slaves. This became the subject of the first feature length film ever made using the motion picture camera. President Lincoln was assassinated after passing the emancipation proclamation and paying union soldiers in government printed paper money while paying the financiers of the union army, European banking conglomerates, the interest on the loans they had made to them for weapons during the war, which had become known as the national debt, from the federal reserve's gold standard.

b. industrial fallout and the turn of the twentieth century

It was around this time in Russia that the depressed potentate Potemkin was causing a backlog of documents mandated by the vassal legislature requiring his signature before they could be passed into law. An eager young clerk named Chuvalkin responded in earnest to the needs of the governing body and approached the ruler on demand that he sign the appointed documents. Potemkin began signing them, and one after another he signed, "Chuvalkin, Chuvalkin, Chuvalkin." This moral of bureaucracy we owe to a confident scribe named Franz Kafka.

In 1905 Alexander "Aleister" Crowley, who had already studied many of the greater mysteries of the universe, visited the pyramids of Giza with his wife Rose and showed her the "ayres," or spiraling tachyonic light that is still visible in the king's chamber of the great pyramid even without the presence of a burning torch or other form of light source. The next day she showed him the stele of Ra - Hoor - Khuit, who Crowley identified as an Egyptian version of Harpocrates, that was catalogue numbered 666 in the British museum. His translation and elaboration of this, called the Book of the Law, ushered in what would come to be called the New Age, and was made to magickally commemorate an event to which Crowley would devote most of his writings on magick for the rest of his life, the change of the Equinoctical constellation due to the effect of the precession of the poles on the ecliptic zodiac.

Every 108 years the orbit of Mars brings the red planet close enough to the warmth of the sun to melt its polar ice caps, and this effect was observed at the time by Heinrich Schlieman as the presence of vast seas and canals on the surface of our nearest planetary neighbor outward from the sun. Most people mocked and ridiculed him, saying that what he was seeing were merely minute imperfections in the grain of his own reading glasses, however there is some evidence that what appear to be a row of pyramids, with the same alignment to Orion's belt as those in Giza and Teotihuacan, as well as a structure near them shaped like an upward gazing human face, might have been constructed at this time, if not earlier in the same cycle.

The Theosophists, begun by Madame Blavatsky from eastern Europe, claimed to maintain through their exoteric and open to the public organization, a direct link to the Secret Chiefs of the Argentum Astrum, or Inner Order of Illuminism that preserved the ancient esoteric mystery doctrines within the highest grades of craft Masonry. After her death, Samuel Liddell MacGregor Mathers stepped forward and claimed to have preserved this link with the occult divine. He, Israel Regardie, A. E. Waite and William Wynn Westcott formed the Hermetic and Qabalistic Order of the Golden Dawn in the early twentieth century, and accepted public applications for admission, teaching what was described as direct information from the Secret Chiefs according to the binding documentation of their compact charter. Aleister Crowley, who had come to join this sect, would become disillusioned by it and, after lengthy court battles with Mathers over publishing rights to the goetic shemhamforash, would quit the organization and instead go on to reform a German branch of Free Masonry called the Ordo Templi Orientis, or Order of Oriental Templars. The greatest error recorded from the Golden Dawn teachings is the inversion of the left and the right sides of the body during ritual invocation from the right proper attribution to the left and right of the active and passive, yin and yang, pillars of Jachin and Boaz.

c. the first half of the twentieth century

Nikola Tesla was the first man to discover the alternating current used in all electrical wiring of today. A late contemporary of Edison and earlier friend of Einstein, Tesla would go on to invent the hydroelectric turbine and the radio wave broadcast transmission tower. He claimed to be responsible for the large blast in Tunguska Russia caused by a "death-ray" redirected through the crust of the earth using magnetic holographic resonance imaging. Having discovered scalar wave electromagnetic effects he claimed to be in contact with "off-world sources." The cylindrical type ufo seen most commonly in those days does resemble the scalar wave reorienting weather control balloon technology Tesla sold the United States military, and since all scalar wave technology is temporally commutative it is possible that the earlier ufos in fact were these same weather balloons, slipped back through time. Tesla was allegedly also involved in the 1943 Philadelphia Experiment, an attempt to create an electromagnetic bottle around a United States military battleship and render it invisible to radar. In this experiment spherical coils were used to generate the field, however the resonance they created was out of harmony with the earth's 40 megahertz Schumann resonance, and the ship disappeared and reappeared sometime later in a different location. There were unpredicted effects on the crewmen. This began the United States military's research and development of scalar wave technology. Wilhelm Reich, independently of Tesla's research, advocated sensory deprivation long before its time in special types of containment units designed to trap and increase ether energy, or what he called "orgone" energy. It is now believed by many that both the Tesla and Reichian types of esoteric energy were in fact the same as what the US military would later dub "zero point energy."

In America the pharmacology business was booming. Morphine, cocain and marijuana were all legal. Morphine was the first to be criminalized, followed by cocain and alcohol, and much later, marijuana. Alcohol would be re-legalized after prohibition failed due to American mobsters running liquor through import shipping companies. In 1938 Dr. Albert Hoffmann, working at Sandoz laboratory in Basle, Switzerland, isolated the psychoactive properties of ergot, a parasitic fungus that grows on rye. In 1943, he was allowed to repeat the synthesis, and LSD-25 was created. Lysergic acid Diathilamide probably began its life being involuntarily tested on prisoners at the Dachau camp 200 miles away by Allen Dulles for I.G. Farben.

The planet would be shaken by two global wars in rapid succession. The first began when a representative of an occult cabal called the Black Hand, or Carbonari (later la Cosinostra, now known as the mafia) murdered archduke Ferdinand. It was a ground war fought in mainland Europe by entrenched troops using machine guns. The second world war began when Hitler, steeped in occult rhetoric of the lost supercivilization of Thule, invaded Poland and declared himself the Fhurher of France, and this war generated even further armament advances, with bombs being dropped from planes, large tanks, repeating rifles and portable rocket technology. Supposedly the Germans were experimenting with scalar wave technology towards the end of the war, and expected it to be their salvation. The Americans, with the help of Einstein, a German Jewish American immigrant, designed a bomb based on nuclear fission. Werner Von Brom, another German, however an ex-Nazi, had designed the long range rocket technology used to blitzkrieg Britain before being brought to America as part of Project Paperclip, a snatch and grab operation in competition with the Russians over the remainder of the Third Reich's secret science. He helped design the atomic bomb, and later began NASA, America's space program.

d. the 1950's

A long cold war followed this between the capitalist United States and communist Russia, during which they kept intercontinental ballistic missiles, some even thermonuclear, trained on one another at all times. In the 1950's Lockheed aeronautics, under contract to the US government, designed and built the U2 spy plane in only one year in a secret facility, constructed especially for that purpose, called "the skunk shop." Five years after they began a U2 spy plane pilot would be held captive by the Soviet Union after being shot down on a mission over Russia.

President Eisenhower, who was aware that every U2 spy plane was equipped with self destruct and every U2 spy plane pilot given a syringe for lethal injection in the event they must avoid capture, was confident that the Russians were bluffing about having shot down the U2 — until it was made public that the pilot, Francis Gary Powers, was indeed their prisoner.

Among the Illuminati, who are the highest ranking Masons, it is common knowledge that the U2 was based on aerodynamics reverse engineered from flying saucers. Thus, perhaps due to the nature of the technology itself, the design of the 1947 crash at Roswell, New Mexico may have karmically carried through into the U2's fate.

If this is true, as we must assume it may be, then surely it is a bluff that the military recovered the alien pilot. Claiming that the U2 spy plane was a "weather plane" and claiming that the Roswell crash was a "weather balloon" does not necessarily mean the two events are identical in other respects as well — it only proves that "great minds think alike."

The U2 was only the American version of what they had found crashed in their own back yard, which they then crashed in Russia. What the Americans found was merely what our ancestors called an angel, fallen from hyperdimension, within his circular craft.

This, however, was only a messenger who had slipped through a tachyonic wormhole composed of light reflected off the moon across the face of the American desert, that was coming from a sun that was, at that exact moment in time, technically on the Russian's side. Thus, we can deduce the location of the sleeper, with whom the Watcher corresponded, who was actually on the other side of the planet, both times.

For reasons that are extremely personal to the initiated it is important for the air / space craft that humans fly to be of human design. The fact that the United States Air Force was formed in the same year as the Roswell crash and that some of its oldest ongoing projects involve stealth (all acute angles — thus deflecting fewer radar waves) technology, beginning with the flying wing and culminating in the V-wing B2 bomber, only shows that humans could have studied extraterrestrial flight technology, but proves that all of the technology they have produced is inherently human, having been designed and built by normal people, and having been carried through from beginning to end by human hands. It is in this way that the actual government classified vehicles — those designed by human beings of today — can fly about today right before our very eyes. Take, for example, the genetic research being done underground at Dulce on the abducted children of atomic bomb survivors.

The Hollywood star system of America establishes a new royal class, comprised of entertainers, models, and news anchors. This is the sum of the substance of modern American culture, the gift of the 1950's plastic fantastic suburbanites, their simulacrum of historical heritage as it existed before the social compact. This era were trying to tear the social middle class away from rule by a Lord of the land. In America in the 1950's, the middle class were housed in suburbs surrounding the cities in which they worked in high-rise offices. The social destiny that had been given to them they had made manifest with great skyscrapers. Most countries have a folk wisdom founded culture that dates back much further than modern American culture. American culture did not even begin on an international scale until the 1950's, when industrialised manufacturing made our production of stylized consumer goods applicable to the mass market of the entire country, and the new technological communication media were first used to introduce international cultural trade into the global free market of ideas. The only reason I go into the society and culture of America is that it is a young nation that has grown rapidly to become a globally recognized economic and political system.

e. tie dying

At that time many new consumer goods were entering the developed world market, most being manufactured in Japan or Taiwan and exported to America — compact tvs, portable radios, color movie cameras and eventually even color television. President Kennedy was assassinated in Dallas, Texas after the Bay of Pigs invasion had led to the Cuban missile crisis, the FBI had begun to countermand CIA-mafia drug trafficking instead of confronting the red scare menace of naturalized communist spies, and great strides were taken for integration of the races by various social agencies.

Then, in the mid 1960's, the Central Intelligence Agency approved use of psychedelics for public brainwashing following initial tests on soldiers, prisoners and mental patients, and it quickly took the forefront in the student class demographic of the illicit drug market. LSD-25 was extracted from ergot, a wheat mold that had also been responsible for the Greek Golden Age of Democracy and the Salem Witch Trials. The result was the birth of cybernetics as the ultimate mechanical manifestation of the Enochian Communications System. This was passed down to the youth culture of America more or less directly by the Golden Dawn.

The LSD testing continued under the name PROJECT ARTICHOKE under official sanction, as well as the continuing research of several independently contracted scientists such as John Lily and Ewen Cameron. While operational, MK-ULTRA contracted with independent scientists Sidney Gottlieb and George Hunter White allowing them to determine preliminary effects of the drug in different conditions. The Gottlieb research, designed to replicate Soviet research done on hypnosis and trance states for the implantation of suggestion, led to the Edgewood arsenal tests, known as THIRD CHANCE, where LSD was designated EA-1729, and used in interrogation. The White experiments, called Project MIDNIGHT CLIMAX, led to Operation BIG CITY.

In the Gottlieb experiments he posed as an artist, luring people into being involuntarily dosed, which led to the CIA's project name of ARTICHOKE. All of this was to root out Communism, to see what the communists had learned from Nazi scientists, during the Soviet equivalent of project paperclip, about experiments with mind control. The Americans expected the communist threat to present itself to America's liberal, left wing side that is familiar with the liberal arts and sciences, particularly associated with Magick. The American military, through Al Hubbard, father of L. Ron Hubbard, friend of Aleister Crowley, had become aware of a large amount of information about the use and application of magick in the form of mass mind control.

By the time of George Hunter White this had been replaced by an attempt to simulate the simulacrum of the artistic media itself. George Hunter White watched people having sex from behind a two way mirror while on the toilet. This represents the average American watching tv. It is also reminiscent of how Elvis Presley died. The way his agents behaved, running around half-naked with their guns, is reminiscent of the American gangster — more a creation of newspaper media distribution than of the original roots of Italy's Black Hand secret society (the Zorro to the keystone Knights of Malta, the sad clown of Opera) — which has been duly distributed also to Russia and to Japan, as well as into American ghettos, where it has become the dominant slum culture for disenfranchised African Americans. White ended by shooting blank wax slugs at his reflection in the mirror. But this was only during the time of the Rat Pack, during the early 60's.

In the hippy era by the late 60's, when liberalism really passed through America, the experiments of Doctor Ewen Cameron were developing the rules for "tripping." In these people were subjected to all the post-JFK alien-esque and inner earth technology that merely represented the history of the old world that the lurking spectre of multinational socialism had accumulated before it had begun to become transplanted into the backyards of national American corporations on its way to the Orient. These "psychic driving" experiments took place in Canada, but many involved travel between multiple locations, often while under the influence of an intoxicating substance and the combined effect of a strongly reinforced post hypnotic suggestion. Just as the Russians had called sleeper agents "fellow travelers" so had Dr. Frank Olsen died not from LSD making him think he could fly, as the myth is perpetuated, but from Bourbon, Nembutol, and a blow to the head.

This rapidly overran the messianic breeding program with the sensation of urgency and immediatism psychedelics induce. The MK of the CIA's project MK-ULTRA went from meaning Mind Kontrolled (*sic*) Masonic Killer to Mind Kontrolled (*ibid*) Messianic Killer. Perhaps in Plato's ideal Republic this would have been Masonic or Messianic Killer of Mind "Kontrol." In 1973 most of the documents regarding operation MK-ULTRA were destroyed by Richard Helms and the operation's name was changed to MK-SEARCH. MK-SEARCH has been the attempt by the illuminati within the CIA to combine the Enochian Communications

System accessed by psychedelics, which had only been used to train assassins under MK-ULTRA, with the Masonic messianic breeding program. The result was the sleeper agent program.

8. the post-modern era (sleeper agents and telecommunications)

The sleeper agent program conditions us terrestrials to simply "tune out" — preferably by physical stimulus or work — when a stronger broadcast signal is being relayed through our brains. The age-old parental guilt routine has proven highly effective in inputting this response in children. The dark karma accumulated from sales of addictive substances such as cigarettes gives first world Masons a truly vast supply of human brains to relay signals through, of whatever nature message they like, and has enabled the government to use these mindless zombies even to kill.

Indoctrination into the cult of sleep, which is the exoteric wing of the esoteric Order of Death, begins at an early age. The Prussian educational system, imported after World War I during the Alphabet Soup campaign to rebound the economy from the Great Depression and formally institutionalize all the one room, one desk, one stove schools of rural America, is employed at an early age to accustom the child to coping with physical environments where they would be put under psychological pressure, without the right to leave, bound by the responsibility to "better" themselves. While they twist and turn the teachers quietly observe the level of orgone energy rising. It will only result in positive reinforcement for the student if they uttered the correct mantra through this time, reciting to themselves and duly noting all the "right" answers. In this way the teachers absorb their authority from the student body — the vague answers to the deeper questions of life held by the majority being reinforced in their minds by the students' submission to their mandatorally enforced, situationally conditional education.

Sleeper agents have repressed memories, usually of Satanic ritual abuse cults run by the CIA or through CIA domestic front organizations — some covert, like the Finders, some not, like the Christian Church. They do not read minds because abuse freezes their ego at the age they were when first abused, and thus puts them into a minimal lifelong trance regarding the true nature of their potential. This minimal trance is simply ignorance of the thalamus, and it is why sleeper agents are said to sleep. This leaves their thalami, and mental projective capabilities, "up for grabs." At this point the CIA accesses them for the mob (the net gross of the mass populous), according to standard practice. The soviets called these field agents "fellow travelers." They are they who know not what they know. They live events they would prefer to forget. The mark of a sleeper agent is how much they forget. The pawns in game reality are the sleeper agents. "Burn outs" are those agents who get left in the field. They are intended to serve as the company's straw men smokescreen.

These drones are tracked by inversion, which occurs as a slight variation in the vibration of brain waves, a change in the pressure of the fluids in the skull, causing them to suddenly change their minds, and then to forget, often forever, what the other, now discarded, thought was. This can be caused directly by others, in the form of a casual suggestion implanted in conversation, or by a projection of concentration from another person's consciousness. The latter can be accomplished long range as well, such as is evidenced by the old folk proverb of your ears burning when someone is thinking about you. This is caused by an actual change in the tachyonic radiation underlying probabilities which your brain projects holographically onto your ears, making them prick up. This can also be relayed via satellites scanning large areas for just such EM shifts. They think that they are intentionally using inversion as a trick or "glamor" to throw other minds off their track. The entire counter intelligence saturation of culture stems directly from this application of "useful" inversion. This grows with the sunspot cycle. It is lunacy.

Because the brainwaves of sleeper agents are easily altered, as demonstrable by their being constantly prone to suggestion, it can be said that, in the "air" of the electromagnetic spectrum, they are listening in between channels — that is — their soul is no more in their body than anybody else's, and no more in anybody else's' than their own — they are simply: detached.

This sort of listening between channels is similar to the stations on the radio where there is no clear signal, and the broadcasts of, sometimes several, different, distant stations overlap. We know these radio substations to be effected by sunspots. This occurs as patterns of static, or repetitive, often overstimulating, soundwaves known as "white noise." Sunspots, therefore, would be having an effect on the brainwaves of the people of this planet even if most hadn't conditioned themselves to it by becoming sleeper agents and tuning their minds out with technology.

The general quandary of sleeper agents is that they congregate in aggregates, collectively known as pop-culture, that tend to produce stereotypes, who, by the efficacy of the masses, are asymptotically archetypal. At this point humans then project their own concepts of good and evil onto these archetypes. Because of this, the Watchers can body jump through these sleeper agents in the form of archetypes.

a. the brain of the waking sleeper

The human being only uses about ten percent of its brain, that is, the electrical signals active in the brain only utilize ten percent of the electrochemical environment of the brain. Almost all of this is isolated in the left hemisphere of the brain, and is perceived as rational thought. Rarely we also utilize neural pathways in the right hemisphere, and these pertain to creative thought.

Now, the concentration of electricity in the brain that occupies the active ten percent is the same substance as the remaining ninety percent of the unstimulated brain. Freud explained it in these terms. When electricity is passed through a nerve, most of the electrical charge, which, combined with its neurochemical reaction, Freud called phi, is transmitted via the nerve's conduction and neurochemical reaction, a process called cathexis, to another nerve, if it is in a system whereby it is in contact with another nerve. Some of this phi, however, Freud proposed, stays behind, and builds up in the nerve itself. This process leads to hypercathexis, or the delivery by a nerve of more electrical charge and neurochemical than what was transmitted to it. According to Freud this is how Ego accumulates. Here we see the one to nine ratio at its root: most of the electricity in the nervous system is not inherent, and is due to stimuli, while some of it has dug in and is related to perception itself.

We know that, while the human will is not being consciously imposed upon it, the electricity active in the brain will fall into regular waveforms that will cycle themselves through in a regular pattern, sustaining all the autonomic neurological functions necessary for the preservation of the inert physical body.

The grey matter of the cerebrum is comprised of an interlocking network of neurons that are made of axons and dendrites connected by a myelin sheath. This is where all the electrochemical interactions associated with free thoughts occur. The cerebrum has no nerves of the somatosensory system inside it, and hence the one thing this part of the brain can never feel is itself.

The grey matter of the thalamus is comprised differently. Although it is also composed of nerve cells, it is more dense and compact. It is not arranged in layers as the cells of the cerebral cortex appear to be. Also it acts holographically, with single neurons relaying sensory information from multiple sources in the nervous system to multiple areas of the cortex.

One of the more meaningful distinctions between these two, however, is that one can have brain waves of a different frequency than the other — one can be "more conscious" and the other "less conscious." This is true between the left and right thalami as well. The thalami themselves, as part of the forebrain, are just as mysterious to consciousness bound up in the tissues of the cerebellum, pons, and medulla oblongata, the parts of the hind brain, as the cerebrum is to the thalami.

For example, at the time when dinosaurs had reached the point in the spiral macro-fractal evolutionary path, approximately pi at the brink of becoming phi, when the thalami had developed thoroughly as their primary cerebral structure, there was a sudden, global disaster, curbing their evolutionary tendencies of size into smaller species such as modern reptiles and birds.

Now the human brain is at the same moment in its evolutionary developmental curve, and our thalami are well developed, but not so well as our cerebrum, and we, like the dinosaurs before us with their stadium sized reptilian forms, have populated the entire planet with our upright, mammalian bodies. If the catastrophe that killed off the large reptiles was related to the development of their thalami, then we can, with all due moral justification, imagine ourselves to be saved from such an outcome by the balancing in development between our thalami and our cerebrum.

Reptile brains are essentially identical to those of birds and fish since those are the three primary branches of species into which dinosaurs evolved. The Enochian Communications System is predicated upon the projection of this key onto the working parts of the cerebrum by the thalami, or more concisely, the belief that mankind will evolve into angels. We see the evidence for this in the behavior of archetypes, which contribute to macroevolutionary conditions, teaching our species how to survive being itself.

There are three Mayan hieroglyphs for the three parts of the brain. Men represented the human brain. Oc represented the mammalian brain. Chichan represented the reptile brain. It is easy to see these as merely referring to stages of development in the evolution of the brain stem, but they are archetypal also.

The reptilian brain, comprised of the medulla oblongata, pons, and pituitary gland, accesses all manner of matter, from the least tachyon to the greatest dimensional extrapolations of the universe, however it possesses, or rather, seems to posses, only enough intelligence to have a very strong opinion. The mammalian brain, comprised of the corpus collosum and the thalamus, accesses the realm of archetypes through the Enochian Communications System, and comprehends the mechanism of manifestation. The human brain, comprised of the cerebrum and cerebellum, comprehends pure dimension in its higher geometric forms, such as potential light, potential energy, potential spin and potential information, in the principles of Light, Love and Life.

What the mammalian mind perceives as the archetypes in the heavens are only the ennegrams in the cerebrum, and the concept of God is the ego. All that is the Enochian Communications System is the perception by consciousness bound up in the tissue of the thalami of the biological functioning of consciousness bound up in the more complex tissue of the cerebrum above and around it. Understand, of course, that this particular deduction is being made by a gland that translates tachyonic holographically concentrated consciousness into chemical neurotransmitters released into the brain in essentially the same way as the heart pumps oxygen exchanging blood in the body. The entire concept of time as an absolute is derived from the production of alpha waves there.

i. biological projection

The prophase of manifestation is projection.

The keeping of sacred objects is not equivalent, as it was considered in the latter ancient times — after the advent of socializing monotheism, to idolatry. It is merely the earlier, no less sophisticated, associative technique of the organized mental cellular structure of the brain, common to all forms of self-motivating biological organism. The only difference is that of scale and degree of delegation.

For elder vertebrate life forms of minimal mental development this process manifests itself in memory storage and mental mapping. It would be impossible for fish, for example, to navigate in isolation without a form of the same power that humans attach to religious practice — namely, the essentially holographic superimposition of their internal reality with the external environment through a softening of the sensory field dividing the nous from the logos.

This process, in organisms of more highly evolved mental structure such as mammalian vertebrates, displays itself in the attachment of particular attributions to specific locations and/or objects, climbing up through a gradient of quantifiable displays of affection essentially initiatory in nature to culminate in the primate branch with the use of tools connecting preconceived intention to the accomplishment of specific endeavors.

Finally, in humans, we find the same process. Although tempered by the strengthening of the ego resultant from the awareness of the cultivation of awareness, which, it can be said, is what this so-called process ultimately results to, and which places humanity, in the hierarchy of our own self-created understanding, at the middle point between the objective animal and the subjective eternal aspect anthropomorphised as God, this same process is present both in the domestication of animals as surrogate self-complimentarity in the form of living examples of our own ability to subjugate reality to our holographic superimpositions, and in our equally proportioned sensations of overwhelmedness when confronted in our personal experiences with ultimate unknowns. It is the same force within us that precategorizes objects of affection for potential usage in moments of emotional crisis as seeks out causes for supersaturation of sensory stimulus.

One of the primary hitherto identified differences in the human as opposed to all other animals behavioral utilization of this process is the seeming necessity for recognition of mortality connected to the human utilization. The comparatively longer life span of the human seems to allow the recognition of changes to a being over time as occurrences in an identifiably constant pattern, and this recognition leads both to an attachment of dominance to the design based on the same attribution of final importance to duration of survival, and to a desire to attain to permanent or at least practice the acquirement of gradual changes to the form of the being which serves as the vessel of transmission for this motivating potential which is realized both in terms of the temporal template, and the raw resource of energy provided by the interaction of consciousness with material reality through the senses.

Animals seem to have much less forethought in their collection of stimuli, due, at least partially, to their having fewer items on their list of lifelong goals than the average human. With every world, philosophy, or, synonymously, lifetime created by a member of humanity there arise resultantly a certain number of new potential opportunities for the furtherance of interest along lines defined by the parameters of subsets to the temporal pattern. In this way humans can produce a unique breed of life which animals cannot, namely, intellectual offspring. On the other hand, so long as humans and animals share the same amount of overall space and continue to have interaction, the attribution of superfluously identifiable traits to animals by human intelligence seems to compensate for the difference in the pace of otherwise common evolutionary development.

It is this conflict across levels of development within the most highly structurally evolved animal minds between awareness of mortality and interest only in the immediate that leads to distraction by differences from the essentially intuitive attention to similarities that is truly their shared inherent trait.

It is because of the long term psychological effects of projection on the weak-willed that manifestation has come to be viewed by authorities as "off-limits" except to the divine. Projection in and of itself is harmless, and actually feels quite good. Its emotional counterpart is empathy, and this emotion is, at least exoterically, encouraged by those same authorities. It has even been proven by scientists that projection, which is also the basis of religion, can occur between animals of two different species other than humans. Their example is Koko's kitten, but anyone who has ever survived riding a horse past a rattle snake knows this in even less uncertain terms.

Projection occurs when holographic ennegrams are externalized via the thalamus, and meaningful value becomes attached to something external to the nervous system as a memory storage referential. Because alpha waves are produced in the thalamus, the concept of divided, or differential, time becomes a factor. The result, and the cause of the psychological impact on the weaker-willed, is consciousness of the eventual loss of the external as memory referential, at which time one will have to legislate between all the things one wishes to remember about the external, and all the things one wishes to forget about the external. This is the religious necessity of cloning, for example, for — even if you cloned some body — you would still have to summon their departed soul into the new form.

b. mechanical projection

From the 1970's - 80's many new consumer goods were entering the developed world market, most being manufactured in Japan or Taiwan and exported to America — video cameras and videotape recorders that could be connected to the tv, portable tape players, portable sound systems, portable televisions, microwave ovens and, perhaps most importantly, the personal computer. The military was also developing the internet, then known as AARPA net, microchips and night vision goggles. Scientists were conducting frequent experiments with lasers, light absorption spectral emission reflectors, or electromagnetically charged gas particle beams.

During the end of the twentieth century America, the Soviet Union and other developed nations raced to install a vast satellite system in orbit high above their heads. As the satellites went up, empires rose. As the satellites were turned on and began to be used, the empires began to fall as individuals began mutating into advertising corporations. The military black helicopters may have been the first branch of this operation and are also the product of stealth technology. They use scalar wave sound mufflers that can be used for mind control when targeted on people.

The Enochian Communications System exists as the unification of the Russian Project Woodpecker, and the American Projects ELF and HAARP (High Frequency Active Auroral Research Program). These create a global shield of electrons as an imposed perimeter on information. This shield acts as a large screen, upon which messages are broadcast. The result is the "painted sky" effect for data dispersal. Every electron acts as a fractal of the full field, and thus can be used as an access port to the patterns of information contained in or surrounding any other electron in the set.

Just above this field are the satellites. These are like sentinels of propaganda. They have been sent up by humans serving the Watchers for the purpose of relaying manifest information bounced off them to remote locations on the surface of the earth, heightening the illusion of our soul being in "two places at once."

i. life in the simulacrum

Within the field there is chaos in virtual reality. President Nixon took America's printed money off the gold standard, and President Reagan bankrupted the fed on big budget international defense systems. We are told that only by trading gold for oil can we keep the Muslims of the middle east from terrorizing developed nations. The United States of America has been in a state of federal emergency since the passage of the federal emergency management act (FEMA), and at any second can declare a state of Marshal Law. The main goal of the system is to utilize the remaining 90% of your individual brain capacity to transmit controlled information. "It becomes gradually more difficult to have thoughts without manifesting consequences" is the dominant paradigm. This, combined with the immune system degeneracy caused by fluoridation of public water, causes a state of light body free fall, where our tachyonic selves, our spirits, repel just above the surface of the earth, much as though in a dream. This contributes to a weakness in the knees, because our tachyonic spirits are gravitational.

Compared to ancient times, when the soul was more ethereal and the body more solid, this sort of feeling of "one foot above the ground" causing weakness or trembling in the knees was so rare as to induce devout prayer. Now it is so common it is the other way around. The monkey primitive enough to lack sensitivity to the ambient level of radiation is taken to be a great leader, and he quickly has the first scapegoat he can find for that effect of weakness crucified — which, sensibly, only worsens the situation. Thus psychiatrists, who differ from psychologists in that they prescribe drugs, are clearly overlooking the taking of drugs as the primary obvious cause of behavioral pattern, because if they allowed the populous to realize this fact, they would be exposed as addictive-substance merchants peddling lethal poisons. Instead we are given the red herring of tobacco, which is only a good workout for the lungs, and yet, when combined with aluminum byproduct lined, immune system weakening tap water, spiritually and physically unhealthy exercise diets, fast food high in starch and useless, fatty carbohydrates such as grease, and trendy designer prescription pharmaceuticals to further weaken the immune system and make it dependent on external sources for support, tobacco is blamed for all cancer.

9. the present (cultural symptoms and social cures)
It is really the fluctuation of frequency in subspace called entropy that cause the divisions we call society and culture. Each of these describes the true condition of the other. What we do socially is called culture, and what pertains to the genetic culture of all mankind they call civilization or society. What we can conclude from this is that culture moves eastward around our planet while society moves westward. In modern politics there are only two forms of control of populations. Those are federalist governmentally liberal, military-industrial complex Right Wing centralized and local governmentally conservative, culturally populist Left Wing decentralized empire.

a. centralized empire and decentralized empire
Centralized empire goes through three common stages: war or revolution; patriarchy; infiltration by interested parties and dissolution. The best examples of centralized empire are either cloaked in religion, such as Catholicism, Judaism, or Islam, while politically centralized empires have predominantly been too recognizably totalitarian to survive free trade without initiating mass conflict. The only politically centralized empires to survive the twentieth century are communist masked patriarchies based on the false belief that 200 year old Jacobin humanism is in any way stronger or more durable than 2000 year old religious affiliation.

Decentralized empires are what centralized empires tend to become. These are still imperial, insofar as they advocate worldwide control by their chosen political system, they have simply adopted multiple parties (as in America) or international governmental organizations (as with the voting blocks in the UN) in accordance with a check-and-balance system that makes them more user friendly to the people.

Covert organizations have been acting within larger organizations at least since the time of the Crusades to destabilize centralized empires and render them more decentralized. The intelligence/counterintelligence department has become so common a part of democracy that it has almost become its own centralized empire within the decentralized empires wherein it flourishes best. There is no question, for example, that the nonmilitary cold war between the USA and the USSR was won entirely by information/disinformation strategies effected upon the opposing citizenry in the guise of (arbitrarily) supply versus demand side economics.

It is usually through the establishment of an agent community that interested parties come to undermine centralized governments. At this point they may still appear to be firmly centralized patriarchies, such as Russia under Gorbechev, Cuba under Castro, China under Mao, Palestine under Arafhat, or even America under any of its puppet dictators, however the true order of the government by this time is economic espionage based on viable resource accumulation/allocation.

In an ideal decentralized empire, such as the withering dictatorship of the proletariat represented by technological capitalism, all of this is common knowledge to the man on the street. Unfortunately, no ideal decentralized empires exist. Instead, what the developed nations of the world have become today are different versions of realist decentralized empires.

These include the fallen Soviet states battling totalitarianism, terrorism and organized crime with Democracy, the European bureaucratic technocracies stabilizing a unified market economy to maintain international economic competition, the corporations of Japan recovering from the late nineties pull out of foreign investors, the Muslim Sheiks remaining above middle class while trafficking in international trade, South American dictatorships being militarily overthrown and drug cartels being kept in check by federal investigation agencies, Canadian and British socialist utopianism seeking to provide better health and social welfare at higher taxes to the people, and the American republic, where two parties represent the dominant and submissive traits of an authoritative system.

These two American political parties are each populated by people whose personal karmic auras are inversions of their party's line, such that disenfranchised Democrats are closet fascists and even only moderately bourgeois Republicans are elitist socialists. The elected president represents a political party figurehead: the least Democratic Democrat or the least Republican Republican. However, to admit to inversion is tantamount to admitting responsibility for manifestation, and because manifestation is considered sacred and holy it is not practised by the majority of members of the political system, i.e. the voters. It is only practised in the form of the token exchange economy system, and this yields no eternal rewards.

b. right proper manifestation

The reason manifestation is considered too great a power to be wielded by the common man, besides the evidence of the destruction of all those peoples who have wielded it in the past, is that it creates a consciousness within God other than God. Let me put that this way: ordinarily tachyons fall into a regular pattern. When manifestation occurs, one of these tachyons swells and opens up into a wormhole, allowing the additional meaning projected by the perceiver's thalamus through into the context of material reality. This is a temporal distortion to an eternal continuum.

What I am about to describe is best done in space. Doing so on earth, on any large scale, is probably contraindicated. This is not to say that this has not been done before, and yet we are still alive. It is simply that myths of a strange energy source are associated with the downfall of an ancient lost supercivilization. Mythology is a strange beast in itself. It is usually only a court jester's perspective on the affairs of the court and health of the royal family, however when it speaks of the human ability to harness energy that can be misused, even if accidentally, and cause a massive effect, it addresses one of the greater issues of the human species.

This being said: by projecting a faster microwave frequency into a slower microwave tachyon, it is possible to expand this tachyon into a wormhole. By mapping scalar wave frequencies over top of the microwave signal it is possible to access any destination from the holographic electromagnetic radiation background. This can be done as easily as by projecting a visual thought of where you want to go onto the wormhole.

When the destination is projected, by whatever medium — mechanical or biological, it creates a particular harmony with the faster microwave gravity, and this creates null space, or zero-point energy — the electromagnetic result of combining radiation and gravity. Another name for this is a temporal singularity. Just as the background radiation of the electromagnetic spectrum serves as a fractal through which any single point in space may be linked to any other, so does the sum over histories of probabilities in the universe serve as the gnomon from which we may extract our temporal destinations.

Once you have traveled through a wormhole you have entered into the multiverse. Things will not be the same, and all things will continually change. It is possible to meet yourself in this way, and there are fewer experiences more alienating from reality.

The stealth bomber is a triangle shape because it uses three microwave gravity generators to expand a wormhole and break the light barrier, while the public of today is unaware that this craft can even leave the atmosphere. There are, however, several possible designs and methods available for it to do so.

Outside the Enochian Communications system, and containing it, are the Akashic Records. Despite the fact that these provide access to past life memories, they are not housed, as mystics such as Edgar Cayce and Madame Blavatsky may have allowed their followers to believe, in the genetic coding of our DNA. They are more universal than this, being comprised of microwave gravitational tachyonic fifth dimensional Light, and comprising the background radiation of the electromagnetic spectrum of the continuum of the universal singularity in which are all the other lesser potential singularities, evolved from stars, that are the wellsprings of our spirit — doorways outside the universe. Between these singularities is the history of photons, the illusion of material time, the lesser light, by which we perceive third dimensional reality. As the earth's electromagnetic field is an Enochian system, part of a solar/galactic/universal, similar system, so are the Akashic records equal to our fully evolved consciousness as part of the communication of ideological singularities, or rhetorical points, in a universal singularity continuum.

c. signs of the times

Now we have come to a time again when we begin to see meaningful events occurring in the heavens, so sudden, and so rich in meaning, that we bow down before knowing their full limit. This began with the passing of the comet Hale Bop, and with the crashing of comet Shoemaker-Levy 9 into the visible face of Jupiter in 1994. This was followed by a solar eclipse between leo and cancer, visible only to the eastern hemisphere, on August 11th, 1999. This was followed on May 5th, 2000 with the alignment of the earth, the moon, the sun, mercury, venus, mars, jupiter and saturn in the constellations taurus and aries, as well as the alignment perpendicular to this of uranus and neptune. This was followed by another partial solar eclipse in sagittarius on December 25th, 2000, visible to the north half of North America alone — what I call a "crescent" eclipse. On January 20th, 2001 there was a lunar eclipse between gemini and cancer, when the penumbra of earth's shadow painted the moon wine red. There was then a full lunar eclipse in gemini on January 9, 2001. The last alignment for a while then is the partial lunar eclipse in sagittarius, June 24, 2002, visible only to the eastern hemisphere. In September of 2004 there will be alignments of the moon with Saturn and Venus in Gemini and then with Mars, Mercury, and Jupiter in Leo. Then there is the sunspot cycle, which will be at its peak between 2001 and 2012, culminating in the alignment of the ecliptic with galactic core on Dec. 12, 2012, the date given by the Mayan Tzolkin or Long Count for the end of the world.

10. the future (post 9-11-2001)

Everything is always in synchronization. Some things are simply synchronized to the past (which things we know by the name of "memories") while certain other things are synchronized to the future (what is "prophecy" in idea and "omen" in form). Prophecy, according to current authorities, can only exist under scrutiny. Omens, on the other hand, are common and trivial, but should not be at all.

It is not that priests can say things like this, but that I, an ordinary person, can, that is biblical power. Biblical power is temporal power, power to warp time with the mind. This is most commonly applied via mass hypnosis of congregations, while they meanwhile mindlessly express their archetypal gestures without having to see them for themselves. Most religious people are either ashamed of being proud or proud of being ashamed; and of course this is an alternating cycle, which only amounts to enslavement to sorrow. I do not bow down to worship religion. I believe in God Above.

This is the power of the I AM presence described by St. Germain.

We are biologically slaves of the I AM presence in a similarly archetypal way. It is present in the dilation of the pupil, as the eye lets in more light. It encourages us to move forward. We live in its fractalized hyper-realities. I have met myself there.

a. the prophecy of a judgment and resurrection

We know the naturally occurring division between differing manifestations of the same source as the exchange of matter - energy, in the "forward" temporal flow we call entropy. This is a single, tuned vibrational frequency that permeates the entire universe. A fine tuning of this naturally occurs through the polarization of the wavelength creating microwave gravity and drawing forth concentrations towards the four elemental harmonic chords (which the Egyptians "built in" to the great pyramid as f-sharp). The electromagnetic force generates centers or fields of vibrational frequencies. Our modern words for aura and hour both derive from the ancient word for the hawk, Horus, called the king of the sky and the son of the sun, Ra, the son of Osiris, God-king of night, dweller beyond the Duat, the Egyptian word for Death. It is these fluctuations of frequency in subspace that cause the divisions we call society and culture. Each of these describes the true condition of the other. What we do socially is called culture, and what pertains to the genetic culture of all mankind they call civilization or society. The brotherhood of the Church is based on the fellowship that began in early smoking circles, however it has become corrupt by housing what it hears of sin and fortune only to rent this ambiance out to its parishioners, who then become hungry for more, as though it meant salvation from the rat race of survival rather than a cancerous sore spreading ineffable memetics that compounds itself upon it. Teaching read backwards in time is theft.

When we do not play the roles created for us by the expectations of others, the labor will become displaced onto another body, and this will represent the part of ourselves that we have assigned to be exiled by projection into manifestation. This allows us to play the role of God. Whenever there is more than one person around playing the role of God, this devaluates the role, and causes the entire paradigm to crumble into the dimension of the magic theater, where all are demigod-kings, representative of the universal harmonic vibrational frequencies of potential energy in hyperspace. It is these strands alone that the multiverse is comprised of, since all the matter-energy absorbed by the black hole and filtered through the wormholes is converted through the inversion engine of the singularity into the baby universe, and the wormholes are only filaments of potential energy themselves. In this level people are more or less psychic, their own roles, and thus their own appearance, being distributed out amongst the masses statistically, while they come to resemble the statistically averaged appearance desired for them by the souls displaced from the bodies forced into secondary roles. None of this amounts to much. Go forth and be merry.

What the left hemisphere knows pales before the right hemisphere's capacity for imagination. Without manifestation there would be no inversion (polarity), and without inversion there would be no deterioration (entropy) of manifestation. Where do you go to find tachyons if you do not want to make one by microwave radiation? To a place where there is a large quantity of photonic radiation, such as a sun. In the case of a sun it would not even be necessary to impose an electromagnetic field, for one already exists. The light that photons emit is microwave radiation. Because the frequency of this is narrowed enough it will flow opposite the flow of entropy, that is, move faster than the speed of the photon. This is a tachyon. Expand this to make a wormhole in spacetime. You can go anywhere you want to. Do anything you want to do. Before beginning a journey, take the journey mentally. See the way clear in your mind's eye. Then, all you have to do is take the first step. The rest of the journey is already there, laid out before you. It is easy enough using reincarnation as a rationalization for resurrection to make a sequel to a movie even in which the characters have died. How much more than this can the media simulacrum teach its militant, mind washed, trench-coat mafia soldiers?

Between 1994 and 2000 the hardware and software that the American military had tested in the the Persian Gulf war against Iraq began saturating the free market through privatised companies in the first world nations. Cellular phones, palm pilot computers, laptops and modem cards were all manufactured in the Orient and exported to America for mass market consumption, and the internet boomed with startup "dot com" entrepreneurial venture capitalism. Between 1998 and 2001 foreign investors pulled out of Asian markets and withdrew support for nationalized and privatized production companies, beginning what would come to be called the Asian Contagion. The bankrupting effects this had on Oriental markets swept westward, through the European economies, and finally hit America with the crash of the plateaued tech market, when all the overextended venture capital entrepreneurs failed to repay their startup loans from national and international banks and businesses and most were forced to declare bankruptcy, putting the American market into a free fall. At the same time the United Nations, along with the World Bank and World Trade Organization, have been pushing first world countries toward third world debt relief. The currency of the budding European Union and the United States is essentially paper money that has been floated off the gold standard and the value of which is largely based on the interest of loans these first world nations have made to their impoverished brethren in South America, Africa and the Middle East. The strategy of collecting interest from loans to third world powers in order to put them in debt to the first world is only the nationalization of the exact same tactic used by multinational banks during the last century through war loans to put first world nations in debt to them, so it is not surprising that the World Bank and World Trade Organization, through the beneficent, purely political, United Nations are pushing the developed nations into third world debt relief at the same time as they collect interest payments on war loans directly from the gold standard.

Then, on September 11, 2001, four commercial airliners were hijacked and crashed on American soil. One was crashed into the countryside, one into the pentagon, and two, one each, into the twin towers of the World Trade center in New York. The President of America, who had been appointed by the chief justices of the Supreme Court in a trial to determine the legitimacy of the election results, swore revenge against the Al Qeida network of terrorist training camps, called "cells," and began dropping bombs from planes flying over Afghanistan, where the terrorist network was said to have been centered. Although it is unknown if the supposed leader of the cabal was killed in these attacks, there have been no further reprisals.

There will be another world war soon. It will be triggered by a nuclear explosion between India and Pakistan that will be seen by the security council of the united nations to impinge environmentally on the sovereignty of Russia. The Green voting block of the U.N. (the Muslim nations) will use biochemical weapons against the developed nations' troops who will be sent into the Holy Lands with energy weapons based on scalar wave technology neutron guns similar to the phasers on Star Trek. No more hiding holes of sacred scrolls will be destroyed, though many innocent Muslims will be killed over the international affairs of their figure head leaders such as Saddam Hussein and Osama Bin Ladden, both of whom are only pawns of the CIA. Furthermore many troops from the developed nations will come home sick, instituting a greater socialization of health insurance and social welfare systems to support the sunspot weakened immune systems externally and artificially. Just as the war on drugs was based on the manipulations of the central intelligence agency under Allen Dulles, when massive shipments of drugs were imported to the united states by undercover agents through front organizations and distributed by american gangsters to the ghettos, so is the subsequent war on terror conducted by president George Bush the second based on the manipulations of the central intelligence agency under his father, former president George Bush the first, when terrorist cabals in Afghanistan and guerillas in south america were trained to overthrow communist invasions in favor of military totalitarian or oligarchical rule in accordance with the international interests of American businesses. While Osama Bin Laden is loose he "represents" the literal resurrection of Jesus Christ, and the representative reincarnation of Yeheshua Ben Padiah, author of the Enochian angel scroll (thought to represent a complex calendar) being translated contemporaneously by the minions and underlings of the Grand Architect and Good Shepherd Pope John Paul the second, head of both the Catholic Church and the Holy See of the Vatican Two council. Just as the Dalai Lama has forewarned, if his successor is not born in Tibet then Matrieya, the Last Buddha, might come from anywhere all over the world, so does Hymaneus Beta, the current outer head of the Order of Oriental Templars forewarn that the current Pope is the last of the Antichrist popes before the second coming of the Alpha and Omega pope of the Christian Church, the actual reincarnation of the historical person upon whom the character of Jesus in the New Testament was based.

I. Dimensions

A. subspace quantum information vortices (qliphotic shells)

Our five senses and our mind inhabit our physical environment, and this is three dimensional, with the fourth dimension acting as motivator. This is the world that we know, and our knowledge of this world falls into the phylum of ontology: the study of the nature of being, and the search for its ultimate constitutional unit.

There are two philosophical schools of ontology: the atomists (including all the greeks from the pre-Socratics to Plato) and the existentialists (who study the essence of the human being as the fundamental recombinatory social unit, and believe that man evolved after the idealizable elements).

The atomists included Thales — who believed the fundamental element was water, Heraclitus — who believed the fundamental element was fire, and Democritis — who established the belief that reality was composed of indivisible units called atoms.

Socrates, a character described by Plato, taught of the differentiation between the ideal realm, attainable mentally, and the realm of the physically real, and of the ratio between these alternate realities that should be held as the standard. Plato went on to describe five rhombic solids — the only five regular solids possible composed of homogenous geometrical shapes — that represent the perfect mathematics of the ideal realm in a way that we can grasp and comprehend through the senses.

The next western philosopher to address ontology was DesCartes, who sought to chart the exact border between the realm of the ideal (or the mental) and the real (or the physical). He postulated that it is possible, without doubting the existence of the materially factual reality, to believe that all of this is merely the bad dream in the mind of a sleeping monster. This would spark the beginning of existentialist thought.

Existentialists study the mind and emotions of the human being in order to reach conclusions about the origin of ethics, religion, and society. They derive a great amount from both transcendentalist (that the mind is a blank slate, humans essentially brutal and savage, and society an achieved ideal) and behaviorist (that the only way to approach comprehension of the internal thoughts of people is by examination of their interpersonal reactions to stimuli) schools. Jean Paul Sartre differentiated between being in itself and being for others, and Edmund Husserl addressed the border of difference between them as the epoché, an exercise in which an object is examined until all meaning has been completely detached from it and it is seen in its true or base state as exclusively a condition of pure being.

While the existentialists studied the ethics of Plato's student Aristotle, the atomists went on to become scientists in the age of reason that followed DesCartes, and introduced the study of physics that would give rise to the fields of astrophysics, classical physics, particle physics, wave mathematics, and finally quantum mechanics — an attempt to unify particle and wave physics in light of probability.

Both the classical Greeks and the Jews studied the Egyptians. The Jews record study of something called the 'qliphoth,' or "shells." The latin word 'quantum,' meaning "as much as" or "so much," may have similar origins in the arena of weights and measurements, where it meant "a small amount of an unspecified substance," that could refer to shells. The latter word has come to be used to describe the basic unit of probabilistic information that comprises our reality. It is unknown the full extent to which the ancients took their understanding of the meaning of their equivalent term.

The quantum unit can be imagined as a self-connected spiral well of pure probability, a concentration of potential.

B. Specializations of the 4 Elemental Forces

Quantum information units accumulate in three dimensions to form macro-structures that are concentrations or centralizations of one or more of the four elemental energies (the strong or weak nuclear, gravitational and electromagnetic) over time. We know these as all things from stars and planets to apples and oranges. In these concentrations of fields, combinations can occur forming fine structures of macromolecules, and this is the chemistry of life. This depends on these fields as much as the intellect of consciousness depends anthropically upon it.

The Egyptians identified energy with the Ka, and matter with the Kha, the vessels which contained it. This is similar to the oriental concept of chi energy, which is contained in tao, meaning "the way," however the tao is more temporal. The Mayans, who also measured time as the fluctuation of energy, shared with the Egyptians the concept of the universe as a unified field (which the Egyptians called Nuit, the origin of our word for night) with an opposable fundamental unit (which the Egyptians called Hadit). The Egyptians shared with the Chinese and the Indians the belief that the energy of the unified field, Ka, Chi or Mana, could be channeled through the physical body, through the pressure points and the chakras.

1. electromagnetic properties

The best example of these accumulations of probabilities toward task specialized functions is the electromagnetic for it is the least esoteric and most well known. It is the cause that orients compasses and directs electrical currents. Usually it is invisible, but when we use instruments to measure it, the force can be seen.

Some known types of concentrations of the electromagnetic force are fields, such as the orbital shells of the basic electromagnetic quanta, the electron. There is also an electromagnetic field surrounding the earth, caused by the earth's rotation. A unique type of field is the bottle form, which is a force field whose inner and outer surface are contiguous, that is one and the same. This shape derives its name from the klein bottle, a distorted version of the idealized torus shape.

The other basic structure of the electromagnetic force is the waveform, an elongation of a field until it projects entirely in one or both directions. When it flows in opposite directions at once it is called an alternating current because the electrons traveling along the current will alternate magnetic poles in a regulated rhythm. Some specialized types of wave forms are scalar waves, which are two different waveforms pulsed at an unequal or irregular rate to one another, and null waves, produced when these overlapping waveforms are exactly opposite or an even ratio to one another.

2. gravitational properties

Another example of the fluid dynamics of these forces is the gravitational force, which accumulates as a function of the pressure of mass. Gravity is common to all objects and is responsible for the orbits of the planets around the sun as much as the orbits of electrons around an atomic nucleus. This force avidly displays the conundrum of the elemental energies with atomic theory, since no "true graviton" has yet been discovered that acts as a particle for the exchange of force of gravity. This is because gravity is thought by modern scientists to be an attractive force.

It is not considered that gravity might be a force as all the others, producing a repulsive solid particle that carries the energy exchanged, that simply travels faster than the speed of the fastest known particle (the photon) and therefore opposite the direction of the flow of entropy that particle is used to measure. Such, however, is the case, as has been proven by research with microwave electromagnetic radiation which, when propelled through a solid medium, arrives before itself, moves backwards, and generates gravitational attraction. The gravity particle is therefore the tachyon, a faster than light, and therefore unmeasurable, theoretical particle, whose shape is the torus, or hypersphere.

When a gravity field is concentrated upon a smaller and smaller space it generates a black hole. The black hole gets its name from the fact that it absorbs all light around it, causing it to be consumed within itself, where it disappears. Because black holes posses spin they are spheres, in fact, and not holes, and for the same reason a small portion of light does escape them, and so they are not altogether black either. Three dimensional black holes exist for a temporary duration of time, however black holes can also be exclusively fourth dimensional. Whereas a third dimensional black hole has a first dimensional singularity, a second dimensional Schwarzschild radius and accretion disk, and third dimensional spin over time, an exclusively temporal singularity can also occur, and these are known as breakthroughs or quantum leaps, referring to the effect of quantum tunneling. Usually these occur in pairs or aggregates, and this effect is known as synchronicity or coincidence, and is created by the increase of probability due to random fluctuations caused by the uncertainty principle of the same event (the quantum tunneling of information through a certain holognomonic ennegram in the mind usually being either one or both) occurring in two or more areas of space at around the same time.

3. wormholes

When the electromagnetic force is combined with gravity the result is a wormhole. Nuclear fusion, of electrons rather than nuclei, creates microwave tachyons, and nuclear fission, again from electrons, opens these up into wormholes. Nuclear fusion can be accomplished by strong electromagnetic irradiation, alike the smelting of metals with heat. Nuclear fission can be accomplished by the projection of a distortion to a gravitational field, alike the dehydration in a pressure cooker. Thus wormholes are a unification of all the forces.

Wormholes, as they are projected at this time, are also exclusively temporal. They connect two points in the continuum such that to get from point A to point B in spacetime takes x amount of time, to get from point A to point B through the wormhole takes y amount of time, where y is less than x, and asymptotically approaches zero.

To understand the mechanics of a wormhole it is best to understand the dynamics of quantum tunneling, since they are only a magnification of this effect. This is thought to be the doubling of quantum information units through the random fluctuations of quantum foam due to the uncertainty principle. The best current example is the double slit experiment, where photons are passed through two cuts or breaks in a surface, and their overlapping field measured on another surface behind that. The result of this experiment is that the light acted holographically: since it was just as probable for the photon to pass through one of the slits as the other, the sum of all the photons projected sorted statistically out to be not only either half passing through either, but all passing through each, such that much more light ended up reaching the other side of the barrier than was expected. Each photonic unit seemed to be acting as though a hologram or exact reflected duplicate of the whole, and therefore was passing not only through one or the other of the slits, but through both at the same time.

In a wormhole the conditions are essentially the same, only the two slits are the opposite ends of the wormhole, the barrier the distance of the space between them, and time the holographic fractal passing through them from the same source at either end, that being the continuum.

C. hyperspace hyperdimension = hypereality multiverse

The timespace inside a wormhole is outside of the timespace outside of the wormhole wherein is the wormhole. That is to say that inside a wormhole is a universe supernal to our own. This is the realm of hyperspace or hyperdimension.

Hyperspace refers to the acceleration of real particles to a velocity faster than entropy, and hyperdimension refers to the prismatic holographic refraction of energy waves in all directions simultaneously concurrent to that acceleration. In the same way that matter and energy both comprise the continuum of the known universe so hyperspace and hyperdimension comprise the hypereal multiverse.

The spacetime inside a wormhole that moves faster than light escapes the limitation that forms the boundaries of our existing universe, and therefore it is properly considered hyper-real, or of a reality that has been raised beyond average reality. In the same way that a wormhole connects two probabilities in the continuum, so does it access infinite potential points in infinite potential universes. It is unlikely that there are actually an infinite number of universes, however their accessible quantity of potential does increase the probability of multiple universes. These other universes are alternate realities in parallel dimensions to our own, and the sum over histories of all of these and that of our own universe is the multiverse.

1. Wormholes and Quantum Tunneling as Information Transport

The basic unit of the temporal field is the probability well of concentrated potential energy. When one of these passes through another, the two occupy the same place in zero time. That is, the time it takes them to pass one another does not occur within the confines of the measured continuum. In the same way that one exits a wormhole at the same time as one enters it, just as we walk through a doorway, when a tachyon vibrates through an electron it literally goes inside the electron, which occupies simultaneously every position on its orbital shell, and comes out at the same time on the opposite side of the orbital shell.

One of the earliest forms of long distance communication was the smoke signal, where someone would create different kinds of clouds of smoke by holding and removing a blanket from over a fire. In this same way, pulses of energy in an encoded format can transmit a translatable pattern of information. Since the temporal distortion created by quantum tunneling or a wormhole is only external and not internal, it is possible to relay such a signal through either method without degradation of the quality of the message.

Whatever information was broadcast through a wormhole, however, would be refracted holographically inside of it, and therefore generate in the idealized realm outside of the timespace continuum eternally recurring spin in accordance with the characteristics of the transmission. It would therefore be possible to access any such signals that had ever been broadcast through any wormhole. Since it is possible, also, to follow the fractalization of harmonically vibrating multiple dimensions in hyperspace towards the various different possible combinations of the four elements that are the multiple universes, it is also likely that it would be possible to tune these in accord with the frequencies of real, subspace waveforms, and in this way to access all manner of information even if it had not been passed through a wormhole.

2. Why there is not always a party going on in the universe next door

Because it exists outside of finite spacetime, the multiverse asymptotically approaches infinite potential, however because it can only restructure the quantum information that it comes in contact with from the local universe, the multiverse is ultimately confined to reflect our reality. As local spacetime is bent and distorted until it becomes a parallel dimension, it compounds upon the original information until a plenum exists. This continues on in an unending interrelated network, each dimension feeding into the next along harmonic distortions to the fundamental geometries until there is a complete round cycle and all the dimensions connect. Even though each of these dimensions, and therefore the universe they describe, is equally infinite in potential, they can only support the probabilities they consume, and this they accomplish through the distortion to the underlying geometry.

D. Pure Dimension/Pure Geometry

"This sentence is false." An impossible loop in a logical hierarchy such as this constitutes the doorway out of hyperdimension into pure dimension. It is expressed as the final phase of distortion to the original, underlying geometrical, physics of subspace, extruding all expressions from the equation except the most elegant forms.

In subspace it is embodied in the gravitational singularity, which is formed when a temporal singularity implodes. This occurs inside the light absorbing event horizon of a black hole, where the immense gravitational pressure causes quantum foam to quantum tunnel in very short wormholes, or temporal singularities. When one of these reaches its conclusion the zero time information it contained phases into the central gravitational singularity. For an astronaut caught in a black hole this would be the equivalent of losing consciousness or falling asleep.

In this model each miniature wormhole has its own signature geometrical trajectory. This constitutes the dimensional information that comprises hyperspace. The multiverse of infinite potential universes is in this way accessed, and additional probability of one possible universe is brought into the gravitational singularity.

The gravitational singularity itself is the ideal ratio phi/pi in manifest form. This represents the most compact expression of geometry capable of expressing the most, since each of the two numbers given are transcendentals, transfinite decimals. When information phases through this system, that is, dies inside the wormholes and then passes into the singularity, it passes outside of the material universe and enters the realm of the ideal.

The information itself inflates a baby bubble universe that expands on the inside of the universe. As much as the wormholes were phi around the pi singularity, so are such baby universes phi around the edge of the pi universe.

Perpendicular to the history of this offspring universe is the tachyon history of pure dimension and pure geometry. As universal expansion inflates the horizon of the parent universe, it emits tachyons outward into the macroverse whose measure is faster than the speed of light. Each tachyon is a phi/pi hologram, and all of them together are a gnomonic fractal. Thus they are the energy and the particles of pure dimension. Their movement, moreover, is phi away from the pi universe, and therefore their form of time is also a purely idealized geometry. However because the particles and the movement of time are one and the same, all is unity, and there is division between nothing.

1. Where one overlaps the other invisibly

Although phi over pi is meant as a geometrical expression for dimension, it is possible also to replace the operators in the same ratio with geometry and dimension themselves. Here we see that, while phi over pi is a number that underlies everything in the material universe, pure geometry and pure dimension underlie it.

Since tachyons move faster than light we cannot see them. They arrive before the lightwaves emitted from the same event. They are emitted by photons. So, in the same way that tachyons are invisible to the eye, pure dimension and pure geometry are invisible ideals to the equation of phi over pi.

To enter into the gravitational singularity would be physically like what standing inside a temporal singularity would be like to the perception alone. There would be zero probability measuring zero time, and therefore undiminished infinite potential. This means potential spin in a gravitational singularity as much as potential energy in a wormhole. Thus the asymptotically infinite multiverse perceived in a wormhole is slowly being brought forth one possible universe at a time by gravitational singularities throughout our universe. However inside our universe we can no more see these baby universes than we can perceive the multiverse while outside of a wormhole. This goes so far as to mean that whenever we imagine the multiverse we are creating quantum tunneling inside our own brains.

In these ways we can say that pure dimension is invisible to us because it is blocked from our sight, however we must also remember that it is invisible to us because it is a transparent energy — tachyonic astral light — that surrounds us all the time. In fact, it is the substance of time, because its materiality is pure potential, without probability wells to interrupt its asymptotically unending expanse. Temporally, our aura itself is an event well inside this field.

2. The nth Dimension and Infinite Geometries

Just as the sum over histories of tachyons emitted from the universe measures the sum over histories of baby universes, produced from the infinite potential of hyperspace accessed around the edge of a black hole and comprised of a single history of consumed temporal singularities, inside tunnels in the quantum information foam surface of the universe, so may we say that the measure of the universe's external luminescence and the quantity of possible generations contained within it are equivalent. This is merely an idealized way of saying, again, that matter (in this case the internal multiverse) and energy (in this case the tachyonic astral light) are interchangeable within a shared continuum. Our realm of third spatial fourth temporal dimensional, manifest, material reality acts as the balancing point between these two depths. It is because of the increase of information from within the quantum information unit itself that our universe appears to be expanding. Both the three dimensional homogeneity of the expansion and the temporal template are illusions, because the true pulse of the universe flows along channels of current exchanged between galaxies in filaments and walls, vibrating harmonies in voids.

The tachyonic astral light outside our universe, the baby universes inside of gravitational singularities inside of black holes in the thin film of the manifest, the expanding quantum information units, and the resonant harmonic vibration underlying deep space (known as subspace frequencies) are all one and the same. They are all expressions of what is called mathematically the nth dimension. This is a variable meant to represent the last in a transfinite series, and to imply that it exists in a way that contains all the subordinate sums.

The measure of the nth dimension is infinite geometry. The ones that I have listed above are only a small start. Include ones based on temporally synchronistic harmonic resonance manifestation systems and you will be one more step closer to enlightenment.

3. Ylem and the Primary Clear Light

Perception of this is the state of enlightenment the Buddhists call Nirvana. They follow a ritualistic practise of meditative trances on an increasing level of awareness of the ideal and sublime. In this system the trance of Samadhi allows entrance into the realm of Nirvana, wherein one can freely perceive the primary clear light of ylem. According to their belief this light, the light of pure potential, precedes even the big bang and the beginning of our own universe.

The next closest concept to the perception of this is the perception of the self in its presence or tangential proximity. This deifies the Light and gives one the illusion of empowerment. The oldest form of this is the Jewish Nefesh, meaning spirit. This is the first anthropomorphication of the spiritual principle, and brought the soul, which the Jews called the ruach — meaning breath, into contact with the world of discorporeal consciousnesses they had previously believed to be inhabited only by good or wicked djinn, or spirits of the wind, that they believed were alive as much as their own memories of their ancestors, and that they, too, were separated from the mundane world by a veil similar in principle to death itself.

Other systems can account for the universe as an automaton, and believe that science is god, and the only right way or path to life technocratic deism. These are no less based on petro, meaning "stone," or violent blood rites than every early system of religion, yet they claim to be atheists. Accordingly science shuns only liars. Yet still, should the demigod Pan come to bear, the result will be Pan-deism, the opening of Pandora's Box.

II. entropy

Since the first Greek elementalists, people have pondered the composition of the universe, but evidence that they sought for a unifying force underlying all nature and everything known by the mind is present in the even earlier pursuits of religion. Despite all of the efforts of religion, however, whether anthropomorphic, naturalistic, objective, all fail to produce data that is anything more than an induced personal experience, and comes down to a basic choice on the part of the individual to further go along with the group or not despite its lack of conclusive findings. On the other hand, science can find no common ether or fixed constant in the whole of the observable universe that they cannot explain away as entropy. As the old Greek saying put it, "When Zeus is toppled, chaos rules, and whirlwind reigns."

The concept of entropy has been described by mathematicians using fractal or exponential pattern generators on computers, in which it is observed that, where the patterns approach infinite finitude, growing smaller and smaller, the computer can no longer represent them, and only large empty zones appear. According to chaos theory such empty spaces are equivalent to random genetic mutations due to replication errors that do not get subsequently corrected — which nuclear biologists increasingly believe to be the trigger of evolutionary leaps, and the mathematicians say that the same "fuzzy logic" creates weather patterns. Chaos theory — which states that cause and effect may be separated by myriad complex systems, fuzzy logic — the concept that natural systems have vast, variable fudge factors for information translation, and the dead zones in fractals do not begin to describe the reality of entropy, though. They are only studies of data methodology and artificial systems. Entropy is the cause of death and decay for all biological tissue, and therefore interfaces with living patterns and organic forms as well.

If entropy could have a personality it would appear little different than the spectre of death. It isn't the word most people would want used to describe their deity, but it fits all the job requirements, being, also, how we measure time.

Entropy is described as the gradual breaking down of ordered systems in increasing states of disorder and chaos. It is this force that causes all change in the universe, and against which biological life struggles to survive. According to the theory of entropy, the big bang was only the entropy of the first quantum fluctuation, and still expands the universe. Think of this as like the first test of the atom bomb, which keeps exploding by causing radioactive particle decay to this day.

It is quite right to call entropy the background radiation of the continuum, because it unifies the four elemental forces and is the cause of their functionality. Entropy is even thought to have preexisted the division between the four forces, at a time when all there was in the universe was radiated heat. Thus it is thought that heat is the basic unit of entropy. Therefore the process of the matter-energy exchange defined as the function of entropy acting upon and working through the four elemental forces is measured according to thermodynamic radiation.

The loss of energy as heat is considered the constant by which the difference between the moments may be measured in all things. Anything that gives off heat or is subject to the presence of heat in its external environment is being restructured by it asymptotically distrophically. This is the liberation of energy from the order of form, and it is a process that will continue as long as there is energy to be liberated from material form.

Astrophysicists predict that the universe will age according to one of three ways. If the universe has an open geometry, it will continue to expand indefinitely. If the universe has a flat geometry, it will continue to expand until it reaches a certain constant point of equilibrium between matter and energy. If the universe has a closed geometry, it will expand until it reaches a critical density, and then it will collapse.

It has been observed that all intergalactic bodies in the visible universe display a red shift to their gaseous emission spectrums, or the measure of light being emitted by stars. This red shift occurs due to the doppler effect expanding the wavelengths of light as the galaxies move away from us, lengthening the amount of time it takes the photons to get here for us to observe them. In other words, we know that the universe is expanding now, or rather, was expanding at the time when the furthest, and therefore oldest, galaxies cast the light that we can observe today. Thus it is the speed of light that has come to be considered the fixed constant, or fudge factor, of entropy, being considered the pivotal factor in the function of matter- energy exchange represented by special relativity. Therefore, if the speed of light is more or less constant, then the most recent observed data indicates that the universe is not only expanding, but expanding at a rate much faster than previously recorded.

This seems to imply that a variable must be wrong, unless the universe's rate of expansion is actually speeding up. Theoreticians are debating these days over whether we are moving between matter-energy level shells in a universal quantum bubble or whether Planck's constant, a truly naturally occurring average electrical velocity, might be questionable. It is more likely that scientists are simply using different values for variables now than they did before, and therefore that the universe is expanding at a constant rate, however that rate is variable within a finite range of differing degree.

Based on these findings let us consider the different possible geometries underlying entropy.

If the universe is speeding up as it expands, then it is likely that we are living in an open geometry universe. This is not necessarily as good a deal as it sounds. In an open geometry universe the fabric of spacetime would continue to expand forever outward, carrying all the material bodies within it along the way, and this expansion would not be one directional, but in all directions at once, thus moving all material bodies further and further away from each other. This would go on forever, and eventually, though long before the universe ever ended expanding, there would be a virtually infinite distance between everything and everything else. First this would happen to the galaxies. Then it would happen to the stars. Then it would happen to the planets and asteroids. Then it would happen to living cells, their nuclei and DNA. Then it would happen to the spaces between atoms and between quanta. Entropy would never end, and the fabric of spacetime would be stretched out and bent parabolically, tearing it to shreds. However if this were the case then all the galaxies wouldn't just be red shifted relative to us, but to each other as well, as the homogenous expansion of the universe pulled them apart; instead what is observed is that some of the galaxies are moving toward one another, forming filaments, walls and intergalactic voids in the universe.

Another theory is that the geometry of spacetime is flat. This may sound amusing as one can look about themselves in all directions and know that they are not living in a two-dimensional universe. Add to this the knee-jerk reaction that people used to think that the world was flat before Columbus discovered America and you immediately find yourself in flatland. However the theory does not address the dimensionality of the universe, but the function of spacetime. Einstein discovered that gravity does act as a well around very massive objects, capable of bending and distorting light rays. These wells comprise curvature of spacetime, however other than this the fabric of the universe does not display any dimensionality at all, and may easily be considered as a plane. This is not a plane such as a flat two-dimensional field, but a plane of reality, such as the mindset of consciousness that has evolved in a universe governed by entropy, and cannot comprehend a universe where space ever stopped or time ever came to an end.

The last theory is the one most widely accepted by astrophysicists today. According to this theory, the geometry of spacetime does loop back around on itself in an enormous cycle, and there is sufficient mass in the universe to create a greater gravitational collapse than the expansion of radiation in the form of the fixed rate of photon propagation and the universe will eventually eat itself through entropic gravity wells such as supermassive black holes. Ironically, the faster the universe expands, the more mass it has pushing outward, and the more mass it has pushing outward, the more gravity it will have pulling inward, and the more likely it will be to contract. Although there is evidence that black holes have been gradually consuming the gasses and radiation of the universe for at least as long as there have been galaxies of stars gathered around them, this geometry has spawned another theory of entropy: the big crunch. This event would be the big bang in reverse, and it is a popular concept among modern physicists due to its similarity to Eastern karma wheels and their cycling cosmologies.

A. thermodynamics

While much else in physics is complex to comprehend, thermodynamics is as easy to understand as feeling temperature with the skin. Therefore it was probably in reaction to this that that first form of sensation evolved in the original life form.

Thermodynamics is hot and cold. Hot and cold are simple concepts: energy moves from cold to hot as it moves from slow to fast, and it moves only between these two, and only in this direction. When something is cold, its particles are moving very slowly. As radiation is introduced, the particles begin to move about. Some become free particles and their energy leaves the system. In this way all energy breaks down from matter, and in this way all energy asymptotically approaches the speed of the fastest real particle, the photon moving at the speed of light.

The best way to describe this process is by looking at the three forms of water. Water can be frozen solid into ice, and here we see that its molecules are almost entirely stationary, except at the surface where they are exposed to radiation and begin to melt and give off steam. In its natural state, that is, at an averaged temperature, water exists in liquid form, and its particles obey fluid mechanics. This is more or less randomized, however it is always contained within the minimal amount of space, given room for the random reactions. When water is evaporated by exposure to radiation it becomes gaseous steam and airy mist. In vapor form the water becomes lighter than air microdroplet macromolecules, and this is fundamentally what clouds are made of.

1. convection currents

Heat is transported in waves called convection currents. Similar to the bending of light by gravity wells, convection currents can be observed as a distortion to the radiation of photons. They can be seen in the distance rising off the surface of paved road in the sunlight, where they appear as a clear liquid fire that melts the view. They can also be observed in infrared spectroscopy, emanating from any living source like a pulsating aura that surrounds us with an invisible, undulating flame.

Convection currents always flow in the same direction, from the hot to the cold, and these are centers or areas where the pressure is different. Heat flows from high pressure centers, cold flows to low pressure centers. In the same way that degree of irradiation determines density cohesion, so are centers of high and low pressure the highs and lows followed by temperature.

2. isobars

Isobars are the features of the terrain defined by pressure. Like convection currents they run along and around pressure centers, but define a dimension perpendicular to the convection currents. Even though hot and cold and high and low pressure are measures of changes induced by energy liberation, such as radiation, their orientation to one another is itself one of a polarity similar to electromagnetism.

The orientation of pressure bars and convection currents relative to one another is more than just caused by electromagnetic radiation creating high and low pressure centers. It is caused directly by electromagnetic field lines surrounding these mobile concentrations of energy. While the high energy, hot, high pressure centers and low energy, cold, low pressure centers move around, for example, in the atmosphere, as the particles between them are moved about by the stirring of radiation, there is increased potential for electromagnetic friction and, for example, in the atmosphere, storms arise.

B. fluid mechanics

Fluid mechanics govern the propagation of force for the convection currents of temperature over the isobars of pressure as waveforms in two dimensional plane fields. The variable representing the movement of force in a fluid continuum is given as gamma, and it is represented such that gamma is a function of solid reflection interference patterns and potential induction propagation patterns. It operates similarly to a gaseous state, only denser and slower, however in both we see complex natural patterns arising from simple external disturbances.

These natural patterns take the form of both fractals and gnomons. The fractals represent the type of nonliving patterns found as the transverse wave fronts of the small waves arcing up onto the shore, or the artificially created smoke ring. The gnomons are living patterns such as the whirlpools of the water or the tornadoes of the air.

1. fractal propagation

Entropy in the form of heat produced by matter-energy exchange moves through things changing them over time according to patterns that are both "living" and "dead." The "dead" sort of patterns are simply higher dimensional forms of harmonic resonance such as the three dimensional Platonic solids, which are the only five regular solid polyhedra that can exist in three dimensions. Such forms are fourth dimensional metaforms like the hypercube and hypersphere, however when they are viewed as fourth temporal rather than fourth spatial dimensional models they display a completely different type of pattern of propagation.

Modern fractals include the Mandelbrot set, named after its program creator, and which resembles a bug or a meditating Buddha, and the Julia set, which is an expansion of a fibonacci spiral that self-replicates on a finer and finer scale until finally disappearing into large blank gaps that surpass the computer's display capabilities. These can also also be mapped three dimensionally, and excellently replicate the natural terrain patterns of landscapes. By turning the fractals around, or giving them spin, quaternions can be extracted from the dark spaces that look like gummy shells of cotton candy.

2. gnomonic propagation

Entropy also propagates in "living" patterns, which is lucky for us anthropically. These types of patterns are like pi, which self-replicates in either a self-terminating, inward cycle, or a self-expanding, outward cycle, and, under the conditions of polarity, alternates these in a double helix, and like phi, which is pi seen at an angle and is extractable from the diagonal of a cube. Phi is also present as the involution of tachyons, and these generate gravity, which propagates entropy.

3. the illusion of chaos

Some systems of entropic propagation have yet to be discovered. It is possible that, there, hiding in the static, is the hologram of a picture. All that is required to be able to see it is the ability to view it simultaneously from multiple points of view. Such optical illusions are common nowadays and known as stereoscopic art. These apparent fields of white noise, or static pixels, are actually arranged in a very specific order so that, when viewed through two eyes the distance apart of the human eyes, the individual pixels will become blurred together and a holographic projection will leap out into sculptured three dimensional vision at the viewer. So far what we know to be true of entropy, that it propagates as gamma energy flow according to highly advanced and relatively simple geometric forms, it is entirely possible to conceive of it functioning according to the geometry of some even higher dimension as well.

C. on the tables of the eight and the ninth

table of eights:

01 * 8 = 08	(0 + 8 = 08)	
02 * 8 = 16	(1 + 6 = 07)	
03 * 8 = 24	(2 + 4 = 06)	
04 * 8 = 32	(3 + 2 = 05)	
05 * 8 = 40	(4 + 0 = 04)	
06 * 8 = 48	(4 + 8 = 12)	(1 + 2 = 3)
07 * 8 = 56	(5 + 6 = 11)	(1 + 1 = 2)
08 * 8 = 64	(6 + 4 = 10)	(1 + 0 = 1)
09 * 8 = 72	(7 + 2 = 09)	
10 * 8 = 80	(8 + 0 = 08)	
11 * 8 = 88	(8 + 8 = 16)	(1 + 6 = 7)
12 * 8 = 96	(9 + 6 = 15)	(1 + 5 = 6)
13 * 8 = 104	(1 + 0 + 4 = 05)	
14 * 8 = 112	(1 + 1 + 2 = 04)	
15 * 8 = 120	(1 + 2 + 0 = 03)	
16 * 8 = 128	(1 + 2 + 8 = 11)	(1 + 1 = 2)
17 * 8 = 136	(1 + 3 + 6 = 10)	(1 + 0 = 1)
18 * 8 = 144	(1 + 4 + 4 = 09)	
19 * 8 = 152	(1 + 5 + 2 = 08)	
20 * 8 = 160	(1 + 6 + 0 = 07)	
21 * 8 = 168	(1 + 6 + 8 = 15)	(1 + 5 = 6)
22 * 8 = 176	(1 + 7 + 6 = 14)	(1 + 4 = 5)
23 * 8 = 184	(1 + 8 + 4 = 13)	(1 + 3 = 4)
24 * 8 = 192	(1 + 9 + 2 = 12)	(1 + 2 = 3)
25 * 8 = 200	(2 + 0 + 0 = 02)	
26 * 8 = 208	(2 + 0 + 8 = 10)	(1 + 0 = 1)
27 * 8 = 216	(2 + 1 + 6 = 09)	
28 * 8 = 224	(2 + 2 + 4 = 08)	
29 * 8 = 232	(2 + 3 + 2 = 07)	
30 * 8 = 240	(2 + 4 + 0 = 06)	
31 * 8 = 248	(2 + 4 + 8 = 14)	(1 + 4 = 5)
32 * 8 = 256	(2 + 5 + 6 = 13)	(1 + 3 = 4)
33 * 8 = 264	(2 + 6 + 4 = 12)	(1 + 2 = 3)
34 * 8 = 272	(2 + 7 + 2 = 11)	(1 + 1 = 2)
35 * 8 = 280	(2 + 8 + 0 = 10)	(1 + 0 = 1)
36 * 8 = 288	(2 + 8 + 8 = 18)	(1 + 8 = 9)
37 * 8 = 296	(2 + 9 + 6 = 17)	(1 + 7 = 8)
38 * 8 = 304	(3 + 0 + 4 = 07)	
39 * 8 = 312	(3 + 1 + 2 = 06)	
40 * 8 = 320	(3 + 2 + 0 = 05)	
41 * 8 = 328	(3 + 2 + 8 = 13)	(1 + 3 = 4)
42 * 8 = 336	(3 + 3 + 6 = 12)	(1 + 2 = 3)
43 * 8 = 344	(3 + 4 + 4 = 11)	(1 + 1 = 2)
44 * 8 = 352	(3 + 5 + 2 = 10)	(1 + 0 = 1)
45 * 8 = 360	(3 + 6 + 0 = 09)	

table of nines:

01 * 9 = 09	(0 + 9 = 09)	
02 * 9 = 18	(1 + 8 = 09)	
03 * 9 = 27	(2 + 7 = 09)	
04 * 9 = 36	(3 + 6 = 09)	
05 * 9 = 45	(4 + 5 = 09)	
06 * 9 = 54	(5 + 4 = 09)	
07 * 9 = 63	(6 + 3 = 09)	
08 * 9 = 72	(7 + 2 = 09)	
09 * 9 = 81	(8 + 1 = 09)	
10 * 9 = 90	(9 + 0 = 09)	
11 * 9 = 99	(9 + 9 = 18)	(1 + 8 = 9)
12 * 9 = 108	(1 + 0 + 8 = 09)	
13 * 9 = 117	(1 + 1 + 7 = 09)	
14 * 9 = 126	(1 + 2 + 6 = 09)	
15 * 9 = 135	(1 + 3 + 5 = 09)	
16 * 9 = 144	(1 + 4 + 4 = 09)	
17 * 9 = 153	(1 + 5 + 3 = 09)	
18 * 9 = 162	(1 + 6 + 2 = 09)	
19 * 9 = 171	(1 + 7 + 1 = 09)	
20 * 9 = 180	(1 + 8 + 0 = 09)	
22 * 9 = 198	(1 + 9 + 8 = 18)	(1 + 8 = 9)
27 * 9 = 243	(2 + 4 + 3 = 09)	
33 * 9 = 297	(2 + 9 + 7 = 18)	(1 + 8 = 9)
36 * 9 = 324	(3 + 2 + 4 = 09)	
40 * 9 = 360	(3 + 6 + 0 = 09)	
44 * 9 = 396	(3 + 9 + 6 = 18)	(1 + 8 = 9)
45 * 9 = 405	(4 + 0 + 5 = 09)	
54 * 9 = 486	(4 + 8 + 6 = 18)	(1 + 8 = 9)
63 * 9 = 567	(5 + 6 + 7 = 18)	(1 + 8 = 9)
72 * 9 = 648	(6 + 4 + 8 = 18)	(1 + 8 = 9)
81 * 9 = 729	(7 + 2 + 9 = 18)	(1 + 8 = 9)
90 * 9 = 810	(8 + 1 + 0 = 09)	
99 * 9 = 891	(8 + 9 + 1 = 18)	(1 + 8 = 9)
108 * 9 = 972	(9 + 7 + 2 = 18)	(1 + 8 = 9)
117 * 9 = 1053	(1 + 0 + 5 + 3 = 9) or	
	(10 + 5 + 3 = 18)	
126 * 9 = 1134	(1 + 1 + 3 + 4 = 9) or	
	(11 + 3 + 4 = 18)	
135 * 9 = 1215	(1 + 2 + 1 + 5 = 9) or	
	(12 + 1 + 5 = 18)	

and thus (18) through 180

198 * 9 = 1782	(1 + 7 + 8 + 2 = 18) or	
	(17 + 8 + 2 = 27)	
243 * 9 = 2187	(2 + 1 + 8 + 7 = 18) or	
	(21 + 8 + 7 = 36)	

and thus (36) through 360

396 * 9 = 3564	(3+5+6+4 = 18) or

$$(35 + 6 + 4 = 45)$$
it can be shown that this pattern of increasing
multiples repeats itself, as demonstrable by
taking the first 2 digits as 1 integer in all sums

The tables of the eighth and the ninth tell us many things about the physical construction of our universe on a purely mathematical level, the level which serves as the bridge between the external formation of the material world and our relatively internal domain of consciousness. My arrangement of these tables therefore allows for the application of numerology, or the realization of relationships between the digits *en-soi*, or in themselves. This existentialist approach to computation is as ancient as the human practice of collecting objects in countable sets, and constitutes an esoteric equivalency to the exoteric fact of pure traditional mathematics. Regardless of the excuse of its origins, this practice's credentials are frowned upon in the light of pure mathematics as an escape from the understanding provided by methodological calculation. They are, quite to the contrary, no more than a reapplication of methodological computation on an entirely other level, self-contained and non-threatening to the approach of exoteric mathematica. Thus, while their examination may be seen as untraditional, it is at least not unacceptable.

One of the fundamental insights of the tables is provided by comparison between the two. Their primary similarity is the repetition of pattern; their primary difference being the nature of these patterns. As the eighth demonstrates quantifiable decline in sequence, so the ninth yields first exact self-replication, and then increasing self-referential sequentialism. These patterns are apparent and undeniable. If the calculations are repeated in any other setting the conclusions will be exactly the same, and therefore the patterns displayed in their relationships will be identical. Numbers do not lie. They lack that motivation for symmetry.

It is possible to see these sequences of both as related to one another dimensionally. The perpetual numerological decline of the eighth and the perpetual numerological duplication and factorial increase of the ninth may be seen as ascent through the first three dimensions respectively.

The diminishing table of eights constitutes collapse into a singular point in space. Naturally one can question how there can be any differentiation at all in the measurement of a point, but the answer is a simple one indeed, for without it there would be no scale-correspondent maps.

The repetitive aspect in the nines resembles the extension of a line in space. At all points along the line two of its dimensions are canceled, leaving only the third behind to mark its position. This is quite obvious in the graphing of a straight vertical or horizontal line in a two dimensional Cartesian coordinate system, where either the x or the y coordinate pairing remains undefined. It may be less obvious in a diagonal, where every point on the line has a defined x and y coordinate pair that differs from every other coordinate pairing of a point along the line. However the distance formula shows that any two points on a graphed straight diagonal will cancel one another out leaving only one integer behind, that being the value of the line itself. This may seem trivial now, but it is essential for understanding the next comparison of the ninth table to dimensionality. One need only to consider that a line in a two dimensional coordinate system, straight or diagonal, is equivalent to a plane in a three dimensional coordinate system to begin to apprehend why.

Although it is a confirmable fact that the integer sums of the multiplicative quantities produce a doubling of results along factorially related lines throughout the entire ninth table, this only becomes really evident after the process has entered its third repetition, when the quantities involved are of such an amount that their sums render divergent factors. What this process is in fact describing is the event of entrance into the third dimension from the second. At first, as the plane is defined as two lines of value nine, the factorial sums elevate rapidly through the established sequence. As the shape becomes clearer while the coordinate system is rotated the number of quantities between the factorial transitions becomes greater, allowing, as it were, more time to pass between phase shifts. The key to understanding the shape that is described is contained within the non-numerological pattern of table nine.

The factorials diverge first after product 180. That is, their sums if taken by pure integer alone or by combined integer render different results, although both results that occur elsewhere as quantities within the initial multiplicative table. At quantity 108 the products break from the ascending sequence and linger at sum 18, or, if they are taken as a pair of paired integers, begin to decrease along the same factorial lines, beginning at 63, but skipping every other factor, such that the next result yielded is 45, rather than 54, etc. This is the case until 180 appears again as the multiplicative function. The result for 198, that immediately following 180, is peculiar, as it constitutes a different rate of change than has been previously established. Although it describes the results for all the quantities of multiplication between 181 and 240 (interrupted only by 200 * 9 = 1800), this still describes a much shorter set of numbers than are contained within the 36 factorial grouping from 241 to 390 which follows, 58 as opposed to 148 respectively. This anomaly also helps to point out that the phase shifts in sums don't occur cleanly at factorially defined breaks, but are governed only by the dictates of the digits themselves. Although this opens up the realization of still another, more subtle pattern — the difference between the clean factorial 243 and the true break of 240 for the 27 factorial set being 3, and the difference between the clean factorial 396 and the true break of 390 for the 36 factorial set being 6 — it is unnecessary at this point to go into it in detail.

It is more instructional, from a dimensional emergence perspective, to examine the factorial breaks that are clean, and those are 180 and 360. These numbers are most immediately recognizable as the definitions of the circumference in degrees of a half circle (the angle measure of a straight line) and a full circle (or, if you like, the angle measure of a straight line that reverses its own direction). But they have more, deeper connotations than this.

The most fundamental polygons are the triangle, the square, and the pentagon, composed of 3, 4, and 5 angles respectively. The hexagon, with 6 interior angles, is somewhat more complex, and can actually be tallied to be the sum of 2 triangles. The sum of the angles of a triangle is 180. It is always, exactly, and only 180. The sum of the angles of a square is always 360 (90 + 90 + 90 + 90). This is true for any 4 sided rhombus, according to the formula that $(n - 2)180$ = sum of the interior angles for any object with *n* number of sides. Applying this same formula we can conclude that the sum of the angles of a pentagon is 540, or 3 * 180, 1 & 1/2 times around the circumference of the unit circle. These numbers are as old as the act of measurement itself, and they are absolute. So we can see how the clean fractional breaks in the ninth table pertain not only to angles that describe arcs, but to those describing well defined shapes, particularly the fundamental polygons, as well.

Moreover these fundamental polygons comprise the sides of the only five regular solid polygons that can exist in three dimensions. All other solid polygons, like the hexagon in two dimensions, are only combinations of these first five. The Platonic solids consist of the tetrahedron, comprised of triangles, the cube, composed of squares, the octahedron, comprised also of triangles, the icosahedron, also composed of triangles, and the dodecahedron, comprised of pentagons. So we have not only the description of angle measures on the unit circle, but also the sums of the angles of the faces on each of the five platonic solids, and all in a pure numerical form. It is easy to follow the progression of factorial breaks up through their ascending sequence and see how the sums constitute measures within a deepening three dimensional space, describing the unfolding of the platonic solids according to the sums of the interior angle measures of their faces multiplied by their facial sum.

It would be more probable, at this point, for a practitioner of calculatory mathematica to caution that the bad habits of a writer may be transferred to their readers, than it would be that they could provide hard evidence that this data is more conjecture than implicit organization. In either event it cannot be doubted that the dimensional bridge between function and form is crossed in the table of nines.

The bridge described herein is not altogether complicated, but is increasingly complex as more governing rules are discovered to determine each additional dimension. For example, the point may be seen as of any size under magnification; the line as definitive of angle, and implying the edge of a perfectly flat plane. Three dimensional objects are formed simultaneously of matter and energy — comprised not only of charged particles, but of waves with measurable frequency related to this charge. This relativity of fundamental components and distance becomes substantial in the fourth dimension, the final to be considered here, where an object has form in both space and time, according to the measurement of intervals. All of these are descriptive of the same process, differing only in the complexity of dimension.

To understand the relationship between the table of the nines and the fourth dimension, it is necessary to lay a minimal foundation. Some formulae and models describing progression should be given, as it is by progression that time is measured. It is this process that has thus far been described, and which constitutes our bridge. Growth in biological organisms is measurable according to an exponential law for equiangular spirals that gradually approach ratio based orientation. An example of this is that governing the formation of a nautilus shell, given as $r = ae^{k\theta}$, where r is the radius of curvature, a the area, e the natural number (2.71) found in exponential balances, k the kinetic energy, and θ the polar angle of predicted curve continuation. Although it seems completely unfamiliar, this formula has been underlying our progress all along. It describes the function of the ninth table whereby the lower dimensions are evoked rapidly, and the forms of the higher dimensions more slowly. What this formula means is that the spiral of the nautilus shell is curled more tightly around its origin, rapidly forming a circularly bound core. As it continues, the arc of the shell expands to break its circular condition, and rate of growth slows. This is the same exact process as described in the numbers of the ninth table.

Now, what does this similarity have to do with the fourth spatial dimension? If the rate of growth of either a nautilus shell or the ninth table's factorial breaks were plotted as a sine wave, it would have high initial frequency and minimal oscillation, followed by lower gradual frequency and more pronounced oscillation. This chart describes the unfoldment of progression for either equally well. It is, itself, also a three dimensional shape. If it were plotted in a three coordinate system the sine wave would orbit about the x-axis, alternating positive and negative in the y-axis, with an expanding radius in the z-axis. If it were displayed with its compliment as well, numerically negative products of all the multiplicatives or geometrically the cosine wave, it would take the familiar form of the double helix of DNA.

Before we go into the implications of the gnomon described in the ninth table, let us first pause briefly and consider the construction of a sinusoidal wave. In trigonometry, the study of triangles, lies the basis of wave measurement. A wavelength, λ, may be measured as two right triangles extending up to the endpoints from the base line of the standing wave, sharing either a common point, where the wave form crosses the base line, or the vertical leg, where the wave form reaches its peak or trough. These triangles are equivalent for a sine wave depicting circularly bounded progression, and isosceles — possessing two equal length legs apiece. The method of creating the sine wave itself comes from a technique for deriving a right triangle's non-right angle measurements from the measures of its legs and hypotenuse. The sine function for an angle is the opposite leg over the hypotenuse; the cosine function for the same angle being the adjacent leg divided by the hypotenuse. These two functions, as it was stated before, are complimentary. This is expressed in the relationships between sine and cosine for a triangle ABC whose right angle is C: $\sin A = \cos B$ and $\cos A = \sin B$. When the numerical solutions of these functions are graphed, a sine wave appears, which is a spiral in three dimensions. As a measurement of interval, this is also a fourth dimensional shape.

Now we may consider the progress through dimensions of the described spiral as it breaks from circular boundary conditions and what form it takes after it does so. The first point that should be made is the distinction between degrees and radians. Both are measurements of the arc of a circle, or the radial expansion of a spiral bound by circular or exponential conditions, but the former is fixed and the latter is open. Radians, unlike degrees, are dependent upon the transcendental measurement of pi, an irrational integer whose value is roughly 22/7 or 3.1415926. Pi, or π, is an expression for the ratio of the circumference of a circle to its diameter. The relationship between degrees and radians is such that each πradian $= 180°$. It is interesting to note that a spiral column of tetrahedrons, such as Buckminster Fuller modeled for the geometry of a double helix, undergoes 1/3 full rotation while 22 of its faces are exposed. Thus π is the limiting factor for the spiral in its early stages of progression, as it crosses the threshold of the first two dimensions. As it enters into the third, the forms it describes become exponentially complex, and the time interval elapsed between factorial breaks therefore begins to widen.

The limiting factor remains a transcendental number, an irrational integer similar to π, but one that describes an open set for growth along dimensional lines. This integer is phi, ϕ, and will be shown to occur in pure mathematics, quantum mechanics, and the helialical pattern of DNA. It is this number that governs the third dimensional forms derived from the ninth table, as well as the fourth dimensional pattern of their progress. In order to decant ϕ from the table of nines we need only convert the products given to degree measurements, a process already implied by the clean factorial breaks occurring at the 180 and 360 multiplicatives. Once this is done a chart for the sine and cosine functions of these angles may be assembled, and, using larger, later occurring sums for these, ϕ will be revealed. The $(2 \sin)^2$ and $(2 \cos)^2$ sums are preferred, and yield results containing both phi and phi prime, ϕ^1, that is the reciprocal of phi, or $\phi/1$. Some of the most notable results of this table are, for 9° or $\pi/20$ radians: 2 - (ϕ + 2), 2 + (ϕ +2), sin and cosine functions respectively; for 18°, $\pi/10$: ϕ^1 + 1, ϕ + 2; for 45°, $\pi/4$: ϕ + ϕ^1 for both; for 72°, $2\pi/5$: ϕ + 2, ϕ^1 + 1; and for 81°, $9\pi/20$: 2 + (ϕ + 2), 2 - (ϕ + 2). The complimentarity of the sin and cosine functions is readily apparent here. These figures represent distance relationships between points on what is called the Golden Triangle of 36° by 72° by 72°, after ϕ, which is itself the Golden Ratio of $(1 + 5)/2$, with approximated value of 1.61803. Some other properties of note possessed by ϕ are that multiplied by its reciprocal, the value of which is -0.61803, its product is -1, and that squared it is equal to itself plus 1. The spiral described by ϕ is thus the so-called Golden Spiral, and is exponentially bounded rather than circularly, its points being ϕ, ϕ^2, ϕ^3, ϕ^4, ϕ^5, and so on. These constitute the bridging of dimensional gaps described by the tables of eights and nines, and the slower, more numerous three dimensional forms that arise by angle sum recurrence during the later stages. It is interesting to note that a line connecting two points on opposite sides of a third point in a pentagon forms a line that, intersected by another such line drawn from the third point, forms the Golden division. Thus, not only is ϕ determinant of the rate of third and fourth dimensional progression, but a fundamental building block in the Platonic forms themselves.

Before going further a brief clarification between the circularly bound and ϕ bound spirals should be considered. In the circularly bound spiral the juncture points of the sine and cosine waves do not necessarily occur at the base line of the standing wave form, nor are both their apex and nadir points compatible with opposing peaks and troughs. Because the oscillations are so small, however, this factor is compensated somewhat, so that they may appear so. In the ϕ bound spiral the conditions are more alike the traditional depictions for DNA if the exponential function is doubled. It is at the third repetitive juncture that the circular becomes ϕ.

We have seen how ϕ effects the factorial breaks in the numerical sums of the products of the ninth table, resulting in first a rapid expansion through π limited dimensions and later in gradual progression through forms in an open dimension that are, themselves, the results of its interactions in that dimension. The hypothesis that, as a measure of interval, the φ/π spiral itself is a governing factor of a still higher dimension has also been forwarded. Next let us consider the applications of this revealed pattern of forms in the real sciences of math, physics and biology.

It has already been stated that the π or circularly bound spiral in its initial stages of formation yields markedly different results than the φ binding does later. This difference is the gap between the second and third dimensions bridged by its progression. The π constraint means that the spiral's growth is arithmetic, and that only the sine and cosine waves can be plotted to measure its factorial breaks. The larger numbers of the later products in the ninth table, allowing for as many as three different simultaneous solutions for a single product's digital value — all representing products already given earlier in the table, are governed by exponential expansion, where only φ and its reciprocal compliment phi prime determine the location, or rather the moment, of each factorial. The difference in a chart plotted for these two different types of growth is the difference between a horizontal line and a diagonal, respectively. Another, more statistically appropriate, depiction of exponential growth is the asymptote, an upward curve approaching infinity. These averages for the variance in the spiral may be taken to represent its standing wave state, or the base line between itself and its compliment. The most significant distinction between the arithmetic progression of the second dimension and the exponential progression of the third dimension is the graph of the relative locations of the factorial breaks, occurring at the complimentary juncture points. The juncture points of arithmetic and exponentially bound spirals differ, as has already been demonstrated by the distinction between the sine and cosine wave forms and the phi and phi prime wave forms. In the later system they will always occur at the event of intersection with the base line given for the underlying growth pattern. This would seem to have little relevance, as the spiral is three dimensional, and thus has no real juncture points between either sine and cosine waves or phi and phi prime waves, both merely arcing around each other along the base line x-axis. But these points are where the factorial breaks occur, and have relevance in the fields of quantum mechanics and biology.

In quantum mechanics particles are so small that there is no way to observe them without disturbing their behavior by doing so. For example, sending a photon into the probability field of an electron cloud changes the path of the electron to be measured. Thus, either position or velocity for a particle can be measured at one time, and this is what is known as the uncertainty principle. What is really implied here is a change in probability itself. The probability of the electron in the example following its given course before being struck by the photon is a certainty; the probability of it following that same course after being struck by the photon is an impossibility. The inverse of these statements is also true. In the context of table nine this translates to the clean factorial breaks occurring at the phi and phi prime intersections, and the rest of the factorial breaks occurring at the sine and cosine intersections or at later points along the opposing peaks and troughs determined by φ. In the practice of constructing histories, the positions of a particle over a period of time are considered. The spiral may be thought of as a graphic depiction of a history. Quantum mechanically, the numerical factorial breaks constitute positions where probability undergoes inversion. Three terms govern the behavior of the particle at these positions, r (radius), θ (theta, polar angle), and φ (azimuthal angle); r allows the emission of a photon, releasing energy, or one's absorption; θ measures precession; and φ is the spiral path taken between orbital energy shells, whereby the electron will lose energy approaching the nucleus, or gain energy retreating it.

We can easily observe the same pattern being repeated on a larger scale in Astronomy. The most obvious would be a plotting of the movement of the heavenly bodies about their elliptical orbits over a period or duration, such that the spiral may be viewed as their positions over time, similar to the history constructed for the position of an electron. An even closer examination of the interaction and structures of the fields that cause this yields an even more detailed discovery as to φ's presence.

The solid state planets, those within the asteroid barrier of our solar system just to name a few, obviously obey only the π binding for their formation. The result is that they are centered approximations of spheres, whose φ, or gravitational force breaking, traits are limited as much as possible, resulting in the equatorial bulge for example, and, by interaction with our moon, a π cyclical timetable of the tides. The atmospheric and electromagnetic forces surrounding these solid state planets are a somewhat different matter, but this is best explained by looking at the planets that are, themselves, more entirely governed by these forces respectively.

The gas giants, therefore, represent the point at which the π internally oriented and φ externally oriented spirals most actively interchange. Their atmospheres manifest violent and extremely mobile centers of opposing pressure, creating immense, ionized, semi-acidic storms. The most notable of these is the red spot on the surface of Jupiter, which has been modeled on a computer using supersymmetric statistical averaging. Saturn represents the most π bound, with its ejected rings of neatly circularly situated frozen materials, and Pluto the most φ bound, with its erratic solar orbit and its nearly horizontal rotational axis.

For both of these planets it is the electromagnetic field that is the suspension where the spiral's transformation plays itself out with such astonishing affects. The best location for an observation of the spiraling nature of this field is the sun, directly at the center of our system, whose atmosphere is entirely dependent upon the mathematical structuring of its behavior. While it is Jupiter's atmosphere that is the most clearly at the cusp of the spiral's third permutation atmospherically, the sun is already there entirely electromagnetically. In short, what occurs on Jupiter within its atmosphere is nearly mathematically identical to what is occurring in the outer corona of the sun. The only significant difference is between the π bound temporal pattern of Jupiter's atmospheric disturbance and the φ bound temporal pattern of disturbance for the sun's electromagnetic field.

The best way to understand this pattern is to glance over first the evidence for its existence, and then explain its cause. The evidence consists of the eleven year sunspot cycle, during which the occurrence of solar flares and prominences alternately increases and decreases. The model that has best, thus far accounted for this process is the winding up of the sun's electromagnetic field currents due to the differential rotation of the hot gas of its body at different latitudes. This occurs for eleven years, causing outbursts where the currents overlap, until finally these number so many that they short-circuit the entire ball of string and it returns overall to a base state, at which point the process can begin again. Because the timing of the cycle does not increase exponentially, it is clearly π bound in total. However the occurrence of solar activity during the eleven years does increase with time, and therefore the rate of flares, prominences and sunspots is ϕ bound.

It is demonstrable that it is at the peaks and troughs of the temporal sine wave that the ϕ, or algorithmic, spiral takes root. It can also be stated with a degree of certainty that it is at the juncture points of this algorithmic spiral that the solar supersaturation event transpires and the entire electromagnetic field resets itself.

Lastly, it may be noted that the practice of using histories for particles is utilized Astronomically in the theory of quantum gravity to study a similar supersaturation in the very young universe, where ϕ bound "bubbles" are thought to have formed within the π bound symmetry of the early universe as the third dimensional factorial phase shift occurred.

In biology ϕ/π occurs as the emphasized function in two areas pertaining to the reproductive process of cells. The first is in the determination of time intervals in the cell cycle, or life-span of a cell. The second is in the actual physical composition of DNA, as well as the interaction of that structure with RNA in protein production.

The cell cycle is divided into four spans: G_1 (first growth), S (synthesis), G_2 (second growth), and M (mitosis) phases. During the first growth stage a "trigger protein" accumulates to determine if the cell will engage in replication. Some cells, such as nerve, muscle and red blood cells in animals, and roots, stems, leaves and flowers in plants, never leave this phase. During the synthesis phase polymerases separate the helix of a DNA strand by breaking their hydrogen bonds. Two new complete strands are then created as these polymerases attach corresponding nucleotides with complimentary nitrogen bases. These will be discussed presently. Replication occurs at several places along the DNA strand at once, its helix being opened up like a zipper into "replication forks," where the opposite directionally oriented copying of DNA occurs simultaneously — the leading strand continuously adds nucleotides in one direction while the lagging strand synthesizes short, discontinuous segments that are joined together by other enzymes which also serve as proof readers. In the final DNA strand there is one in 10^5 to a billion (10^9) errors or mutation factor. Mutations that do occur are corrected by nucleotide complimentary enzymes and sealed back together with DNA ligase. Within minutes of its synthesis new DNA wraps around nuclear proteins called histones. Eight molecules of histone (two each of four types) make a nucleosome core, around which the DNA loops twice. Nucleosomes are linked by DNA bound to a fifth type of histone, and all of this together makes up chromotin fibers. If there are three pairs of chromosomes (called chromotids) in a nucleus, the cell is diploid. If there are only two it is haploid. During the second growth phase of two to five hours only minimal RNA, protein and macromolecule production occurs. When the cell enters its final reproductive stage, the mitosis phase, chromotids are collected and held together at a centromere. There are four spans of mitosis. The first is prophase, during which chromosomes condense and the nuclear envelope begins to disappear,

Microtubules form two diamond shaped spindles and attach fibers from them to the centromeres. Next is metaphase, where the chromotids are pulled into a planar arrangement across the cell center perpendicular to the spindles. Then anaphase occurs; the centromeres divide and chromotids separate into chromosomes. The final phase is telophase — the chromosomes gather at opposite ends of the cell and a new nuclear envelop forms between them. In animal cells this is the plasma membrane, in plants the cell wall.

We should also consider the similar process of meiosis, or the replication of genetic material in the combination of cells from parental sources as occurs in the reproductive function. It is essentially the same as mitosis, but because the cells start out with three pairs of chromotids, they must undergo division twice before their full amount of genetic material has been haploid integrated. Thus they have twice the phases of mitosis, but undergo the same process. The only significant difference occurs in the very first state, prophase 1. During this phase each chromotid in a homologous pair (of which there are six total) exchanges parts of its replicated chromosomes for their identically placed counterparts with their paired chromotids. The new chromosomes are a genetic recombination. This is called "crossing over."

The time interval of the first and second growth phases in the cell cycle is clearly determined by a sine wave. As will be demonstrated shortly, in these cases it is the amount of labor and not the length of time which is exponential. Thus they may be plotted as points where the wave curves into the lower x values in our three dimensional coordinate system. The spans of the cycle allotted for the replication and reproduction functions, however, are temporally fixed by ϕ. They sweep upwards into the greater integer values and remain there longer.

Now let us consider ϕ as it relates to the structure of DNA itself, and to rate of processing in the relationship of DNA and RNA in building proteins. The backbone of DNA is a congregation of deoxyribose (sugar) phosphates. This forms the exterior pillars of the strand. Between these, like steps on a ladder, are four types of nitrogen bases that fall under two classification categories, united by a hydrogen bond. The classes of these bases are the single, hexagonal ringed pyrimidines, and the double (one hexagonal, one pentagonal) ringed purines. Thymine and cytosine are the bases belonging to the former class; adenine and guanine those belonging to the later. One purine always binds with one pyrimidine: thymine to adenine, cytosine to guanine, according to complexity of structure. The formation of the helix itself is not exactly ϕ, but more closely resembles the complimentarity of the sine and cosine waves. The axis does not penetrate the center at the hydrogen bonds, but is rotated about by the sugar phosphate backbone. This is due to the heterogenous pairing of the nitrogen bases. If the same effect were produced binarily, the image of the medical cauduces would be exactly evoked. ϕ itself only comes into play when the strands are separated, at which point it serves as the angle governing their divergence, much as in quantum mechanics. It is also according to a ϕ determined number of rotations that the replication routine will begin and end. This is best examined in the RNA process.

There are three types of RNA, with purposes implied by their titles: mRNA (messenger), tRNA (transfer), and rRNA (ribosomal). These are involved in the three consecutive phases of protein production from the coding of nitrogen bases along the DNA double helix. mRNA polymerase produces exons, copies of regions of DNA that are expressed as polypeptide chains, or meaningful instructions, and throws away introns, replicated non-coding DNA. tRNA is comprised of codons (promoters) and anti-codons (terminators) that begin and end this copying. Through tRNA charging, L shaped triple looped rRNA binds to ribosomal subunits to create proteins from DNA coding sequences. Certain codon genes, called transposons, jump around to cause mutation. These alleles are alternative forms of a gene that have slightly different base sequences as a result of mutation and cause the genetic variation on which natural selection acts. The simultaneous existence of tRNA and the ribosomal subunits that act as factories for protein production begs the question of the chicken and the egg, but a solution is readily available in the primacy of DNA and analysis of its mathematical implications for congruent progression. All of these purposes occur during the first and second growth phase in the cell cycle. The replication process undergone by tRNA is somewhat different from that in mitosis, as it only represents the replication of abbreviated segments of the DNA which are specifically designed to promote protein generation. As such it breaks with the time interval coding of these phases, extrapolating a π sine wave predicted amount of material and integrating it into a ϕ regenerative system in a π factored span of time. It is at the beginnings and ends of these coding sections that mutation is most likely to occur.

These examples from some fields of hard science can be called by one group term: Spontaneous Symmetry Breaking. In each of them the occurrence of a given and predictable number in the progression represents a point at which change occurs, where the existing pattern is broken and a new, similar pattern is established, just as throughout the factorially identifiable sums of the products in table nine. In short, with each dimensional gap it crosses, our bridge changes form.

The spiral itself, particularly the unique multidimensional nautilus-like spiral described by the ninth table factorial breaks, is only one form of fourth dimensional shape. The Platonic solids all have their correspondences in the fourth dimension as hypershapes, of nested like-within-like forms. As we have already seen however, the number of definite coordinates increases for each additional level of dimension, and so there are many more basic polygons that can occur in the fourth dimension than in the third. The number of these, as well probably as the structure, is a factor of the interval proportion of the underlying pattern, that is, the ϕ spiral.

In closing it is perhaps best to clear up why the spiral bound by pi (whose value is 3.14159) is actually less than the spiral bound by phi (the value of which is 1.61803) despite the balance of their values being tipped more in the former's favor. It is best, for this exploration, to perceive the spiral in organic terms. The singularity is alike the seed, from which the structure springs. The second dimension is the roots and the third dimension the trunk, offering stability and durability. They are both bound by pi in this respect, the greater quotient, providing fuller form. The fourth dimension presents the shape of the fractal branching pattern that yields sticks and stems and finally leaves, down to the veins. The leaves themselves are as myriad as the solids of the fourth dimension, and arise, thrive, deteriorate and drop away as living intervals of time. And these are bound by phi, the more delicate and diverse function. Now we may consider theories that substantiate this model.

All structures that thrive in nature do so by establishing a firm foundation. The φ/π spiral shows us this has been accomplished by diversification. The circular binding becomes spherical, and subsequently, omnidirectional. Not only are the later forms built from those of earlier dimension, each greater dimension is itself a bigger picture of the same hologram. Pi represents the base of the pyramid, which, although in our case more compact in terms of interval, is the keystone of the arch. It manifests this best in the singularity seed, the alpha and omega point. The latter lattices of the fourth dimension are merely elaborations upon this element.

There are two types of complexity at work in our construct. That of pi is digital; geometrically the form it expresses is simple and precise. That of phi, however, is irrational; its structure is elegant unto divine, and yet only rendered with expertise. One can infer a great many things from complexity of structure. As I have already stated, the closed and open states described by these shapes are inherent within their expression, and yet of so pronounced a magnitude is this difference that herein lies their entire relationship. They define the very boundary of internal and external.

The best explanation for the dynamic interaction of pi and phi is the simplest. At each point along their prospective arcs, it is their digital value that defines the degree of correction for their progress to the next point. Thus, pi is circularly closed, because each point is connected to the last by an angle such that all of these angles combined totals a full rotation after a certain number of points. So, phi is spirally open because its points follow the same relationship with a different digital code, such that a complete rotation is only accomplished after a greater number of points than that already established as comprising a circle.

Finally let it be said that it is not at precisely the third juncture that the pi spiral is transformed into the phi spiral, because this does not occur directly at a clean factorial break. The closest factorial break listed in the table of nines is multiplicative 117, but the true break occurs between multiplicative 112 product 1008, and multiplicative 113 product 1017. It is only at this break that the numbers again begin their repetitive climbing through subset integer sums between clean factorial sets, as already evidenced in the second dimensional set. Thus it is not exactly at the gap between the single and double digit sum sets that the dimensional leap occurs, but rather, a little after. It is, in fact, exactly 14% between the first complete repetition of factorial cycle in the second dimension quantities and the first transition in the second subset of factorial breaks in the third dimension that the leap occurs; in other words, between 99, the fractal redoubling of the line, and 198, just after the formation of the holographic triangle, both the work of the interaction of time. The φ/π transition occurs at dimension 3.14. It is likely that a parity to this computation is evidenced in the termination of the multiplicative set of table eight, where it transits to the beginning of table nine. As the eighth table is necessarily infinite, its conclusion must be transfinite. Thus its transdimensional crossing must supersede singularity by a slight, though incalculable degree. As neither of these gaps technically exist, I will hitherto refer to them as positive zero and negative zero.

D. Mathematical Proof for Inversion

Begin by imagining a straight line. Divide the line into regular segments, and number these segments, rendering a number line. Call the center of the line Zero, and imagine that the line stretches out as far as Infinity in either direction.

Even though the name of the number is the same in either direction, the value of one is positive and the value of the other is negative. Whether the number is positive or negative is entirely arbitrary, except in the context of their relationship, where it is convention to render as opposing the polarity of the two extents of Infinity.

Next imagine the intersection at Zero of two more identical number lines corresponding to the graph of three dimensions. This is the structure of metaphysics. Here we see that Zero represents potential energy contained within the unit sphere of probability, and the six infinities represent potential spin, where three are positive and represent movement towards, and three are negative and represent movement away from.

Traditionally these were rendered as a series of ten ideas, the sefirot, where the cardinal directions are taken as certain characteristic attributes, expressed at a lower energy level, of the central three concepts. Here we see probability described as fire from water, potential as water from breath, and singularity as breath from breath.

We see from the number line that probability is in three ways One and in three ways negative One. Potential is Zero from all directions, because it is intangible. A singularity, however, equals the spherical sum of the number line in all directions at once, and we know the number line to reach Infinity; but the Infinity of the singularity is an Infinity that doesn't exist, for the singularity is a function of Zero. In its mathematical formula it is Zero / Zero = Infinity, because any number divided by Zero is Infinity. The singularity, therefore, though a function of Zero, can be anywhere on the number line, and is, invisibly, already everywhere.

Zero, as we can see from all directions, is Infinity plus negative Infinity, or, in other words, the complete measure of the number line. Thus, what the idea of Infinity in all directions is describing is potential. For the sake of convenience later, let me introduce the idea also of seeing potential spin as positive potential, and potential energy as negative potential. This potential is only the uncertainty contained within the wells of probability that comprise physical reality.

One, also, is equivalent to Infinity in all directions, insofar as it is divisible into infinite fractions. This microcosmic continuum is the measure probabilistically between Zero, statistically impossible, and One, statistically certain. On the surface of the well of One, those points that are positive impact upon the future, and those that are negative reflect upon the past. The most profound insight about One is observing the difference between the unit sphere and the unit cube. One is the measure of the thought at the point of consciousness on the sphere of the mind, although this too is traditionally reckoned as a singularity. The point of concentrated consciousness is archetypal, and the proof of this is that the measure of its pattern over time is ϕ/π.

Now let us look at why zero should be as negative pure potential, and why Infinity should be seen as positive pure potential. Each Infinity, because it is also equal to One, is archetypal, and because it can itself be divided by Zero, is as well a singularity. Thus we may say that the value of an archetypal singularity is Infinity; yet still the exact opposite archetypal singularity is negative Infinity — it is exactly identical to, and completely excludes, its counterpart measure of distance on the number line. Directly between them is their combination and absence.

However, the inversion between them occurs at One, or in reality, rather than only in potential. This is because Zero cannot act as the lowest common denominator or the greatest common factor, and these functions are necessary to translate between the sets involved. One is the smallest common multiple of the set of all integers, including Zero, expressed as rational numbers, which set is equal to infinity, and it also is the highest denominator less than the set of all prime numbers greater than it, which set is equal to infinity. This is the mathematical proof of inversion between the entire number line and a pattern within the number line, as above so below, both equaling One, and One equaling Infinity.

Having done with this imagine folding up our measuring apparatus. Traditionally, each of the six directions is sealed with a different combination of potential energy, probability, and potential spin (represented by the letters Yod, Heh, and Vah from the Tetragrammation), thus rendering the traditional attributes (the seven central sefirot) associated with each. Only the three directions of probable spin measuring positive values produce cubes in which for these changes to be calculated, however, as their negative counterparts represent the past, and together these two halves comprise the thirty-two verticed shape of 2^5; the shape of 2^4 is the sixteen verticed hypercube, or the measure of the distance of the unit cube from itself, which is temporal, rendered by using the unit sphere to separate a cube internal to it from the unit cube external to it. Continuing to count down our dimensional exponents, the cube is eight vertices, each the intersection of three lines of two points each; the square is four vertices, 2^2; the line itself 2^1; and each point on the line is a well of probability (One) containing infinite potential (Zero).

1. inversion of inversion

When inversions occur actively they occur in more rapid succession. When inversions occur passively they occur in a slower sequence. The proof of this is given by the formula where the greater number of steps in an impossible feedback loop, the longer the amount of time it takes to filter through them, given a fixed measure of the properties of time; and where the fewer number of steps in an impossible feedback loop, the less time it takes to filter through them, given a standardized definition of time.

The shortest form of the impossible feedback loop is one step, and it has special properties. Like all other impossible feedback loops, it is a hologram of all other possible multiple step impossible feedback loops. This property has a specific name, and this name is applied to the one step impossible feedback loop itself; thus, also, is the transition between steps in a multiple step impossible feedback loop known as inversion.

Inversion is thus the active/passive principle of its own autocorrelation. When we describe this as having been between steps over fixed, regular time in the context of an impossible feedback loop, it is the same as saying the inversions have occurred between sheets of neighborhoods on a manifold. Such inversions in set theory usually only represent binary spin symmetry, however in the simulacrum of the font of consciousness that is the impossible feedback loop they may be seen as visual-verbal inversions that occur faster or slower, and are of varying length.

2. average law of depleting resources

The average law of depleting resources dictates that more an effort becomes displaced, the less likely for its effect to a random event. When someone does something at the behest of another, it depletes the pure probability by causing it to be a factored probability. Therefore, when production demands 90, consumerism dictates at least 45. When 45, 22.5, when 22.5, 11.25, when 11.25, 5.625. However, the opposite would be false — at 0.703125, 1.40625, when 1.40625, 2.8125, when 2.8125, 5.625, etc. Since available resources would deplete. Therefore, for the sake of one, there are two, and thus. However as the many increase, so does effort necessarily become displaced, from one head to the next even. And therefore the great work is opposed to that essentially random quantum functioning that is equivalent to the Great Work of the spirit. As the spirit moves the probabilities about in ways we can only call random due to our lack of understanding so does the mind's action govern that of he hand, the hand the tool, and so on.

III. Manifestation

The modern practice of manifestation began with the Old Man of the Mountain. He was the leader of an Ishmaeli Muslim sect called the Hashishins, who took their name from the hashish plant that they smoked to put them into an altered state of consciousness, and from which we get our modern word for a political murderer: an assassin. This stems from the Hashishins' conspiracy to infiltrate every surrounding and distant foreign government with a covert espionage operative close enough to a person in an office of great authority to be able to kill them upon command.

Members were recruited to the order by being drugged with hashish and carried up to Hassan I Sabba's mountain fortress, guarded by a pit of snakes, inside the walls of which was a palatial garden equipt with all amenities and luxuries that anyone would desire: wine, women and song. The initiate, once their thirst had been piqued by relaxation, was then brought into a room and shown a beheaded body. He was told to choose between serving the sect and living in the secret paradise or betraying them and being decapitated. This usually converted them quickly enough, but if an added incentive was needed the candidate would be shown the same beheaded man alive and whole and told that the order could resurrect him to life. His men were known to be so loyal that when he commanded one of them to jump off a castle turret for a visiting dignitary, the man did not hesitate but to obey.

Thus was born the simple act of conversion by conjuring. The allegedly beheaded man was hiding his body beneath the floor boards and wasn't dead. When he reappeared, "resurrected," he was merely standing on a trapdoor. This was an act of macromanifestation, however it appears as if micromanifestation may be much, much older.

A. The Ethical Apple and Moral Wormholes

The mythology of the Jews tells us that Moses spoke to a bush that burned in the wilderness but which was not consumed. He went on to lead his people out of Egypt and into the Promised Land of Milk and Honey. The Greek mythology concludes the story of the theft of this holy fire from Olympus by Prometheus by describing his punishment as similar to that of Sisyphus, eternal and self-regenerating. While the tribe of Hebrew speaking nomads wandered in the Sinai desert waiting to enter Canaan, Moses compiled the Pentateuch, the first five books of the Old Testament. In the first chapter of this, which he called Genesis, he described the formation of Adama, the first man, from the red clay of the earth.

The moral of this is manifestation. It is born from the flame of the will of the self, and molds us into who and what we are and how we see ourselves. In this way the concept of Original Sin represented by the theft of the apple of knowledge over good and evil from the tree of life in the garden of Eden is guilt for existence itself. This has carried itself out in all the subsequent attempts by humanity to civilize the animal self, and to bring their instincts under the influence of their reason, and has led to the institutionalization of better judgment and higher learning. In short, the negative declarations of original religions against taboos or sins are the key stone of civilization — the ultimate macromanifestation.

B. Concentrating Trance Through Will Power

The goal of trance meditation is to transcend mundane reality and attain prolonged unity with the most high within the self. The ways of inducing trance are too numerous to list, and therefore I will not go into them here. All of the most popular are the basis of religions, however the institutionalization of the experience inevitably leads to more ritual than education.

The basic formula is for the raising and cleansing of the aura in a single sitting. In this the electromagnetic field is concentrated upon in the mind's eye until it is possible to feel the energy of it in the air. Then there is a ritual which differs in every faith that is meant to clarify this aura, and set it in an even balance with the external continuum. This is practised repetitively to keep the soul in good faith.

The method of raising the aura is based on the generation within the self of an increase of energy. This should be drawn in through the lungs, follow the paths of the blood vessels outward from the heart, and the fibers of the nervous system upwards toward the brain. Ancient methods for awakening awareness included skrying, or the zoning out over various types of meditative object.

The method of concentration depends upon the free will of the individual being brought into play. If the mind is easily distracted, as by an ever flowing, never ending internal monologue such as the font of consciousness itself, then it will diverge from purgation for the duration of the diversion. Enlightenment is achieved only by rigorously screening out not only of the thoughts that we think, but the thoughts that we think others might be thinking as well.

For the aura to be clear, the mind must be clear.

The world does not change when we meditate.

Only our perception of the world changes.

1. Opening the Ajna

See what you want to see.

Do you see something that is not really there?

Do you really believe that anyone cares?

Often the hardest step of a long journey is the first.

For the projection and cleansing of the aura the first step is seeing the aura around yourself with the mind's eye. The Hindu name for the mind's eye was the ajna, and they marked it with a red dot in the center of the forehead. The Egyptians also believed in a third eye seen through by the spirit, and associated with it a serpent standing up from the forehead, featured on the crown of Upper Egypt. It has also been said that the eyes are the windows to the soul.

It is more or less necessary to use the mind's eye because it may not always be convenient to rely upon technology as a medium whereby to see the aura. Also, because the aura is invisible to the naked eye, it is necessary to use the eye of the full imagination, tapping all resources of information, legislating intelligently. This is the technique of passive reasoning, allowing all thoughts to come and bow down before the central silence.

Therefore in order to be able to see complete clarity through the mind's eye the last lid one must open is the sleep of the self, or the ego. To accomplish this it is necessary to invert the idea of the self into the presence of nothingness. One must seek nothingness out, for it can be measured by various containers. The cumulative goal in this is that you will be able to see the presence and movement of forces where there are none to be seen by the common eyes.

Although the third eye as the concentration of pure consciousness is usually associated with the forehead, and therefore the frontal lobes of the brain that lie immediately behind the skull there, one must not discount the importance of the left and right thalami, collectively called the thalamus. This is often mistaken by the lay person for a gland, because of its position directly superior to the pineal gland and the pons of the medulla oblongata. It is however, only fueled by the chemical neurotransmitters produced in these subjacent areas, as is the rest of the brain and the nervous system, and is not itself part of their glandular secretive process. The thalamus is, instead, a holographic depot from one or more incoming sources to one or more outgoing receivers in the neuron columns of the cerebrum.

2. alternate possible perspectives

In order to suspend rational disbelief, which is based on our being accustomed to the realm that we perceive with the five senses, stimulated by inspiration and distraction, legislated by logic and faith, it is necessary to accept that all possible lines of reasoning are equally real at least in the realm of potential, and that the realm of potential is equally real as the realm of ideal in the mind. It is only after doing this that we may begin to separate probable fact from probable fiction, and to discern for ourselves what is true and what is false, thus determining what is real and what is not, and in this way, answering questions along the way, deduce the ends to the history of any event. However without first accepting that anything might be, then we cannot fully and firmly follow from one end to another, and thus never know beyond doubt that our conclusions are in harmony with aggregate reality.

Neither has the average mind been prepared to deal with any infinite concepts by non-trance, social exchange, what modern meditators call game reality. So, in order to realize the existence of infinite possible right answers, one must grasp the concept of infinity. This is a sum, such as one, that is at the end of a number line. Since it would take forever to count to this sum, it is considered equivalent to saying that the number line goes on forever.

When we imagine infinite possible solutions it multiplies this number line of sums by infinity, as though the sums were hollow vessels filled with additional meaning by infinite explanations. This squares infinity because it doubles the dimension of the number line in depth. This can be referred to as the exponential expansion of consciousness. Applying this same effect to a timeline displays the reason: our ability to see all possible explanations for events as equally true in theory. This is the last vestige of the self to be shed — the locus of external control.

3. dimensional levels of truth (socially unacceptable archetypes)

As we relate to our reality we determine what is real and what is not. We come to understand that it is possible to convey greater amounts of information by an appeal to more than one of our senses at the same time, and in this way to know that there are some things that exist entirely beyond our senses as well. This causes us to become aware of ourselves as an ego inside of a mind, and the ability to imagine the absence of this usually associated by polite society with fear of death. Since it is this very self-negation that deep meditative trance hopes to accomplish, it is necessary in order to engage in it to know this ideal conceptual realm of things beyond our senses with the mind.

First one must see what there is not to see.
Then one must know what there is not to know.
Then one can begin to decide for themselves what is truly reality.

This means utilizing the mental realm in a formally systematized way — storing memories and categorizing perceptions according to structural hierarchy. Naturally, the dimensional model presents itself. We may say that, where the dimensions represent parallel alternate possible realities, that each may posses its own truth in potential, and that therefore they again compound infinity with infinity, thus expanding our consciousness into the third dimension, and giving us the form of a rational cube of infinite implications to our infinite spherical consciousness.

Just because the human mind can transcend the focal ego, does not mean that is has seen its way free of all the limitations of human instinct, however. There will always be different goals to be gained by different souls, and it is important not to let another's path become your own without your knowing. Take for example the competition between humandroids, who always want to keep everything "real" (meaning formalized) and humanimals, who believe that reality is an illusion, and want to let everything "go with the flow" (meaning relative chaos). The former diagnose the latter with various types of mental disease. What follows delves deep into the conflict over the manifestation of socially unacceptable archetypes.

a. the right diagnosis of depression or melancholy

Just as the mother may feel post-pardem depression after giving birth to her child, so does every single one of us feel homesick for the womb. This is the emotional reaction to the birth experience, which imprints upon the brain developed to the human stage the first impression of death. For the rest of our lives we try to recreate the perfect safety of the conditions of the womb, until we cannot fend off the forces of other forms of life anymore and finally give in to death. The human condition capitulates this in its strivation for the ideal, the supernatural and the extraterrestrial, because this is not simply a behavioral condition implanted on the individual human organism, but the culmination of millions of years of co-evolution between autotroph and heterotroph microorganisms as small as our DNA and RNA.

b. the misdiagnosis of manic-depressive bipolar cycling

Get married; otherwise you have turned your unique personal genetic heritage of millions of years against itself and deserve to die. Get a job. Otherwise you are not part of the human race, which struggles and strives heroically against entropy. Die of old age; otherwise you will die unpleasantly and be trapped with a bad death imprint experience in the next form of life, which will cycle unendingly without your knowing of it just as does the birth trauma in the background of the mind in this life. Never abort fetuses: otherwise you are killing an intelligent soul. Always abort fetuses: there are too many intelligent souls already. If you want some examples of untrue things that many millions of people believe anyway, here are a few. I am no better, though, I have to admit: I believe people are hopeless.

c. the misdiagnosis of mania-induced obsessive compulsion

As soon as the idea that by following patterns we can control the cycles of time is introduced it is enforced. Since this is an untrue premise to begin with, the modern heirs to the knowledge of this fact originally held by the first medicine men, that is, psychiatrists, identify this with physical gestural behaviors as being symptomatic of different lines of reasoning, and begin to use different prescription drugs to erase such different symptoms that are seen as contraindicated by the social patterns established as the law of the land by the local chief of governing bodies. Interestingly enough these same psychiatrists, who differ from psychologists in that they prescribe drugs, are clearly overlooking the taking of drugs as the primary obvious cause of pattern, because if they allowed the populous to realize this fact, they would be exposed as addictive-substance merchants peddling lethal poisons. Instead we are given the red herring of tobacco, which is only a good workout for the lungs, and yet, when combined with aluminum byproduct lined, immune system weakening tap water, spiritually and physically unhealthy exercise diets, fast food high in starch and useless, fatty carbohydrates such as grease, and trendy designer prescription pharmaceuticals to further weaken the immune system and make it dependent on external sources for support, tobacco is blamed for all cancer.

d. the misdiagnosis of obsessive-compulsive induced paranoia

This being the case it follows naturally to accuse anybody even only half aware of this degree or level of social manipulation of their individual biological systems with being too scared by it to be able to accept that it exists and find substantial evidence for its proof, which makes them a target for being seen as a threat to the national security of the collective unconscious, a danger to themselves and others. This only represents another gradual step in the marginalizing of intelligent citizens who, for reasons perhaps of neurotransmitter deficiency induced by a mother's lifestyle while the fetus was in the womb, cannot or do not partake of the vast and ample fruits of the social system. At this time in history more and more doctors are diagnosing people as paranoid. The question remains unanswered by the experts if this is because there are more paranoid people now than before, or if there has always been this same percentage of paranoia per population and it has simply gone unidentified as such, or treated differently.

e. the misdiagnosis of paranoia induced schizophrenia

Historically the realm of infinite dimensions where all outcomes are possible has been associated with the djinn, or discorporeal spirits of our ancestral dead. This has led on the one hand to the comparison of idealized hyperspace to utopian heaven and on the other to the persecution, on religiously anthropomorphic personal psychological grounds, of anyone outside traditional religions who explores consciousness expansion. The weapon of choice against any such hyper-rational heretic has been to accuse them of living in a fantasy world created out of fear of living in the real world. This is a dressed down version of the same "truth" preached by the church — that the suffering of material reality is penance for transcendence.

The examination of paranoia has even infiltrated the art world in the past one hundred years. Salvador Dali, of the surrealist school, wrote of his own "paranoid critical method" for inducing states of heightened nervous excitation for the purpose of energizing self expression. This same "hyper-ration" is associated with mania, hyperactivity and attention deficit disorder by modern psychiatrists and treated with medicine to return the brain to a slower, more "normalized" rate of thinking. The operative clause in the persecution of schizophrenia however is that it is based upon a preexistent emotional state, usually extreme fear such as occurs during prolonged trauma. However in some there appears to be no such experience in their past background, unless of course one counts the trauma of being born to begin with.

Many latter day philosophers have gone much further in endorsing schizophrenia as a desirable and controllable mental state. While Foucault wrote of the appalling conditions of mental institutions, where the most common patient is the person diagnosed schizophrenic, Baudrillard wrote of the technological simulacrum (meaning) and simulacra (appearance) of the mass media in the terms of it imitating the gradual increase of psychic tension associated with paranoid schizophrenia and its application in the brainwashing of our entire culture.

f. the misdiagnosis of multiple personality disorder

The djinn of our ancestors became the angels of the church and the voices in the heads of many modern mentally disturbed people. These, too, are as natural as talking to yourself, however they are more or less independent personalities that are being perceived by the mind. Just as talking to ourselves is merely taking conscious, lucid control of our stream of consciousness by interjecting more concentration into the train of thoughts, so to are the full personalities that we perpetually pass through emotionally merely extra egos from which we can access information. The survival instinct applies to these as much as it does to us, although the resource that other consciousnesses compete over is the central concentration of attention.

Our own memories are often perceived as through the appearance of others or of things or places, as is evidenced by their eventual difference from formal reality, so that, where we perceive no change in ourselves, our memory comes to be deceived into its existence. This alienation of the event continuum usually does not follow as a result from anything to do with the mind, rather with things changing around it with age, and therefore to say it is triggered by the transcendence of the ego (or self-concept) simply because the consciousness relates to its root environment instinctually, and can only be metaprogrammed when the ego is obliterated (preferably by the self itself), and in this way open the mind's eye up to the consciousness plenum of the multiverse, is to miss the very obvious fact that this is the very truest reality of the situation all of the time.

4. astral traveling between dimensions (socially acceptable archetypes)

When this first began it was simply a detached mental state associated with trance meditation, often over a skrying object such as a pool of water or a mirror, resulting in a virtual free fall sensation, a state that has come to be associated with levitation. Then it was associated with the subject of the meditative trance, or meta-object, and at first with the ka as the vessel of the light body that transported Osiris, the king of the underworld, dead through the dreams of others' sleep. Immediately after this it was anthropomorphised somewhat by the Jews, who divided the astral form into the ruach (soul) and nefesh (spirit) and the astral realm into the formative and creative, aethyreal and material. After the Crusades the Rosicrucians, particularly John Dee, devoted much esoteric alchemy to the recreation of the body without organs (simulacra or simulacrum) on the astral plane. The astral body has come to mimic the physical shell and not the electromagnetic aura usually associated with the soul. The astral realm is merely another name, again, for the realm of the ideal. Two modern forms of astral projection are the potential temporal or gravitational singularity and body-jumping, that is, moving about in the short-range between different peoples' bodies. These are both popular topics in such genres as science-fiction, fantasy and gothic suspense horror, where the cosmological or supernatural aspects come into play.

a. dream archetypes: mother moon, star children

All existing religions are ways to hide from children their innate psychic potential. They all originated from Mother Goddess cults that were violent and bloody. This was at the time when the megaliths were still new, but their builders had been lost, and all lay in ruin and the tribes of the world ate each other. The concept of Time weighed heavily on everybody's mind, and frayed the edges of their sanity. All subsequent orders, rituals, observances and formalizations have been an attempt by the patriarchy to hide the existence of such a time from the records of all the taught histories of the world. Figures from the late stone age of voluptuous pregnant women are associated variously with Venus, Inanna, and Gaia.

b. working stiffs: god the father and the prodigal sun

It is held by all native peoples that once, a very long time ago, the entire world had reached the same level of collective intelligence as it has now. It is the belief of these native people — who have never been any less capable of developing industrial technological civilization at any time than those cultures that did and have developed it when they did, at any time, only less willing, less interested — that this first great global civilization was less structurally organized than materialism, and was devoted more to the study of spiritual or trance consciousness experiences. Many modern legends hold that it was because of this, non-materialist, focus that humanity fell from grace, and this is associated with stories of the world flood. According to these same traditional mythologies, before this great catastrophe there was little distinction between the inherent potential of humanity and the plants and animals, and thus all levels of awareness inherent to this earth were in communal harmony.

The result of the loss of this was the beginning of the study of the passage of time. This was the era of the Mother Goddess the moon and the womb sky of Nuit, or night. This was thought to, subsequently, "give birth" to the next great era, just as day follows night with the rising of the sun. Early mythologies about this kind of cosmology began to be recorded at the same time as the study of spiritual metaphysics became formalized. God became associated with the invisible atmosphere, that was present during both night and day, and in this way divided the sun from the zodiac.

C. Practising Manifestation in Subspace (modern)

At this time we are encouraged to begin to grasp the concept of manifestation in the material realm. This hinges upon the Manichean division of spirit and flesh, which needlessly associates these two with finite and therefore insufficient value judgments. This doctrine is one of the last modifications made to collective dogma on the nature of the aethyreal or astral realm, and states that spirit and flesh exist on opposite ends of a gradient the midpoint of which is the soul. Thus we see that, to move beyond the obliteration of the self and into the realm of pure idea is to become aware of the principals by which all things operate in the realm of base matter. This is the principle of recovered knowledge that can exist in zero time.

There are two principle types of material manifestation, and these are on a scale of smaller to larger. It would be wrong to categorize these as true or false, as these are terms defined by relative perception, and manifestation of any sort is a trick played upon perception, usually to divert the primary attention while the rule of a sensation is being broken subliminally. Micromanifestation and macromanifestation do not necessarily depend on one another, but are both effects of the same cause. Consciousness itself defines reality between these two, and as well depends upon their cause.

It would be right to call the original cause of manifestation an event only because this unit is generalizable to a whole universal set comprised thereof. This universal set is Time, or the fourth dimension, and the unit of event is probability.

1. micromanifestation (magick)

The optical obstacle course leads to rhyming ideas. In other words we can only be as inspired by what we see as what we see can be inspiring. It is thought, therefore, to be healthy to believe, and unhealthy to doubt — however, the truth is that no amount of energy anyone alone puts out through their belief can change the fact of the continual fluctuation of energy that is manifestation, and therefore it is irrelevant to the survival of manifestation whether we do or do not believe in it.

The prophase of manifestation is projection. It is because of the long term psychological effects of projection on the weak-willed that manifestation has come to be viewed by authorities as "off-limits" except to the divine.

Projection in and of itself is harmless, and actually feels quite good. Its emotional counterpart is empathy, and this emotion is, at least exoterically, encouraged by those same authorities.

Projection occurs when holographic ennegrams are externalized via the thalamus, and meaningful value becomes attached to something external to the nervous system as a memory storage referential. Because alpha waves are produced in the thalamus, the concept of divided, or differential, time becomes a factor. The result, and the cause of the psychological impact on the weaker-willed, is consciousness of the eventual loss of the external as memory referential, at which time one will have to legislate between all the things one wishes to remember about the external, and all the things one wishes to forget about the external. (This is the religious necessity of cloning, for example, for — even if you cloned some body — you would still have to summon their departed soul into the new form.)

It has even been proven by scientists that projection, which is also the basis of religion, can occur between animals of two different species other than humans. Their example is Koko's kitten, but anyone who has ever survived riding a horse past a rattle snake knows this in even less uncertain terms.

The keeping of sacred objects is not equivalent, as it was considered in the latter ancient times — after the advent of socializing monotheism, to idolatry. It is merely the earlier, no less sophisticated, associative technique of the organized mental cellular structure of the brain, common to all forms of self-motivating biological organism. The only difference is that of scale and degree of delegation.

For elder vertebrate life forms of minimal mental development this process manifests itself in memory storage and mental mapping. It would be impossible for fish, for example, to navigate in isolation without a form of the same power that humans attach to religious practice — namely, the essentially holographic superimposition of their internal reality with the external environment through a softening of the sensory field dividing the nous from the logos.

This process, in organisms of more highly evolved mental structure such as mammalian vertebrates, displays itself in the attachment of particular attributions to specific locations and/or objects, climbing up through a gradient of quantifiable displays of affection essentially initiatory in nature to culminate in the primate branch with the use of tools connecting preconceived intention to the accomplishment of specific endeavors.

Finally, in humans, we find the same process. Although tempered by the strengthening of the ego resultant from the awareness of the cultivation of awareness, which, it can be said, is what this so-called process ultimately results to, and which places humanity, in the hierarchy of our own self-created understanding, at the middle point between the objective animal and the subjective eternal aspect anthropomorphised as God, this same process is present both in the domestication of animals as surrogate self-complimentarity in the form of living examples of our own ability to subjugate reality to our holographic superimpositions, and in our equally proportioned sensations of overwhelmedness when confronted in our personal experiences with ultimate unknowns. It is the same force within us that precategorizes objects of affection for potential usage in moments of emotional crisis as seeks out causes for supersaturation of sensory stimulus.

One of the primary hitherto identified differences in the human as opposed to all other animals behavioral utilization of this process is the seeming necessity for recognition of mortality connected to the human utilization. The comparatively longer life span of the human seems to allow the recognition of changes to a being over time as occurrences in an identifiably constant pattern, and this recognition leads both to an attachment of dominance to the design based on the same attribution of final importance to duration of survival, and to a desire to attain to permanent or at least practice the acquirement of gradual changes to the form of the being which serves as the vessel of transmission for this motivating potential which is realized both in terms of the temporal template, and the raw resource of energy provided by the interaction of consciousness with material reality through the senses.

Animals seem to have much less forethought in their collection of stimuli, due, at least partially, to their having fewer items on their list of lifelong goals than the average human. With every world, philosophy, or, synonymously, lifetime created by a member of humanity there arise resultantly a certain number of new potential opportunities for the furtherance of interest along lines defined by the parameters of subsets to the temporal pattern. In this way humans can produce a unique breed of life which animals cannot, namely, intellectual offspring. On the other hand, so long as humans and animals share the same amount of overall space and continue to have interaction, the attribution of superfluously identifiable traits to animals by human intelligence seems to compensate for the difference in the pace of otherwise common evolutionary development.

It is this conflict across levels of development within the most highly structurally evolved animal minds between awareness of mortality and interest only in the immediate that leads to distraction by differences from the essentially intuitive attention to similarities that is truly their shared inherent trait.

There are three essential phases of micromanifestation. A quantum tunneling wormhole, a synchronicity or coincidence, and the assertion of will upon cause and effect. Because we know of more than we are responsible for, all we must do to cause something to come into being because of us is take responsibility for it by claiming that it is ours. Therefore controlling the process, which is the last step, is actually the simplest and easiest as well.

a. random information transport

Just as a random fluctuation in pure potential caused by uncertainty created the first probability in the history of our universe, so does this same process continue today in the form of random quantum tunneling through the existent quantum foam. Each of these quantum tunnels is a temporal singularity creating an interior zero-time, hyperspace event. If this event is expanded upon in space by supersaturation of the surrounding quanta it absorbs and projects at two different, distant points in spacetime, it becomes a wormhole. The quantum information that passes through such a temporal tunnel or wormhole can comprise the natural ambiance of the vacuum of deep space, a message coded into a pulsed wavelength, or even biological organisms. The geodesic influx of gravitational magnetic energy causes these to occur everywhere all over the globe, and as we pass through them we react emotionally because our perceptions of what we see differ.

b. acceptable qualities, quantities and velocities of meaning

We must always beware what we mean, and be aware of what we mean at all times. To this extent our measure is logic, by which we measure factual evidence against belief, faith and degree of trust. This determines the scale upon which these two parallel realities come into existence before the perception of the ego, the conscience and idea of self. The meanings we accept in this way are units of karma that accumulate into the guiding gravity of our fate. The trade in them unwinds the lifeline, making the individual perception of time "turn around."

What we see will be a mirror of what we can expect, which is only a fraction of what we know, and we know this to be much less in turn than what we do not know. Thus any form of seemingly meaningful "sign" that we perceive is in truth a co-creation of reality and the mind. Its significance is dependent upon the presence of an aware observer, otherwise it is meaningless and therefore non-referential. In different words, the mirror does not reflect my face if I am not facing it.

The concept of the wishing well is a good example of early, possibly prehistoric in origin, human conception of information exchange through a wormhole.

c. inversion

Whenever manifestation occurs there is a build up of potential energy preceding it that is exchanged through the wormhole into the manifested material. This is the concentration of energy in one place in spacetime as another point in timespace is distorted towards it through hyperdimension. At the moment when it occurs a subspace quantity of quantum information passes through the temporal singularity. This information is the manifestation. By bending it into the material reality of spacetime, the manifest information is inverting the substance of the wormhole from its normal state of potential energy in the fourth dimension.

A good example of inversion occurs in social interchange theory. In this area of human development the rhetorical argument has been refined into the format of: introductory statement; thesis statement (main body); conclusion. In this practise the use of inversion has become common between either the introduction or conclusion — whichever contains the antithesis — and the thesis, leaving the other to represent the synthesis, or manifest resolution to the conflict between the two ideas.

Because it occurs in arguments even when they are being made with great emotional energy, inversion has come to be associated with the sudden break from tension into disunity described variously as "the situation coming to a head" and "the straw that broke the camel's back" and is associated with a pressure drop and feeling of compression. (The oldest existing description of this is as zimzum in Hebrew QBLH.) By the same reasoning, because inversion is the reversal of matter and energy, like fact and fiction, from virtual to real particles it can be achieved by epoché, or diminution, upon the dasein, or pure being — in the here and now. Both of these perceptions are dependent upon variable conditions of the inversion, though.

Although the original masonry of QBLH in pure dimension has been "lost" (according to modern Masons), the body of literature associated with what has come to be considered "pure" QBLH continues to represent a vast collection of possible techniques for inversion. The languages of QBLH are based on the encoding of vibrational frequencies, associated with their equivalent factors according to alphabetical number sums. Much of QBLH is verbal, although, originally esoteric, it pertains more to written concepts associated with the spoken sounds in theory, and to intonations without sufficient referentiality in practical craft. These ceremonies and codes serve as an automatic inversion between the realm of the ideal and that of the manifest, whose full effect is determined only by the initiate's degree of open mindedness. The literature of QBLH serves only in lieu of a theory for true esp.

2. macromanifestation (magic)

When my generation was young and our imaginations were at their peak potential for solution seeking, our elders behaved towards us and tried to reinforce for us through ritual the roles of blank slates, in order to pass on the idea that the only cause of entitlement is timing. While this was done out of the best intentions, attempting to impart their highest wisdom — that there is no such thing in reality as responsibility, only association ante facto — it nonetheless came to be only as valuable to us as one more distraction in our event history. This has apparently been a problem in education for at least as long as there has been domestic child abuse, for both are mentioned in the bible, where it says that we must "spare the rod and spoil the child," that we must "enter into the kingdom of Heaven like a child," and where it also says "an eye for an eye, a tooth for a tooth."

The result of this type of socially indoctrinating interaction mechanism overlaying the process of education is the imitation of the official version of global history: where the pre-homo sapien humans and early Atlanteans are considered the youth of humanity, the flood a coming of age, and industrial technological development work. In short, they give you credit for having already known none of the recovered information you are reawakened to at school, and encourage you to believe that you are only aware of as much information as you can consciously (meaning against the opposition of social distractions) account for beyond the shadow of a doubt, wherein their doubts in you are tested out against you so that you will grow to fit the mold they have created for the individual based on the official version of history, which constitutes the entirety of their expectations for you.

When we do not play the roles created for us by the expectations of others, the labor will become displaced onto another body, and this will represent the part of ourselves that we have assigned to be exiled by projection into manifestation. This allows us to play the role of God. Whenever there is more than one person around playing the role of God, this devaluates the role, and causes the entire paradigm to crumble into the dimension of the magic theater, where all are demigod-kings, representative of the universal harmonic vibrational frequencies of potential energy in hyperspace. It is these strands alone that the multiverse is comprised of, since all the matter-energy absorbed by the black hole and filtered through the wormholes is converted through the inversion engine of the singularity into the baby universe, and the wormholes are only filaments of potential energy themselves. In this level people are more or less psychic, their own roles, and thus their own appearance, being distributed out amongst the masses statistically, while they come to resemble the statistically averaged appearance desired for them by the souls displaced from the bodies forced into secondary roles.

a. duping somebody

The mind is a five dimensional hand. It is capable of accomplishing all the same feats as we are accustomed to performing ourselves in the physical world. Its strength is the spirit. It can also palm objects and misdirect attention from a sleight of hand, so to speak.

A mental conceptual magic trick is simply a conceptual lie — a false premise based on unjustifiable and wrong reasoning. These are acts of bad karma, even the little "white" lies. These are considered psychically unhealthy because they introduce dead ends into the flowing progress of consciousness down the infinite forking pathways of the transdimensional evolutionary holognomonic system.

Get what you can out of life: this premise is agreed upon by all philosophies. Good karma is giving back all of what you don't want and half of whatever you have left over that you don't need. This is living comfortably, and this increases the probabilities of dying comfortably — hence morality has become associated with a rewarded afterlife. In this way, by getting all you can, and by giving back five eights of it while still alive, or else setting the process in motion for perpetuity, philosophy and religion coincide. However religion gives all the glory to God, and philosophy prefers to contribute to more or less humanistic pursuits. Survival is reasoning thus.

Any use of reasoning can become corrupt by false premises. These are simply wrong information. Often times wrong or misinformation is created on purpose, as with disinformation leaked to intermediaries in order to confuse the heads of competing groups, but most of the time misinformation arises accidentally. The machine was introduced initially to limit misinformation, however as long as it is under human control it is limited itself in how much it is being applied to its goal.

b. duping everyone

It is possible to induce results in reality through the consciousnesses of others by convincing them of the solidity of your conviction in the concluding inversion at the end of a stream of consciousness that has been gradually accumulating energy within your nervous system. Although it should not shock you to learn this, it probably will — just as it should not shock the audience to see the magician pull a white rabbit out of his black top hat, though it always does.

Manifestation, the practical exoteric application of speculative esoteric materialism, has become intertwined with the barter exchange system that has evolved into the valueless token exchange system of capital today. In this economy, manifestations accumulate in aggregates of stereotypes in genres and in merchandise. Usually manifestation is not associated with the performance of a service; this is the purely ritualistic side of money.

The world of finance is a complex system of impossible loops that is wholly self referential. This is not to say that one cannot get in or out, for this is done all the time. Nor would it be fair to class this creature under the same phylum as any form of biological organism that has ever existed on this planet before. It is, nevertheless, what it is: a closed system. All access points are finite, but there is infinite potential within. Richard Nixon obviously agreed with this philosophy, himself a freemason who knew that taxation was entirely premised upon disinformation of the American public, when he took the American dollar off the gold standard, thus floating the loans of the Nation to every country to whom we owed outstanding debt, and declaring our statements of debit and credit invalid and therefore null and void.

c. duping yourself

Whenever there is an unpleasant reality, the mind's instinct automatically creates an alternate reality for the personality to retreat into. This is simply a coping mechanism that naturally accesses the multiverse, but the same effect can be accomplished by a calm, rational choice more easily. Here we see that the mind will obey all the same rules as the dimensional geometries formed from quantum foam — up to and including the creation of alternate personalities as baby universes. In the same way that all baby universes are virgin births so is each mind its own unique kind of singularity — neither exclusively gravitational nor temporal.

It is in this way that the personality first begins to form self referentials or role model archetypes. These can take any and all forms that they like, and operate in an ideal state more or less independently of the self, only answering immediately when called upon and the rest of the time at utter liberty. We consider these spirits of our own higher being that exist in the aethyreal realm just as much, and in much the same ways, as we do in the material realm. Our ancestors like to use ritual magick to summon the ghostly memories of their own ancestral dead, however our knowledge of this remains cold and detached, while our use of the mental projective space is free to wander with the will power of our imaginations.

The interconnectivity amongst past generations seen by the shamanic elders of the yet primitive tribes is much alike the interconnectivity of suburban yuppies on the internet, or rather like the social party rule of six degrees of separation applied to zen Buddhism in which we are all just reflections of one another, therefore it is very tempting for us to lie to ourselves about such things. To have contradictory ideas, modes, paradigms, or even religious beliefs is quite common now, and there are fewer and fewer authorities who are worthy of answering for the very decrease in deserving authority. Take for example those authorities who feed off the repressed desire of their subjects to leave and you will see the quantities of injustice.

Some say that inversion — the necessary transference between opposing viewpoints, their union-intersection, like an all purpose right answer at any given wrong time, is necessary for the rolling of probability over from one event to the next. This, in turn, constitutes the forward flow of entropy, measuring the unfolding passage of time, only insofar as this effect on probability is in the nature of disruption and disorder. It turns out, it is not only necessary that it be thus, but a given as well, for it is random quantum fluctuations, due to the uncertainty principle, that caused the universe to come into creation, and that continue to transport information inside our brains, or even through hyperspace across the universe, to this very date.

Still, it is worth noting that the real measure by which we perceive time is by space (distance) plumbed by the speed of light, which is energy upon the distortion of gravity by the substantive matter of the universe, which itself is a field comprised of microwave tachyons involuting opposite the universe's entropic expansion — thus against time.

3. common odds

It is safe to say anthropically that there would be no magic without magicians, just as it is safe to assume that most, if not all magicians, are more or less like you and me, and that, moreover, for them to share a magic trick with us, it is necessary that they inhabit the same continuum. This is the same continuum shared by the mind through manifestation. It is a continuum of pure probability.

Pure probability is the substance of prediction in quantum mechanics, which is the attempt by applied mathematics to bring the workings of subatomic spinning particles and force carrying waves into the grasp of the mind. It is more commonly known to the gambler as chance, and control of it, which they seem to think is random, they call "luck." Science has given us other names for it, the best of which is "probability." Either simply means the likelihood of an event happening or not.

It is difficult for the modern mind to fully accept that our entire universe, the timespace continuum itself, the fourth dimensional motivation of our third dimensional reality, is no more than a fluid math of randomly fluctuating measures of "probability" (manifest nothing) arising from and within pure potential (ideal nothing). But this is the way it is, mathematically.

Now this "continuum" is not constant. The particles of the atom themselves resemble a miniature replica of a solar system gaping with empty space, while even the highest frequency microwave background radiation only vibrates its gravitational waves through subspace in the pattern of the dispersion of galaxies in the walls, filaments and voids of our universe. However, it is important to note that the manifestation (the probability or magic "trick") is merely a concentration of potential (or that which we know is between the subatomic particles and in the universal voids, i.e. nothing, for such as do we know) that amounts to no effect other than the creation of spin in particles or the stirring of force carrying waves — for if it did, all probability as such would be revealed for misdirection (and the "magic" would be taken out of the trick).

a. dimensional molecules

Fermions and bosons are molecules of dimensions. Their spin potential variability is determined by transdimensional interaction. The dimensions themselves remain outside of and beyond the forms they generate, as the nuclei of two covalently bonded atoms. They are as the philosophical definition of the ideal.

Before we can consider the specific cases of fermions and bosons we must delve a little deeper into the nature of the dimensions. It cannot be argued that the first dimension is a singularity, the ultimate embodiment of symmetry. The second dimension is, equivocally, planar geometry. These are ideal forms, impossible to express in our material reality, and exist as realms of pure equation. Without them, however, we could understand nothing of our own condition, and would posses no mathematics by which to quantify nature. In fact, without them, nature itself could not have assumed the forms and functions which we now know. However, even the simplest understanding we posses cannot belong to one of these dimensions alone. Consider the point for example. Its properties are compression and scale correspondence, where the former applies to a singularity and the latter to planar geometry. Neither has any meaning except in relationship with the other, and what meaning we do comprehend of either is therefore a creation of both. Without the point, a Cartesian coordinate system could not exist, but without a singularity, there would be no point.

This natural progression continues upward through the dimensions even up to light and beyond, which I have here classified as the fifth dimension. This is due to the unique properties of light which seem to imply a domain unto itself, particularly as it relates to time; the primary example of this being the seeming transtemporal nature of tachyons.

The fourth dimension should be set aside for special consideration as well, as doing so now will relieve the pressures of later describing the particles it helps to form. The fourth dimension is simultaneously spatial and temporal, meaning that it generates highly complicated structures over fixed intervals. The complexity of its spatial component is due to its having as its foundation an open geometry, in progression from the flat second and closed third dimensional geometries. Its temporal component, however, is semi-second dimensional, as it is traditionally observed to unfold binarily between past and future. I have demonstrated elsewhere that this relationship is indeed far more complex than this, but in the context of this exposition we need only remember these two defining traits for this dimension.

Now we may consider the dimensional molecular cases of fermions and bosons.

Bosons ("real" particles) posses fractional spin as they occur between the third and the fourth dimensions. They arise as the result of the interaction of matter and energy between the pure, formal geometry of the third dimension and the pure, formal time of the fourth dimension. They begin as ultimate, unformed material chaos. As time passes this mass becomes motivated, and an exchange of matter and energy begins to take place, constructing a framework of preexistent geometric relationship to govern the materialization of real particles. Bosons' symmetry is quarterly in three dimensional coordinate space, meaning in any direction it is rotated a boson will look the same after less than one complete turn. In other words, its matter-energy structure is determined by geometric (symmetry) and temporal (spin) interaction. These may be measured as either particle or wave.

Fermions ("virtual" or force-carrying particles) arise similarly, but posses integer spin potential variability. The fourth dimension exists in field form, and the fifth dimension in no solid form at all, and therefore there are no geometric restrictions placed on symmetry. Fermions thus existing between pure Time and pure Light can only be measured as waves, and as such obey spin in accordance with oscillation interval, or frequency constant wavelength.

The nature of further dimensions, and therefore implications for the particles they may generate through interaction, is discussed elsewhere.

i. dimensional relativity

Just as we observe that on the scale of atoms there is a structure which mirrors the arrangement of the planetary spheres in the heavens, with the electron orbiting around the nucleus as does a moon around a planet, or a planet around a sun, but that the structure of the subatomic particles themselves is probabilistically regular, alike that of the planets, moons and stars, each having its own set of characteristics and rules for behavior, and no two classes being identical in either, so must it also follow that the patterns governing over the smaller aspects than the subatomic particles should follow the same structural organization as that which is of the lowest order for the arrangement of information in these larger scales. Thus, for the properties of dimensionality itself, it should be observed the same ordering as occurs with and for the heavenly spheres and for the electrons of atoms, that they, being of an order beneath the spheres of the subatomic particles, just as are the subatomic particles exponentially one progression beneath that of the heavenly bodies, should obey the same rules for behavior as the orbits of electrons, just as the electrons obey the same rules for behavior as do the orbits of the satellites of space. The force that acts upon the satellites is gravity, and its constituent particle is the hypothetical graviton. So, the force over the smaller electrons is microgravity, and its constituent particle is thought to be the tachyon. As explained elsewhere, the graviton and the tachyon are really the same particle, carrying the same negative entropic force on two different size scales. Though even smaller than any particle, even the superstrings of tachyons which are thought to be infinitesimally small, the same force still applies to the substance of space itself, and in this capacity it is called time. Time is a movement which warps space into the shapes of the smallest particles, tachyons, which shape is a torus. Therefore it is this shape that we must look for in the raw substance of the continuum in ts most fundamental form, that of dimensionality itself. And here we see that dimensionality is, indeed, a unity that is divided into a plurality, and that this is because its levels of plurality are orbital paths, like those of electrons or satellites, occurring within a single pattern of a torus.

Thus, we see that manifestation occurs as the orbital vector of dimension. As dimensionality itself is revolved, manifestation occurs. In the first revolution dimensionality manifests the singularity. In the second it manifests the plane. In the third it manifests solid shape. With the fourth revolution dimensionality manifests time itself. Beyond this is not known to modern physics, but it is held within the esoteric tradition that the fifth emanation is pure light. This is in keeping with gnostic doctrine, which expostulates that before the beginning of creation there was chaos, and this chaos was a darkness. This darkness was a shadow beside the Light of God. That is that, the act of creation itself, what I have been referring to as the act of manifestation, was itself a solid thing, itself a manifestation, and that between God's Light and the unformed chaos this created a shadow which was the measure of nothingness, known as the primordial abyss of Ain. Therefore manifestation is said to have two sides, a light aspect and a dark aspect, and between them an inversion. But we have also seen that this first manifestation is the manifestation of dimension itself, since there is no smaller measure for the continuum to set it aside from the void, and that it is by the rotation of this first manifestation, that of dimensionality itself, that the manifestation of the subsequent known dimensions comes into being.

Now just as the darkness and the Light were the two loci of the diameter of the universe when it first began to spin into form through dimensionality, and the darkness was but a shadow of that dimensionality cast off by the divine Light, so did these loci become separated by the rotation of dimension along its orbit manifesting; the darkness rising up to become the matter of space, and the Light sinking down to become the progression of time. Thus, just as the particles of matter and energy occur as though shared electrons between dimensions like as in a molecule, so does each dimension revolve in its way around the two loci of space and time here within our universe.

Now time is a greater force than the matter-energy substance of space. Together they are a unified continuum, but it is time which is outside of and above space that causes it to be moved in the way of dimensionality. The course of the manifestation of dimensionality itself, thus, because the two loci of its orbit are the substance of space and time, must pass around and beyond them, so that part of the time pure dimension penetrates the divided plurality of the manifest universe, and part of the time passes beyond its confines and outside of it into and through the multiverse of refractions of the one real timestream into the hyperdimensions of alternate possible timestreams.

The shape of the torus becomes apparent when it is reckoned that, while dimensionality itself is orbiting in its course of manifestation between the divided dimensions of the local universe, between which orbital rotations the particles of matter and energy, among others, occur, the two loci of space and time are themselves perpetually changing place with each other, so that, while around them pure dimension makes an orbital process, within this perimeter also there is another orbital loop being made by the two loci around which the greater orbit revolves.

Because of this precession of the inner orbit we can therefore further subdivide the two loci. That which began as Light greater than the whole of all that is, became time outside of and motivating all that is, and can now be known as timespace, or the external surface of the continuum that is all that is, that which is also called the multiverse or hyperdimension. That which began as darkness and unformed chaos before the creation of the universe, and the volume of space thereof upon its creation, may similarly be called spacetime, or the substance of the continuum, space infused with time, the interior surface of the one universe. Thus, also, it should be noted that it is always spacetime that is nearest to the manifestation of multiple dimensionality that is pure dimension on its orbital course of manifestation. The reason for this shall be expounded upon presently, for it is a series of shadows.

Now dimensionality within space is as one sphere within another, and as dimensionality passes on its circuit of manifestation, so space is changed over time, such that the sum over histories of space, or the multiverse, assumes a form to encompass this orbit, and thus takes on and is bound to the shape of the torus. So too does time, which remains a force outside of space that can only be observed as the speed of light squared, thus double itself about the diameter of the interior tube of the multiverse, wherein is the space of this universe, and, it is known by the forty five degree angle of its diagonal, takes up to itself the shape of a tesseract, or hypercube, of which much more shall be discussed duly. Thusly, spacetime is the interior surface of the sum over histories torus of the multiverse, and timespace is the exterior surface of the tesseract that comprises the force of hyperdimension. The relativity of dimensionality proceeds, therefore, thus: manifestation (the orbit of dimension) is the shadow cast by (pure) dimension; (multiple) dimension is the shadow cast by space; (space is the shadow cast by spacetime; spacetime is the shadow cast by timespace; timespace is the shadow cast by hyperdimension; hyperdimension is the shadow cast by time ... etc. shortened to...) space is the shadow cast by time; time is the shadow cast by manifestation. God is above and beyond, the source of the Light.

ii. conservation of dimension

Pure dimension orbits elliptically the double loci of space (volume) and time (motion), which themselves orbit each other so that space always faces dimension.

At different points in its orbit pure dimension casts the appearance of multiple, different dimensions. This is called manifestation and is measured by geometry.

Manifestation is dually active and passive. As a verb it represents the motion of time. As a noun it represents the measure of space. When pure dimension is at the antipode point in the orbit of manifestation space and time overlap to create the volume and motion of the local universe.

As pure dimension orbits in manifestation its poles precess, such that they alternate in their orientation. Thus the precession of manifestation leads to the conservation of dimensions such that plural dimensionality scale compresses to the original singularity of ylem at the time of the big bang.

Manifestation is the vector of dimension in the same way as time is the vector of space, hyperdimension (n-dimensional multidimension) is the vector of subspace (the infinite volume zero mass vacuum), and as the multiverse (sum) is the vector (over) of the local universe (histories).

The precession of pure dimension in subspace (the apparently flat surface of gravity bent by mass) equals the lightcone history (past and future) of the universe in the hyperdimension sum/history multiverse outside of timespace, which is the exterior surface of spacetime, whose geometry is different (flat, closed or open) in different local areas due to the conservation of dimension.

As pure dimension orbits through manifestation perpendicular to the temporal vector of hyperdimension in the multiverse (that is, the precession of manifestation), the past and future light cones revolve around through the torus of the sum/history multiverse outside of and around the local space-time universe.

The measure of the lightcones of the antipode point of pure dimension, where time and space overlap, is phi, or a regular conic spiral turned at a forty five degree angle. Here we see that the lightcone does not represent a past and future projection of vector, but the regular degree of curvature to the background radiation in the subspace of the local universe. The measure of the lightcones for the projected past and future vectors in the hyperdimensional tachyonic multiverse outside of, containing and surrounding the subspace local universe is the same, but because they are relative to one another along the precessing orbital vector of manifestation as pure dimension casts the geometric appearance of multiple dimensions, the measure of the multiversal torus comprised of the sum over histories of these projected tachyonic light cones in hyperdimension is relative to the precession of manifestation, and therefore its curvature is pi, a complete, cyclical, circle. Thus, the measure of the lightcones of spacetime and timespace of the finite dimensionality of the subspace local universe is the sum over the histories of the projected lightcones of future and past trajectories of tachyons in the multiverse, or phi/pi.

b. instantaneous manifestation

Instantaneous manifestation is defined as the focus of time when potential energy expenditure and information being delivered are exactly equivalent, as in the case of perfect clarity upon some point, or the passing of a thought perpendicular to that point, but of great elegance.

Instantaneous manifestation means equalization of energy expenditure in the realm of potential and information being delivered. This pertains to the duty of the messenger. There is much literature on the synchronization of one's mental projective waves of probability to the natural, inherent vibrational frequencies of solid matter. There is a strong belief, rarely realized, among researchers that time goes only as slowly as you let it.

The relationship of the mind to time is one of the most complex geometries of pure dimension, in potential — for what is not defined specifically remains present in potential — and thus all the dimensions remain balanced between their outward pattern and their movement of potential. Here we find nothing to confound ylem, or pixelate the subordinate hyperdimension; only pure potential. When we say that it is complex in potential, it only means that it is simple in fact, though the simple units we are using to measure complexity are themselves potential. Here we find that, as in all pure dimensions, the relationship described between consciousness and potential is one of energy exchange.

Thus we can recognize patterns in event, for here, when the concentration of active consciousness is applied to it, it mirrors the hypercubic form of the fourth dimension. We perceive and catalogue the data that some patterns of energy change more than others over time.

This is true both for the patterns that arise in the dimensional subspaces where manifestation occurs and for the fields of potential that occur between these realms. Therefore instantaneous manifestation occurs when active consciousness is aligned directly with the fourth dimensional event.

It is caused by the transmission of potential energy from one pure dimension to another in a route circuitous to the subspace between them. This allows for a certain amount of subspace to be left over in the multidimensional elemental situation of an event, and this results in one of two things happening, each in one of two ways. Either the extra subspace manifests hyperdimensionally, in a way either obvious or oblivious to extent, or in accordance with the argument that all matter is potential energy, in again a direct or abstruse way.

In the first case it would be in a quantum information orbital state close to the data pattern of which it has been shed, and therefore may result in manifestation similar to a thought's line known to be occurring on one channel of your bio-electromagnetic aura. In this way you could say that the manifestation was "close" or was "aimed at" event. This is the cause of both all miracles and all accidents.

In the latter case then we are in a situation of chaos in closed systems, where there is no hope of predicting the manifestation's relative position of occurrence, unless it were plotted as upon a phi-bound projection from a pi-bound orbit as pure information units in hyperdimensions, or as a phi/pi spiral, assuming one has found sufficient signs by which to measure the positions of these orbits of pure information.

The question becomes — which of these perspectives came first? Is it true that, for someone who drew before they spoke (or who perceives instantaneous manifestation as hyperdimensional before material), one could expect that they had actually seen images in images with their minds before having conceived of in words descriptions of the impossible?

In any event instantaneous manifestation, being somewhat of an ideal meteor, usually is found to occur in the form of the appearance of objects that coincides with a thought-line. It may be speculated that this is a type of proto-telekinesis, proving that the Great Creator preceded the Primary Motivator.

c. random quantum tunneling and the uncertainty principle

It is said by the bravest minds of science today that the universe began with the first random quantum fluctuation, and I have no doubt that we are still living in this same expanded singularity today. Theologically this displaces the prime motivator as Grand Creator with the uncertainty principle being the only primarily preexistent condition. The uncertainty principle itself is a factor of the speed of light, which is the inversion between manifestation and entropy in this third dimension. According to this theory it was the force carrying photon that was the first particle — a quantum tunnel that was alone without any other points to connect, closed up into a sphere, like real numbers inside and like imaginary numbers outside.

Now, however, the gnomonic and fractal expansion of fluctuation within that original seed of the first Light has progenated the quantum foam — a sea of various frequencies, some combining into harmonic melodies — the background microwave radiation continuum. Thus when a quantum forms the interior of a tunnel it can connect various points in spacetime. However this still occurs mostly, if not entirely, at random and, it is thought, due to the uncertainty principle. Admittedly random fluctuation and the uncertainty principle alone do not account for the myriad of patterns that are formed in the universe.

d. continuum singularitas

If we accept that the universe is a unified field such as the vibrational model or the multidimensional model, then we can say that it is altogether apprehensible by the mind as itself a single unit. One such model of the universe is the Jachin and Boaz model, known in the east as the yin and yang model, which combines two inverse fields — one more manifest, and one more aethyreal; one representing beneficial karma, one representing unpleasant karma. It is thought that, though opposite, the two forces are one. Buddhists, who study this model, give the name ylem for that which existed before the big bang, and Zen Buddhists call the energy underlying the universe chi and the guidance of this energy tao. If, as I have described as being indicated by the theory, we say that the uncertainty principle accounts for ylem and tao, then quanta themselves comprise chi. However another perspective on these same factors is the concept of the Primary Clear Light. This is the state of enlightenment (or absence of neuralectric kinetic energy in the nervous system) associated with Nibana or Nirvana achieved by the trance of Samadhi or Samasamadhi.

Thus we have in our hands all the pieces, both modern and ancient, of the cosmological puzzle. Let us see how they fit together. We may associate ylem with the Primary Clear Light as a precedent has been set in previous consciousness research. Also we can see that chi possibly describes waves and microwave frequencies of electromagnetic radiation in a field of ambient potential energy created by random quantum fluctuation. Therefore tao would be, by process of elimination, quantum probability wells. Tao and Chi are Yin and Yang — good and bad karma, over time. Yin and Yang blur out of ylem in the beginning and into it again at the end.

e. the prediction of events based on random information theory

A good example of inversion occurs in social interchange theory. In order to understand the inversion inherent in social interchange systems we must understand that there is an exoteric and an esoteric side to all social interaction. On the one level we are communicating only what we are communicating consciously. However beyond this level are countless others, where information as idea is constantly being exchanged, that we can only sense emotionally as the shadow of our social conscience and relative light of social spirit, and that remain obvious to us at all times, and yet of which we remain unconscious.

If we adopt the notion of ownership over the full capacity for awareness of our mind, as philosophy suggests as the course it has followed since the age of Reason, then we must accept that it is only as ordered as we have allowed it to become. If, however, we reserve rights to a "higher power," such as any of the versions of God offered by institutionalized religions, or if we give these rights away to a spouse, the state or the devil (all karma adds up to the same sum, whether "good" or "bad," "black" or "white," namely: nothing), then we can assume that the world of which we know nothing and to which we continue to remain blind is founded upon whatever form or degree of order we imagine associates to this nobody of nothing.

In either event the only measure we are given to plumb the depths is inversion between the end pointing at us, of which we know "something" and the other end disappearing into the void, of which we know "nothing." When we communicate with another, this inversion ceases to be the inversion between "something" and "nothing" and becomes the inversion between "something" and "something else."

In this area of human development the rhetorical argument has been refined into the format of: introductory statement; thesis statement (main body); conclusion, with the inversion between the thesis and the antithesis. At first, philosophy had adopted this form verbatim to express itself, however, it eventually sided entirely with the dramatic chorus against the dialogue of the bit players in favor of attributing both thesis and antithesis to the same voice, thus delimiting possible reaction, and gaining better target audience control. This represents a devolution from the communion of the double helix (inversion between two "somethings") towards dictatorship of force according to singular wave function (inversion between peak and trough of one "something" and the surrounding relative "nothingness.")

Although the original masonry of QBLH in pure dimension has been "lost" (according to the publicly released literature of modern Masons), the body of literature associated with what has come to be considered "pure" QBLH continues to represent a vast collection of possible techniques for inversion. The literature of QBLH serves only in lieu of a theory for true anomalous cognition by which to communicate the "tradition" in a way meaning "to sing a song in silence." Originally QBLH was all of the esoterica, the unifying geometry of the multiverse, and in this way is originally representative of the twin helix of Life.

One of the finer points of QBLH rhetoric is the distinction between fact and fiction. This dates back to ancient bigotry between the initiated and the uninitiated. The initiated were horrible know-it-alls who were, moreover, rich, while the masses of humanity toiled in debt, starved and went ignorant. Much of history is the story of the reversal of this fortune up to the modern day — the uninitiated slowly coming to grips with all the secret facts of history, usually only as conditions for their continued loyalty to substandard living status. The terms of fact and fiction constitute the entire language of the social compact as it effects change in the society of its citizens, which it calls culture in order to depower it, calling itself society. According to this agreement, facts move upward through hierarchical systems, while fiction moves downward. Since the concept of a naturally formed hierarchical system (such as the gnomonic growth pattern of a tree or the double helix of DNA, both factors of the negentropic evolution of life) is contrary to the relatively heathen belief in the uncertainty principle as being the only guiding light for all universal being, the full implications of this agreed upon association go generally unrecognized by the uninitiated.

D. practising manifestation of pure geometry in pure dimension (ancient)

It is perpetually becoming easier, and therefore wiser, to accept that our common ancient ancestors had a much greater knowledge of the stars than their modern uninitiated antecedents. The popular paradigm dogma has become the rule by the elders of the children, and this is becoming safer to apply holographically to human history, perhaps, than it is easy to carry out in actual life. What the forefathers of our civilizations concealed esoterically in their descriptions of the cosmos may itself only be the tip of the iceberg of our ancient ancestor's knowledge.

One such governing theory, understood by the initiated but not understood by the uninitiated commoner, is Astrology. It is a systemization of the heavens according to temporal differentials, however the given explanations of it are needlessly complex and convoluted, and have historically distorted the underlying and fundamental practical astronomy of it. It apparently derives from an ancient art known of in great detail at the lost roots of all existing peoples and civilizations. An interesting application of Astrology often overlooked is the possible correspondence of data about people born under the same conditions of timing, determined by astronomical alignments. (If such a record were compiled over the ages it would produce a very interesting tapestry woven of many various people's lives and bearing an image that might offer better insight into the overall shape of the human mind than any method before.)

We know that the ancient people of the earth, whose history is largely otherwise lost, erected enormous stone megaliths in perfect astronomical alignments. Our history of civilization begins with the creation of the zodiac, and constellations formed by imagining connecting lines between certain stars in the sky in patterns reminiscent of natural or otherwise known objects. Their exact observations were based on the simplest of means, and though monumental, their works are aesthetically minimalist and ergonomically utilitarian. They have however endured for all time, and many maintain even to this day.

Because of the ancient's observation of the stars being relative to our own temporally, and because the events observed by our ancestors and the events that we observe are also relative to the distance from us at which they occurred temporally, we can extrapolate the formula that our ancestors were observing events at a time when events other than those that they were observing were actually occurring at the distance to their origin, and that these events will be observed before the events that are going on right now at the distance to the origin, which are also different from the events that we can observe now. An application of this would be realizing that what we are seeing now for a star 2000 light years away is relative to what was going on on earth 2000 years ago, and that what we are seeing now for a star 4000 light years away is relative to what was going on on earth 4000 years ago.

This is the measure of inversion our ancient ancestors used between their own social "something" and the "something else" of the stars in the heavens. Because what they knew of the stars was exclusively idealized, and what they knew of the ways of the earth was exclusively material, this inverse relativity is in hyperspace.

1. the theory of Atlantis: a lost supercivilization

It is thought by the modern occult that, just before the beginning of the era of the Mother Goddess, and ending with the menstrual flood associated with her divine birth rite, there was a civilization that was similar to our own, and in very many ways surpassed it, particularly in fields of technology. This thought has been disseminated by making the writings of the author of the original description of the lost world required reading in the very same educational university system he had helped to create. Therefore, although his own words tell us otherwise, the entirety of the world's ancient, lost civilizations have come to fall under the nomenclature he gave to that one land, and any differences between areas and eras is on the brink of becoming hopelessly lost to a semantics battle between yuppie and hippie new agers, inheritors of the eternally unworthy role of torch bearer to Atlantean illuminism.

It is widely known today that around the world there are cities that lie off the shores of many countries, below the level of the waves on the surface of the sea. The most probable explanation for such cities is that they were built during the last ice age, when global sea levels were lower. However this would mean establishing a much earlier date for the beginning of civilization than is traditionally accepted. Currently it is taught in our schools that civilization first arose in Sumer, also called Mesopotamia, between the Tigris and Euphrates rivers, as well as, around the same time, along the Nile river to the southwest and the Indus river to the east. The oldest known artifacts of civilization in existence date from this time, about 9000 - 7000 years ago. The peak of the last ice age, however, was closer to 20,000 years ago.

These are only some of the unexplained artifacts of the late stone age, however. Clearly a great art of stone masonry has been lost to our times, for beyond our means are the constructions of many ancient megalithic centers, shrines, temples and monuments, involving the quarrying and transport of solid stones, some weighing over 100 tons, most weighing over ten tons, and their exact placement in alignment with astronomical signs on the horizon and at angles in the sky. Since we also lack the ability to determine when they were moved into place, we do not even know when the hundreds of thousands of sacred stone sites throughout the world were erected, and therefore we know absolutely nothing about the people who did so.

We have only recently discovered the worldwide knowledge of the hunting weapon, the atlatl, and now believe that it was already known of globally during the last ice age. The atlatl itself is no more than a wooden spear propeller. Nonetheless before its discovery, no such tool or implement was imagined by the greatest minds of modern men when they imagined back on our common ancient ancestors; and it is in fact a marvelous adaptation to the cumbersome spear that reveals a level of knowledge of physics that anthropologists still consider to be as much of an anachronous advancement at the time as was the atomic bomb in the 20th century.

While for the uninitiated Atlantis remains a pleasant myth, for the initiated the fact of ancient lost highly technologically advanced civilizations is an ever present reality. It is equated with the fundamental masonic practise, as well as fundamental religious doctrine, of confronting one's own mortal finitude. Its inevitable ending has become so wrapped up with the mortal or living flesh that the very word that describes it, "mortality," has come to mean instead only death. So, with the world's great civilizations, have all of their efforts amounted only to the myth of the moment of their loss.

a. megaliths and underground rivers

Many stone monuments are not only aligned with celestial events, but are centered over sources, junctures or near-surface points of underground rivers. The only known way finding underground waterways today is by dowsing, which involves using a wishbone shaped stick held lightly and allowed to droop. There is no scientific method to dowsing, and it has not found any basis in scientific work, however its success rate can be measured against the success rate of remote viewing and the temporal fluctuation of the earth's electromagnetic field and found to correspond. This indicates that dowsing may have more to do with neuralectric sensitivity to earth's electromagnetic field, and that this electromagnetic field may have to do with the paths taken by underground rivers. It is not known if dowsing is the descendent of any ancient art, but according to the legends associated with many ancient megalithic sites by the indigenous people, especially when these same people believe the megaliths to have been built supernaturally by an alien race of gods, the megalithic sites are associated with early wells during the age of the Mother Goddess, and subsequently with blood rituals at the beginning of the age of the Father God.

b. underground rivers and ley lines

Just as the various megaliths plot the progress of the various underground rivers throughout the land, the alignments between them are known as electromagnetic current ley lines. Just as they had discovered vast underground caverns and lakes during the ice age, when the ground water level dropped along with sea level (Tennessee remains perched atop a network of deep caves and subterranean waterways and lakes), so did the ancients establish a web of energy channels guided by the megalithic conduits, all across the face of the land. This energy was called the Vril or Orgone energy by German occultists immediately prior to World War II, although it was also identified earlier by Hebrew scholars as Ophanic energy associated with the Ophanim, cycles or "wheels." It is essentially similar to biophysical energy present in the nervous system, which Freud called phi, however it is equivalent to a very great amount of this, such as Count Orgaz identified with the Orgasm. The control of this energy in the confines of the biophysical system is the long time practise of the east, where Tantric Kundalini channels the chi through the chakras, and where acupuncture and acupressure free the energy paths between pressure points. It is probably not very far off the original intention of both to compare these two schools on the subject of the electromagnetic force: the biophysical and the geographical, however while the intended effect of biophysical energy channeling is ultimately enlightenment and transcendence of the ego, or self-actualization, it is unknown what the effect would be of establishing a working megalithic electromagnetic current circuit system across the face of the globe.

c. ancient ley lines and modern churches

While the simple solutions of the domed ceiling and the support frame modify ancient megalithic architecture very little, the practise of placement above underground water and in geoglyphic alignments of holy sites for shrines, churches, mosques and temples has been systematically adhered to by all subsequent faiths. Modern churches can be found arranged in various mystical geometric forms and patterns relative to one another and to the ancient megalithic sites. It is believed by some very recent thinkers that all of these alignments are relative to different positions of the geographic north pole over the past 80,000 years.

2. flood myth/sacred geometry school cults spread in India

The first calendrical system was that of the yugas of kalpas in India. According to this system there were three yugas in each kalpa, one the creator (Brahma), one the maintainer (Vishnu) and Kali-Yuga, the time of the destroyer (Shiva). Along with this system came the description of an immortal soul that reincarnated through a myriad of lifetimes seeking enlightenment and perfection. Thus were begun the world's first sects, or religious cults, devoted to the study of pure number, and naturalistic allegorical encryption systems. They describe a blue skinned race called the Aryans, who have long since vanished from the region, from whom they inherited their kingship by divine right, and from whom they inherited their system of laws, both natural and social.

a. the Rig Vedas and the Hindu Caste system

This was the system described in the Rig Vedas. Of course, as the world was short on civilization at the time, a large part of attaining "perfection" was associated with civic duty, and thus the same manual also created the class system, where the slave caste served the land owning caste, and the land owning caste served the kings. In exchange for a life of servitude, the elderly at the time of retirement were exiled to live in aescetic meditative purgative contemplation of their value to the world and brace themselves for the impact of their eventual demise. It was these people who, in their aged wisdom, began the original orders of spiritual attainment, and outlined the parameters of much subsequent research.

b. the cult of thugees and petro rites

The thugees cult was the first to perform ritual human blood letting as part of a sacrificial ceremony to Shiva. Because this was done over stone idols consecrated to the deity, these became known as "petro" rites (for "stone"). This would not be the last of blood letting cults, which continue to this day. The ancient Jews sacrificed animals, the Aztecs gave up their own young, and modern voodoo places hexes using blood as a medium. All people throughout history have shed blood for war, and almost all if not all have put prisoners of such wars up for public execution; even in cases where the war is a civil war or only a political revolution this has been the case.

c. Buddhism and the transcendence of Samsara

The melancholia described variously by all great minds throughout all time was the subject of the meditations of a particular Hindu prince turned monk. He came to believe that the material world was an ever changing illusion, and that the depression and futility felt by most people with their status in the social hierarchy would be dealt with best by greater spiritual benefits than were offered by traditional reincarnation. The Buddha taught that the world could be transcended by degree, and that, at a state of ultimate transcendence, it was possible to leave reality altogether and to ascend beyond the physical form for as long as one wanted.

d. the sunken city off the coast of Japan

There is a city off the coast of Yonaguni island in the islands of Japan that is believed to date back 10,000 years at least. It has evidence of the same, smooth surfaced, several ton stone masonry known from megalithic sites throughout the world. There is no reason to doubt that civilization beginning some 7000 - 9000 years ago in India could have come directly from this large site in nearby coastal Japan. It is known that during the Ice Age, the islands of Indonesia between the Indian peninsula and New Zealand, were all connected forming one giant land bridge uniting Asia to Australia. The only question is why a technologically advanced people who had migrated far to the east would migrate back to the west and up the Indus river when the end of the ice age returned sea levels to their present height. There is evidence in Siberia, Russia, northern Asia that thousands of huge wooly mammoths from this time were flash frozen with tropical vegetation undigested in their stomachs. This opens up the possibility that this group of Atlanteans might not have had any choice in to where they were navigated, since it may have been the only hospitable environment at the time.

3. flood myth/sacred geometry cults spread from Sumeria

While the Rig Vedas were still a primarily oral tradition Sargon the Great unified upper and lower Egypt and created the first system of writing — hieroglyphics, which, it has only recently been discovered, predates even linear a and linear b, the predecessors to cuneiform, which would lead in turn to Sanskrit, he language in which the Rig Vedas would eventually be written down. At this same time this great ruler, who came out of Sumeria, developed the system of representing counting by numbers, for the purpose of stocking inventories of staples for the purpose of trade-based distribution amongst the people. This would eventually lead to the creation of the concept of money, or arbitrary value. Sargon spread the myth of Gilgamesh, the survivor of a great flood, and this was the world's first written work of literature. It is particularly astounding because it is held exclusively as a work of fiction, without any sort of cult attached to it at all, such as is the case with many other accounts of the world flood. Like Plato's account of Atlantis itself, Gilgamesh remains held as mythological speculation, accepted even by the people of the time as a flight of fancy, falsity, fallacy, fraud or fiction.

a. Babylonian astronomy

It is known that, while the Hindus followed cycles of many hundreds of thousands of years, the Sumerians were primarily interested in yearly cycles and cycles related to crop seasons for planting and harvest. To this end, it is thought by modern scholars, they created the twelve month zodiac. For whatever their reasons they were the ones to originate, 8000 years ago, all of the signs of the houses as we know them today. Some of them seem natural — Sargon was known as the Scorpion King, and hence Scorpius the Scorpion; simply an indigenous animal.

Others, however, are quite arcane, such as Aquarius the water bearer, or Sagittarius the philosopher — part man, part goat, part fish. These have been, however, faithfully preserved through the ages, in this exact form, to today. Though they were life or death to ancient mariners, they are almost totally ignored by the uninitiated of today.

b. Egyptian architecture

While the zodiac of constellations given for the solar system's ecliptic orbital plane is credited to Babylonia, the first full chart of all the animisms for the different parts of the night sky in the whole year's northern hemisphere comes to us from New Kingdom Dendera, in Egypt. The Egyptians carried astronomy into far more than this single, simple motif however, incorporating divine measurements and observations of the heavens into every single work of art and craft they produced. The most recognized of these, even in ancient times, are the great pyramids of Giza. These are three enormous, stone, megalithic structures constructed to align with the belt of the constellation Osiris (modern Orion) as part of a massive scale geographic representation of the constellations on the surface of earth. The tips of the pyramids also align according to a phi spiral, a shape unfolded from the expansion of a transcendental number. Buried next to the great pyramid, and on display there today, is a large, beautiful, seafaring boat.

c. Essenes and Nazarenes

The cult that most profoundly embraced the harmonious parity of Yin and Yang, or Jachin and Boaz, was that of the Baptist Essenes, known also as the Poor Nazarenes. The Essenes of the order wore white robes, and the Nazarenes of the order wore black robes, however they merely represented the left and right hand of the same organization, the Nosrei-ha-brith, or "keepers of the covenant," the esoteric masonic Juwes of the exoteric Hebraic Jews. This was the inspiration for the later order of the Knights Templar, whose emblem was two brothers on one horse. Though both the Templars and the Essenes were practitioners of communal property, the Templars began the system of banking that is still in use today. The Essenes, on the other hand, dealt in esoteric doctrines, and were fundamentally determined to form a unified description to fit all existing dogma. The brothers of the path of the Black Hand went on to form the cult of the Assassins, who inspired the Templars.

d. the sunken city off the coast of Spain

It is thought by some that the blood line of Jesus, the Essene Buddha, moved to France and eventually to Scotland near stone henge, a famous megalithic site. During this same time the Muslims of the middle east spread west across Africa as far as the Atlantic, then crossed the straits of Gibraltar and took took Spain in Europe. What motive would either of these peoples have had to seek territory in Europe, a miserable land populated by heathen barbarians, the Gaels of France, the Anglo-Saxons of Brittany and the Normans of Germany. Aside from the treasures of many thousands of megalithic sites on land, there is a city off the coast of Spain that would have been common knowledge to the utopian, map-preserving, astronomer Moors. The only real question that remains is whether or not Julius, ignorant swine that he may have been, knew of the significance of these sorts of sites during his Gaelic campaign just before crossing the Rubicon, reentering Rome and declaring himself Caesar.

4. flood myth/sacred geometry cults spread in America

Long before Columbus and the Conquistadors of Spain landed in America to colonize it for Christianity, the indigenous peoples of the region maintained oral and sometimes petroglyphic depictions of ancient times according to astronomy, and preserved traditions of a recent cataclysmic flood that had destroyed the original peoples of the area and which only a few families, high atop various mountains, survived. Since by now we can see that this myth is so common, we can see that it must date back to a time when all the peoples of the earth were unified. When comparing the sports games played by the Maya with a rubber ball to the contemporary gladiator games, chariot races and mock sea battles at the Colosseum in Rome, it becomes easy to see that the world is always, and will always be, "unified" — on the same wavelength, the Schumann resonance of the earth's electromagnetic field, which is about 40 hertz. The indigenous people even to this day preserve the oral traditions of their heritage, including various spiritually complex cosmologies.

a. the Anasazi (astronomer) and the Chocco (cannibal)

In north America the first peoples to develop civilization were a tribe of cliff dwellers called the Anasazi, who would go on to build the first cubical astronomical observatory in human history. Their ways were peaceful and they knew no war. Then the neighboring clan of the Chocco valley Grand Lodges came in and devoured them. Their culture was literally consumed by the Chocco, who built a great round observatory that, from the similarity of which to a Masonic lodge, earned their houses their name. The likely mechanism of the original turn in both cultures of attention to the heavens was the psychoactive mushroom psilosibin. Similarly it may have been peyotl which awoke the cannibalist impulse in the Chocco.

b. the Olmec (stone masons) and the Maya (sports)

In south America civilization evolved much sooner, about 4000 years ago. The oldest civilization is thought by modern anthropologists to be the Olmec, who carved remarkably African heads out of giant stones in the jungles of south America. Almost nothing is known about them, though they were thought to exist for so long. They left few megalithic sites, and these were all pyramids similar to the ziggurats in Sumeria. They are usually compared by modern scholars to the pyramids of Egypt — but these, the same scholars think, were built up as monuments on top of or around underground or containing tombs, whereas both the Sumerian ziggurats and the south American temples have steps leading up to the top, at which is a small hut wherein lived the ruler or shaman, and before which there was a small altar for making a blood sacrifice. This identifies the origin of recognized civilization in south America clearly with the beginning of the age of the Father God.

The second civilization to arise in south America were the peaceful empire of the Maya, who venerated above all things astronomical events represented by megalithic monuments, such as pyramids in the same Orion arrangement as found in Egypt, along a great flooded roadway known as the Way of the Dead. It is thought that their culture also venerated the "seven sisters" of Pleiades because of a peculiar offset in the alignment of many of their sacred sites, however this difference from orientation to modern true north is only meant to signify orientation to a more ancient true north, when the north pole was located in a different place on earth.

c. easter island, the bimini road and the Nasca lines

In the Pacific ocean on easter island at an unknown time over 600 monolithic heads were set upright, each weighing several tons. These provide a double edged mystery for, assuming the first people to arrive on the island, whether they arrived 6000 years ago or 600 years ago, had the technology to raise the monolithic edifices is entirely contingent upon admitting that they had the technology to built sea faring boats out of thatched river reeds to get there in the first place. These ancient people, now forgotten by the indigenous people of today, left behind also a pictographic language called rongo-rongo writing, the ideograms of which are strikingly similar to later Chinese characters.

In the Atlantic ocean near the Bimini islands near Bermuda there is what appears to be a road or the top of a wall under the surface of the shallow sea. The blocks are divided regularly and evenly, although some people argue that this could be a natural occurrence, only a formation of the waves. This particular area was predicted by psychic Edgar Cayce only some 20 years before as being the area where we would begin to see Atlantis "reemerge from the ocean." It is also at the southern most tip of the Bermuda triangle, an area in the Atlantic where more boats and planes have disappeared than any other spot anywhere else in all the seas.

Recent discoveries by ardent researches indicate that the Nasca natives were in fact the first civilizers in south America, earlier even then the Olmecs. They traced out perfect geoglyphs of their zodiac and ley lines miles long in the nasca desert, large enough to be seen from space, and their skulls have all been lengthened presumably by some form of artificial means.

5. the cult of sacred geometry turns to addressing the matter of time

We have seen throughout history how the lost cities have guided the evolution of societies. Remember that it is only the modern occult (in particular the O.T.O. as revised and reconstituted right proper free and associated masonry according to Aleister Crowley) that traces back the lineage of the age of the Mother Goddess (beginning 6000 years ago), the Father God (beginning 4000 years ago), and the Crowned and Conquering Child (beginning 2000 years ago). The 2000 year cycle is between the sun's alignment with the earth during the solstices or during the equinoxes, and this determines weather on earth. This is only part of the much larger, 26,000 year precessional cycle, during which the pole pivots 360 degrees about the polar regions of the earth, causing the sun at earth equinox to seem to precess through the constellations along the entire ecliptic; this cycle is in turn in flux with the alignment of the ecliptic, which appears like a coin coming to rest after spinning on a tabletop in extremely slow motion, with the center of the galaxy. This latter cycle interacts with the sun's electromagnetic field. The combination of these cycles results in the 41,000 year ice age cycle for earth. It is believed by all of the few who know of this that all of it was known in ancient times as well.

a. Thoth/Trismegestus/Templarism

The first task of the ancient order was to anthropomorphise the alignment of the point (in Sagittarius) where the ecliptic overlaps galactic core and the sun in various seasons. It was thought that the changes in the electromagnetic field of the sun and the earth could be used to predict climate changes in the polar regions of earth, which would precipitate change to global coastlines. As the plane of the solar system gradually rotated around, its only spin the orbiting planets, slowly wobbling back and forth relative to the center of the milky way, and the earth precessed in its rotational axis relative to the plane of its orbit, earth's axis of orientation to the sun gradually changed relative to the time it was aligned with the sun and galactic core, such that, it was thought, when the electromagnetic effects coincided with a winter season, there was an increased possibility of an ice age. Since the alignment alternated between equinoxes and solstices every 2000 years the ice age cycle took nearly twice the full precessional cycle before the signs began to enter the winter months. Some of the suspected electromagnetic effects include reversal of magnetic polarity between positive and negative.

The earliest as well as the greatest civilizers were deified by the classes they created, and all were made to stand for anthropomorphications of the precession of the solar alignment with galactic core, thus being given the same history and duties. One of the earliest of these is the archetype of Thoth, who recurs in the alchemy of the middle ages as Hermes Trismegestus, who knows the secret of how to transmute metal into gold. It is believed that this is what started the quest of the Templars as well — who believed that orichalc, the fabled metal of the Atlanteans, was obtained in ancient times beneath the temple of Solomon by some extreme, concentrated heat source. This may make them sound crazy, but they did subsequently get rich, and it may have only been for the knowledge of the very same Dead Sea scrolls available to all today, and they may have never sought anything other than these scrolls in Jerusalem (the city of Lucifer, the Rising Star). These scrolls do contain fragments of the Book of Enoch, believed to be attributable to Thoth.

b. the hypercube of time

The corpus of Thoth is QBLH, the hypercube of time. Today time is considered a fourth dimension, however the temporal force, whose agent is entropy, is not weighed like the other spatial dimensions. There is also thought to be a fourth spatial dimension, and we use constructs from it to stand for concepts of time only because the geometries overlap conventionally, according to folk wisdom or common sense. From this we derive the hypercube, or third spatial dimensional shadow of a fourth spatial dimensional cube, properly called an octaholohedron. The hypercube is a cube within a cube, and when we weigh the outer cube as the continuum of time (Nuit) and the inner cube as the moment of an event (Hadit) then we reopen the ancient wisdom of the QBLH, associating the temporal dimension with a higher spatial dimension that can be perceived only through geometry.

c. different existing frequencies of space (rhomboid crystals)

The QBLH is considered a right proper meditative object upon which to perform an epoché, and thus invert dasein from being for others to being in itself, and in this way transcend the ego and achieve enlightenment. The finest equivalent object in the lower third dimension of manifest subspace, a wholly imperfect world, is the nearly precisely aligned molecules of certain natural rhomboid crystals. These have become known as shew stones and associated with the prediction of the future. They establish a nearly perfect harmony of frequencies aligned along their linear arrangements of molecules.

d. the methods of skrying according to mystics

First it should be noted that skrying is the attempt to instantaneously manifest an accurate reflection of the future of a fractal time pattern in the form of a micromanifestation that serves to suspend the observer's disbelief. It mainly uses some form of medium as a tool, though the experience is ultimately neither finite nor bounded by any delimiting measure. What we call life is only a form of skrying that our spirit is doing to our body through our soul. Similarly what we call communicating is only skrying of idealized concepts between the bodies of two souls.

Skrying is the universe looking at itself.

i. Nephotes and Narcissus

The first recorded skryer in history was an Egyptian of little note named Nephotes. His accounts make it obvious that he was familiar with the art only as if one among many who had some practical knowledge and could call themselves skilled craftsmen. His tone goes so far as to imply an entire class of skryers, sooth sayers and fortune tellers in ancient Egypt. This may have represented the first trade guild unionism by working class men, and the first step towards commerce in the trade of goods and services already perfected during the age of the Mother Goddess in Babylonian temple brothels by the priestess prostitutes. If this is true then Nephotes can rightly be credited, not single handedly, but as among a class of men, with the pre-Socratic elementalist and atomist movements (which have become what we know of quantum mechanics today), and deserves more credit, therefore, than to be canonized in Greek mythology as the boy who turned into stone looking at his own reflection in a running brook.

ii. Dee's birdcage

The system of the Enochian hosts inhabiting the 30 ayres created by John Dee and Sir Edward Kelley in the fifteenth century is the most complicated endeavor to date derived from skrying. According to Dee, Madimi, his Holy Guardian Angel, was contacted, and guided him through the realms of the angels, and showed him the right proper construction of his Enochian system. It bears little in resemblance to the Watchers described by Enoch, from whom Dee's heavenly characters took their collective name, except for a twelve month calendar with only given names of the months, the guardian angel for each (given by Dee), and the dekans. This may have represented the lunar zodiac, but if it were applied to the sun then it could also be set to measure precession. The system Dee created is an unfolding of the ruling planets in the three dekans of the twelve houses of the signs, multiplying this pattern out to a factor of two by the four elements. When it is taken to form a cube, as it was by Aleister Crowley in his design for an altar for the Golden Dawn, it should be thought of as a fractal hypercube. Inside of it are the Watchers, guarded by the seven angels that watch the watchers. The name of the woman that gave them all wings was Becka.

iii. Quetzalcoatl's cat and mouse hunt

The demiurge of the sexual instinct, known as the great serpent Kundalini, wraps itself around the auric egg and grows wings, one of light feathers, one of dark. The great serpent force was known in ancient south America as the plumed serpent or feathered snake, Quetzalcoatl — representing the movement of the aura over time. This force upon perception is undoubtedly the fading afterimage upon our genetic memory hologram of yet more ancient dinosaurs — the enormous "thunder lizard" predecessors to all modern forms of reptile and bird, and even mammal. This strange ghost, often known as Choronzon, the discorporeal essence of the aeyther, still haunts us in our perception, and nothing is clear.

iv. remote viewing

Distance learning. Diversion of spin onto a different dimensional level, directing the concentration of perception in active consciousness toward a certain person, object or location. In these times of product branding, youth market angling and target audience product placement, and of the body-without-organs in virtual reality in general, the focal point for trance meditation is no longer restricted by any bounds — it can be near or far, stationary or moving, past, present or future, fact or fiction. It is as easy to channel the living as the dead now, courtesy of up to the minute live coverage of breaking events by the all too "in-your-face" news media. This is one good yoga move away from being just lucid dreaming, so detached is it from the manifest corpuscle. Taking society as the macrocosm and bands of individuals as the microcosm it pits the one against the other as the old humandroids and young humanimals.

e. the end of time

The universe resembles the nervous system, the walls and filaments of galaxies strung throughout the empty voids resemble the neurons linked at the axon-dendrite gap branching out in synapses in all directions.

Freud described the accumulation, which he called hypercathexis, of the electrochemical energy of the nervous system, or phi, as being the source and cause of ego or personality. Not all of the phi in each neuron would transmit during regular cathexis and then some would be available for use in making conscious decisions, or hypercathexis.

This is similar to the way the universe works. The many millions of galaxies variously align with each other, effecting their electromagnetic polarity, and this causes the transport of condition from one location in the system to another. These alignments cause a finite number of gravitational reorientations, and this effects change in the universe.

I have no doubt that the universe will end due to an effect essentially similar to Freud's conception of hypercathexis.

IV. Time

We have already seen that duality is inherent in everything in the universe. To understand our relationship with time, however, it might be best to consider one pair of opposites in particular. The cell structure of the bacteria — usually a formless protozoa with a coat of sensitive fibers, and that of the virus — a finely honed hexagonal killer designed to seek and destroy other living cells, with a distinct lower body equipt with gripping limbs capable of anchoring it to its prey. Here we see that humanity truly is a hybrid species of bacteria (the body of flesh, muscle, organs) and virus (the brain stem, the eyes, the limbs and digits, the lungs, the pathways through the systems). It is for this reason that we are taylor made to confront the concept of time, because its effect upon the universe is like that of a disease in a human, and the universe has yet evolved in symbiosis with it, just as has the human body unified the bacterial and viral forms.

A. Time Travel

From birth until death we are time travelers visiting this dimension. In this field of phased matter, time passes through everything. It seems to be coming from nowhere, and causing the universe to expand. It is really coming from outside of the universe, and merely causing the universe to attract. Time is made up of microwave gravitons known as torus shaped tachyons — light faster than light. These cause gravity as a repulsive force faster than light to flow opposite entropy, which on the level of subspace below photonic velocity this acts as the attractive force. The greatest centers of gravity, or the oldest probabilities to begin to spin in the universe, now black holes at the centers of galaxies, emit these from a higher dimension into our own. This higher dimension is the space of the fourth temporal dimension. Our souls are born through these sun eaters the same as our bodies are born from our mothers womb. The first portrayal of the concept of time was as a storm over the grass fields and a flood of the rivers and lakes, followed, according to scripture, by a rainbow to symbolize that such a storm would never happen again. This was during the time of the first hominids, and subsequently associated with the era of the Mother Goddess. It may have been time itself that caused her change of face from Isis (the twin stars of Sirius), to the moon, to Venus, the morning star.

Time changes everything. It is the sum effect of all the forces. It is a shape in hyperdimension that creates the points of emotions as it moves through the aura of perception in our mind. We are a living, organic, fractal refraction of it as it is reflected off the surface of itself. However all of this is only time moving through us.

The goal of human life, and of the Great Work of the occult, is to reverse the flow of entropy. The evidence for this is how the common man, through the millennia, has been gradually manipulated by the various orders into working for his welfare, fighting upstream of entropy. Meanwhile politics only exists as the argument over what to do with those people unwilling or unable to do likewise, who float with the flow of entropy. Meanwhile entropy itself is misrepresented as chaos, when really it is simply temporally inverted order of a higher dimension.

1. ships

When people first began writing, in ancient Egypt at the time of Nephotes, they described the material universe as comprised of vessels housing energy. Now we know that what they were describing was the holographic fractal torus shape of the tachyon and the microwave vibrational frequencies that caused them to assume this form. However the ancient comparison of the kha to ships for the greater aethyreal energy of the lesser electromagnetic soul should be taken for noting when considering the human relation to time in the light of travel.

The system passed down to us from the ancients describes a spirit, which is eternal formlessness outside the temporal universe, a soul, which is immortal energy inside the temporal universe, and a body or self, which is mortal matter inside the temporal universe. In truth the only fact the people of this planet have ever known is the mortality of the corporeal form, while they pray for their souls, still playing make believe like little children hoping the impossible come true, and revel in the communal spirit, calling it the Holy Ghost, making any young fool scapegoat for the millennial in-joke of the occult.

a. the eight circuits of consciousness — unleashing the atman

Scientists of today are more interested in what parts of the brain control what functions of action and reaction than ever, and this is in large part due to a heavy reliance on the part of neurologists on the encroaching application of wave technology to the brain, as laser technology has already been applied to neurosurgery and eye surgery, and it is possible to eliminate tumors in the brain using resonant sonic frequencies. Thus there is much hope for the future of neuroscience, which remains today a largely pharmacological field, and little attention paid to the curriculum vitae of thoughts our minds accumulate, except by advertisers, who are responsible for most of our man-made sensory stimulus. According to a doctor named Timothy Leary, who studied states of consciousness relative to the evolution of human cognition, there are eight primary circuits, or "ennegrams" that are predetermined levels of awareness associated with the timed function of genetic adaptation, where the survival instinct is inherent to the single nerve, and where survival of the fittest is considered the only governing principle.

The first is the bio-survival sense, often called consciousness, which evolved 2 - 3 billion years ago with the rise of invertebrates and remains visible in the individual human at the stage of infancy. Here, an either/or circuit establishes itself, that will govern most of the rest of the third dimensionally relative habitual patterns of survival. This determines the mechanism of approach nurturing/helpful things, avoid noxious/dangerous things. This is water.

The second is the emotional sense, or ego, beginning 500,000,000 years ago with the birth of mammals and apparent for the individual human at the stage of upright walking, both due to the start of competition for territory. This activates the up/down binary circuit. This is air.

The third is dexterity-symbolism sense, or mind, arising 4 - 5 million years ago when hominids entered the paleolithic age, and began making stone tools, and is equivalent for the individual with the developmental stage of using toys, tools and weapons. Also, the beginning of speech, for the toddler as well as the early human branch of the primate family tree, is credited to this neuro-genetic encoding. It activates the left/right circuit of dominant handedness. This is earth.

Fourth is the socio-sexual sense, which awakened for mankind 32,000 years ago when the first hominids penetrated India, and is found in the puberty of the individual person. This is thought to also be the first beginning of civilization, for it is associated with hunting in packs among animals. This is fire.

Fifth is the neuro-somatic sense, and it is first to occur in the right hemisphere. Leary, drawn towards both scholasticism and the populous of the student class, balanced the scale of controversially ascribing the use of certain different psychoactive drugs, some illegal at the time, for the extension of the sense of self into these different conditions of intellect, with the staunchly authoritarian doctrine of the division of the brain into functional areas. One of Leary's students, Stan Grof, holds a view of the brain as a holognomonic whole, which he calls holotropic, perhaps in apology for his predecessor's antiquarian affiliations. This sense of the creative hemisphere, which we only seem to send electrical currents through while on certain chemical substances, was first activated 4,000 years ago, and distinguished the difference between visual, Euclidian space and sensory, multidimensional space. This is gravity.

The sixth sense is neuralectric, and is 2,500 years old. In this condition the nervous system becomes actively aware of itself, and free will determines our way through different tunnel realities, each leading to its own goal, by metaprograming. This is commonly associated with anomalous cognition or extra sensory perception. It allows comprehension of the magnetic force and electron scale. This is the electromagnetic force.

Seventh is the neuro-genetic, which is 2,200 years old. It is associated with the concepts of reincarnation, immortality, past life memories, as well as with the akashic records and the collective unconscious. Here we discover the marvels of our own cellular deoxyribonucleic acids, and inherit control of the mechanism by which these senses have become activated, namely free will governed by the survival instinct in a game reality the only rule of which is survival of the fittest. This is the force of nuclear fusion.

The last is the neuro-atomic sense, associated with astral projection, out of body experiences, lucid dreaming and remote viewing. When this sense becomes aware we gain access in our mind's eye to perception of the behavior of particles and waves on a quantum level, as we are actually thinking about what we are thinking — and discover by doing so that our thoughts are subatomic electrical impulses inside the neuro-cellular columns of the grey matter of our brains. This is the force of nuclear fission.

The philosophy Leary used this marvelously open minded and broad sweeping data to justify was, however, cyberpunk — a lifestyle unaccepted by the community yet today. It is clear that this was in keeping with the agenda of the same level of thinkers as released the comic book inspired movie, the Matrix, which equates metaphysicians with terrorists in the eyes of a predatory authority. Leary himself recognized that we are, all of us humans on planet earth that is, involved in a war between the levels of this hierarchy. I would only go so far as to add that I believe it is in essence a war over whether or not we will be allowed to manifest.

I have listed along with these the relative elements. Cosmos, or spirit, moves into and passes through this system.

b. actual government classified vehicles and popular phonies

There are centers for souls to gather that have not yet been locked down by the institutionalized religious system. These are called the Traveling Lodge. They have been described in publicised occult literature since the medieval Rosicrucians. Now they are associated by the populous with such sites as crop circles and possible UFO crash sites such as Roswell. The true Traveling Lodge bears more similarity to the brainwashing cult of Rosalind Air Force Base, however such truths will never be pursued. This demonstrates the difference between what is known esoterically and what may be revealed exoterically.

For reasons that are extremely personal to the initiated it is important for the air / space craft that humans fly to be of human design. The fact that the United States Air Force was formed in the same year as the Roswell crash and that some of its oldest ongoing projects involve stealth (all acute angles — thus deflecting fewer radar waves) technology, beginning with the flying wing and culminating in the V-wing B2 bomber, only shows that humans could have studied extraterrestrial flight technology, but proves that all of the technology they have produced is inherently human, having been designed and built by normal people, and having been carried through from beginning to end by human hands.

It is in this way that the actual government classified vehicles — those designed by human beings of today — can fly about today right before our very eyes. The black helicopters are also the product of stealth technology. They use scalar wave sound mufflers that can be used for mind control. Similarly the stealth bomber is a triangle shape because it uses three microwave gravity generators to expand a wormhole and break the light barrier, while the public of today is unaware that this craft can even leave the atmosphere. There are, however, several possible designs and methods available for it to do so. A double rocket propulsion system is commonly used by NASA to send the space shuttle into orbit, however this is little more than a freight vessel rented out to the highest bidding multinational conglomerates for putting their satellite into space or doing some worthless waste of research or development, more for a public relations show than for pure science.

The rule of society is rule by the law, where justice is only what can be proven, and where it is easy enough for those with money to determine which version of reality there will be evidence for. In this way all actual UFOs and extraterrestrial encounters are usually counted among the number of intentional human frauds, and in this way "sneak by the radar" of the public eye. These crafts are only our future selves returning to our shared home planet in time machines, just as we have been doing since the sudden and as yet unexplained appearance of our species.

c. builders, pilots and passengers

As I have described, the U.S.A.F. have contributed greatly to the number of identified flying objects that defy what we expect for earth bound air craft. However they were not the first to do so. Orville and Wilbur Wright, the two brother co-owners of a bicycle shop, built the first human air craft in recorded history as recently as the nineteenth century. Unidentified flying objects date back to the beginning of this same history, and are recorded all throughout the texts of time. Cigar shaped scalar wave weather balloons were the first modern UFOs, appearing in the skies of the late 1800's — although they may have originated later, during the 1950's weather balloon experimentation phase of the U.S.A.F. and slid backwards in time as an effect of the scalar waves distorting the temporal synchronicity of the balloon's molecular frequencies.

The German National Socialists were probably the first people to propel a rocket into space — as they were responsible for all the early solid fuel propulsion systems used to break earth's gravity. It is known that they had access to scalar wave technology, although, because most of their research and development facilities were bombed during World War II (which they started and lost), it is unknown much of what their application projects with scalar wave technology might have been.

Nikola Tesla, who first invented scalar wave technology around the turn of the twentieth century and applied it to a death ray that was directed through the surface of the earth and which was responsible for devastating tunguska, Russia, claimed to be in contact with "off world sources." Another group that would make a similar claim was the Golden Dawn, who associated with the legacy of the true and invisible college of Rosicrucianism by way of the Secret Chiefs of the Great White Brotherhood of illumined seers and ascended masters.

The Russians, contemporary to the Americans, developed both solid fuel rocket propulsion systems and scalar wave technology. Because the two super powers' competition was to dominate their entire drive and motivation for research and development, both countries suffered down to the lowest common denominator of weapons manufacture. While intercontinental ballistic missiles became a new reality both sides directed enormous scalar wave programs against one another and against themselves for mind control of the mass populous. The Russians were the first to put a probe in orbit around the earth that sent a radio signal back to the ground, and they were the first to put a man in space, that is — in a pod outside the atmosphere, and they retrieved him happily. The Americans were the first to land on and return a man from the moon, however, because of superior, Japanese manufactured broadcast media, were able to make this seem like a bigger accomplishment. This competition between the two greatest nations in all the land did not end until while I was a child, and I can attest to the stress that this Cold War produced for me even then.

d. the time capsule itself — tachyonic technology

What I am about to describe is best done in space. Doing so on earth, on any large scale, is probably contraindicated. This is not to say that this has not been done before, and yet we are still alive. It is simply that myths of a strange energy source are associated with the downfall of an ancient lost supercivilization. Mythology is a strange beast in itself. It is usually only a court jester's perspective on the affairs of the court and health of the royal family, however when it speaks of the human ability to harness energy that can be misused, even if accidentally, and cause a massive effect it addresses one of the greater issues of the human species.

This being said: by projecting a faster microwave frequency into a slower microwave tachyon, it is possible to expand this tachyon into a wormhole. By mapping scalar wave frequencies over top of the microwave signal it is possible to access any destination from the holographic electromagnetic radiation background. This can be done as easily as by projecting a visual thought of where you want to go onto the wormhole.

When the destination is projected, by whatever medium — mechanical or biological, it creates a particular harmony with the faster microwave gravity, and this creates null space, or zero-point energy — the electromagnetic result of combining radiation and gravity. Another name for this is a temporal singularity. Just as the background radiation of the electromagnetic spectrum serves as a fractal through which any single point in space my be linked to any other, so does the sum over histories of probabilities in the universe serve as the gnomon from which we may extract our temporal destinations.

Once you have traveled through a wormhole you have entered into the multiverse. Things will not be the same, and all things will continually change. It is possible to meet yourself in this way, and there are fewer experiences more alienating from reality.

e. accidents and insurance empires

Wormholes are like wishing wells. Wishing you were home from a wormhole is like falling down a probability well. Many people in the present and near future will find this to be true, because if you wish you weren't brought before the most high, then low you will be cast down. So it is with hyperspace — if you try to turn it inside out and use it to get where you already are, it will put you down in another time.

Paranoia is a kind of insurance paid out in karma.

2. civilizations

The prevalence of evidence shows us beyond equivalence that the full extent of advancement in the natural sciences and in technology of ancient, lost civilizations is now beyond our comprehension. Every artifact in existence seems to add a stroke to the blank canvas, painting a picture of humanity as a more and more adaptively developed species at an earlier and earlier time. Far from wandering blindly before the creation of civilization, which, according to scholars, is barely a second ago on the smallest cog of the cosmic clock, the modern time traveler, a historian in this case, exploring the past, can now carry with them all the artifacts essential to survival throughout the lifetime of humanity.

There can be no doubt that ancient civilizations, such as the early dynastic Chinese and the Mayan, evolved highly complicated calendrical systems based on astronomical observations that modern scholars think would have required modern computers to calculate, however there is a building amount of evidence to indicate that even the localized tribes who remained technologically primitive, such as the Aborigines and the Dogons, also possessed this level of astronomical knowledge. It is perhaps easier to imagine a more complex civilization discovering more complex data, however since people of all lands and affiliations possessed the same skill, it can only be explained by a highly evolved common source.

This point of origin is in our future. Ancient ruins are all that remain of what we will have built in the past. The technology of today can get you to yesterday through tomorrow.

a. our future (our origin)

The mind, when seen as a holographic imprint overlaying the brain, created by the brain, is seen to exist in a continuum unique from that of the fleshly body. There are billions of interacting neurons in the brain, each comprised of a fiber that delivers an electrical signal, and connected by gaps crossed by neurotransmitters. If you could take the mind out of the brain, as John Dee extrapolated his angelic hosts from the four elements, it would be comprised by an asymptotically infinite number of such connections, however it would no longer be responsible for triggering the nervous signals to the body. It could be transferred into an electrical machine, or into the ambient electromagnetic field, or the background microwave radiation, it is really irrelevant where it goes outside the body.

These connections are thought to be the wells from which emotions arise, and so we must see that our mind goes with our emotions. People in times of yore romanticized about their feeling coming from the heart, but modern biology's guess is the brain. The nerves in this way spread throughout the system of the entire body, and so the phantom sensation of the entire form may come along. This has been rigorously researched, from primitive natural birthing and early tantra yoga to modern advanced pharmaceutical experimentation on animals and by people on themselves, all of which is in all honesty unethical only in that it is pro-entropic.

b. our past (the destination)

Atlantis is a timed utopia. You could travel back on one day, live there a year, and arrive back in the present the next day. Assuming that you established a temporal signature by returning to the same timeline in such a way, that is, by being in one timeline day after day, and being gone in between these days for undefined durations, then you would establish a referential evolutionary cue to determine life-span. This would be an artificial temporal pattern, however once one enters the multiverse, there is no going back to the way things were before, so higher natural and new artificial patterns alone can serve as moral compass there.

When navigating the timestream it is much like an electrical impulse traveling upon nerves, and this experience stays with a person such that it is difficult to know when it has fully stopped and one has arrived at their destination, because the gravity to which this disorientation is relative is Time, or the forward flow of entropy defined by the spin of a photon measured as the speed of light.

c. our present — (slow, old) socio / culture (fast, young)

While society appears like a giant dinosaur to culture, culture appears like a microscopic fly to society. Culture is the poor man's society. There is no such thing as an exclusively bourgeois fraternal order, unless related to a job, like police officers.

The star system of America establishes a new royal class, comprised of entertainers, models, and news anchors. This is the sum of the substance of modern American culture, the gift of the 1950's plastic fantastic suburbanites, their simulacrum of historical heritage as it existed before our forefathers established our more perfect than thou union with the social compact. This era of free thinkers, who were jacobite Masons from Rousseau to Castro, including Washington, were trying to tear the social middle class away from rule by a Lord of the land. In America in the 1950's, the middle class were housed in suburbs surrounding the cities in which they worked in high-rise offices. The social destiny that had been given to them they had made manifest with great skyscrapers. Most countries have a folk wisdom founded culture that dates back much further than modern American culture. American culture did not even begin on an international scale until the 1950's, when industrialised manufacturing made our production of stylized consumer goods applicable to the mass market of the entire country, and the new technological communication media were first used to introduce international cultural trade into the global free market of ideas. The only reason I go into the society and culture of America is that it is a young nation that has grown rapidly to become a globally recognized economy and political system.

Human cultures have grown with human migrational patterns and evolved from the same origins differently in different regions. Society, on the other hand, derives from the tribal hierarchy of chief and medicine man, and from the first codification of law, in Sumeria at the same time as the beginning of a written record of history, which was simply a generally binding agreement to provide services to one another. Society did not begin until people were spread throughout the world, and had already established what would become their traditional cultural customs. Society formed in two camps — the middle-eastern and the Indian. The middle-eastern rule has been passed to us directly in the form of the ten commandments. The Indian rule established the class system. There can be little doubt that — while the Samurai warriors of later China and Japan, who are thought to be inheritors of an originally Indian cultural heritage, represent a totalitarian, or dictatorial, though essentially otherwise classless, society — the class system of India spread to the west and influenced the minds of even the early Greek philosophers, considered to be the first social politicians, as opposed to laborers or lords, and it was upon the framework of a class hierarchy that the first utopias were speculated. This would lead to the Roman empire conquering Europe, and Christianity — the idea of the anthropomorphic deity as discorporeal ruler — conquering the entire western hemisphere. Insofar as it was cultural, Christianity spread to the east as well, though this caused a split in the church over the doctrine of the trinity in the west and that of the ascended human in the east. In the west the class system would evolve through the dialectic (thesis-antithesis-synthesis), with America arising as the synthesis of a capitalist free market and a representative democratic republic. What we can conclude from this is that culture moves eastward around our planet while society moves westward.

d. aliens throughout history

First of all let me assure you that there are no aliens walking the earth at this time, nor have there been at any time during our recorded history. There has not been any genetic interbreeding. There is no technology that was not made by humans on the planet nor has there ever been. No one else controls our fate. Human beings have free will, and this controls all human outcomes. It alone always has.

What have been called fairies, elves, goblins or monsters throughout history, particularly among the fairy tales of the European middle ages, are usually only either wishful anthropomorphications of natural forces and events, or actually deformed people forced to live outside of towns due to scientific misunderstanding. But these are definitively not exclusively extraterrestrial forces, no matter how alien they may have seemed to people. Similarly the concept of the djinn, or discorporeal ghosts — probably the hallucinations caused by drinking the first artificial alcohols, as well as the elemental pantheon should both indicate to us that the Nefilim of the Sumerians who came in "vehicles from the sky" probably refers to events of weather or astronomy, and should not be taken as entirely and exclusively a literal description. Another example of such a vehicle that appears from the air is given by Ezekial and recorded in the bible, however a technologically advanced time machine from our future or a drug induced experience could both account for this as well.

More recently the reports of alien abductions all involve beings that are thin and grey, with large heads, large black eyes and small mouths. Visual hallucinations of such beings — presumably doctors during a birth experience flashback — have been documented as being produced by the strong electromagnetic fields given off near areas of tectonic pressure, such as fault lines and volcanoes. It is likely that the original inspiration for beings of this appearance was that of the earliest mummified corpses, who look like the state between health and the skull. Natural mummies — such as the one frozen in ice on a mountain top and some of those buried in the peat bogs of England — date back to the last ice age and to the stone age.

The concept of the djinn, whose astral realm is the motivator, source and intersection of the four elemental forces, however, introduces the applicability of trans-dimensionality, as the ghosts of the dead are thought to be beyond time. Whether such spirits exist or not it is clear that they would exist according to rules that would not be confined to the earth. Being beyond even necessity of incarnation such concentrated intelligences are eternal within the universe of the four forces. They may be raised on the sounds of the sunspots and solar flares, and heard in the earth's electromagnetic field, but the realm of the djinn is hyperdimension, the same as that accessed by wormholes. The djinn have been seen as the peaceful angels and the wrathful demons of various world religions who inhabit the same heavens as the spirits and ghosts of the dead. In the same pages of the first record of the djinn are descriptions of human giants and men of renown who were of the first generation following human interaction with the aethyreal spirits. It is from these people that we derive all the subsequent arts of metaphysics, mysticism and magic. They were the first to be alienated by their communities, and this is why these arts are secret.

3. the intersorting of subspace matter through hyperspace

It is difficult, often, to address the cause of something if it generates many, and / or various, different types of effect. It is tempting to rhetoricians to focus too much on examining the difference between effects than to cut to the core of the cause, particularly if the cause encompasses a greater number of outcomes. However it is in fact much more fruitful to study the cycle of time that allows the harvest than to simply reap the rewards of nature's workings. This is illustrated by Jesus in a parable as the difference between feeding a man a fish and teaching him to fish. Because of this proclivity of the analytical mind a single, ongoing cause may remain undiscovered for any finite number of effects throughout time, even while each of these effects is studied in detail by specialists.

Such is the case today with the effects on third dimensional material forms caused by their interaction with higher dimensions. The best, or at least most immediately prevalent, example for this is the mystery of death caused by third dimensional spatial interaction with the fourth temporal dimension. We believe that death is an inevitable end of any and all forms or patterns of matter that exchange energy with their surrounding environment, however by this criteria the fact that life exists at all is an unprecedented miracle, since the fragile biological organism is an engine for energy-exchange that functions at ratios equivalent in other physical bodies with much shorter life spans. In fact, the biological form of life is that which has been chosen by survival of the fittest attributes of atomic chemistry to exist in direct opposition to death. This is proved anthropically by our species being the only one with a working concept of finite time.

On the most open scope, all that exists is the interaction of dimensions. The matter-energy we call the third dimension is the effect of the presence of time. Down to the first dimensional singularity well in a brane, plane or event field of spacetime we see that the known dimensions are entirely interdependent, and so we must allow that this be true for any number more dimensions that might exist undiscovered. When we imagine these dimensions as distanced from each other, such as the measure given by whole numbers, they comprise the multiverse — each potential universe with a different vibrational set of laws of physics relative to its neighbors. When we imagine these dimensions all bunched up close to one another, as with the measure given by fractions between whole numbers, they comprise hyperdimension — each different dimension distorted with its own unique underlying geometry.

a. effects of radiation on the genes

Some vibrational frequencies are beneficial to the life of the microorganisms that comprise the fibers of our genes. Others are harmful. The same common knowledge is associated with the modern folk wisdom of pregnant mothers putting small speakers playing music upon the outside of their growing uterus to expose the foetus, while still developing as en embryo in the womb, to the external world's sensory stimuli. This is thought to enhance the baby's appreciation for composition later in its adult life.

The beneficial vibrations of electromagnetic radiation, which includes the visible spectrum and is of a similar form to sound waves, are those equal to the angstroms of the gene — that tend to encourage or advance it along its existing pattern of growth. Also, just as lasers can now be used for cutting, the right frequency of light can aid in the splicing of different genetic fragments, and activation of different inherent traits, while they are still alive in the cells.

Harmful frequencies are all the rest of the frequencies that do not exactly match the criteria of the existing measurements of the genes. It is the sad truth that the life line followed by our genetic sequences and sequencers — birth, mate, death, the same holographic pattern we follow in our own lives — is more influenced by these negative influences than by the scarcer positive ones, but only according to the statistics, which show that all things that are born die, but that only as many things are born as mates have found each other to parent them, and that fewer things find mates than are born.

The result of exposure only to those frequencies which stimulate our growth as a genetically unique species would be an appearance similar to that which people appear to have to a person being exposed to these frequencies. This has been described by people living near the electromagnetic fields and flux beside the geophysical friction of a fault line, and I have already mentioned it in the section on the hallucinatory appearance of aliens.

b. effects of radiation on the mind

There remains a debate about the mind over whether it is more aethyreal or substantive, and this debate has gone on for almost as long as there have been minds to debate with about minds in the first place. However when considering the effects of radiation exposure, nowhere is the line more clearly drawn between the two opposing concepts. We see that the debate hinges upon the concepts of aethyr manifesting (motivating) matter and matter representing (reflecting) aethyr. The earlier argument is then usually called moral or ethical, though it is really impossible according to matter/energy, and the latter obfuscated by the division between the body and brain, when it is really moral or ethical. At the two polarized ends of the argument are the concepts of mind as aethyreal, eternal spirit or as the mortal self-cognition of organic matter. In truth it is the biological circuit which is the temptation, though anyone subject to natural law will eventually find themselves having believed in both.

When the brain itself is exposed to electromagnetic radiation, it has different effects according to the different wavelengths of the vibrational frequencies. Very short wavelength, high frequency microwaves of the scale of the diameter of a single electron cloud penetrate to the inner neuron pillars and columns of the cerebellum, as well as to the holographic neuralectric projection tissue of the thalamus, stimulating quantum fluctuation there, and thus increasing the kinetic potential of the brain overall, that is, it stimulates thoughts. However, microwave frequencies of a slightly longer wavelength — that of the diameter of a nucleus or an atom — will disintegrate the brain tissue at an atomic scale. One method of introducing electromagnetic fields into the brain is by creating a gravity band or gravity well, and it is known that magnetic elements were used in the creation of some Mayan crowns, possibly for the purpose of elongation of the wearer's skull.

The thoughts that arise in the environment of the mind, in the context of the ego or the recognition of self, due to exposure to electromagnetic radiation are similar to those induced with certain psychotropic or psychoactive chemical substances, and this is not surprising, since the brain, along with the entire nervous system, is an electrochemical hydraulic turbine. These thoughts are usually distortions to the type of sensations our sensory organs ordinarily perceive. It is usually thoughts such as these that are associated with mystical experiences of the aethyreal spirit, and therefore the primary goal of adapting to them is the differentiation of fact from fiction.

c. you can never go home again

Changing is an ending of one thing and the beginning of another, and everything is forever changing. In this context you cannot hold onto anything, and may even let go of your sense of self, for it can only serve to anchor your soul to one region in space-time. Here is where we experience the astral body of the free spirit. This is the idea of the second self, and even it can be shed for the greater Light.

Permanence is an ending of changes, the fixing of all potential future histories for a subject into singular form. It is associated with the edifying of idea, however it is hypothesized that, in so far as it cannot be accomplished artificially, it can be accomplished naturally at least as well. This amounts to the idiom "dust to dust, ashes to ashes," which refers to the original creation of our internal carbon compounds in the nuclear furnace of the sun, and to the continual radioactive decay of carbon element fourteen in all living and once living things.

Thus we see that both change and permanence constitute their own form of ending — change the ever ending of everything, permanence the end of change.

i. infinitely changing parallel universes

When we begin to see time as the history of a wormhole, the measure of the moments will be infinitely changing parallel universes. These are equivalent to the infinitely different possible outcomes in the potential of every event probability well. Thus each different possible universe measures as the intersection of two distinct dimensions — infinite depth within and infinite length in sequence.

Inside the wormhole history of time we experience temporal gravity towards the past, where our memories tie the ego down to the experience of material reality in the present moment, and invert these into concerns (intellectual and emotional) for the ego's future. However this is no more than a sign of the times, erected into the freefall limbo of the future. It is used as a guidepost and for ascension, it is the subject of gospel and gossip and derision. As I have said before, our concerns for the future of what we believe to be our selves are irrelevant relative to death. This does not mean people don't think that these two things go together. On the up side, by letting go of the constraints tying one to the sensation of the here and now, it is possible for the center of awareness to rise up into the future.

ii. permanent timelessness

With no fixed focal point for the self in the temporal continuum, one can explore, detached, wherever they will. At this point we can stop time, fast forward, rewind, do all these things within our minds. Meanwhile an infinite number of possible options surround us at all times, populating the multiverse with pathways connecting to singularities like us. Singularities, like all things, are attracted to one another — temporal and gravitational both and alike, and so the wormholes comprise potential histories between these points. To be drawn into and caught up by this web is to be beyond the reaches of time. Singularities are permanent more or less relative to one another, and wormholes only connect them as long as they are apart.

The only equivalent concept to permanent time distortion, or time-space disorientation, is the concept of hypersleep, the sleep of ages. In this the subject body is put into unconsciousness in suspended animation and reawakened, usually many light years later. A parallel theme to this is that of post mortem cryogenic preservation and resurrection, or even the potential extrapolations of transplanting.

d. the way out

It is a popular saying among metaphysicians that "the only way out is through." This is a half truth, and unlike many of my colleagues, I do not wish to appear false or misleading. It is generally accepted that the only way to the singularity is through the wormholes, and the singularity itself is the doorway out of the universe and its internal, ever expanding multidimensional multiverse. However going through it leads into the baby universe created outside the universe by the alignment of the temporal singularity wormholes as they feed the substance and laws of physics from the parent universe to the child.

To come to grips with the popular misconception of "getting out" by "going through," we must extrapolate what is indicated from what is observed. In this case we see that, as a body travels through the wormholes inside the event horizon of a black hole and outside the singularity at its core, it is divided by the temporal singularities it passes through on its way between them to the central gravitational singularity according to the law of infinitely repeating halves. Thus, what is apparently theorized by the esteemed past scholars is that, to finally exit the universe and the expanding informational dimension multiverse it contains, one must first pass through an infinite number of baby universes containing smaller black holes. As ritualized and formalized as this doctrine has become in the modern occult, it really amounts to little more than this.

While there is an absurd consensus — in light of naked singularities, that to get to a singularity it is necessary to travel through quantum foam with random quantum fluctuations caused by the uncertainty principle (the wormholes), it is generally not considered to move temporally perpendicular to the history of the singularity after entering into the realm of microwave gravitational tachyonic ylem.

e. archetypes

Archetypes are transtemporal — they exist throughout time. Commonly they are called stereotypes, and, while valued, also made the subject of much abuse. The name "archetype" is a crude play on words, doubling the arc of the rainbow with the ark of the covenant. The true ideological entity this asymptote of meaning describes is perhaps better thought of these days as a "messaliengel" — a triple play on the words "messenger" (in that it transports information), "alien" (in that it is not confined to the earth) and "angel" (in that it is spiritual and ideal in essence).

Perhaps the reason that stereotypes are so abused as a social statistic is the ongoing debate between the body and brain and the mind and spirit. They represent a blending of the self-identity with some greater purpose as their guide, and thus make highly recognizable minorities and often too totalitarian majorities. The people who do not wish to be seen as stereotypes, on the other hand, see themselves as being somehow more spiritual and thusly more righteous. The truth is that, as long as there is analytical observation, nothing can remain long unclassified.

The social movements — usually themselves the cumulative effect of combined social, political and economic factors that give rise to these stereotypes as connections along the "grapevine" joining mainstream and underground culture — are thought constructs of information systemization that naturally organize these statistical aggregates into relationship with one another. It is because of the imagined similarity between social institutions coagulating stereotypes and seraphic hosts and orders of angels and the djinn that they are thought to be justified.

It should be worth noting that these social institutions and classificational orders of archetypes began at the same time and in the same place as the institutionalizations of State and Church. However besides that they are literally as different as these meta-entities.

The angles, demons, spirits and ghosts thought to populate the afterlife are all either originally anthropomorphications of discorporeal forces or founded on the personal memories of departed ancestors. The synthesis of these two realms is the key of all religion, and it continues to remain a struggle in the minds of holy men today. In the far east, where Shinto is prevalent, the people believe that even the trees have guardian spirits, and Buddhism shows how the soul passes from one body to the next on a journey towards ascension into the primary clear light. In the middle-east and Europe they have made equally complicated explanations unifying the realm of the forces and that of pure consciousness without a living shell. In the west the various religions of the native Americans succeed best at understanding that these two underworlds are one, and the rest focus on extraterrestrials and the Apocalypse.

None of these clearly specify or agree as to the exact details of the nature of the archetype. Using elements from all the existing systems, as we see that it must do — since these are temporal systems, and archetypes are transtemporal — an archetype might best be described as a reincarnative holognomonic interpersonal social mold.

It is reincarnative in that it transcends its representative manifestations in the eternal duration of ideal or potential. It does this from moment to moment, event to event, stereotype to stereotype, and so, gradually taking every possible road in the garden of forking paths, tracing out its parent pattern like pixels on a tv screen.

In this way it is imitating the holographic behavior of photons in the double slit experiment. Here we see that the individual photons themselves follow the same fractalization, (in this case binary) being equally likely to pass through either slit, as a hologram, wherein each part carries potential for the pattern of the whole field. However the difference between a fractal and a gnomon is that a fractal has a closed or finite set of parameters, and a gnomon has an open set. Therefore all fractals are self terminating in the course of their permutations, however gnomons continue to replicate exponentially. Because our human population grows in this same way, as I have said, opposite the curve of death, the gnomon is thought to represent a more life like form of mathematical expansion. Thus to say that archetypes are holognomonic implies that their existence is in some way more like our own than otherwise.

Insofar as it is relative in fields, the archetype is commutative in the same way that humans communicate interpersonally. We cross gaps between one another with our ideas, and in this same way a particle, such as a photon, transmits a force from one event location in spacetime to another. Thus, also, matter and energy, though inversions of one another, are both subject to interaction temporally. We see that the more we interact with others, the more affirmed we feel that all is one.

Society is but a vague shadow cast by the desire to preserve a deteriorating knowledge of an ancient technology, which is based on the science of time. It is hints of this great Light that peak through now and then to bestow revelations and spiritual experiences that foster and engender and encourage systemizations. These are only glimpses by the left hemisphere of the potential for reunion with the right.

Archetypes are artificial. Stereotypes are man made, copyrighted, advertised and merchandised every day, and so, it is thought: God controls the angels. Einstein, one of the greatest minds to grace this globe, said, "God does not role dice." Man is given free will, but this right is reserved for him alone, as punishment, along with death. The angels have no rights of their own and the demons have no release. This little bit of metaphysical legislation sounds about accurate as the result of millennia of human relations. It leaves alot to be said for the description of the actual conditions of archetypes as I have described them by unifying their attributes.

B. the multivibrational universe as time capsule

What we are seeing with our eyes are photons. They enter through the lens, reorient the rods and cones, and are translated into electrochemical signals in the nerves connecting the eyes to the brain. Photons are real particles that transmit radiation from a source to a receiver. It takes different amounts of time for a photon to travel different distances, since all photons travel at the same speed, known as the speed of light. A photon reflecting from the surface of a flower to our eye, and carrying a distortion to its wavelength that translates as the flower's colors, travels across that distance almost immediately. It would have taken that same photon six minutes to get from the surface of the sun to where it was reflected off the surface of the flower. It takes light 100 million years to travel across our galaxy, and only about 11 to 15 billion years to get to earth from the furthest edges of the known universe. What we are seeing in the night sky, for example, is not the light given off by the stars as we are watching them, but the light they gave off millions and millions of years ago. Therefore it is not exactly accurate to consider anything that photons are conveying to our eyes as reliably real. Just as the one color wavelength the flower reflects is the only one it didn't absorb, so are the very heavens above an illusion.

Samsara is a recursive cycle of events.

The most of the populous is given blindly and bodily over to Samsara. We learn what behaviors are appropriate and what not, and eventually become completely distracted by our own lives.

Then there are those who leave Samsara. Of these some leave only to return again. Of these some return and forget, and some return and remember.

A few leave completely. Of these most leave at the moment of death. The only humans known to have transcended Samsara while still alive all became great teachers and healers, because only those outside Samsara can change it, and only those who are completely free of it can guide the rest of the populous closer to its conclusion.

In Samsara, though, everyone thinks they are a fit guide. This is how they forget their own suffering. When suffering is ignored, it worsens. When internal suffering worsens, it becomes external suffering.

For Buddhists, only the trance of nibana can free us from external suffering. Then, in the trance of Samadhi and Samasamadhi one sees their own internal suffering, and by seeing it, can pass through it — for it is only an illusion. In the trance of nibana one sees the veil of external suffering as well, and, by seeing it, can pass through it — for it too is only an illusion.

Suffering is an illusion because it occurs to the self. The self is an illusion. Both the self and its sufferings, or what comprise Samsara, are an illusion of time. As time passes, events change, and within these events are physical patterns, such as ourselves, trees, computers, etc. When events change, these change, causing patterns of event for physical form over time.

Samsara is a very specific such cycle of events that applies on a wide scale because it governs very general patterns. It has been known of since the creation of civilization, and has existed at least since the birth of consciousness. There is no way of knowing whether or not, say, an electron suffers in its orbit around a nucleus, or whether the stars can know that they are doomed.

It has been hypothesized that, according to the law of natural similarities, the consciousness by which we know Samsara is probably a pattern that we have found in other forms, so it is possible that the suffering is of the pattern, and that consciousness is the mechanism by which to pass through external suffering without internal suffering.

The trances used by Buddhists are methods of concentrating consciousness so that one can see the physical pattern of their own existence as the source of internal suffering, and perceive that they are more than this alone, and see the cycling pattern of existent events as the source of external suffering, and perceive that the universe is more than this alone, and in so doing, find peace.

The pattern of samara is quite simple. Two forms of it are the Buddhist Wheel of Becoming and the hexagrams of the I Ching.

You should notice that the wheel has spherical application and the hexagrams have cubical application. This only arises because the mind, the source of both forms, is represented as the sphere inside the cube, where consciousness occurs as a point shared by the sphere and the cube, the sphere is an infinity of thoughts, and the cube is the five senses and the six fundamentals of reasoning. This shape is representative of a ϕ / π spiral — the sphere being π bound, and ϕ being extractable from the cube as the height of one of the tetrahedrons it contains if the leg of the cube is one unit, or as a flat spiral on the surface of half of any one of its sides. This is the universally occurring pattern that has been postulated for consciousness, or the pattern of concentration of mind over time.

By applying this pattern to the Wheel of Becoming and the I Ching we can see how simple the pattern of events called Samsara actually is. In the I Ching there are eight trigrams that compose the hexagrams, and consequently there are eight double or inverse hexagrams, which look the same above and below. These are the corners of the cube. The sides of the cube are the six lokas, or realms of reincarnation, between the spokes of the Wheel of Becoming, yet these cardinal directions are eight when taken with the ascending and descending paths through them, and ten when counted with consciousness, at the center, and time, the mandala's outline.

The eight trigrams can be arranged either according to the early heaven or later heaven arrangements. The proper arrangement for the eight "doubled" hexagrams is determined by a pattern that balances their numerical values from the context of the King Wen sequence of 64 hexagrams, given in the Mandala of Time. The eight trigrams usually represent characteristics that govern the eight events represented by the eight doubles.

What is described by Samsara is the fluctuation of a field of energy. The 64 hexagrams represent the regular functioning of this field, represented by their first order of difference, or number of tao lines changed between them, and the eight doubles the regular fluctuation, or wobble, within that function, similar to precession. The inventors of the I Ching called this energy chi, and housed it within tao, meaning "the way." Another way of expressing it would be to call the energy potential, and tao probability. Here we see that probability is the substance of time.

1. the geometrical spirit and the electromagnetic soul

Regardless of how we describe the states of others that dwell before God, we describe ourselves as having souls that become spirits when they are raised before the lord. This has been described variously by all the different institutionalized religious cultures, and it is a belief that traces its roots back to the birth of the djinn and civilization. In the far east the internal energy, or chi, is studied as it flows in the kundalini path through the chakras and manifold pressure centers, generating the person's karmic aura, visible when they open the ajna, or third eye. In the middle east the Hebrews study the Ruach (breath or soul) and neschamah (spirit) in the context of the four worlds of action, formation, creation and emanation; Christianity expands upon this esoterically by way of astral travel and indulgences; and the

Muslims offer grand promises of paradise beyond death. The native Americans in the north and south continent, even as far as the western most islands of oceania, believe in a force called mana which is responsible for all things, and is the substance of the power that the gods have over the heavens and nature, and by which they effect the laws of men through their appointed administrators.

The subtle distinction between the soul and the spirit has been argued philosophically for equally as long, and "insanity" and "intent" are the equivalent modern legalese. According to current legislations, the letter of the law is replaced by plea bargaining and the spirit of the law revolves around proving preconception. all this means in straight talk is that modern pop culture media, whose target market is rebellious teens, is itself the child of election politics and national populist religion and that the protection of the inherent human trait of free will is guaranteed by written law in the American constitution. Pop culture represents the golden dawn of artificial intelligence, the media itself replacing the soul, or link between the astral and material realms, and the spirit becoming a simulacrum body without organs in virtual reality, possessed of the absolute, dictatorial freedom to impose the full limits of their will upon the seeming substance and the very essence of their environment.

As I have described in the context of hyperdimension, pure mathematical geometry underlies the laws of physics. It has long been held that pure math is on a different level of reality, that of pure ideal, for it can be done even in the context of only the mind. Dreaming was thought to be the degeneration of meaning of mental images from this realm when it was first conceived. Thus it was thought that the soul descends from this realm, like a dream, while the spirit raises upward toward it.

a. the corporeal vessel

This is that shell for consciousness that acts as the umbilical chord connecting the soul to the earthly environment. When we die this chord is cut and our spirit is born into the next world. This world after death is thought to be the aethyreal astral realm, that of the ideal, where exist distilled to their essentials all patterns that give rise to shape in manifest forms. Just as the body manipulates objects on the terrestrial plane, so does the free spirit apprehend and shape ideas from the astral aethyr. However when the body dies, and the spirit leaves it for the last time, it does not die completely. The spark of life remains, in the form of the genetic coding of the body's DNA and the ambient electromagnetic signature that it accumulates in life. The result of this is seen in the continued growth of hair and nails after death, caused by the aura's afterimage stimulating the tougher follicles and tissue to progress along the lifeline's patterns of growth, and in the decay of radiocarbon 14, which acts as an electromagnetic pulse, the only radioactive element present in organic matter, and thus serving as both an elemental frequency and a biorhythm. In this way even the dead are not dead, only waiting for their voices, carried on their continuing frequencies, to be listened to and understood.

i. practising for death

The secret doctrine of the Therapeutae, contemporaries of the Pythagoreans of early Greece, is that ultimately it is better for man to die at the hands of man than through the intervention of illness. Thus they began the practise of deriving medicines from poisons, in order to treat fire with fire, and therefore insure different effects of preservation of the corpse after death with drugs that treat incurable diseases with death. This continued on through the Roman empire, when lead, a deadly toxin that assists in post mortem preservation by killing off the bacterial microorganisms that line the walls of the intestines, was used to line aqueducts and thus taint the public drinking water supply. Lead based medicines continued to be used throughout the European Middle Ages, as well as various other types of torture, such as bleedings with leeches. In modern times the equivalent of this practise is vaccinations with small dead doses of the virus to boost up the immune system against them should they encounter them again. Another modern ingredient of this foul plot is tobacco, which, as it is packaged and sold to the consumers in the medium of cigarettes, contains arsenic, tar and other deadly toxins, all which serve the same function as lead — to kill off the intestinal bacteria that cause the body to decay after death. Traces of tobacco have even been found in the mummies of Egypt, indicating their ancient knowledge of its use in preserving the genetic integrity of the fleshy tissues after their life has ceased. The practise of mummification is much older than either the Therapeutae or the mummies of ancient Egypt, and is probably a millennial practise in preparation for the eventual end of the world. The most recent mummies are, or were, Buddhist monks who drank lead based lacquer and arsenic laced spring water to stop their autotrophic life functions, after surviving on nuts and tree bark for six years to thin and harden the tissues.

Mummification is only one branch of the wellspring of spiritual legislation on the preparation for the moment of death and the unknown duration to follow. As I have said, much of humanity's metaphysic for the afterlife is founded on the soul or aura and astral body or spirit. The soul is attached to the body in the form of the aura, comprised of our accumulations of karma, which usually follow an identical pattern along a path of ascension. The spirit detaches from the corporeal vessel, and rises into a clear light, as described by all mystics and those with near death experiences. According to the doctrines of all the various mystery schools it is important, almost necessary, to practise leaving the body in preparation for the moment of death, when the spirit will be severed from its customary control of the corporeal vessel. In the highest levels of gnosis the process of self-actualization is identified with Satan, or the concept of cognitive dissonance in the universal mind.

ii. dying

Strong believers in anthropomorphism might perceive the moment of death as a familiar personage, and, just as indicated in Greek theater, these psychodramas can play out over the entire course of someone's lifetime leading up to their final moments. The reasoning for this is that the mind is in a highly deluded state relative to death. Blood bearing oxygen remains in the brain several minutes after the fibrillator has stopped pumping blood through the heart, and this oxygen allows neuralectric activity to continue for some time. Because people usually do not return from such a state, little observational data regarding it has been collected, and so it is unknown, for example, if the perceptions of the sensory mechanisms continue to function or not. In many people whose near death experiences involved very sudden or shockingly transmundane experiences they "see their life flash before their eyes," becoming cognizant in microseconds of macropatterns underlying their lifeline that had eluded them their whole life long. It is possible that such a surge of neural kinetics and the continued hum of the brain as it feeds off the last of its air produce equivalent internal experiences, and these types of personal transcendent events are associated by modern psychologists with visions of a Final Judgment. The moral of all historical judgment visions is to know yourself while you are alive. This is not contingent on being understood by others, and the road less traveled is always the only one that does not lead to Rome.

b. the electromagnetic aura

Once one has begun to learn about the physical part of their self, one will come to learn that it functions on an electrical charge, that flows through all the flesh, controls the heart, and runs the brain. It is fact, if one wishes to learn it, that the neuralkinetic reactions in our brain are synchronous with our thoughts, such that they are thought to be mutually causal, and that different frequency brain waves are produced in different states of waking consciousness, unconscious sleep and dreaming. The entire physical corpuscle contains this flow, called chi by the Chinese, who taught with martial arts how to channel and harness it for physical health, strength and endurance. The duality of chemical (soma) induced neuralectric activity as external motivator and naturally occurring inspiration as internal cause of thoughts predates the beginning of recorded history, where we find the roles of chief or elder hunter and shaman or medicine gatherer established possibly as the original hominid division of labor between the sexes, however it is an issue that continues to this day involving scholars of all fields. While the right and wrong of it are debated, by their compromises it gradually becomes a medical reality as all of our inherent physical systems are augmented by advances in scientific biotechnology and we are indoctrinated to the rules of the state. Drugs are about as equivalent to the soul as the state is to the realm of spirit, but all schools that have derived wisdom from them teach us that there is much more karma in life than in any one style of it.

The earliest professional philosophers were the Greek elementalists or pre-Socratics. They were the first to begin to describe the aethyr as more or less alike certain forms of natural phenomenon, namely the four elements. They were, in doing this, opening up the watchtowers to what would become an established spiritual realm in institutionalized religion, though at the time they were seeking to describe the essence of nature. Plato followed this with his strict segregation of the ideal realm, where the forms or phylum were found, and the real realm, where the phenomenon were observed. Little progress has been made in humanity's subsequent effort to tear down this wall from the side of the real even to the modern day existentialists, who believe that existence, or base matter (subject), preceded essence, or the potential for spiritual ideal (object). The diamond key to cutting through the glass ceiling of the Golden Division is the pythagorean extrapolation of the vesica (or heart) and the cross (or helix) proportions, representing the realm of the ideal, from the pentagram, representing the realm of the real. There remains much to be said about this relationship, however the aura is the energy of the pentagram, just as the potential for the release of kinetic energy is represented by the five digits of our hand.

i. the ajna

As I have already described, the ajna is the mind's eye. It represents the vista of mentation occurring in the mind, just as brainwaves represent the neuralectric activity occurring in the brain at the same time. Enlightenment, the goal of all seers, is a brainwave flatline while still alive and consciously awake. Since alpha waves, those produced during the dream state, are the slowest frequency and longest wavelength, such a state is pursued by attempting states of lucid dreaming and states of conscious dreaming, usually achieved through prolonged sleep deprivation. It is thought that, in this state of total mental detachment, the concept of time becomes meaningless, and this is the ultimate achievement of all mystery schools forever. The state of pure illumination is also associated with levitation. It is unknown why this would occur, even by many of the monks who attain it. However one theory is that, when their brainwaves stop creating a pulse regulated field around the body, its electromagnetic signature begins to become distorted by the ambient background radiation of the environment, and the quantum frequencies are no longer in temporal synchronization with the temporal electromagnetic imprint of the situation. A similar technology used in aircraft antigravity propulsion systems is the null wave zero point projecting triple scalar wave generators of flying saucers and the stealth bomber. The null wave field, or zero point (without established temporally synchronous context), is the result of crossing scalar waves, which are themselves overlapping fields of wavelengths from multiple sources. In this way the flatline history is derived from a scattered field, just as pure enlightenment may be achieved after waking dreams.

ii. projection of the astral body (as a conscious singularity)

In the same way that the loss of the controlling rhythm of the personal electromagnetic field causes it to fall into dystrophy, and this may stimulate a physical phenomenon as the field becomes disoriented to its polarized environment, so does this produce for the field itself the effect of generating a light body sensation or feeling of free fall. This is the astral body, that naturally occurring electromagnetic signature of the soul, as it begins to de-tune itself from reality, and fall out of phase with the comprehensible dimensions.

The ambient shape of the aura surrounding the chakra path of the 33 spinal vertebrae is that of a torus, or hypersphere at antipode. When the pathway of kundalini is aligned and the chi energy rises unimpeded, the aura involutes. Although it is represented by a light ionization of electrons, most of the aura is given off in the form of heat, or pure thermal radiation, the essence of which is tachyons. When the brain stops sending brain waves to control the regulation of the body's biorhythms, the aura begins to involute at the velocity of these underlying tachyons, becoming a potential energy dynamo, creating a temporal singularity.

When this happens consciousness flows through the mind in free association, without singular origin, being omnidirectional, and without end, being unimpeded by the interjection of the ego. Here one perceives everything as nothing. This is the ascension of the soul to the astral, spiritual, or ideal realm. From this point one identifies the self with a conscious singularity in the realm of potential that can manifest itself where and whenever it chooses.

iii. exploration of the astral plane (hyperdimension)
The soul, when elevated before the Most High, becomes the spirit. The aura becomes the astral body. It is not the vessel that is changing. It is only its environment that is different. It is thought that the gross body is in accord with this same harmonic. The form of flesh, the electromagnetic aura, and the geometrical spirit are all one. When the electromagnetic aura is active there is a physical reaction. When the aura leaves the host and travels throughout the universe and beyond the multiverse it must adapt in its own way. Thus we see that the vessel responds to its environment, whether this is physical, psychological or heavenly. We must assume that it is therefore equally able to manipulate outcomes in the astral and karmic worlds as it is in the physical.

As I have already described, the spirit is thought to be most "at home" in the idea of heaven as a community of other spirits, and this is why the realms of the unknown afterlife and the drug-induced spirit realm have been so judiciously legislated as similar, when in truth, since little serious research has been done on either openly and in a rigorous scientific manner, no such connection can rightly be inferred. The home of the angels in the heavens has been identified with various locations, both real and imaginary, ranging from the more realistic Hebrew zohar on the ten sephira (senses) and Buddhist six lokas (archetypal realms) in the reincarnative cycle, to the less realistic Mormon interplanetary aliens (although I do like the Star Trek interstellar and Star Wars intergalactic ones). It is thought that the home of our ancestral dead is in the same location.

In truth it is clear that, if projected in the from of a potential, that is, temporal, conscious singularity, a departed spirit could go anywhere in the universe instantaneously by traveling through the wormhole multiverse surrounding gravitational singularities. It would even be convenient in these locations to leave the universe altogether, if one so willed it.

c. the geometrical spirit
By existing in the universe, in any form whatsoever, one is subject to the laws of physics. When one travels through higher dimensions to move from one place in the material universe to another through potential, one is subject to the laws of geometry. This is because geometry underlies the laws of physics in the universe, and higher dimensions tend to break down material forms into less manifest, more ideal systems according to the framework of this underlying geometry. In hyperdimension, in the light of geometry, spirit is like a pattern through the dimensions. Finally, in the presence of the primary clear light, all that is left of spirit is the pure geometry of this history without dimensional context. This is the highest extent of the existent gnosis, and it is why there is a "G" for geometry between the

Masonic compass and square, both of which are, themselves, instruments for making geometrical measurements.

The spirit can be free even of geometry, however, while it is in the presence of the divine, geometry is how spirit will know the divine. When the spirit frees itself of the trappings of geometry, there is no longer any difference between it and the essence of the deity, and they become in essence one and the same. However if the spirit wishes to know itself as apart from the divine, geometry becomes key. In other words, pure geometry and pure spirit are one and the same in pure dimension, and pure spirit and the presence of deity are also the same in essence.

i. free will

As geometry underlies the laws of physics, and these are thought to be universal, and because pure geometry can be accessed through hyperspace by a projection of consciousness, there is literally nowhere that a spirit cannot go, and nothing that a spirit cannot do. It has utter free will, as it alone in the universe is capable of doing anything simply by instantaneous manifestation.

However we are told exactly the opposite is true. Just as ghosts are bound to a hoary netherworld of karmic limbo, angels are segregated to the clouds of heaven and demons to the cavernous fires of hell. In Heaven the angels obey God, and God acts on behalf of man bending his utter freedom to the bow of the covenant. In Hell there is a similar hierarchy — demons answering to Satan, and Satan the slave dog of God. Finally the same scriptural style sources proclaim that humans alone are free. This last seems to be the missing capstone, since until very recently, wageless slavery was common throughout the world, and many of the same legislators who fought to abolish it owned slaves themselves. Now there is only extreme economic imbalance between the less than one percent of the world's population who control over seventy percent of the world's wealth and the child labor sweat shops, environmentally polluting factories and starving, diseased tribespeople they are responsible for. Still, if this is meant to represent a hologram of the mind of the archetypal human, it falls far short of the original lawmakers' specified standards.

Too much importance has been placed on the brain in the legislation of spiritual matters. When it dies apparently the free will given to man by God is thrown out the window, just as in the story of Adam and Eve being banished into labor. In the west it is believed that no spirit that has lived in a body has free will without one, and in the east it is believed that to be free of the spirit is to be one with the divine. Thus it means that to give up the spirit is to give up the free will, and thus to join with the divine. However there is ample evidence of spirits that have given up their free will. This can occur for a person in their life, and then, when they die, their spirit will have no free will. According to eastern tradition this is the pattern of karma that results in reincarnation, or the same spirit being born again into a different body. However, in the west Judeo-Christians sacrifice free will in this life for a better afterlife, building up karma for their future. This is commonly called the "work ethic" and is seen pushed to its furthest limit by Moslem extremist suicide bombers. Thus it seems that it is possible for the spirit to regain free will after death without necessarily reincarnating.

It should only be important that the brain, as an organ, contains as much potential as the spirit, as an essence, can imbue, and that the spirit, connected to the divine by geometry, is infinite in potential.

ii. the geometrical system unifying the multiverse

Just as the form of each living being has a signature electromagnetic aura, so the earth and all the celestial bodies also have electromagnetic fields, and these are all arranged to one another in a webwork that generates gravitational spin in the universe by microwave electromagnetic fields, and also the universe itself emits an illumination of tachyonic astral light. All these things obey the definition of a soul in the sense described by all mystics of all lands throughout time. It is easy enough at this point to see that all these souls are only one soul over time. This should not be taken entirely in the hierarchical sense, as alike offices held by bureaucrats. These spatial levels can be seen as more alike parts of a whole, like organs in an organism. In the same way that the strong nuclear force (nuclear fusion), the weak nuclear force (nuclear fission) and the gravitational force (microwave tachyons) are all peripheral radiations within the electromagnetic spectrum, geometry underlies all the laws of physics.

The study of classical physics began with the study of shapes and measurements, in ancient Greece. At every step in the scientific progress of classical physics it has been supported by its geometry, from ancient public works to Newton's astrophysical theory of gravity. The study of modern quantum mechanics or particle physics is derived from the study of classical physics. The concepts of vector and spin, wavelength and frequency, while expressed as algebraic variables in the axiomatic equations, are advanced derivations of geometry. So, just as there may be many levels to the soul, there is only one spirit: math.

Black holes are the nuclei of galactic cells, and the singularities are its DNA. Just as mitochondrial DNA exists outside the nucleus of a cell, singularities can exist outside of black holes. Here we find the physical equivalent to the temporal patterns governing the evolution of the universe. Singularities cause geometric distortions to the laws of physics. Temporal, or naked, singularities are probability wells causing temporal distortions to fields of potential energy. Gravitational singularities, inside black holes, are potential fields causing gravitational distortion to wells of probability. The result of both is wormholes. In the case of naked singularities these wormholes lead to other naked singularities in the universe. In the case of black holes all the wormholes lead into the singularity, through the fabric of spacetime and outside the local universe into a baby universe. The sum of all the parent universes and baby universes is called the multiverse, and comprised of infinite dimensions accessed inside the wormhole. In the same way that the laws of physics governing the matter passing through the black hole into the baby universe are slightly distorted by being passed through the wormholes to the gravitational singularity, so are the physical laws also distorted for the matter passing through a naked singularity wormhole. In both cases the distortion is geometric, and therefore can be predicted.

2. the macroverse and the microverse

Long ago, an unknown personage of ancient origin, according to the histories out of southern Persia, came into the Greek and Nile areas and began writing divine instructions for understanding the cosmos. These rules were filtered down throughout most of the rest of western culture, and we are told that a similar school began in the far east oriented around the concepts of light and dark. Although the name of this person has been forgotten by the ages, his association with certain local concepts of deity, such as the baboon or ibis in Egypt and the messenger in Greece, has preserved the tradition he long ago began. The most famous of his dictums was, "as above, so below." During the Roman era this philosophy was present in the economic division of patrician and plebian, while the essenes and nazorenes described themselves as "children of light and darkness." However during the European dark ages this concept was rediscovered by alchemists, and the terms "macrocosm" (describing the cosmos) and "microcosm" (the unit), were derived. Only now are we finding similarities to the cauduces or staff of the healer and the double helix of genetic DNA as the microcosm and the zen garden of forking paths and the filaments, walls and voids of the universe as the macrocosm. The pattern of phi/pi clearly emerges, where phi represents division and pi unification. Since division (as between filaments, nerves, genes during RNA replication or quantum waves) and unification (as for individual celestial spheres, the gamete progenerative cells, DNA packing inside of the nucleus in ribosomes or quantum particles) are obviously inverse of one another, like dark and light, it is cognitively dissonant for the unenlightened mind to perceive them as equivalent. Therefore, the idea accepted by the Greek philosophers, who began the first philosophical colleges in light of the wisdom of the ancient law giver, was that man must be separated from the cosmos in the same way that an idea itself — such as the idea of self— is separate from the ideal realm of the mind. This makes sense, because a light source, even one as prevalent as the sun, is only emitting rays into the darkness of the void, in the same way that ideas generate lines of reasoning that progress into unknown dreams. In this way the macroverse became associated with the realm of ideal — where phi and pi are one geometry, and the microverse with the real, given or material.

Currently, scientists are seeking to discover a unified field theory, or grand unified theory (GUT) of everything. Until chaos theory became popular in capitalist culture from the mid 1970s to the early 1990's these theories consisted of debates between MACHOs and WIMPs, or strong and weak centers of gravity. The increased government funding of superconducting supercolliders and magnetic particle accelerators following the mideast oil embargo, however, allowed early computer program using, fractal generating mathematicians to exchange ideas with quantum physicists and this is how string theory began. According to string theory, the universe is comprised of a finite number of vibrating frequencies, and these are hyperspace dimensions, coiled up very tightly, that cause all pattern in matter. Chaos theory, which is the attempt to make the mathematically abstract axiom of the uncertainty principle more elegant using fuzzy logic systems, and string theory both miss the mark of Einstein's own attempts at a unified field theory involving gravity, however Einstein was quick to admit that his own theories, too, were imperfect, and remained unconvinced of a gravitational constant as unifying ether.

Combined, these theories would spin off into the even more modern theories of the holographic universe, based on the Hindu Maya — an energy illusion, and quantum consciousness, attempting to account for esp. Another modern theory, called M-theory, doubles the dimensionality of the string frequency fabric of spacetime, making the chaos-fractal vibrations into waves of potential, passing effect through probabilities, and calling them dimensional branes. Each one of these theories is satisfactory in its data, however each is incomplete. The longer scientists wait to reveal to themselves what some ancients already knew long ago, the more these GUTs will pile up, until eventually we will need a GUT of GUTs. It is within the reach of every one of us to know that everything is one, and this generates wide belief in one god, but the mechanisms necessary to prove in order to convince others of the ways in which different things are similar elude the grasp of most, tainted as this is by the superstition of good routine and evil change.

The six fundamental questions of rhetorical reasoning lead us inevitably to a description of the universe as a dimensionally resonant, holognomonic metaform, an effect being the inverse reaction to a cause on each swing of the temporally motivated pendulum of quantum uncertainty.

a. consciousness as a fractal represented by society

It is as easy to anthropomorphise as it is to sympathize, or empathize symbolically. The most common token exchange unit used for information transfer in this market is social statistics, thought to be an idealized, purely mathematical reflection for aggregate psychology. This implies that buying power accurately accounts for all quantifiable self expression in a society, such that what is available in terms of goods and services meets all of the consumer's needs, amenities and luxuries to satisfaction. This is the modus of totalitarian supply-side economics, however its philosophical apology is a rhetorical trojan horse, revealing demand-side economics as doomed to dependence on over-analysis of market research on target audiences, an effect already responsible for the social subclass of producers and executives in the strata of the capitalist cultural entertainment industry. We feel the effect of this struggle every time we shell out five dollars for a bad movie.

Because both supply side and demand side economics resort to the use of social statistics to understand the will of the masses, it could be imagined that the use of social statistics preexisted economics, however this is not accurate. Agrarian barter, or trade, economies flourished very well the world over using only folklore as culture, and it was not until the nineteenth century that the first industrial based economies began opening up global commerce with the standardized token exchange system created long before in old Sumeria, and thus began again compiling detailed data on social statistics as had been done in the great civilizations since the invention of money. Social statistics are no older than the covenant called society, but statistics themselves are much older, and underlie all probabilities that exist in potential.

Statistics are the math upon which is based all cause and effect, that is, all physical behaviors, the human (guided through fate by free will) as well as the gross or base subjectivity of existence. It is always a factor of whether it is more or less likely that an event happens that causes an event to happen or not, and at this scale the whole makeup of the universe is as relatively binary as the code in a computer program. It is thought to be at this level that quantum consciousness, that which interacts with the substance of the cosmos to co-create interactive reality, occurs.

i. taking out the mental garbage piles up

If Freud's theory of hypercathexis is accurate, then ego is something that accumulates electrochemically in the nervous system. If the theory of a holographic continuum unifying man, society and the laws of physics is accurate, then social consciousness, awareness or civic spirit can be thought to build up to a point of hypercathexis in a similar way. It is easy for the eye to see this as possible in the filaments, walls and voids of galaxies in the universe, because they resemble the human nervous system, however this may prove to be an illusory similarity as to the amassing of sufficient probabilities to perform electromagnetic hypercathexis on a universal scale. By this same reasoning the social structures of man, which do not physically resemble nerves, but mimic them on the ideal level in the order of ranks in the branches of their hierarchies, might be capable of spiritual hypercathexis.

These are all the hopes of many, that there may be scientific proof for God, the ego of the universe, and that this may bring true justice to our society and equality to our world. However it is a house of cards comprised of axioms and dependent premises. The truth is that the winds of death long ago began to blow through the universe. If the galaxies furthest away are comprised of stars as mortal as ours, they burned out long ago, and hypercathexis even on a social scale only takes one person.

The truth is that ego is not the goal. The accumulation of pseudo-self definitive belongings is counterproductive to the ultimate sacrifice of leaving the dying universe. However it effects self-replication. It engenders dopplegangers, who in turn duplicate themselves with each other. This is looking at the spiral from an angle that accents the phi element. This is looking at the hypercube of time from above one of its corners, and seeing it as an octaholohedron. This is the universe on its ear.

ii. the point system, the buddy system and the end game

According to scholastic consciousness researchers of the twentieth century, the mind that is trapped in reincarnative Samsara, the cycle of the soul reincarnating through manvantara, often turns to the playing out of roles relative to social situations. We see that this is true throughout all the more sentient animal species, and, in so far as pollination is a mating ritual, may account for the entire variety of appearance in plants. The study of how these social roles are transferred from one generation to another is currently called, by these same scholars, game theory, because in the higher mammals it is through game playing in the young that their place in the pack is established. This returns us to the concept of the social sciences, which in no way differ from an order of anarchy.

Here we see that the finer relationships, such as that between when and where, that between how and what, and that between who and why, elude the gross populous. In the place of these fine facts are fostered to fester rancorous and odious, repetitious rituals of all sorts to engender our imposed roles. This is done because speculative philosophy is not considered as necessary as craft guildsmanship for survival. This is probably due to an old, pre-Egyptian custom of human cannibalism. Social class researches have also studied the difference between these in terms of infinite and finite games. As the saying goes, "when I was a child, I behaved as a child, but when I became a man, I put away childish things." So it is with the finite games of guild craft and the infinite games of speculative philosophy.

Of the finite type of game is paramount the point system. This serves only one function for the brain: to teach it math. Doing math is for the brain much like eating vegetables is for the body. However, in conveying the ideal key of mathematics to the gross gray matter of the brain, society includes the hidden compounding interest fee of morality, and in the west greatly promotes the Epicurean idea that more is good and less is bad. Originally the barter economy was based on conservation of resources, however with the introduction of the token exchange system, the value placed on material goods has successfully been driven down as the token of exchange is made increasingly ideal, that is less and less substantial. The concept of technological capitalism as representative of the withering dictatorship by the proletariat has already taken the dollar off the gold standard and floated international credit. We get paid placebo points for work to placate us with playthings we must identify with. This type of game is a lifelong game, and an everyday, common game, and nowadays considered necessary for survival.

Another like this is the buddy system. Just as the ownership of private property became the twisted means by which to teach the karmic point system, so is the idea of collective property derived from naturally occurring friendship. As we grow we will find ourselves near others that are in some way like ourselves, and if we are attached enough to that part of us that is similar, we will be pulled towards these people. In the same way, we will gradually grow apart, the similarities we had changing in different ways, until only different things are similar, and we will have already begun to turn away from the other. Our best friend is usually a person with whom we choose to share very personal information, and this is often because we see them as like a part of ourselves that we like. Buddies, chums and pals are a lot of the worst sort, since they will always be aware of the karmic score in a way that a best friend would ignore. Best friends go on to make excellent spouses, should they fall in love and get married, because they will see reproduction as an extension of their combined self, just as a carpenter sees the hammer as an extension of the hand.

Both the point system and the buddy system are games. They create the sort of existence known as virtual reality, tunnel reality, or game reality, where an organism contributes as little of their inner psyche as possible except in exchange for the necessities of survival. When observing this archetypally, that is, manifest in someone, the common sufferer and contributor of this syndrome may feel repulsed, however this is only because they are not seeing their own face when they look in the mirror — only a person who is either more or less like what their parents, their loved ones and themselves had hoped for and expected.

On the other hand, the end game is a system. This is an infinite game, because it is a zero sum game, and depends on the absence of all resources for conclusion. The end game is the knowledge to end all games.

iii. the invisible world and the leviatan

Cultural self-referentiality is artificial intelligence. Perhaps because of the aluminum byproduct lined water people go about their daily lives as though they were already cyborg, part organic and part machine. That which they have cultivated within themselves as civility is like an intangible tax on their immunity, and like dinosaurs, the toiling masses wither beneath their share of days. Rather than our actually being anything like dinosaurs, it is only our dependence on the fossil fuels produced by the resin of ancient plants and the fats of the dinosaurs' skins that has brought us so low, as well as that has elevated us to such heights. In the same way that many biblical scholars associate the realm of tempest and flood with the chaos and luxury of politics, so can this ghost of a great beast, the thunder lizard, be ideally seen as equivalent to the demiurge of the leviathan. Hobbes described the leviathan as the political hierarchy ruled by the monarchy. Considering that the dinosaurs died, the idea that their spirit ruling over us in the form of our nascent capacity for social administration is in some way potentially threatening is altogether understandable, however they merely represent an earlier form of life also evolved by the interaction of bacteria and viruses, so we see that our system of checks and balances is inherently the same, though guided by free will. To this end it becomes increasingly easy to find examples of how the reptilian mental characteristics, or the blue bloods, have always turned survival into their gain, from the story of the dinosaur snake tempting mitochondrial eve into the trees to evolve thumbs for the apple of civilization and banishing monkey Adam to evolve feet wandering the grasslands of the tropical savannah, through Quetzalcoatl as Teli, to the promised return of the King of the Jews to rule the whole of the Godly people.

b. thinking outside the box

Another, more spiritual than geometric, symbol of masonry is the coffin. Its importance in ritual is similar to the womb, but its significance socially is best evinced in the Russian fable "how much Land does a man need?" This is the area that a large part of our unconscious energy is being driven towards, that directly surrounding the corporeal body, and all life has evolved this way for survival. Because a similar thought bubble surrounds various other forms, it is tempting to discard one appeal for another, however this does not follow, and one will not find satisfaction unless they part with the ways of all the vessels. Just as the coffin is the box for the body, the hypercube of time is the cube root over cube of the universe.

i. the hypercube of time

Imagine the universe as an expanding sphere, with the fabric of spacetime as its surface. Inside this sphere is infinite potential dimension. Outside of it are tachyons traveling from the future to the past. This is the hypersphere of the multiverse. Now

250

imagine a unit of time, an event, as a cube. Another cube can in the same way be drawn around the entire universe, and this represents time. Just as the universe becomes the multiverse, sphere expanding into hypersphere, through the force of time, so does each individual unit of information in the universe shift about and change with time. Time changes as the temporal hypercube slowly turns around in the fourth spatial dimension, causing spatial involution on the surface of spacetime, and slowly bringing the third dimensional representation, or shadow, of the fourth spatial dimensional object of the temporal force through a cycle from cube to antipode to osciholohedron to nested hypercube to osciholohedron to antipode to cube. These are only the names of the points for alignment, and there are many other lattices that represent potential view points on the shape that go unnamed.

The reason that the hypercube is the cube root over the cube is the same reason that the hypersphere is phi/pi. Just as phi/pi can be represented as an oscillating spiral, so can the cube root over the cube be represented by a hexagram within a square, representing a cube within a cube. If they were only arrayed upon the spiral, the ten emanations would have no apparent order, but because the pattern of the alternating hexagram and square form a smaller fractal hypercube at antipode, they take on the familiar pattern of the Kaballah.

ii. the holognomonic projection of cyclical events
The emanations, or corners, on the actual cube root over cube fractal pattern inside the hypercube of time are best understood as cyclical events, or a series of similar settings and situations that recurs repeatedly. Another example of a temporal measurement, that to derive the cube of the sphere, stems from the Orient, and is a set of 64 different event outcomes arranged in a predictive, oracular system known as the I Ching. Both the Kaballah and the I Ching are temporal systems, and represent the flow of the unified field, or time, through the universe. The I Ching has been decoded as a lunar calendar, ending in the year 2012.

A Long count in the ancient Mayan calendar also ends in 2012. The Mayan calendar was based on observations of the solar system's ecliptic orbital plane relative to the center of the galaxy, which they apparently derived from the retrograde cycles of Venus, and of sunspot cycles that rival those made with the scientific advances of modern technology.

Samsara, or the six lokas of reincarnative realms, derive from the eight doubles representing the regular variation of the temporal fluctuation ($E = M c$ sq.) of the I Ching.

iii. why some thoughts come back to haunt you
Phi/pi and the hypercube at antipode are both living patterns, the former the growth signature of bacterial DNA, the latter the shape of the ribosomal virus. Therefore they are alive relative to one another. The fact that these forms, which are the geometry underlying the laws of physics, comprise the two, intertwining histories of a living double helix relative to one another proves that electromagnetic entities such as archetypal spirits are as much alive as you or me. If free will is a right only of the living, then free will to all. In this way the body of God is a living organism of dissonant and resonant tonalities. Just because you are only learning

251

this now does not mean that it has not always been. Angels are only points of reference bent toward us by the karmic signature of our aura carried over from all of our past lives.

iv. Asymptotic Time

First we must agree on the direction of the flow of time. This can be in any combination of the three recognized essential components, Past, Present and Future, so that time flows from one to the next, or from one to the last, or from both to the next and to the last, or to neither at all. What our senses compel us to recognize is the particular arrangement of these components we have come to call the "forward-flowing arrow of time." This is basically a measure of matter-energy exchange dictating that as things continue they decongeal.

The best device created by the hand of man for common interaction with this all-powerful force is the clock, a sequence of numbers arranged in a circle around a set of three hands that move from one point on the circle to another at regular rates. The most powerful type of clock is the atomic clock, the rhythm of which is regulated by radioactive decay internal to the mechanism. This makes it accurate at varying altitudes, where one must travel greater distances in a shorter amount of time to keep up with a fixed spot on the globe below. It renders time then relative to the speed of light, since the particles ejected by the unstable element will appear as a doppler shifted wavelength of photons. The distortion will occur between the atomic clock and the position of an observer either before or behind it on its path above the planet, and this is a very nice little model for time.

Now imagine that you wish to count numbers; say, perhaps, you have grown tired of the atomic clock counting all the numbers. Nor is there any reason for you to count the numbers just like an atomic clock would, since you are not an atomic clock, you are a quite sane and rational human being, so you decide to count the numbers as fast as you can. You will quickly see that you can count faster than the atomic clock.

Now imagine that you can count asymptotically fast. Of course we know that this is impossible, but pause for a moment to consider why we think it so. I propose that the location of this particular idea is between the full set of ideas developed by and for our evolution, and the abyss of what is not known even in idea. Thus, it is possible, although it has never presented itself as an option in such a way as to impinge upon our survival selectivity and become a more common probability. On the other hand, for example, if we happened to build space crafts that were piloted by just such a method of counting, then it could be argued that, at least after several generations, we would begin to spontaneously mutate towards a mode of thinking more compatible with that method of comprehension, and would then be as comfortable with it as we are now with our atomic clock.

So, begin in this manner, then: watch the seconds tick past on the atomic clock and start to count asymptotically fast. At some point you will find that you are continuing to count, and the seconds on the atomic clock have wound down and stopped.

Well what has happened here? It would be reasonable to hypothesize that you had gone off onto on alternate timeline, leaving the timeline counted off by the atomic clock behind — if you can still call what you are in a timeline. You would have to say that it was an offshoot of the other timeline similar to the measure of depth transecting the measure of length, with one exception. Since your timeline is counted off in an asymptote, it only lasts one moment of atomic time, but since the asymptote can never reach an end, neither does that moment ever end.

What we are perceiving now is time flowing in a different direction. Both time lines count the same ultimate measure; it will be infinity on the clock before time runs out. The clock stops, your counting speeds up. Therefore they may just swap over, from one direction to the next, and continue on without more than a momentary break.

This, it seems, is similar to the geometric distortions to gravity around massive or phenomenally dense objects. If time-space encounters an impediment to its "forward" flowing momentum, it simply adjusts the angles of its coordinate system around it. Of course, what we are observing in the case of gravity is only the refraction of photons due to their impact with the force fields called gravitons that surround all physical objects prone to entropy, and it is, in that context, convenient to see relativity between the curvature to photon trajectories and the emanation of virtual particles by real particles as they contact other real particles or other virtual particles. What, however, does this mean in our asymptotic counting experiment?

The idea of an alternate timeline having been introduced as a potentially common occurrence in the twisting and turning of the timestream, we see immediately the ultimate implication such a concession to gravity lets in: the existence of infinite potential timelines, each leading to a different future, all occurring simultaneously within the moment, and, while this seems an accurate depiction of an event from the perspective of potential, we also know that probability intercedes to manifest possibilities which delimit the event and put it into a context which we then call time. Without possibilities such as the atomic clock, we would not understand time in the linear way that we do. The asymptotic counting experiment has allowed us also to conceive of time as being infinite in potential.

I ought to further stress the fact that one need not necessarily count asymptotically to accomplish the affects we are discussing here. One could recite a chant, even hum. In fact, when you see that your mind itself is only a conduit whereby for the thought of an asymptote to come into existence, and that it is an idea that can exist independently of your mind as part of nature, then you can remove your focus to one side, if you like, and let this asymptote float in the center of event at the very core of time. It is, afterall, always there in potential. Conveniently, it is the mathematical expression for the infinite potential that constitutes time.

However, the asymptote itself and its potential effect on time must still be seen as separate concepts. Time doesn't stop every time you imagine an asymptote, so what trait of the asymptote makes it possible for time to potentially wind down and stop? We see evidence for the asymptote's physical existence in the gravity of black holes, and the symmetry breaking of certain particles when accelerated towards the speed of light, but we know that the asymptote is somehow removed from and above these artifacts just as we know that time would continue to exist even if we didn't have the atomic clock to measure it. The real question then, is — is time potentially an exclusively mathematical function, like the asymptote, that can be removed from material reality into the same realm as the asymptote? Is there time in the domain of mental projective space in which you created your transtemporal asymptote? Is time nothing more than the relativity of sets of numbers?

If we say that time is something other than death, then we can answer that it might then be nothing more than the relativity of sets of numbers. The atomic clock, then, is the dependent object, rather than its subject; we understand time as the linear unfurlment of a circular cycle. This same observation applies to the orbits of the planets in the heavens. They, too, are only a representation of the idea of change underlying their perpetual rearrangements. And if time is described best only by the relativities of these sets of finite objects, parts, or merely numbers, then we can say that it does exist in the realm of the mental conceptual, and even in such a way that it will allow itself to be imagined as a geometric extrapolation of potential like the asymptote and sustained that way as a visualizable thought.

Now aside from our understanding of time as the linear relativity between circular cycles and our knowledge of the asymptote as a geometric function we cannot say whether these two do or do not exist in the same form, except in potential. We only believe what we do about time because of applying certain tools to it, and not others. As we have seen, you can easily imagine counting faster than the clock, and I can easily imagine you doing this infinitely faster, and from these combinations we can see time as potential. In the future, other people will look at time differently, and see it to be like different things.

In fact, I think our old friends the Ancients may have had something to say about this very thing. It was their observation in the Orient that heart rate often joined with time in a mysterious way such that, when the heart rate was slow, events would seem to move slowly, and when the heart rate was fast events would seem to move rapidly. The Yogic mystics and the Brahmas attained to states of mastery over their emotions, because it was thought that, by binding the movement of the subtle energies within, one could influence their ebb and flow without.

There was a similar belief regarding the eyes that originated in Egypt. It concerned the mythical third eye, the ureaus, located in the center of the forehead, identical to the Vedic ajna, from which it was thought the substance of the human essence flowed down into us from the realm of the mysteries. The best explanation regarding the third eye comes to us from Jose Silva's modern method of mind control, which informs us that a twenty degree angle of inclination for the closed eyes produces Alpha waves in the brain — those found in the minds of all meditators the world over. This has been used more frequently for the purposes of manifestation.

A further note was made that certain nervous involutions contribute directly to the elapse of time. The Ancients of the Orient again sought mastery over these breaks from ideal concentration, believing that the nervous system was the instrument triggering certain possibilities. Particularly the practitioners of the marshal arts and the Dervishes, as well as the American Colonial Shakers, all partook of this fountain of energy.

The biological circuit these comprise is mathematically a phi / pi spiral, the same essential pattern as a flower. The fundamental idea of the phi / pi spiral is symmetry breaking governed by irrational numbers, wherein the irrational numbers are numbers like phi and pi and the symmetry breaking is often in an asymptotical form, from an asymptote, or between two asymptotes. Let us look for a moment at this last example.

In our counting experiment we felt relatively safe assuming that the time line of the clock was linear in so far as it described a sequential progression without retrogression. But were we wrong in doing so? It can equally as easily be imagined as an asymptote itself, if one dwells upon the potential that the instants, as soon as they are measured, disappear. To where do they disappear — when something changes where does what it once was go? And from where do the instants come before they are measured?

Both the Past and the Future are like asymptotes stretching offwards and upwards into infinity, disappearing in the same direction as the asymptote of time measured by your counting. The asymptotes of Past and Future meet, then, at the point where the clock has been stopped by the asymptotical counting experiment. I think this is a satisfactory definition of the Now — the event that can begin but never end. In this capacity we can imagine it as a phi / pi spiral where phi is the future, pi the past, and their ratio the singularity of the moment.

Let me remind you that this is the same as the distortion we saw at the beginning that effects the observation of photons emitted by the radioactive decay of an unstable element in motion depending on whether it is perceived as from before or behind it on its course. If we observe this phenomenon from the side, seeing both the doppler distortions which constitute the Future and the Past for the trajectory of the element, we will again behold the phi / pi spiral, where phi determines a compressed wavelength before it and pi determines an elongated wavelength behind it. This is known commonly as the red and blue shifts to the frequency of light, and this is how it would be observed in the case of the atomic clock.

There is no comparable observation for the change in timelines represented by the asymptotic counting experiment, although we can now see it, at least if we like, as a representation of the same underlying mathematical pattern.

c. filaments and superstrings

It is thought to be the vibrational fluctuations of probabilities in chain reactions of cause and effect from one event to the next in a continuum of infinite potential that cause the function of matter-energy exchange to occur in the forms that it does, that is, as the expansion of the universe (indicated by the homogenous red shifting of galaxies), as the electrogravitational arrangements of galaxies in filaments, walls and voids, as the underlying cause of the different forms of nebulae, galaxies — such as spiral galaxies like our own Milky Way — and clusters, the orbits of the planets around stars, and even as the behavior of quanta, such as the uncertainty principle. The best name for this theory is wave theory, but it would never catch on.

This theory has been advanced as far as this stage publicly, but it has been advanced much further in private. It applies to brainwaves and can be seen in the patterns of traffic in the street.

i. the walls and voids

We are all aware that everyday the sun rises and sets, and we can be confident that it has done this in the same way for many millions of years, and, as scientists predict,will continue to do so for many millions more. The sun is a star, just like the billions and billions that we can see in the night sky after the sun has set, but the sun is close and the other stars are very, very far away. The stars that we can see at night are arranged in nebulae, galaxies and clusters, and our own star is part of a spiral galaxy called the Milky Way. The closest galaxy to our own is the Andromeda galaxy, and both these galaxies are part of the nearest supercluster of galaxies, named Virgo for the constellation of the area in the sky where it is visible. These superclusters are connected to one another by strands of smaller clusters and stray galaxies known as the filaments. A very thick strand of these superclusters runs around the entire universe from above earth's orbital plane to the north to below it in the south, and this is called the great wall. Between these filaments there is empty space, and it is only based on the anthropic argument that we are led to believe the same laws of physics apply in these "dead zones" as in the rest of space. According to recent scientific studies, the universe's rate of expansion cannot be accounted for as being based on the amount of gravity resultant from the mass of known matter, or that which emits or reflects radiation. Therefore something must be out in the voids to account for our apparently accelerating expansion. This has lead astrophysicists to the erroneous postulate of dark matter, and the study of black holes.

Though black holes were postulated as a mathematical probability based on the accepted theory of timespace relativity according to the exchange of matter and energy, they were not observed until they were discovered at the centers of spiral and other flat disk galaxies. Here we see the most ancient of the universe's quantum fluctuations, a spinning wave that acquired polarity and became a particle, that passed through the phases of all the four elemental forces as it became a star, then through all the main sequence of different phases in a star's life, until finally it accumulated enough gravitational distortion to collapse in on itself and rip a hole in the fabric of spacetime with so much gravitational pull that the stars in space all around it began to move towards it and around it, and nebulous gas clouds approached and its gravity started star birth within them, until finally an entire galaxy surrounded it in a vast accretion disk, flattened into orbit by its continuing polarity, and spewing gas jets out of its electromagnetic poles comprised of tachyons, the only particles fast enough to escape its gravitational pull.

These tachyons are the particle carriers of the actual gravitational force of the black hole, moving in the proper direction through time, as measured by entropy, for a force carrying particle. The tachyons involute as they pass through all the larger particles, and this pushes the particles back in the opposite direction the tachyons are traveling, hence creating the force of gravity. They are high frequency, microwavelength electromagnetic radiation, and are only visible when they are passing through a solid body, such as the gas jets of the black hole. When they can be seen, they are observed to move faster than light, or photonic radiation, thought to be the standard maximum velocity for radiation. Entropy, as a measure of time, is fixed to the speed of the photon, and so when we say that tachyons emitted from a light source will arrive before the light, meaning that they will have already delivered their depiction of the carrier information pattern before we can see it, we are saying that tachyons are negentropic, or opposite the standard arrow of time. We also consider ourselves to be negentropic, as we struggle to survive against all odds.

Just as when the orbital plane or pole of a star lines up with that of another there is a slight electromagnetic effect, so when it aligns with a black hole there is a greater effect. Similarly when the galactic accretion disks of black holes align with one another, there is a very strong electromagnetic effect, and so when the gas jets of black holes align with one another there is a tachyonic effect.

ii. random quantum fluctuations produce wavelengths

The singularities inside black holes probably represent the oldest quantum fluctuations in the universe, but they do not represent what will be the fate for most of their offspring reactions. The oldest black holes and any black holes formed by stars they created will eventually merge into super massive black holes. By then there will be no matter left in the universe, and not enough ambient radiation to feed these super giants, who will eventually burn out, taking the last of the light in the universe with them, and leaving our entire multiverse of alternate universes in parallel potential dimensions leveled down to the constituent tachyons. This will not mean the universe has been destroyed. In accordance with the conservation of potential, matter can only be converted to energy, while energy cannot be destroyed. So will it be with our universe, that every last particle will be converted into a wave.

The reason for this is simple. Virtual particles have fractional spin while real particles have whole spin. It simply takes more than one wave to form a particle. Therefore, while wavelengths can arise from random fluctuations in probability, a number of these must align in a combinatory way over time to create a mass particle.

Our universe is already built on laws of physics whose structural framework is a geometry of vibrational dimensions, however eventually these holognomonic vibrations will be all that remains. The potential inversion of the uncertainty principle necessary to continuously manifest this will statistically disperse with the elements of the radiation until eventually this geometry will be entirely unrealized.

iii. the human nervous system

Of all the known structures in the universe, one of the most complex is the human nervous system. Consider the difference between the tissues of the nervous system and the trunk of a tree. While the tree has a homogenous tissue that allows fluid transport, channeling water from the roots to the leaves turning them green and air down to the roots so they can breathe underground, each nerve has a similar structure, with a myelin sheath, made of high acid sugar (similar to our genes), surrounding the nerve like the bark of the tree, and inside which an ionized fluid suspension allows the transport of an electric charge, from the dendrites (roots) to the axon (leaves) where it is translated into chemical neurotransmitters and crosses over to another dendrite. Axons also act as sense receptor nerve centers, electrically transmitting neural sense perception from the skin. In the grey matter of the brain the neurons grow in functionally associative columns outward from the middle. All of the five senses travel through the nervous system, and in the electrical potential of the nervous system lies the capacity for extra sensory perception. When parts of the nervous system are shut down, the parts of the body or the perception of the senses these nerves control are also shut down, and, though still functional in potential, and as alive as any other part of the living body, they will be delivering no information to the rest of the brain and can carry no commands from it.

It is the primary organ of the nervous system, the brain, that has become the primary organ of the human body, governing all of our behaviors, a savior to our survival. The brain created society, and society is responsible for our exponential population growth. As our population grows we imprint ourselves more and more on our environment, and this makes culture comparable to a virus. The nervous system itself, with its five lower branches and the genitals below the brain at the top resembles very much the familiar, hexagonal headed shape of a virus. It is possible that, while the first, most basic bacterial microorganism was busy fighting off its natural predators, that is viruses, the archetypal imprint pattern in DNA that would result in all the later organisms from trilobites on was conceived between them. In this way, at least, humanity is a hybrid species, and in so far as the virus may have originated off world, we may even be a truly alien species in origin. Coincidentally it is entirely bacterial microorganisms that are responsible for decomposition.

iv. is information contagious?

One modern theory for information transfer that has arisen from social exchange theory is memetics, or the theory of information itself as virus-like in its transmission. This would mean that the entire struggle between active society and passive culture is merely the continuation of the war between the two genders of organism, viral and bacterial life forms. Aside from its being generally applied by populist darwinian Malthusians, who believe that society is a popularity contest governed by survival of the fittest, this seems to be a valid theory. According to this theory information is transmitted in units, called memes, although the same conclusion is reached by quantum mechanics, and the information units called quanta. Memes become more or less popular, or "catch on," in a direct proportion to how easy they are to remember. Perhaps the best example of this is early subliminal messages used in muzak played in department stores, containing simple messages, meant to be delivered directly into the subconscious, such as "don't steal," "mommy loves you," and "it's okay to do better than daddy." By this reasoning it is easy to understand why more people know Einstein's famous equation for relativity than the more obfuscated equation for the uncertainty principle. This leads to a general devaluation of information, and has resulted in what has been recently identified by sociologists as a "sound byte" media, "fast food" culture and "hot button" issues. In this media circus, everyone is famous for fifteen minutes, which is never enough time to get anything done. A short cut to saying that mankind is a plague on the environment is saying that intelligence is a disease to humanity.

d. space — what is it good for?

First, come to accept that half the universe is dead: we still see the light from distant stars even though they burned out long ago. Then understand that the universe is a gravitational field bent into an expanding well the surface of which is comprised of matter-energy exchange and is measured by the speed of light. As the light moves inward on the surface of the spherical universe, gravity moves outward. Inside the third dimensional surface of the gravity field is the illusion of light we perceive as our universe. Outside the universe it glows faster than light tachyons. Inside the surface of the sphere are other dimensions, accessed in tachyons, coiled up very tightly and vibrating in superstrings, or microwave gravity, that cause the underlying harmonic resonance, or geometry, of the filaments of superclusters.

The multiverse of infinite potential dimensions inside the expanding third dimensional universe, as well as the matter-energy exchange of the material universe caused by the fourth dimensional force of time, and any baby universes formed through black holes, are the microverse, while the tachyons that shine from the surface of the universe into the greater black hole of the parent universe, that comprise the primary clear light of ylem, the universal singularity, as well as the parent universe that has formed us as a singularity inside a black hole, are the macroverse. The microverse inside the macroverse is spacetime. The macroverse around the microverse is timespace.

i. spacetime

It has been demonstrated that, when two metal plates are held only microns apart from one another, so that no particle of any size should be able to fit between them, there is still a quantum foam, or geometrical kaleidoscope of empty bubbles separated by inversions, from the uncertainty principle causing probabilities to arise at random from potential. This quantum foam effect is stimulated by prismatic refraction of microwavelength probabilities in fields of potential that are thought to be caused by the vibrations of superstrings, or tightly coiled dimensions, and resembles the same randomized pattern of webwork as the filaments of galaxies.

If we say that tachyons, which produce gravity as microwaves opposite the flow of entropy, are the hitherto undiscovered gravitons, or force carrying particles of the gravitational element, then they can be seen to serve as Einstein's elusive gravitational constant, and if gravity is an illusion of light moving faster than light that results in the effect we call time, then we are well on our way to a grand unified theory that unites the gravitational constant with the temporal dimension.

Just as this original and most basic form of matter-energy is brought forth through the quantum foam, that is, over time, so is it the substance of time, that is, probability. However because tachyons move faster than light they are antimatter, and therefore cannot be a probability in this universe, that is, this reality of light.

ii. timespace

Tachyons are an inverse probability because they move faster than photons. Photons themselves are a real particle, and therefore cannot carry the electromagnetic force of their own light. Instead, it is tachyons (as Cerenkov light) that emanate from the photon that give it the appearance of illumination. Therefore we see that the speed of light is irrelevant, since the light from a photon precedes the photon. Thus, it would not be exactly accurate to call tachyons antimatter. They do not, for example, immediately explode on collision with their counterparticle of matter, because they do not have a counterpart particle. On the other hand, if we could see these particles, it would prove they didn't exist as just clear ideal.

We know that there are tachyons outside the universe because the universe is bound by the speed of light and tachyons move faster than light. Therefore we associate them with the holographic hypercube of time. Their frequency vibrations, known as superstrings, are equivalent to the pure geometry that exists in the realm of the ideal. This means the cube root over cube of the turning hypercube and the phi/pi spiral expanding hypersphere interact with one another in the continuum of tachyons.

iii. dimensional molecules

Fermions and bosons are molecules of dimensions. Their spin potential variability is determined by transdimensional interaction. The dimensions themselves remain outside of and beyond the forms they generate, as the nuclei of two covalently bonded atoms. They are as the philosophical definition of the ideal.

Before we can consider the specific cases of fermions and bosons we must delve a little deeper into the nature of the dimensions. It cannot be argued that the first dimension is a singularity, the ultimate embodiment of symmetry. The second dimension is, equivocally, planar geometry. These are ideal forms, impossible to express in our material reality, and exist as realms of pure equation. Without them, however, we could understand nothing of our own condition, and would posses no mathematics by which to quantify nature. In fact, without them, nature itself could not have assumed the forms and functions which we now know. However, even the simplest understanding we posses cannot belong to one of these dimensions alone. Consider the point for example. Its properties are compression and scale correspondence, where the former applies to a singularity and the latter to planar geometry. Neither has any meaning except in relationship with the other, and what meaning we do comprehend of either is therefore a creation of both. Without the point, a Cartesian coordinate system could not exist, but without a singularity, there would be no point.

This natural progression continues upward through the dimensions even up to light and beyond, which I have here classified as the fifth dimension. This is due to the unique properties of light which seem to imply a domain unto itself, particularly as it relates to time; the primary example of this being the seeming transtemporal nature of tachyons.

The fourth dimension should be set aside for special consideration as well, as doing so now will relieve the pressures of later describing the particles it helps to form. The fourth dimension is simultaneously spatial and temporal, meaning that it generates highly complicated structures over fixed intervals. The complexity of its spatial component is due to its having as its foundation an open geometry, in progression from the flat second and closed third dimensional geometries. Its temporal component, however, is semi-second dimensional, as it is traditionally observed to unfold binarily between past and future. I have demonstrated elsewhere that this relationship is indeed far more complex than this, but in the context of this exposition we need only remember these two defining traits for this dimension.

Now we may consider the dimensional molecular cases of fermions and bosons.

Fermions ("real" particles) posses fractional spin as they occur between the third and the fourth dimensions. They arise as the result of the interaction of matter and energy between the pure, formal geometry of the third dimension and the pure, formal time of the fourth dimension. They begin as ultimate, unformed material chaos. As time passes this mass becomes motivated, and an exchange of matter and energy begins to take place, constructing a framework of preexistent geometric relationship to govern the materialization of real particles. Fermions' symmetry is quarterly in three dimensional coordinate space, meaning in any direction it is rotated a fermion will look the same after less than one complete turn. In other words, its matter-energy structure is determined by geometric (symmetry) and temporal (spin) interaction. These may be measured as either particle or wave.

Bosons ("virtual" or force-carrying particles) arise similarly, but posses integer spin potential variability. The fourth dimension exists in field form, and the fifth dimension in no solid form at all, and therefore there are no geometric restrictions placed on symmetry. Bosons thus existing between pure Time and pure Light can only be measured as waves, and as such obey spin in accordance with oscillation interval, or frequency constant wavelength.

The nature of further dimensions, and therefore implications for the particles they may generate through interaction, is discussed elsewhere.

V. Light

If we say that we are beings that are made out of matter-energy in the third dimension that undergoes interchange that we measure as being the fourth dimension, and also if we say that time is like a holographic hypercube underlying the third dimension that is a shadow of a spatial object in the fourth dimension, then we can safely say that light is the fifth dimension, measured as the speed of a photon, and whose substance, just as probability wells in fields of potential are the substance of the fourth dimension, is gravitational tachyonic microwave electromagnetic radiation. We have already established that tachyons move faster than photons, so to say that the speed of the photon is the measure of the fifth dimension is like saying that entropy in the third dimension is the only way to measure the temporal fourth dimension. This introduces an important concept: that time moves faster than matter-energy, and that light moves faster than time. This is to say that matter becomes energy when it reaches the speed of light, and that this is the backbone of the flow of entropy from past to future, or the measure of the concept of time. Because tachyons move faster than this speed, they do not obey the forward flowing arrow of entropy, or what we call time. There is still much debate about the concept of matter-energy interchange being dependent on the speed of the photon.

A. the role of light in the genesis

The universe began in the first dimension. There was a microwave vibration that occurred under the influence of uncertainty and this caused everything. A single amount of the void spun around itself and split off, forming the first particle of karma or quantum information unit. This particle was a singularity compared to nothing, and thus was compelled about itself with the combined weight of the fullness of the abyss, which was a great greatness. It was forced to begin to consume itself by the emptiness, and this it did with such haste that it began to implode with a force greater than that of the darkness, that is, that velocity known as the speed of a photon, and thus to bend the space-time within it, as it had been bent from the null space and zero time of the void when it was conceived. This turned it inside out quickly, and filled it with light so that it shone then in the pitch. But these were not rays of photons, too slow in the darkness and too easily consumed would they be, but the projection of astral light, that is the microwave gravity particle tachyons, and these are projected as an outward rippling orb. It is said then, that the finger of the creator came down and touched the spot on the globe in the heavens from outside the space that was outside our universe, at the moment of the Big Bang, so that it would be swept away with its generis to become proper space-time.

Then it was the time of the second dimension, when the waves of tachyon luminous microwave gravity stirred the void up into action, and caused more reactions that created particles. These are the events when the four forces were set down, and everything had been called into spin. Time began then, as a measurement of the spinning, a speed that could be measured by the velocity of a photon. Space was conception itself. A single point in null space would be drawn out and then turned about itself, creating polarity. A particle would spring into existence as a self-expanding wormhole tachyon torus in the vast expanse of the nether realm and immediately progenate a stream of similar shapes, that would continue on filling in the lightlessness until they were all a solid throng occupying a region, and causing by their continual exchange of motion between them, which asymptotically approached regulation, the oscillation of that great polarizing force we know as time. These tachyons tended to accumulate themselves then in a topical aura, since they were emanating outward from a center, and so their region of most profound discourse was around the edge of their expansion. It is upon the surface of this three dimensional shape, expanding in the fourth spatial dimension, that the story of our universe continues. By this time the four elemental genres of particle had been formed, and this had given the Light a fine quality, invisible to the Darkness, that of all those less intense manifest fluctuations of those particles slower than the speed of light. This was the material universe that was becoming polarized as three dimensional space on the surface of the fourth dimensional inflation of tachyons.

Once the third dimension began to appear out of the pure heat following the Big Bang, it rapidly accumulated masses in space similar to those underlying its own mechanisms of creation. These are, in order ascending outwards from our planet, stars, galaxies, and the walls and voids. The planets and the stars are spheres, the stars emitting light and the planets reflecting it. The orbits of the planets around the stars and the orbits of stars around the centers of galaxies are both planar, that is, purely based on the polarization principle, that elemental and temporal-spatial opposites attract. In the case of stars and planets this means the star is too weak to attract heavier objects then the solidification of the fine layers of the gas cloud that surrounded it before its fire scorched them making them curl up into spheres. In the case of stars in galaxies, that is that those bastions of the lesser light all fall towards and are caught in the wake of a singularity where microwave gravity has torn a hole in space-time leading at the edges into hyperdimension, and in the center to the abyss outside. The walls and voids arrange themselves in random strands and gaps, the extended projection of the first spurts of probability in the infinite field of potential.

The fourth dimension gives us time. This is the surface upon which we measure a beam of light as it is guided. It is homogenous to the very small and the very large, though we recognize these terms to be relative to our perception, and it makes the smaller particles to move faster and the larger sphere to move slower, although we can project our understanding of the relativity of size onto the relativity of temporal durations. Again it is only the measure of the averaged frequency over wavelength for an area given as pi squared, or a factor of the force of the bending of microgravity, the force that causes all points in the universe to expand apart from each other as microgravity is perpetually self-generative and repulsively charged toward matter-energy, being that is on the degree of frequency where it is thought to be so improbable for it to exist in the confines of our universe, in the presence of its finer aspect, the larger solid particles or the longer wavelengths of energy, that the likelihood of it is so infinitesimal that it is considered antimatter, or otherwise, bordering on being opposite possible reality. The formula for time is thus given as phi over pi, that is the formula for a hypercube that is contained within and surrounding a sphere, that it is set to work measuring the difference of that sphere, so that, as the sphere expands, so does the hypercube.

On the day of the fifth dimension let there be Light, for as we are given to know of consciousness and sleep, and of day and night, so too do we know of the nature of these tachyons. In the proper conditions they can be observed in the three dimensional matter-energy universe, where, true to form they can be measured by instruments before the time it would take a photon resultant from the same events from which they derived to arrive. In these cases we see that they are able to utilize the same factor of the uncertainty of existence as a probability in potential to quantum tunnel through solids, moving from one point on the surface of a virtual particle to a point exactly on the opposite side, not by going through the center of the atom, nor by following a curve defined by the orbit of its electron, but by passing into and then out of the electron itself, which can be at all points on its orbital shell at any time, where it does not manifest trajectory spin as a probability like a photon being absorbed or emitted by an electron, but warp spin as it is swallowed up into itself between the two points, consumed in hyperspace where the point it disappeared and the point it reappeared are the same point, and the tachyonic wormhole itself fills the space between them, such that spin is conserved by the tachyon. The realm of hyperdimension, or the hyper-real warping of the fabric of space-time so that it is always consuming and regenerating itself simultaneously, is the surface of a geometry in pure dimension also, and this is the origin of spin-wave mechanics.

In the sixth dimension there is potential Light, that is, the absence of space as a continuum of vortices, and the absence of time as this substance in motion. Here is the dark pit from whence we started. It is the black hole of the larger universe that ours lives in, between which various frequencies of microwave vibration are shared, though it only looks light because the light of spin burning off pure potential that is our universe is so dim compared to the speed and involution of the Greater Light of this field, equivalent to the electromagnetic torus surrounding the singularity of a black hole as we know them, on the inside of which wormholes to alternate universes form. Thus it is truly here, in the quantum foam of spontaneously upsurgant probabilities, that we see the connection points between such wormholes form as a gravitational microwavelength that is the history of a single tachyon, known as a "transcendent" tachyon because it occupies all points in this microwavelength history simultaneously, and thus we see how our own universe formed as well.

The seventh dimension is that of potential information, where all pure data is truly relative and thus it is said to be the dimension of dimensions, that is, the one dimension containing the differing geometries of all the others and providing for them a basis for their continuous contiguity. It is for this reason, for example, that we can say there is no division between multiverses in hyperdimension where the geometry governs fine waveforms, for the same reason there are different divisions in the manifest realm of basic matter-energy exchange governed by entropy. So there is subspace, so there is hyperspace in hyperdimension — the hyper-reality of the multiverse, and so there is the pure dimension of the primary clear light called ylem. To this end they say that the creator rested.

1. Energy = Mass times (the speed of light squared)

It is the standard course of entropy that energy is given off by matter as long as the matter is cohesive. This is an effect of the electrons bonding the nuclei of the atoms that comprise matter together. Molecules are formed by electrons joining the diameters of their orbits, while the nuclei remain far apart. Gradually, random quantum fluctuations break some of these bonds and form new ones. The matter begins to deteriorate as the electron shells tend more to break up than to join.

When energy escapes the macromolecules of a living organism, it does not necessarily do so in the form of raw electrons. Most of the energy is given off in the form of body heat, or thermal radiation. This means basically that the electrons spin off from their orbits and become waves of electromagnetic radiation. These waves are high energy, and therefore warm and heat the air around them, causing miniature isobars of pressure to convect molecules of air asymptotically transversely. Most of the electron shell decay in this category escapes from carbohydrates (ash water), or acidic macromolecules derived from fats and lipids.

A certain proportion of electron orbital bonding decay in matter molecules is due to exposure to wavelengths of radiation. The closer the wavelength of radiation gets to matching the diameter of the constituent particles, the more probable their cohesion will be disturbed. The radioactive decay from nuclear isotopes, because the alpha and beta waves are equal to the size of different, sizes of atoms, particularly common in organic material such as living tissue, is considered to be the most dangerous form of such radiation, though the sun that we rely on for daylight is nothing other than a huge nuclear furnace burning of a gas cloud floating through space. Although this process is very slow and gradual, it is responsible for most of the molecular decomposition that occurs in the universe.

Photons and the orbital shells of electrons are of nearly the same size, however when a photon strikes an electron's orbital shell, it causes the electron to shrink to a singular charge at the point of the photon's impact on the orbital shell. The photon bounces off in an angled trajectory, carrying as a function of its wavelength the holographically stored information of the electron's spin. This has resulted in the modern scientific quagmire of only being able to tell the projected trajectory, or spin, of an electron by measuring its collision with a photon. Whenever this happens, it increases the probability of the electron breaking its orbit and destroying the molecular cohesion between its nucleus and another atom.

There is such a thing as the conservation of angular momentum, and this law of physics states that an object in motion, or set in motion by being acted upon by another object, has a predictable trajectory. There is also the conservation of energy, which states that there is a fixed amount of ambient energy in the universe, and that none is ever created or destroyed. Therefore, when electrons are liberated from their atomic orbits, by whatever means, in whatever way, their transformation from mass into frequency is measured by the geometrical equivalent of their probabilistic dispersal. Thus, the speed of a photon, a real particle that carries a charge and is similar in size to an electron (potential energy) cloud (the orbital shell around the nucleus), is arranged in a second dimensional (the orbit of the singular charge in the potential cloud) square field around the source of the energy's liberation. It is thought that, since it is equally likely for the energy to disperse in any direction, and since photons, as particles, mimic the behavior of waves, as fields, and are equally probable to travel in all directions, the quantity of the electron charge is thought to disperse equally throughout the pattern of radiative particle velocity. So it is said that the two are equivalent in potential.

In natural data we can observe this rule, expressed as mass times the speed of light squared equals energy, to hold true, and thus it has become an accepted axiom of physics, known as the theory of special relativity. The title of relativity derives from the second dimensional geometric relativity between the orbital path of the singular charge in the orbital shell of an electron cloud and the trajectory of the projected photon. The successful proof of this special relativity led to the grander general theory of relativity, which attempted to show that, using the accepted relativity between distance and duration, time and space are relative.

Those people who have not devoted thought to penetrating the full amount of meaningful (true and real) implications there are in this simple formula are, for the most part, happy to accept the only recently blasphemous notion that matter (the self) and energy (the soul or spirit) are really one and the same thing. While it is amusing to allow people this delusion, it is important to remember that this is not an exactly accurate interpretation of the data. It is true, by rigorous extension of the theorem, that earlier in the evolution of the universe matter and energy were unified, in that there was no matter, and because the radiation that there was was so high frequency it surpassed the modern parameters for the definition of energy. Because the theory was strictly derived from energy liberation, it has only been very recently, with the use of particle accelerators, that we began to be able to conceive of the complete inversion of this flow of entropy necessary for the genesis of matter.

a. matter can become energy, but energy cannot become matter

In the same way that a particle accelerator speeds up quanta until they explode, so has the theory responsible for their being built thrown off and exploded many modern myths. One of these is the backbone of the idea of time itself, the standard flowing arrow of entropy, or the measure of molecular decay duration.

According to classical physics before special relativity, electron clouds bond with one another less frequently than when they are separated, and this accounts for the eventual de-cohesion of all existent physical form. It was even known that matter took on different forms, and even had different particles, according to the four elemental forces. The most popular attempt at unification of the four elements was known as cosmos to the ancient Greeks, the philosopher's stone to the alchemists, and aethyr to astrophysicists. This was essentially an additional, unknown force, or possibly a higher dimension. After special relativity doors opened up for cosmologists to root out the source of the division between the elemental forces comprising modern matter and energy, and seek the elusive original ether. While the conservation of energy is seen to apply now, the big bang, or cosmology of relativity, shows mathematically that at some time in our ancient past, the high energy "ether" did appear out of nowhere, and that over time, as the universe has cooled, this slowed down into what we now know as energy, and even slowed down further into what we call matter. Fluid mechanics supports this, showing that the molecules of a liquid will move faster (boil) with more radiation (heat) and more slowly with less radiation.

Most recently quantum mechanics, the offspring particle physics school of special relativity, has unearthed the mechanism by which they believe the energy of the universe first came into existence, and then gradually cooled and became more solid, as well as why this solidity is temporary, and matter eventually turns back into energy. This aptly named uncertainty principle is nothing more than the mathematical description of the original quandary of special relativity: the relationship between the solid particle of the photon and the virtual particle of the electron.

b. fermions (waves) and bosons (particles) agree

The centralized (real) particles of matter and the decentralized (virtual) waves of energy are the only two types for which substantive quantum data have been found studying the four elemental forces arising from the known laws of physics believed to be generalizable universally. Through the functioning of these components in accordance with the geometrical physics of the elemental forces we see that the real particles do gradually degenerate into energy wavelengths, and that the energy is liberated and disperses into a field of potential defined by the parameters of the speed of light. Thus we have seen how matter seems to become energy, though we know that all that is defined as matter due to structure is really a function of energy. Therefore the sum of all particles combined with the sum of all waves is a quantity of potential energy.

Quantum wave packets of energy follow geometric patterns relative to one another, and this force causes the assembly of the elemental atoms as well as their combination in molecules. In the same way an atom is made up of a particle nucleus and a wave electron, so the particle nucleus is made of wave quarks. These quarks have different properties, and perform different functions within the structure of the atomic nucleus, determining, for example, radioactive (particle) decay and combined dimensionality, the latter as measured by spin. Quarks carry an electromagnetic charge, and so they are defined as being themselves part of the electromagnetic energy spectrum, however in the presence of one another they are a single, unified, solid particle. However some of the properties that define the different types of quarks indicate that they are more complex than other waves.

The electron exists as a charged point dispersed in an orbital cloud of potential until acted upon by a particle, such as a photon. Similarly, the quarks obey the expected behaviors of fields as well. The electron cloud is, like the individual photon to the entire wavelength of light, a fractal part that carries the pattern of the whole, in this case an electromagnetic field, such as exists around the earth or sun. Such electromagnetic fields are defined by polarity, and this creates a torus shape. Quarks, also, can appear in the torus field shape.

This is only known because atomic nuclei have been sped up towards the speed of light in particle accelerators until they fell apart, and the trajectories of their constituent components, the short lived quanta called antimatter that trigger electromagnetic energy release cascades when they come in contact with a real particle, are measured as a function of their internal composition.

c. the A-bomb as punctuation to end all war

When the nucleus of an atom is split the wave packets that it emits trigger energy liberation cascades, displacing the internal quanta of surrounding real particles, such as other atomic nuclei, and causing them to become radioactive — so strongly electromagnetically charged that particle decay occurs, emitting wavelengths the size of atomic nuclei that further destabilize molecular cohesion. When special relativity first demonstrated that cause and effect relationships occurred between particles and waves, one of the first uses it was put to was a bomb.

The military mentality — that self-loathing is sufficient motive for murder of those reminiscent of one's own undesired qualities — was quick to apply special relativity to the building of an atomic bomb ignited by a controlled nuclear fission. The destructive capacity of this weapon was expected to ensure the peace by outweighing the impetus for war. Although such weapons of mass destruction have been and are still being developed by multiple nations around the world, the tacit globalism of the economic media has arisen binding the people together in agreement not to destroy the whole earth over their individual differences.

2. where does matter come from?

The attempts at theories of a unified field since special relativity have all been based on tracing the origin of matter, since it is from the time of the first substance that derives the division of the four elemental forces governing the laws of physics. This is the case because fermions (the particles of the electromagnetic and, it is thought, gravitational forces) cannot yet be artificially combined to form bosons (the particles of the strong and weak nuclear forces), only derived from them by particle decay caused by random electromagnetic inversions in probability or particle collision. Therefore, the substance of the unified field, or aethyr, has come to be called space-time, and thought to be best observed in extreme events of gravitational distortion, where the other, surrounding forces, become warped. Here it is seen, or rather, can be measured, that the particles and wave packets of the other elemental forces can all be accelerated to the speed of light, and — it can be mathematically proven — even break the light barrier. According to the increasingly accepted theorems using special relativity describing these "black holes," the singularity at their center is the same substance as the universe before the beginning of the division of the four elements.

Black holes are black because they absorb light, and to do this they must be able to accelerate it beyond the fastest speed of the electromagnetic spectrum. When the geometric fabric of the electromagnetic spectrum itself, such as the speed of photons being a fixed constant, this dissolves all real particles more rapidly than the weak nuclear force, even the photon itself, which has no scholastically known constituent components. Because the gravity well of a singularity can be so strong that it can crush a photon, it has no difficulty reaching photons with its pull and speeding them up towards it, so too does it accelerate other particles to nearly the speed of light, causing nearly infinite particle collisions, or nuclear fusion, stronger than the strong nuclear force.

The only thing that escapes black holes does so due to the uncertainty principle of relativity, stating that the vacuum of pure potential undergoes random quantum inversions producing probabilities. This is thought to be the cause of the big bang. What is emitted is an extremely high frequency microwavelength. It is possible, using the measure of the characteristics of these microwaves, to determine the nature of the singularity emitting them.

a. spin equals energy

The study of quantum mechanics delves into the nature of both particle and wave. While a particle is solid and a wave transmits formless force, they do have some features in common. Both can be measured as statistical probabilities. In the same way that a particle remains a statistical probability for as long as it exists, a wave can appear as a probability or as potential. The wave function, otherwise entirely mathematical, possesses one physical feature, the dimensionality of its wavelength. So do particles posses a combination of directions in which they spin, these can all be combined and their sum expressed in a single dimension as a prime exponent.

In the same way that most real particles are actually comprised of energy waves, so is spin mathematically equivalent to a wavelength of energy. The proof of this is in the disruptive effect of exposure of real particles to virtual radiation.

The spin of real particles is in truth the measure of vector (movement of a surface point over duration) expressed as a function of three planes of dimension. Two of these describe directions a spherical particle can rotate, while the third describes the trajectory of the real particle's angular momentum. The dimensionality derived from vector spin is equivalent to frequency of energy. The dimensionality derived from trajectory spin is equivalent to wavelength of energy.

b. microwave hyperquanta

Just as a photon is the packet of information (frequency) received at the end of a ray of variable distance (wavelength), so too is a tachyon that point of extension of an electromagnetic microwave moving faster than a photon. To call a tachyon electromagnetic would be similar to calling a photon particle the wavelength of a light ray. The spin of the photon is equal to the information of the wavelength light ray, and the spin of a tachyon is equal to the information of the microwavelength.

The electromagnetivity of the microwavelength itself is contingent upon the concept of it bearing a charge arising as a result of its frequency's fluctuation, and therefore a magnetic field arranged perpendicularly to the flow of the electrical current. Thus to call the microwaves electromagnetic energy depends on whether or not they carry an electrical charge that can produce magnetic polarity, which is indicated by their structure as a wavelength, similar to the known wavelengths, waveforms and fields of the electromagnetic spectrum.

The microwavelengths are smaller than the point of charge of the electron when it is singularized, because they comprise the geometric vibrational framework underlying the electron's different possible orbital shell diameters. By establishing orbital shells for the charge to occupy as potential, microwaves act like gravity. Therefore, while the microwaves look electromagnetic, because of their very short wavelengths and very high frequency, they do not so much act like electrons as act on electrons, and thus in the apparent form of gravity.

The prevalent modern theory is that tachyon particles do posses aethyreal energy, or what I have called elsewhere, astral light. The reason for this is that particles, such as photons, when converted into phased waves of pulsed energy through solid media can exhibit faster than light characteristic velocities, and therefore it is assumed that tachyons, which share this class of velocity, might also possess the electromagnetic charge characteristic to the accelerated subluminal particles. It should be remembered however, that tachyons are of a quantifiably different dimension altogether, and should therefore be expected to possess qualitatively different traits.

While modern theory does not argue over the notion of tachyons being commutative with subluminal electromagnetic and atomic radiation, it has postulated mathematically that tachyons accelerate their speed as they lose energy, which is the elegant contra-positive to the constant speed of light in a vacuum rule of relativity, which states that as subluminal matter increases it gains mass. This expectation of tachyons will likely be proven according to countless mathematical applications of the relativistic equations, and during all that time delay the true nature of tachyons being discovered by experimental physics.

It should suffice to say that tachyons are visibly radiative as Cerenkov light emitted by photons. Since the light from one photon alone would not illuminate the entire diameter of the universe, however, it is also safe to say that this illuminating energy wears out. In this regard we can say that the energy of tachyons which is positively charged when emitted from photons is negatively charged when absorbed into atomic nuclei, moving from a dynamic to a static state. The modern theory speculates that when a tachyon has burned off its Cerenkov radiation, what I elsewhere call astral light, it loses all of its energy and achieves infinite velocity, occupying every point on its gravitational microwavelength superstring history simultaneously. Such a tachyon is called a "transcendental" tachyon and, having no energy, is referred to as "zero point energy" — a category shared with certain scalar phased waves.

c. involution

The microwaves do not just travel through the electron cloud without disrupting it, they also move faster than light. Ordinarily they are invisible as a result of this, and though the waves themselves are harmless, they are still believed by theoretical physicists to be associated with the explosive conclusion of the lesser, similar trajectories of antimatter because breaking the light barrier goes against the observable laws of physics simply by being considered improbable.

The only time it is possible to see, or rather, to measure these microwaves is when they are projected through a solid medium. Here it is seen that energy beginning as an electromagnetic charge can travel through the material substance by quantum tunneling faster than the speed of a photon traveling the same distance unimpeded. Here we might predict that the photon would win because it is larger, about the size of the electron cloud. However because the microwave is generated from an originally electromagnetic source, it carries a charge that interacts with the electron. It does not compress the orbital cloud into a singular point of charge, as does the photon, but instead the microwavelength and the electron field inter-phase, and the microwave exits the opposite side of the orbital shell at the same time as it is entering it from the opposite side.

The reason for this complex effect is simple to explain. The particles at the tip of these microwaves, whose history forms the fractal waveform, involute. Tachyons turn inside out, and they do this continuously. They are doing this right now. When one contacts the surface of the electron's potential energy orbital cloud, the tachyon absorbs the pattern of the shell, and moves through this geometry according to its own trajectory. In other words, when the tachyon and electron inter-phase, they combine to form a dimensional warp in the probabilistic continuum of space time.

This is how the tachyons travel faster than photons. Because of this they are considered as moving counter to the flow of entropy, or the increase of disorder. This comparison, again, implies a false similarity between tachyons and antimatter, since the particles and waves that have been classified antimatter were all discovered as being generated by the decomposition of larger particles at speeds approaching the speed of light. However the difference between the world below the speed of light and the world above it is like the difference between the world under the surface of the water in a fish pond as opposed to the vast expanse of the world above the water.

d. the unified temporal-emotional field

The saying goes, "time flies when you're having fun." The conclusion of this by science is that time should always move at the same speed whether you are having fun or not, and that this speed should be roughly equivalent to the average between the rates at which it seems to move when you are having fun and when you are not.

The quantum inversions from potential to probability that occur in the brain simultaneous to thoughts and stimuli interact with fields of tachyons through the left and right thalami of the thalamus, or reptile mind, since it is roughly to this stage in evolution that our reptilian genetic ancestors have evolved their consciousness.

The interaction of mind and hyper-light opens up the doors of perception to the multiverse, and its infinite dimensional planes or levels of reality. Since the interaction occurs in the holographic reception and re-projection area of the central nervous system, and this is common to so many varying different types of life form on this planet, it is safe to say we should naturally be accustomed to dealing with this co-creative, interactive, ideal realm of spiritual experience, and I doubt that anything anybody has ever put in the drinking water would dilute this much.

The idea that the mind and the continuum of the universe are relative does not lead immediately to perfectly honed manifestation. Instead we see that humanity is constantly making sacrifices from the realm of potential to bring into materiality, and that the result of all of history's combined idealism remains relative to the environmental conditions of nature and setting. What this causes in the individual is defined as emotion, or a qualitative holognomonic similitude to the movement of objectivity relative to the fourth temporal dimension. Emotions often cycle in patterns as the electromagnetic ambiance of the world changes around them. So it is thought that the tachyons that interact with their neuralectric stimulation obey fixed quantifiable rules, the same as the quanta of the four elemental forces.

The belief that thoughts effect reality is at the very core of humanity. Our unique ability to grasp object and form tools allows interaction with the environment in forms of direct physical causality, and we see everyday the scale to which this has accumulated effect. The creation of materiality is not believed to be exclusive to our consciousness either. Religions tell us that even our thoughts result in karma, the ancient equivalent of the modern quantitative concept of money in the bank. This may be thought of as like a monkey's understanding of their genitalia.

Elsewhere I describe consciousness as potential extrapolatable to the tenth dimension, at which point it finds itself reflected in the core of external consciousness. This makes tachyons, though they are subatomic, equivalent as the midpoint of dimensional consciousness to the division between the terrestrial, left and spatio-temporal right hemispheres of the brain in Leary's eight circuit model. By applying the seven Hermetic axioms, we can see that tachyons obey the same set of rules as the mind, and that the two are, therefore, relative.

e. back to the second dimension
In order to comprehend why the microwavelengths of tachyons can be measured as an emotional reaction we must examine some examples of how tachyons interact with particles and waves. As I have already defined, the microwavelength histories of the involution of individual tachyon particles is an equivalent ratio of information to frequency of a wavelength of energy, or spin and trajectory of a particle of matter. This does not mean that they are all describing the same thing. Each is a characteristic function of the specific type of quantum information unit serving as the carrier of the macro-pattern, or fractal, relationship to its individual behaviors. However the role that each feature plays to its more general component is comparable to that of the others in this definitive capacity, and therefore the unit of measure for the parameters of the field for all these types of effect is the same: dimension.

Just as an electromagnetic wavelength is composed of the plane of the electrical current, and the perpendicular plane of magnetic polarity, expressed in the individual electron, composed of one rotation of these two dimensions, as the distinction between the polarized cloud and the single point charge, so also can the vector of a real particle be expressed in terms of the polar rotation, precession and axis path, in the absence of externally induced trajectory. It is thought that, insofar as mind seems to be relative to time, tachyons represent two geometrically ideal dimensions as well.

i. photon and electron interaction

The photon is a massless particle about the size of an atom. Because it has no mass, when it strikes an atom it does not cause the atom to move. Instead the result is that the photon freezes the electron's negative charge at the point of contact, and then bounces off of it at an angle determined by the second dimensional projected trajectory of the photon and the first dimensional point vector of the electron.

The reason that the electron is brought to a singular point in its orbital shell is that the spin of the photon is comprised of the dimensions of its particle vector and its projected trajectory, and according to the theory of the conservation of dimensionality, because the event of the photon's impact with the electron is of the minimum possible duration, the electron can only contribute one dimension within three dimensional space. It therefore drops the dimensionality of its orbital trajectory, and reacts only as an electrical charge of magnetic polarity — or as a repulsive point. This is enough for the real particle of the photon to react physically to the electron's orbital cloud of potential energy, but not enough to contribute any significant change to the trajectory besides accelerating or diminishing wavelength frequency, that is, changing the color of the visible light, and even this is only by a factor determined by a function of what the wavelength history of the photon particle's projected trajectory was before striking the electron.

Not all photons are reflected from all electrons. When electrons are built up into different shell numbered atoms, the likelihood varies as to whether they will be reflected or absorbed. When a photon is absorbed, the electron changes its orbital shell level to a higher energy level. Because the photon has no electrical charge, then we must look to tachyons to explain why this effect occurs.

ii. absorption and reflection

Tachyons are the commonality of electrons and photons. In electrons, they occur as the microwavelengths emanating from the nucleus upon the peaks and troughs of which form the electron's energy level orbital shells. In photons tachyons occur as the microwavelengths emanating from it that distort the surrounding space making the photon appear visibly radiant.

It is the information of these microwavelength frequencies that is being exchanged between the dimensions of the photon and the electron, and that determine how both will react. If the microwavelengths are a certain distance from the nucleus of the atom, and carrying a certain amount of electrical charge (determined by the number of electrons occupying each orbital shell), the probability cloud will either reflect or absorb the capacity for potential microwavelength progenation of the photon.

To say that the microwavelengths are generated by the photon is not exactly accurate. Instead what we are seeing in the photon is a lightless, real particle traveling through a continuum of potential microwavelengths that are traveling in all directions. Those that pass through the real particle of the photon become probabilities while inside of it. This has an effect on the substance of the photon similar to heating it, and this causes the photon to glow.

iii. the vibrational continuum of the shells

Other than black holes and particle accelerators it is not known what produces tachyons in the material universe. It is possible that the substance of the universe before the division of the four elemental forces consisted entirely of potential tachyons. This seems in keeping with the Buddhist description of the primary clear light of ylem, the substance they believe preexisted the universe.

These microwavelengths, when looked at head on, are tiny tori whose insides move opposite their outsides. Thus they are moving through time in both directions at once. Their frequency determines the structural geometry of the different dimensions, and these lay down the laws of physics that compile quantum information unit particles for each of the four divisions of their elemental, involuting transdimensional multiversal vibrational force. Thus they can be seen to cause our reality.

f. matter as concentrated energy

We have seen that, for the most part — that is, except in situations where special relativity applies — energy and matter are separate forms of substance that do not combine or subtract from one another, but interact independently of each other. If time itself were stopped in a single event for the entire universe, there would be no matter-energy exchange between real particles and virtual waves. We see also that two spin one half electrons don't make a spin one photon, even though both are forms of microwavelengths, and that even the different types of quarks in a nucleus are bound together by real, whole number spin quanta gluons.

So to say that the photon and the electron, as well as the quarks, pions, mesons and gluons in the nucleus, are all structures founded on microwavelengths of energy is similar to comparing day and night. However, what we are seeing by day or by night is the same scene, we are only seeing it differently. So it is with particles and waves upon the shores of time.

i. Thule

The concept of a unified field has been expressed in as many different terms as drugs in as many different cultures. Both are like an elephant in the living room of society, that everyone knows exists, but for which no one claims responsibility. Both are natural realities irrelevant of the ones who discover them or live by their light.

The Hindus had a concept of energy that they called Maya, measured by karma, which was the earliest known concept of energy. The Chinese called this energy chi, and its cycling tao, and the Egyptians identified it with ptah, the energy unifying the Ka and the Akh in the seven vessels of the Ba, and it is, probably due to their practice of keeping the extruded organs of their mummified dead in cnoptic jars, from this word that we derive the word for spitting. The Hebrews associated such power with the ophanim wheels, and the Catholics with the Holy Ghost, while the

Buddhists compared it to enlightenment. The Aryans of the Caucuses called this energy Thule, and the Maya called it mana, which was also the Hebraic Greek word for bread. At the turn of the twentieth century it was identified as orgone energy by Wilhelm Reich, and associated with scalar waves discovered by Nikola Tesla. Most recently the heads of the ss of the third reich based their research of it on a description of Vril energy in a science fiction novel. Currently such microwavelengths have been said to be the cause of gravity by Robert Lazar, who claims to have learned of this while working on government classified alien aircraft. The term tachyons is common on Star Trek, the science fiction tv series set in a future where many particles that remain theoretical today have been discovered and applied for human space travel. The other most popular science fiction series of modern times are the Star Wars movies, where the characters can perform telekinesis using "the Force" — an all-powerful energy field that surrounds all things in the universe, binds them together and makes them grow.

ii. what is left of consciousness yet to explore
What is our future is guided by the light of what is the universe's past. What one person thinks another person is just going to have to learn is the same as what the other person thinks that person is going to have to learn. Other than this there is no psychological structure to time. People's minds think like they are stuck in tunnel realities, and measure the amount of similarity from moment to moment as karma, burdened by their own prejudice against themselves. There is little enjoyable about beating oneself over the head with the weight of the world.

We should not allow ourselves, in light of tachyons, to see light as equivalent to gravity. Matters of enlightenment are not the same topics as those considered gravitas by the old Roman prelates. Tachyons, a particle of radiation off the scale of the electromagnetic spectrum, may carry the force of gravity, and they may be wound up so tightly that they generate the emission of luminosity, but the ideals of light and gravity are as different as the Buddhist conception of the primary clear light, utterly free of all conscience, and the Judeo-Christian God, the anthropomorphication of that conscience. I say this because I do not wish to foresee a future in which ideation, which occurs on an equivalent scale to microwavelength tachyons, is considered the result of pressurization.

This is for the reason that, as people progress closer to a complete understanding of genetics, and immortality becomes an increasing possibility, the function of the birth experience, as described by Stan Grof, a student of consciousness researcher Timothy Leary, should no longer be considered the imprint of the mental ennegram of the experience of mortality as a tunnel reality. If we become immune to death, we will only lose the eternal changing of our voice, as we begin to speak with the voice of an intelligence beyond the times.

This is why much of metaphysics throughout history has been classified by the occult orders that have kept their scientific legislations secret. For example, we know that an atomic clock outside the atmosphere and one inside it at ground level run at different speeds, due to the force of negentropic microwave gravity, and we know that the United States military stealth bomber can fly at altitudes above the end of the oxygenated area of the atmosphere. The stealth bomber is little more than a very expensive boomerang going through temporal wormholes in space-time. On the other hand, it would do the uninitiated little good to scoff at real spacecraft, even if only as a sign of modernity. Similarly it is important to embrace the craft signs of the builders, and to see them as alike keys to unlocking the greater understanding of all.

At the turn of the twentieth century a young occultist named Aleister Crowley translated and elaborated upon an Egyptian stele from the British museum collection. Although what probably caught the young man's attention was that its catalogue number was 666, and he may have been attracted to its apocalyptic Christian numerology, he would claim that his expanded translation was dictated to him by an audible voice of which he could not distinguish the source. He was not alone in this experience at that time, as Nikola Tesla would also claim to be in contact with "off world sources" through his first radio transmission towers, and it is a proven fact today that even the radiation of sunspots effects broadband transmissions.

In this text, called the Book of the Law, the original Egyptian author, a scribe named Ra-Hoor-Khuit, writing for Hoor-pa-krat, who Crowley believed to be Harpocrates, describes three ages of humanity: the age of the mother, the age of the father and the age of the son. The age of the mother was said to have ended with the beginning of civilization. The age of the father was marked by paternal authorities. According to Crowley the age of the crowned and conquering child began in 1905, with his transcription of the book. This work would set the tone for the twentieth century as an age of more rapid technological progress in a shorter amount of time than ever before in recorded history. This is truly a new age.

In Crowley's lifetime the world would be unified as social movements swept across the face of the globe like changing expressions of the emotions of the masses. Two world wars would bring the planet almost to the brink of total destruction, and the special theory of relativity was applied to intercontinental thermonuclear missiles, and humanity introduced to the fear of "the man" pushing "the button." Crowley himself would go on to become one of the heads of the Golden Dawn, a new age occult order that resurrected Rosicrucianist alchemy and applied it to the Enochian system of John Dee. Crowley himself was active in an oriental influx into the order, introducing the practise of kundalini tantric yoga as useful in ritual invocation of the angelic hosts. He would go on to join and restructure a Free Masonic lodge called the ordo templi orientis, and arise to the degree of Ipsissimus, or outer head of the order. Towards the end of his lifetime Crowley had written the Book of Thoth, illustrating the shemhamforash with the tarot, or rotas, wheel of karma. Using this temporal system, the precession of the equinoxes can be measured. Thus, by the end of his life, Crowley had unified the theory of the ages of humanity with the 2000 year long ages of each sign of the zodiac as the sun rises in a different one then at the same time of year.

According to the precession of the equinoxes, in the amount of time it takes the earth's axis to be changed relative to the direction of the sun from one point in its orbit to the opposite point in its orbit, there is an ice age for either the southern or the northern hemisphere, depending on which pole faces away from the sun as the planet's axis moves relative to its orbit. Currently sunrise on the winter solstice is aligning with galactic core, as visible from the north hemisphere. Since the earth's axis precesses relative to the plane of the galaxy, this means we are exactly at the beginning of a north hemisphere ice age.

The primary moral of the Book of the Law is "Do What Thou Wilt," and because of this unorthodox hinge pin of his philosophy, which resembled the will to power described by Friedrich Nietzsche, he, as had Nietzsche before him, proclaimed himself the living embodiment of the Anti-Christ. This was probably a vanity, but his principle of the law, called thelema, Greek for the will, accounts for this sufficiently.

The Roman word, thalamus, meaning bedchamber or sanctuary, is probably a derivation of thelema, since the Romanization of Greek culture was equivalent in many ways to DeSade's "philosophy in the bedroom." It was for this part of the Roman home that the portion of the brain was named wherein chemical electrical nervous impulses are relayed holographically to and from multiple nerve sources through single nerves. This holographic projection is very important, for it is the prophase of manifestation, and this may be the same channeling of energy described by Nietzsche as the will to power.

In fact it is the philosophies of DeSade and Masoch that demonstrate the bitter ends of human will, particularly in amorous matters. Therefore it is important to distinguish between the just wills of Nietzsche and Crowley as Anti-Christ, as opposed to the corrupt ones of sadomasochists as the archetypes of Satan and Maloch.

After Crowley's lifetime scientists applied genetic research sufficiently to produce an exactly replicated clone of a sheep, which I'm sure he would have smiled upon. The further possibilities for genetic research open up endlessly before our eyes, allowing us free will over biological matters previously considered the domain only of doctors. For example, cures for many modern epidemic diseases will appear within less time than new epidemic diseases can naturally evolve or change form.

The modern method for the treatment of cancer, or the rapid growth of a tumor, or lifeless mass, from cells that are deprived of oxygen, involves multiple source lasers, focused on the tumor using magnetic resonance imaging, to create a scalar wave field, or area of overlapping rays of extremely high frequency electromagnetic radiation. This is a far better technique than the primitive practise of giving patients radioactive barium drinks, effectively giving them cancer to treat their cancer. The laser itself derives its name from its function: light absorbing spectral emission radiation, that is — a mechanical reaction inside the laser source generates a wave of probabilities to move through a wavelength of electrons absorbing photons along the way to propagate itself. When such a ray reaches the tumor and bounces off other rays like itself, it breaks apart the molecular cohesions of the cells' atoms by causing the electron shells to emit photons. This is a sound cure, though only half as likely to cause radiation induced cancer relapse as barium, and can be accomplished even more easily and with less risk by mentation through the hands. It is not, however, a prevention of malignant, or even benign, tumors from forming. A genetic cure for cancer could come from the chromosomes from the pubic hair follicle stimulating hormone (fsh) that cause inherent stoppage of growth at a predetermined length being implanted in (combined with) the genetic material responsible for the creation (reproduction) of the cell wall, such that, should the cell reach a certain predetermined, maximum tolerable amount of oxygen retention by the cell wall, the nucleus would react by stopping the thickening of the cell wall as it regenerates it, allowing oxygen to evacuate the cell regularly, and preventing preemptively the stagnation leading to malignancy.

One of the main causes of cancer is cigarette smoking. What would have been fatal doses of tobacco and cocain were found to have been used in the mummification process of ancient Egypt. Cigarettes also contain arsenic, which is known to kill off the micro-bacteria in the intestines that cause internal decay after death, as well as being deadly to humans. If there were a cure for cancer there would be no social movement against cigarettes.

Cigarettes are not the only cause of cancer. The dropping of the atomic bomb only marked the beginning of the age of atomic radioactivity. There is more ambient radiation trapped inside the atmosphere by the greenhouse effect now than there has been since the times of the earliest dinosaurs. Much of this is due, ironically, to smog produced by the burning of fossil fuels, comprised of the liquified body fats of the dinosaurs. This has produced a hole in the ozone layer, letting in even more ultraviolet radiation. In addition to this the sun is now producing more radiation than it has in the last 2000 years, as it is at the peak of its sunspot cycle. All of these combined factors make cancer a risk even from only prolonged exposure to daylight.

An ancient technique for life extension involved drinking water with trace amounts of gold in it. These gold atoms were thought to bond with the cell walls of the red and white blood cells and increase their efficacy. The idea of bonding human cells to metal atoms may also be behind the fluoridation, a byproduct of aluminum manufacture, of public water supplies. It is known that the ancient Romans used aqueducts lined with lead, and this lead to lead-poisoning, an effect similar to arsenic. Many modern scientists predict that in the future much of human biological dependencies will be compensated for by nanites, or microscopic robots the size of cells, with atomic mechanics, that would inhabit the human host's living systems.

Karma accumulates in centers of electromagnetic gravitational influx, which often occur along ley geodesics. As we pass between these we find our emotions change, our mood changes, and our perception of the world changes as well. This fractures the central personality into the stream of consciousness, and when the brain waves are altered into normalized wave functions it is easy to go with the flow. These are also the origins of the various different schools of religious thought, as humans migrated along trade routes and moved ideas across the land.

So have we learned how tachyons, like the essence of Otherness described by Sartre or the spirits described by world religions, reflect in the form of ideological ennegrams in the neurons of the brain, upon the surface of and effecting the genetic code.

B. known particles

Karma is associated with many things, chief amongst them good and evil, but also with the flow of time. It provides us with an excellent example of the same principle behind the anthropomorphication of God, as karma was the first concept to be quantified symbolically. Just as the end all, be all idea of God is humanity's own idea of itself, so karma was measured as beads, or spheres, usually on a string. While this idea was busy leading to the abacus in the orient, it led to the Last Judgment scenario through Egypt into the west, where all of our good deeds and all of our bad deeds are measured against one another, and this determines if we spend eternity in a paradise or being tortured. Even though it was Jesus himself who overturned the tables of the money changers in the Temple of God, Christians unanimously view him as the one to return and judge them according to their accumulations of karma. In truth, there is no real, human God, and there is no ideal particle. Other than these two concepts, religion has no basis at all in all the history of humanity. What there is known to be, however, has only been proven by other peoples' obsession with the opposite concept, a real universe and an ideal God, an agnostic spirit that eludes humanity's grasp, and at the same time is capable of fulfilling humanity's wishes. This is the same concept as karma, only with a psychological personality, and either a deistic faith or a belief in cosmically nascent emotions is only another foil of genocentric anthropomorphism. Even the concept that time is the blood of God still implies that God bleeds, and can therefore be killed, which is impossible, since there is no such thing in reality. Similarly the concept of God as the rule of irony leaves open the legal loophole that even the law of irony can be broken, and is thus not absolute.

The fact that there are real particles, as first predicted by the Classical Greek atomist philosophers, at least proves that karma, in an amoral, non-ideal sense, is a reality, and that even human beings are composed of very small particles and waves of pure probabilistic information. In that it is taken as the mechanism for cause and effect, karma cannot be considered a moral ideal, though, because it is equivalent in quantum mechanics as a rule over time to the uncertainty principle, or the axiom that cause and effect occurs at random, and due to what is, quantifiably, nothing.

According to an application of irony the existentialists believed that existence preceded essence, and only then could people use their free will to determine for each individual person whether God exists. At the same time as this great modern philosophy was being cranked out, the existence of real particles of karma was being used to argue that there is no order to the universe, ergo God. To put this in terms both the clergy and ley-scientists can understand, the randomness of real particles of karma as being the lesser, or more base and profane, aspect of the law of God, disproves the existence of God as primary motivator or intervening legislator.

1. electrons

Quantum mechanics delves even deeper into existentialism, though, disagreeing with its central premise by saying that the essence of the continuum, quantum waves, did in fact preexist solid particles that form the nuclei of atoms, which form the macromolecules of matter. Therefore it is safe for modern scientists to say that electrons are older than photons, in the same way that astrophysicists have determined that black holes are older than living stars. This is because, just as a black hole once was a star, an electron goes through many different energy levels in its history as it absorbs and emits photons.

If electrons are, indeed, older than other, solid quanta, this may explain their state of flux between particle and wave. Electrons are raw potential energy, and therefore can be either a probability cloud in which the charge is dispersed as an electrical field, or it can react as a particle, reflecting another solid particle off its surface. It is important to remember that electrons are both particle and wave, because one is a spike in the wave function representing probability, and the other is a waveform frequency of pure potential.

a. the orbital shell as probability cloud

While it is in the form of an orbital shell around a nucleus, an electron occupies all points on all orbital paths simultaneously. Imagine a soap bubble. It is the same inside and outside, air, but the surface is a seething fractal of patterns and colors. This is like an electron. An electron is like a bubble, the same inside and out, but in the case of the electron this is electromagnetic energy. Just as the bubble's surface integrity is maintained by air pressure, so to do the conditions on the inside and the outside of an electron equal one another quantitatively. However here is where the electron differs from a bubble: inside of an electron bubble the electromagnetic radiation is a negatively charged probability, while outside the electron bubble the electromagnetic radiation is positively charged potential. Because its surface is acting as an inversion between these two descriptions of the same microwave gravitational force, the electron manifests electrical charge.

As I said, the charge of the electron occupies all points on the orbital shell, or bubble, ubiquitously, but the electrical force carries with it the magnetic force. Because the charge of the electron can be at any point, and the axis of orbital polarity is always perpendicular to it, then there are infinite possible positions for the magnetic poles of an electron, and we can also say that the electron therefore possesses infinite potential spin, or vector.

In the metaphor of the bubble, the fluid dynamics of the oil on the surface of the sphere would represent the electron clouds probability. If you poke something through the surface of a soap bubble, you will see the effervescent oil slick cohere homogeneously towards that point. This is similar to the reaction of the electron cloud when acted upon by a real particle, such as a photon.

b. the temporally singular electron

Ordinarily the probability curve of the electron cloud would measure something like a square well, in other words, as existing for a fixed duration of time. However we have no way of knowing if this is the case or not because of the uncertainty principle. According to the laws of the uncertainty principle it is possible to measure only either an electron's trajectory (spin) or velocity (frequency), because which ever one you are observing, and can factor in geometric conservation for, the other is changed relative to the conservation effect of the first. What this means in effect is that, when a real particle strikes the surface of the electron cloud, the electrical charge coalesces into one more probable orbital history than any other, so that it has an effect on the frequency and the reflected trajectory of the photon equivalent to its own frequency and trajectory, however in that those of the electron change those of the photon, so do those of the photon change those of the electron, and since the charge immediately fades back into a cloud of potential energy, there is no way to measure any such predictable change in the orbital path of the electron.

The reason for this is that the charge of the electron is one dimensional. When it is struck by a three dimensional real particle, an orbital plane makes it second dimensional. Because of its minimum possible duration, it is a temporally singular event, and this adds the dimension of time as the electron's third dimension, allowing it to act as a real particle opposite the three dimensional real particle of a photon. The temporal singularity of the electron mirrors its singularity as a point, and this wave function of probability can be measured, as a short, sharp spike rather than a square well.

i. the nonsensical concept of the point

Because the electron's charge is massless it is the same size as a singularity. This is a difficult concept for many modern scholars to grasp, since we are told that a singularity is a point so infinitely small that it bends spacetime in upon itself and produces the massive gravitational effect of black holes, and we do not see this same effect occurring with electrons. Here we must remember that if a singularity had spin it would have mass, while the electron has spin and yet has no mass. Instead it has an electromagnetic charge. While the singularity operates as a quantum dynamo for the gravitational force, each electron carries an infinite potential for energy.

ii. the extremely small size of the electron itself

If it weren't for the mathematics telling us there was something there, an electron might as well not exist for us, as there is no way that we could ever perceive it. For this reason, based on their two dimensionality as charge and orbital trajectory, electrons are considered virtual particles, that is: waveforms of energy with fractional spin, existing only halfway in the material universe. As a cloud of potential they are not even measurable probability.

The best way for people to understand the electron is to visualize how the gas cloud around stars gradually, over millions of years, coalesces into the spherical planets orbiting the star at their different orbital distances and speeds. This is the change an electron undergoes whenever it is struck by a photon.

iii. the temporally singular duration of the electron itself

What this spike well waveform means is that the duration of the electron's concentration into a point is of the minimum possible duration. In other words, it is only while the photon and the electron are actually in contact with one another that the existence of the electron can even be measured. During this time it can be expressed as several different reactions to the photon's internal spin and projected trajectory. The only one I have described is the effect of the electron coming to a point equivalent in size to a singularity, except containing an electrical charge.

Also as I have said, according to the uncertainty principle only either the frequency of the wavelength history of the electrical charge or the projected trajectory of the magnetic polar orbital rotation of the electron point can be measured at one time by the impact of a photon, because the impact of the photon changes these characteristics of the electron after reflection. The way that these are measured is by the amount of change to the opposite characteristics of the photon, where the trajectory spin of the electron changes the photon's wavelength, and the wavelength of the electron changes the photon's spin, or vice versa.

The change to an electron's spin is therefore a precise geometric function relative to the spin of a photon colliding with it, and these changes in trajectory can therefore be measured using quantum mechanics wave mathematics.

iv. the pi orbital trajectory

Ordinarily, when the electron is in its square well state as a cloud of potential energy with the electrical charge dispersed throughout the surface of its orbital shell, there are infinite potential magnetic poles for it to be orbiting around. However, if the electron is brought into relationship with other electrons by a magnetic field, such that its magnetic polarity becomes a fixed function relative to all the other electrons around it, then it becomes possible to observe a single orbital trajectory of the electron charge as a point on the surface of the orbital shell. However, while it is still a cloud of potential, the electrical charge will seem to occupy all points on the surface of the orbital shell at once, because of the velocity at which it is traveling around that orbital trajectory. In this case if the electron cloud were struck by a photon only a small fraction of the electron's orbital trajectory could be projected, because it would not be known if the orbit of the electrical charge would change relative to the magnetic pole.

The best way to answer this question is to study electrons in aggregate groups called fields. We find electromagnetic fields around every living thing, around the earth, and around the sun. While all of these share polarity, the orbital trajectories the electrons take as they cycle through them are determined primarily by external factors. The aura of living things is fluid and obtuse, while of those artificial things imbued with charge, rigid and precise. The ionosphere of the earth seems more or less stationary, but that of the sun coils up due to the differential rotation of the gaseous fusion engine. It is probably this last model that best reflect the behavior of the electron charge in its orbital shell.

The pattern of the electromagnetic field lines of the sun, and of the projected trajectory of an electron charge in its orbital shell around the nucleus, begins at one pole and spirals around the sphere in concentric circles until it gets to the other pole. It is not known what the electrical charge does when it gets this point, but it is known from the passing of an electrical charge as a current through electrons aligned by a magnetic field that this is how the force of electromagnetic radiation is transported.

c. absorption and reflection of photons

The dependence on the use of a photon as activator for the measurement of the properties of an electron is similar to the fetish of modern astrophysicists insisting on describing only the effects of a black hole on a biological physical body, rather than performing a thought experiment on them in the laboratory of the mind. It is a malfunction of modern machinery that cannot be expected to endure as a restriction.

The human mind is a very sensitive entity, and this means that it empathizes, even if only unconsciously, with every change in the environment it is within. The description of an electron being struck by a photon, therefore equates in the emotional mind to the stoning of sinners, and there is a knee jerk reaction to the nascent enlightenment involved on the one side with prayer upon some moral cliché like "people in glass houses shouldn't throw stones."

Neither of these viewpoints is correct — either the dependence upon the use of photon intervention to realize the properties of an electron, nor the idea that there is implied by this dependency the inverse necessity of prohibition. Photons are going to come into contact with electrons as long as there are photons and electrons, and when they do they will always have the same, basically predictable, results. We are not responsible for having first caused this condition of the universe to come about simply due to our meager understanding of it, and what universal effect our knowledge of the use, and abuse, of it may have remains to be seen.

The observed effects researchers have collected data for using quantum probability as the unit of measurement are that when a photon strikes an electron it can do one of two things. It can be reflected, that is, the electron can act as another three dimensional real particle and bounce the photon's projected trajectory off its orbital shell's surface, or it can be absorbed, causing the orbital shell of the electron to expand to a higher energy level. If an electron's orbital shell level is not on the base line frequency of the lowest energy shell level, then it can also emit a photon if struck by another photon.

i. absorption and dissolution between real particles

The idea that one real particle that has whole number spin in three dimensions and another particle that can act as a real particle, a solid against which another solid can reflect, can also dissolve together into a new function that acts entirely like a wave is an astounding concept. However, according to the principle of relativity, which states that matter and energy are relative, and that of uncertainty, which states that potential energy and probability wells are interchangeable over time, this is a simple fact of natural science. As I described earlier, the photon carries no charge, and therefore what it is contributing to the energy of the electron is the radiation of tachyons. When these are absorbed into the field of potential of the electron cloud, it increases its distance from the nucleus, and maintains the same velocity (or increases its velocity and keeps the same distance, but to find out a scientist would have to measure it again, and to do so would disturb it, and thus change its conditions). Because the electron's positions of orbital shells are determined by the microwavelengths of tachyons emanating into the nucleus, when more are combined with them from the potential of a photon, it simply serves to lengthen the measure of the wavelengths of tachyons, and move the electron's orbital shell out further from the nucleus. In order to compensate for the additional energy, the electron moves over a greater distance in its orbital trajectory in the same amount of time. However if there is not another electron that stays as a place holder in the base energy level state, the electron will emit the photon and fall back down to the lower energy level shell.

ii. energy shells and the conservation of reverse spin

The reason quantum mathematics indicates the electron changes orbital energy level shell when it absorbs a photon is the conservation of spin. When the electron absorbs the photon it absorbs its particle spin and wave trajectory components, and this additional information expands the parameters of the orbital shell. This demonstrates a rule with two particles quantum physicists believe still applies even if there is only one particle or wave contributing effect: the law of reverse spin. According to this outlandish axiom, in potential every particle is moving in the exact opposite way as probability allows us to observe. The reason we are observing things spinning the directions they do is because this reverse spin is conserved on a higher dimensional level. When two particles with different spins, such as, in particular, the photon and the electron, interact, they represent two reverse spins combining to form a higher dimensional spin, and this accounts for the effect of the increased information expanding the electron's orbital shell.

iii. reflection and trajectory spin

When a photon is absorbed and emitted from an electron cloud, it happens so fast it is virtually simultaneous. This gives the effect that the photon, as a real particle, is bouncing off the electron, which it is thought would have to be acting like a real particle. I have demonstrated how this is merely a function of the factoring in of information from various different dimensions of spin, and we see that it is these same factors of spin that are changed by the collision and reflect combined trajectory spin. These change for both the photon and the electron, however it is only by weighing the measurements of spin as dimensions for a photon and then bouncing it off of an electron and subtracting the change that we can currently comprehend the characteristics of the electron beforehand, and doing so changes these characteristics for the electron. So, while our knowledge of the whole number spin, real particle photon can be more or less absolute, our knowledge of the electron is confined by the uncertainty principle to be incomplete.

What we can observe are the three precise factors of spin that can be catalogued according to the first three dimensions, and relate these two between the photon and the electron. Therefore we can make predictions for both based on the measurements obtained from either.

2. photons

Two fractional spin virtual electrons do not make a whole spin real photon — unless they are both orbiting the same nucleus and one is emitted as a photon. Of the quarks and gluons and other quanta inside an atomic nucleus are held together very much like the electrons in orbital shells surrounding them that can combine with other atoms to form molecules. Other than this the various gauge bosons, mesons, protons and neutrinos are all more or less free particles that do not bind together except as upon wavelengths to transmit a carrier force. Photons are one such boson, or large, real particle, with long vectors, and wavelength histories so long they comprise the spectrum of visible light, which is only a little bit shorter than audible sound waves, which are so long in frequency that they actually disrupt the particles of air, which are macromolecules of oxygen atoms and water molecules.

In other words, even though it is of a size that it can interact with an electron, which is the same, massless, size as a singularity — or hole in spacetime, a photon is still one of the larger quantum particles that occupy Hilbert space. In the light of relativity, physicists seeking a general unified field theory concluded that the emission of very large particles from radioactive elements undergoing decay is equivalent to the propagation of force through the continuum of smaller elements such as the large photons and the very small electrons. However, following this, their attempts to find a very large particle responsible for the propagation of the force of gravity over vast distances has failed. This, it is believed, would unify both the gravitational and the weak nuclear force to the electromagnetic spectrum.

However there already is a form of radiation that, according to superstring theory, some scientists are coming to believe may be responsible for the propagation of the gravitational force. It may be because of this form of radiation that the large, real photon can interact with the infinitely small, massless electron.

a. a particle of radiation: "real" energy

In the same way that an electron is a force carrying wave form of the electromagnetic spectrum, so is a photon a force carrying particle of that same form of energy. Just as an electron carries a charge, a photon carries light. Just as a charge can be passed along a series of electrons arranged in a magnetic field, so is it by the motions of photons along wavelengths of light rays that everything we see is brought to our eyes. Just as radiowaves are actuated by the propagation of an electrical charge (and so is it now thought for large particle radioactive decay), so are photons the carrier particles for the propagation of radiation. Many photons begin their lives as the potential electrical charge of an electron.

b. the mystery of histories

In the same way that an electron is represented over time as a wavelength of energy, so also is the trajectory of a photon or other real particle over time measured as a wavelength. In both cases this is a two dimensional wavelength unless it is a factor of vector in three dimensions, including particle spin (polar displacement), orbital rotation (surface vector) and projected trajectory (wavelength history). If the electron cloud is moving, it will posses these traits, and if the photon is moving, it will posses these traits.

These wavelength histories are associated with the electromagnetic spectrum, but they are the measure of the force carrying particles of all four elemental forces, and these are all the particles that are known; therefore the wavelengths represent a sort of unification of the four elemental forces, and in a way a transcendence, revealing that underlying all of these is dimensionality itself.

i. the coexistence of particle spin and wave emanation

While cosmologists now believe that the four elemental forces originally diverged from one another due to the separation of matter and energy, it is also believed that energy preexisted matter, and that, through the process of matter-energy exchange, all existing matter will eventually return to energy. In fact, even the matter that does exist, does not exist for very long, and it is only because matter is being continually passed through phases of energy that it is created more often than it is destroyed now.

While both exist they interact. The waves of emanation form fields of potential while the spinning real particles form probability wells. Everything looks much the same on the atomic scale as on the interstellar scale, although much, much more active. The difference in levels of activity between the two determines the number of dimensions there are in the multiverse. In the same way that there are no patterns of similarity on the intermediary dimensional levels of the multiverse because there are only ordered forms occurring as patterns of vibrational frequency on certain harmonic scales, and on these levels we see these similar forms occurring.

So it is with matter and electromagnetic energy and the greater wavelengths of dimensional planes comprising the higher function of their geometrical organization. Such wavelengths are best evidenced in tachyons, the microwavelengths that compress the vacuum of uncertainty into a solid particle of light.

ii. strobing light reveals droplets from a waterfall

Just like the water from a waterfall appears continuous, so to do the wavelengths of the electromagnetic spectrum appear continuous. However, when the water in a waterfall is exposed to strobe light, it becomes clear that the water is comprised of discontinuous droplets, each one self contained and separate from the others. Similarly, if you reduce the frequency of electromagnetic radiation to a single pulse, the spirals of the dimensional waves will combine to form the spin of a singular quantum particle.

This is the way to distinguish between the wavelengths of matter, which exist only as the measure at any one time from one peak to the next, or from one trough to the next, from the wavelengths of energy, that takes up a whole series of peaks and troughs in potential.

c. the illumination of a single photon

Science has discovered what mathematics predicted: that inside each quantum packet of dimensional spin wavelengths, there are smaller such wavelengths acting upon and permeating the quantum information units, and these comprise individual particles of even smaller quanta, the microwave tachyons.

In the case of photons what we are actually seeing is the concentration of tachyons in a vortex at such a frequency that they actually solidify uncertainty, causing it to radiate, or emit, tachyons. When the photon moves through space it is really these tachyon microwavelengths transporting the probability for such a spherical solid along an emanation of force propagation.

i. microwave radiation

In the same way that the radioactivity of large particle elements, the radio-transmission on electromagnetic bandwidths, and the radiation of emitted photons are all considered synonymous, so may microwavelength tachyons be considered gravitational radiation. These are equivalent to superstrings, that propagate in dense, ultra-fine wavelengths, upon dimensional planes, in resonant frequency forms, and therefore manifest a particle force carrier.

3. tachyons

My theories regarding tachyons are based on the comparison of three separate theories and studies. These are, in no particular order: a study conducted by a laboratory in America that showed that, when projected through a solid medium, it was possible to propel an electromagnetic wavelength faster than the speed of light; the description by Robert Lazar of lenticular antigravity craft that use microwave technology propulsion systems; and the theory of tachyons as a faster than light particle. The rest of my descriptions are based on extrapolative interpolations of geometry on the predicted behaviors of such a theoretical particle.

Let me repeat this: the tachyon is a theoretical particle. Despite the study of microwavelengths, there is no public scientific link between the concept of microwave energy and the scientifically theoretical faster than light particle popularly known as the tachyon. To say this goes against the scientific community, because it has only been recently that the weak nuclear force has been associated with the electromagnetic force, and this has been a concession on the side of very large particles, rather than ultra-fine energy. Furthermore the search for a graviton as a force carrying particle for gravity has been based on its being thought to be a very large particle as well, since the force that it propagates effects very large scales.

In other words I represent a very special link through which these three fields of research can come together, however to me they are only one idea out of many, many millions more in potential. The stage is set for much bolder research to come, however now the limitations are largely mechanical. Because of the nature of our own minds we are being forced to confront the constituent components of the more and more finite dimensions as pure data, for we have no other method with which to perceive beyond the lowest levels.

Therefore, tachyons can remain a theory, irregardless of my facts and research, just as well as they can remain a fact, regardless of all the failed attempts at proving them or disproving them thus far by the rest of the scientific community.

a. temporally inverted energy

What it means for a tachyon to move faster than the speed of light is that the information unit of its carrier particle would reach a destination before a wavelength of photons emitted at the same time from the same source from the same distance through any medium. If you consider the fact that photons are the carrier particles for the entire electromagnetic spectrum, from very slow ultraviolet, to very fast infrared, at a speed faster than known to be traveled by any other particle or form of radiated energy, then to say that tachyons travel faster isn't really like saying that tachyons just break the light barrier as much as like saying that tachyons break the law of entropy, or the forward flow of timespace, since photons are the fastest form of particle decay and energy redistribution. In reality this is not the case — since as a form of microwave energy tachyons merely represent a faster form of entropy, but according to general relativity, which also introduced the gravitational constant, this was considered very much the case for anything that challenged the fixed constant of the speed of light.

In other words, to say that tachyons move backward in time might be considered at best an illusion and at worst an anachronism of absurd axioms. However this is in fact exactly what they do, since they involute upon themselves in a torus shape, tunneling through from the top pole to the bottom pole as they are projected through space. At the same time that their outside moves forward in time along its projected trajectory, their inside moves backward. When one of these is expanded into a wormhole, it is possible for someone to enter it and travel through timespace.

i. microwave radiation and the material horizon

Although it is only due to a quirk of its design that an individual tachyon can be thought of as traveling simultaneously forwards and backwards in time, it is a very real fact that this theoretical particle, because it has not been proven, exists outside the material domain. No known particle can be projected faster than a photon. When any other form of particle is sped up towards the speed of light, it disintegrates, its cinders impacting with equivalent sized particles with opposite spin. The idea that a particle is traveling faster than a photon is derived entirely from the ancient belief in an astral light and spiritual realm.

ii. beyond the speed of light

Can you see by the light of a single photon? It is a packet of information, a collection in three dimensions of different spin vectors, whose combined effect measures as its projected trajectory and illumination. This combined unit of information is the "Light" by which we see a photon, and it is by the light of photons that we know the world. Each one of the vectors of the three separate dimensions of the real photon comprises its own microwavelength, and each one of these is the history of a tachyon. It is probably the same tachyon that we are seeing, only intermittently hidden from view in another dimension, appearing like a pulsar.

A tachyon is more than just a particle traveling faster than light, it represents a particle that can flow opposite the direction of the projected trajectory of the particle emitting it and therefore opposite the flow of entropy as defined by a photon. Therefore it is a particle of time, that is, it is a meta-object of the fourth dimension.

iii. the gravitational pull of these wavelengths

Because they are moving against the wake of fluid dynamic entropy, the theoretical particle of the tachyon would create a counter spin equal and opposite to that force in which it was in suspension. This would mean that it would be a force that moved opposite the standard cause and effect progression of the rest of the entropic forces, which all radiate force outward from a central point. This force would have to attract and pull exactly opposite these other forces' standard flow of progressive entropy. The force that we can see doing this now is gravity.

It is still generally thought that the graviton particle would be a large particle, since its force is propagated on the scale of planets, stars and galaxies. However, as an energy, such as a microwavelength between quanta, all that a tachyon would need to be able to propagate force on very large scales would be to act upon a fractal framework, or pattern of gnomonic or exponential expansion. And in fact we do see all of these things happening.

iv. on their rapid projection outside the universe

Because tachyons would, theoretically, propagate faster than the speed of light that defines the material horizon of our three dimensional universe, they involute faster than the parameters of the real universe and exist outside of and beyond it. Most attempts at accelerating particles to the speed of light result in explosion, an effect known as antimatter, as probabilities and their exact inversions collide. Tachyons would, theoretically, implode so fast that they would finish before they started, and would thus be invisible. All the lesser results were called antimatter, and it is important to remember that tachyons are an anti-energy. Mathematically they are still more complex than this, as their insides are moving backwards through time at the combined speed of both their external influx and their frequency of projected trajectory, which is obviously faster than their external surface involution alone when they are in motion. There is a specific bending of the quantum uncertainty of the fabric of the spacetime continuum into a phi over pi spiral on the surface of a torus, and in this we see that the phi involution moves a greater distance in the same amount of time, and therefore is moving faster, than the outside revolution of the pi surface. Therefore, even if what we were seeing were only a photon with an electromagnetic charge, and the surface of the "tachyon" were traveling at the speed of light, the inside of it would still be traveling faster than the speed of light, and therefore entropy, and therefore would be moving backward in time.

b. the torus shape

The torus is the shape of a hypersphere at antipode. If a tachyon is involuting backwards through time, and is therefore like a particle of time, then its shape must be a metaform, or fourth dimensional solid. These are known by their third dimensional shadows, such as the hypercube, or cube within a cube, and hypersphere, or sphere within a sphere. When these fourth dimensional shapes are moved through the virtual plane of the third dimension, the metaforms of their shadows in three dimensions change. The hypercube becomes the cube, the double cube, the octaholohedron, the hypercross and the hypercube; the hypersphere becomes the sphere and torus. The pattern of a vector on the surface of such a particle is a phi over pi spiral as it follows pi upward and around the outside of the surface and then phi down into the center of the upper pole and through to below.

i. a sphere from the side

If it were seen from the side a tachyon would probably appear like a sphere, if it even appeared at all. Like the electron it would be comprised of a singularity sized point with a polarity fixed by trajectory in a differentially rotating pi orbital path around the surface of a spherical probability well. However, while the very large photon and very small electron are still of relative size, what they share in common is their substance — tachyons. The photon emits tachyons and the electron's orbital energy level shell is based upon them. Therefore it is natural for the tachyons to have the same behavior as both the photon as a real particle and the electron as a wave of potential. This is what is expressed in their combined external spin, projected trajectory, and involuting influx.

ii. the hole through the poles

As the whole of the tachyon's outer surface differentially rotates in a tightly coiled pi spiral from the past to the future pole, it simultaneously slices through the center of the tachyon in a phi spiral to emerge from the past pole. Similarly, if a particle possessing exact counter spin to a tachyon were created, it would lead into the future. This is more theoretical than even the theoretical tachyons themselves, though. Instead the function of the tachyons seems to be to establish and preserve order through gravitation, opposite the dissolution of entropy, so that things stay as much the same from moment to moment as possible. This is the most theoretical extent of all their possible applications, and acts as much the same as Einstein's gravitational constant or the ancient concept of the ether as a unified field, or spirit as the unification of the four elements.

iii. expanding it into a wormhole

The places from which we find tachyons emitting in space are black holes' gas jets, which are ejected from the poles of the spinning black holes as they accelerate all the matter-energy they will consume to asymptotically faster than the speed of light, and only a few tachyons escape. Inside the tachyon surface of the black hole, that makes it absorb photon light and gives black holes their name, is the gravitational singularity that began accumulating mass by spinning. The tachyon surface itself is called the event horizon, because inside of it is an event that is taking place outside the confines of the visible material universe. It is defined by the Schwarzschild radius, which is a function of the spin of the singularity. However the surface of it is similar to a soap bubble, just like an electron. Here we see the infinite potential dimensions of the multiverse, each leading to its own universe, all gravity distortions to the surface of spacetime by the singularity hole in spacetime that leads to the one real baby universe. To enter into any one of these universes is to be pulled into the singularity, and the matter energy of your mass to be converted into its inversion in the baby universe, unless you are made of tachyons and can escape the gravitational pull, in which case it is possible to find oneself exiting from another point in the same universe, far away from the tachyons around the black hole. It is also possible, again by using tachyons, to create such a dimensional warping effect in space without the presence of a gravitational singularity.

c. flatline history

Ironically, because tachyons are contrary to uncertainty as having constant characteristics, they are considered improbable, and therefore remain theoretical. Because of the fact that they cannot exist according to the laws in this reality, they have no history. This is merely a further characteristic of their dimensionality, in that a particle that cannot exist in the moment cannot exist from one moment to the next. Therefore tachyons are considered to be removed from and beyond the confines of not only the first three dimensions, being its polarity, orbit and involution, but from the fourth temporal dimension as well. Because its shape is based on a metaform, or the third dimensional shadow of a fourth dimensional form, and the particle itself is removed from even the fourth dimension, then we are truly looking at something fifth dimensional here, and that seems to agree with the harmonic octave and divided decade forms of the elements of the material universe.

To imagine this form of anti-history, think of a standard spiraling wavelength measuring the forward flow of change to the vector of a particle over time. Now imagine that this is an exactly flat electromagnetic field line. This is where tachyons occur, where we see microwavelength electromagnetic radiation sped up beyond the speed of light, such as in the gas jets of a black hole.

i. contained temporal singularity

In the same way that the electron warps magnetism, the spinning singularity warps spacetime, and tachyons warp timespace. The electron warps magnetism through tachyons, the singularity warps spacetime through gravity, and tachyons warp timespace through the matter-energy exchange that is reality's cause and effect. In that they are free of the rules of the fourth dimension, they are bound exclusively to it for their form, which is expressed as potential radiation.

Tachyons are exclusively waves, in the same way that photons are exclusively particles, and electrons can act as either. For this reason they propagate force according to the underlying geometries of the fourth dimension. Electrons, being half like them in this respect, but confined to temporal finitude by their third dimensional solid counterparts, are also fields of pure potential energy. In the fourth dimension there is infinite potential energy. In the fifth dimension there is infinite potential spin. The manifestation of such spin within the realm of potential energy is a tachyon. The manifestation of such energy in the realm of the real third dimension is an electron.

d. emission in the known universe from black holes

As I have already described, the best known source of faster than light radiation in the universe is in the gas jets emitting from the poles of rotating black holes. Black holes rotate, and as they do they speed up all the matter-energy approaching them to the speed of light, and convert it all into ultraviolet light. Behind this the light is accelerated even further, however this is the limit of the Schwarzschild radius, and we can see no more beyond it with modern instruments. If we could we would see that inside is a mass of churning wormholes — tachyons perpetually inflating and deflating, all leading to the singularity, where a baby universe inflated by the consumed mass bubbles off from the surface of spacetime. The tachyons' rotation is counter the rotation of the photons being absorbed because it inverts the information of their spin from matter into energy, and for this reason the tachyon surface of the event horizon rotates differentially around the black hole like the electromagnetic field of a star. When it gets to the poles, however, it cannot tunnel through to the pole at the opposite side, as the torus pattern of the individual tachyon seems to indicate fractally, and it is no more likely for the tachyons on the surface to open up and fall in towards the singularity at these points than at any other, and so the effect is that some of the tachyons escape. The electromagnetic field lines invert their direction and extend out in perfectly straight lines from the poles.

What we are seeing in the gas jet of a black hole is much like what we see in the accretion disk of the black hole, which is, ironically, alot of iron. It is not because of the mass of the iron atoms that they are prevalent in the ultraviolet corona of the black hole and may also be common in the gas jets, but because they can carry an electromagnetic charge and may therefore be acting as a conductor for tachyons. Then we would see their behavior as similar to ashes dancing around a fire, carried in the convection currents. Further we see that a solid medium is necessary to conduct microwavelength energy faster than the speed of light.

It is in this form that the black hole vents its excess energy, that is, whatever it consumes more than it can digest at the same time, and stabilizes itself relative to the progressive flow of entropy.

i. wormholes and black holes

The event horizon is the boundary of the mass of the black hole defined by the barrier of its gravitational inversion of matter-energy to beyond the speed of light known as the Schwarzschild radius. The surface of this is black because it even absorbs photons, but just above this is a photosphere where the matter it is consuming is being spun up and dragged around it at asymptotically the speed of light. This light is transparent however, since it is all being dragged back down to the surface of the black hole. Around the black hole is an ultraviolet corona of particles being sped up to the speed of light and pulled very rapidly into it. It is by this light that it is possible to see the effects the gravitational inversion beyond light are having on the clear light halo of the very dark sphere. On the black hole's surface are an infinite number of tachyons, each swirling inward toward the black hole and outward from it at the same time. These are warped around the surface of it so that there are an infinite number around the edge of the visible curvature of its face, and only one very large one expanded into a wormhole at the nearest point.

Nothing about this description contradicts anything astrophysicists predict about black holes. In fact it is a confirmation of their repeated descriptions of the effects on a clock of it being dropped into a black hole. According to these descriptions the clock would slow down as it passed the event horizon, until eventually it would stop altogether when it reached the singularity. This is merely their allegory to illustrate that time would have stopped, and in truth the clock would have been smashed into atoms before it could even get close to the event horizon.

e. making tachyons on your own

Tachyons are the product not only of very short wavelength, high frequency microwaves traveling through the electrons of a solid medium. If there were a chain of electrons arranged pole to pole, and a tachyon were traveling through them, it would follow not only the flat line history of jumping from the point of one pole to the next — that is only the internal involution of the tachyon. Its surface would also undulate over and under the external surfaces of all the electron clouds, without disturbing them from their orbital shells. Thus, to create a tachyon, one would have to project not one microwavelength, but two identical, opposite frequencies, in perfect symmetry pairing, from between which the tachyon would emerge. This is called a null wave, and the tachyon the zero point energy. Experimentation of this nature using harmonic scalar waves has been ongoing since the turn of the twentieth century.

Tachyons also occur in the neuron stacks, columns and pillars of the brain. The potential energy carried inside the myelin sheath from axon to dendrites in each nerve of the nervous system also creates tachyons, as does the fibrillator of the heart.

C. known energies

In addition to the particles I have listed there are many others that can be classified as belonging to the electromagnetic spectrum, and these are all force carriers for the fifth force that unifies the other four, the force of light. It is light that causes gravity in the form of tachyons. It is light that strikes and bends electric charge, and the electric charge is light. It is light that is generated when two particles are fused together, and light that knocks the nuclei out of other atoms.

The force of light has been known throughout history as being associated with the concept of the ideal. For example, to carry a torch for someone is still considered noble. The spirit is called the light body, or body of light.

Nowadays science knows of many properties of light, but it is only the tip of the ice berg. The work on resonant harmonic variable frequencies is literally music to the ears. It is also becoming clear that the work on holographs and holograms will yield us representations in virtual reality of the resonant metaforms of higher dimensions, such as fractals already have for quaternions.

1. charged particles from all four forces

The force of light is not just the force of electromagnetic radiation. It is also the large particles of radioactive decay, and the superstrings of gravity. Light in its fine aspect, however, as an electrical charge with magnetic potential, can be found present in particles from every one of the four forces, while it is absent in only a few particles, such as the photon, in which it is replaced by radiation.

2. the entropic bubble

Entropy is often associated with the linear polarity of time, and in this way time is often thought of as being linear and two dimensional. This leads to the concept of the continuum of spacetime being the two dimensional surface of a three dimensional universe expanding in the fourth dimension. This seems like a fairly straight forward description, except that, in three dimensions, the universe appears to be expanding outward in all directions at once. If earth were only a point on the surface of an expanding sphere, then perpendicular to the surface of spacetime there would be no other stellar bodies until the opposite surface of the sphere. This is obviously not the case.

This is because it is not the third dimension that is expanding. This is only what appears to be happening because it is thought that the doppler effect is responsible for the red shifting of the photon emission spectrum of intergalactic light. However this is an optical illusion. The expansion is actually going on in the fourth spatial dimension, which underlies the three dimensions that we perceive, and is the substance of the surface of spacetime that is made visible in photons and measured by the speed of light. The substance is pure potential energy and the motion through it is light, not only in the form of photons, but in the form of tachyons. It is the self-propagation of these tachyons holographically that causes the universe to expand, and is responsible for the effect of entropy, or the fourth temporal dimension's passage through us.

It is now possible to predict, based on different possible geometries for spacetime and the observed mass and age of the universe, according to the laws of physics, if and when our universe will end. This factors into similar eschatological systems. Rather than either expanding indefinitely and fizzling out or collapsing inward and compressing back into nothing, the voids in the walls and filaments might be seen as expanding holognomonic forms, or even that the photons themselves on their carrier waves of superstring tachyons might amount to only Maya, or illusion. One such system is the Mayan, which dates three previous destructions of the world, each by a different element, and another to occur on December 12, 2012 gregorian, the last, by fire. Another is the Hindu system, which describes how the universe is a dream in the mind of God, and ends when He awakens.

3. the cyclotronic vortex

When the energy of light moves through the lower dimensions, from the first dimensional electrical charge to the fourth dimensional probability cloud, it creates a whirling spiral concourse, and this is measured as spin. Spin is measured as an effect in the other dimensions, but, since even fifth dimensional tachyons posses involution spin, it is a cause in a dimension above all the rest. Without light acting as an intermediate carrier dimension between it and the manifest realm it would have no effect, however, and the temporal aspect of the fourth spatial dimension would end. According to the standard aging of stars, those in the galaxies furthest away from us and most of those even only a few hundred million light years away are already dead, and this wave of destruction is sweeping through the universe towards us behind and beneath the illusion of light they gave off so long ago.

4. the time tunnel

As I have described, entropy is a function of the temporal involution of the universe through infinite potential tachyons. History itself is going through this time tunnel, leading as much from the more disordered future to the more ordered past as from the more disordered past to the more ordered present. There is no question as to the history of karma: what goes around, comes around; right before anything became past it was the future. There is no such thing as the here and now.

5. body jumping

Time can be stopped in death. In this state the spirit can perceive itself as apart from the body. It moves freely. The mind is just its shadow. The flowing stream of the font of consciousness is a well at the surface of an underground spring and the spirit is the radiation of ley lines out from it. Nietzsche said, "man is a rope stretched across the abyss." Many people have read Nietzsche, and have found themselves as such. Thus they constitute the fibrous network of the multiverse, a dimension accessible through the relaxed concentration of consciousness.

The only thing hindering the common person from their day dreams is moral guilt for being able to have them. Otherwise we would all be able to freely admit to one another that we can see through each other's eyes. Our hearts beat with one another. We even breathe with one another. Rhythms flow through us and between one another and these are harmonic frequencies that we like to cuddle.

Spirits are living karma that can move about between people in the form of consequence. Our phi/pi genetic coded DNA is manifest spirit. It is both matter and spirit, form and function, monkey and God.

VI. Spin

While the speed of light is the measure of entropy as cause, the spin of all real and virtual particles is the movement of its effect. Since it is velocity of spin by which we measure the speed of light, spin can be classified in a higher dimension than the force of light, which unifies the other four elemental forces.

To say that spin measures a higher dimension when in fact any good modern quantum physicist would be able to tell you that it only measures the three real and one additional potential dimensions of our universe seems a little inverse. This is only due to the very ancient tradition of believing the mysteries of the universe are better off left alone as ineffable. Honestly we are lucky to even know of depth, height and length.

In regular quantum physics, spin is measured in as many as three dimensions for real particles, and as many as two dimensions for virtual particles. This is because a real, or solid, particle can be measured as a wavelength even while in motion, and its particle spin, or vector, and its polarity can also be measured by its being struck by another particle, while a virtual particle can only be measured as the wavelength frequencies of its orbital shell and either velocity or trajectory. In this way we see that the way virtual particles' spin is measured allows the introduction of the variable of duration, and this equates to the pulse duration as well as the length of the wavelength of a potential virtual particle or a moving real particle, and in this way we know that virtual particles exist even when they are not being struck by a photon to be measured, similar to a zero acting as a place holder, and it is by the length of wavelengths of the real particle photons that we measure time.

Therefore, insofar as spin measures the four known dimensions, it is not so unreasonable to say that it may be utilized to measure a higher dimension as well. Since we have seen that tachyons are a particle of a dimension transcendent to time, it would seem only fair to say that their spin would represent an additional dimension to those four that are known. However, rather than this, what I would propose here is that the characteristic of spin itself is actually the substance of a dimension even transcendent to tachyons, since we have seen that tachyons spin is involution, however we have not assigned a complete set of characteristics to the behavior of spin itself.

A. vector

As I have said, vector is usually only used as a measure of particle spin, or the movement of a single point on the surface of a real particle. Since this movement is guided in three directions simultaneously — by orbital spin of polarity, trajectory spin of the whole particle's movement through spacetime, and the spin caused by the inclination of the particle's polar axis — vector can refer to three different dimensions of particle spin. It can also be used as a name for the measure of duration, although this is not a common practice, since spacetime itself is not yet commonly viewed by quantum physicists as a single, unified field upon which each particle itself is acting as a vector governed by the same laws relative to spacetime as a fractal or a hologram.

In this regard a single vector can be multidimensional, and some quantum mathematicians have extrapolated different results as to the maximum quantity of dimensions a single vector can represent. Depending on the geometry of our universe, spacetime may be curved in different ways, and thus effect particular different vectors on the local particles. There may be a finite number of ways in which this occurs, and if there are it would seem to indicate that the relative geometry of curvilinear spacetime must be closed. However there are other theoretical models that extrapolate base vector dimensionality to the nth exponent, representative of a sum that is greater than infinity, and greater than the transfinite variables as well.

It is because of the multidimensionality of the single vector that there is usually a prejudice against classifying vector, as representative of spin, as a dimension in and of itself. However the same scientists would no doubt have to concede that a multidimensional universe, or a multiverse, would necessarily be of a higher dimensional order than the universe of minimal possible dimensions we know today. Therefore the more dimensions a vector has, the more it is in an autonomous dimension, unique from the vectors of all the other particles.

Therefore think of a vector as similar to a person on the surface of the earth. As the earth rotates and the pole precesses, and as earth orbits around the sun, so does the solar system orbit the center of our galaxy, and thus we are moving relative to billions and billions of other galaxies, each formed and guided by its own dimensional geometry. Not only this but we have free will to move around on the surface of the planet wherever we wish, and moreover we are perpetually regenerating our physical form such that there is motion even within the point of the vector of a person on the planet itself, just as a tachyon involutes.

1. first dimensional

If we subscribe to the closed geometry theory for the model of spacetime of our universe, then we must accept that entropy reduces the number of possible dimensions to the minimum number it can. I have already shown you this is not necessarily true, but it is still the prevalent belief, and therefore the one I shall describe. In truth, we negentropic humans have created computing machines through which it is possible to enter into a virtual reality in which the open geometry of n potential dimensions governs the laws of physics, and these are just as real in that they have been made and do exist as is the universe outside of them in which our bodies of flesh exist. However much can be learned from the minimizing method, and we see that it does indeed govern entropy, which is the greater factor in the universe than negentropic evolution. Even though this is the prevalent scientific practice, it remains elusive that the universe itself is and always has been nothing more than a first dimensional singularity.

a. the singularity

Until recently the singularity itself was nothing more than a theoretical concept, expressible utilizing the ideal of mathematics and geometry, however unknown in any natural forms beside the socially independent individual. According to Einstein's equations for special and general relativity, the spacetime singularity was a statistical possibility, and it was not until the end of the last century that the black holes these were thought to cause were finally observed through a telescope, masked by the radiation that they were inside of consuming. Now it is known that there is at least one super massive black hole at the center of every spiral galaxy, and that there may be countless other "naked" singularities, or black holes with no "hair" (electromagnetic field lines acting as attractive gravity), are thought to be the particle unit of superstrings.

i. gravitational (particle exchanging)

The modern model of the black hole is based on a gravitational or rotating singularity that is emitting the attractive force of gravity, and pulling the surrounding material of the continuum towards it. We know from tachyons that gravity is actually a repellent force that moves opposite entropy, and in this way acts attractively on the surrounding particles of the lesser dimensions it encounters. These surrounding particles are pulled towards the singularity, and form a shell of radiation around it. It pulls matter apart further out, reducing it to light, and in the center, it is hypothesized, there is an area that even photonic radiation cannot escape from, and this is called a black hole.

It has also been hypothesized that the singularity in the center of every black hole is itself the medium of another, smaller continuum, or baby universe. However I have not yet specified sufficiently the difference between universes resulting from the temporal singularities or faster than light wormholes between the black hole's event horizon and the spinning singularity, and universes resulting from the spinning gravitational singularity itself. The universes arising from the temporal black holes are illusory refractions of our own universe, parallel worlds. The universe in the central singularity is a gnomonic inverted reflection of the laws of physics of our own universe, as they are inverted beyond the speed of light and then again when they enter the gravitational singularity.

Here, in its raw form is the microwave gravity of the universe itself.

ii. temporal (wave exchanging)

A temporal singularity is a tachyon. The wavelength of its projected trajectory is called a superstring, while we see that the tachyon itself is an involuting torus that propagates more of itself that are smaller and then grow. This classification of various different sized faster than light microfrequencies by particle is in keeping with the standard pursuit of quantum mechanics, to produce tables of quantum particles for the elemental forces alike the tables of atomic particles for the same.

The reason a tachyon is a temporal singularity is that it is moving both backwards and forwards in time at once. This is simply to say that its spin is moving simultaneously in opposite directions, and that, due to its surface area, a vector on one spot on its surface would not necessarily be moving in the same direction as another vector on a different spot on its surface, and in fact, since its internal core is a phi/pi gnomonic spiral, it is actually bending the surface of spacetime quite a bit.

These exist on the very smallest level, and are so small that we cannot even prove them using modern machines. They are still only a theory, existing only in potential. Ironically, however, potential itself is the substance of the fourth dimension, for it is only through the study of statistical probabilities that we can predict the future. Therefore, to say that these particles exist in potential is tantamount to saying that they are only a matter of time, that is, a singular temporal possibility.

It is this kind of singularity that occurs inside the neuro-cellular pillars inside the brain, and why our mind is a singularity. They occur as well in the electric fibrillation of the heart, and, because they occur here as a pulse, configure geometric fourth dimensional hypershape meta-object thoughts perceived as emotions around us forming our aura or electromagnetic soul, which communicates the continually changing contents of our minds as moods.

2. second dimensional ("branes")

Since wavelengths have been observed to propagate along relative planes, it is thought that the singular vector can also be extended onto a relatively flat surface. This theory, called M-theory, describes the surface of spacetime as shaped by several different dimensional branes, each one warped in a different way, and in this way describing the force that is warping them.

All this means is that dimensionality itself is allowed to be thought of as planar, which is a hard conceptual leap to make having just come from thinking of it exclusively as a point. However let us examine the electron. When it is struck it is a temporal singularity. The duration of its existence as a measurably real particle is called an event, and this is used as the standard shortest measure for the progression of time by units. The electron, when it is struck, and has compressed into a singularity sized point of electrical charge, can be measured by its orbital trajectory, which is planar, and its electric charge, which is polar. We are not looking at four dimensions here. We are looking at two dimensions that are both potentially two dimensional. In other words, M-theory is correct: every single dimension can be expressed as like a two dimensional plane with two opposite potential surfaces. In this way, not only does every vector express multiple dimensions, but even more so is every dimension an expression of multiple dimensions.

This leads philosophers easily into the age-old debate of dimensional levels and the hierarchy of the heavens reflected in our consciousnesses. It would be easy these days for a child to reason that we are living in a multiverse of tunnel realities and that the surface of spacetime was little more than a graphic user interface system. Of course, for an adult to do the same would probably get them arrested, however both these facts only serve to indicate that computer technology has, indeed, penetrated the collective unconscious with digital newspeak, and is leading the common consumer down multiple roads at once in the capitalist garden of forking paths. However, any ancient QBLHist can also remind us that the first names for cognitively dissonant thoughts that occurred in the same mind were Satan and Maloch.

a. the electron

Because it is an electrical charge the electron always has polarity, and this means that the charge orbits around the poles from one to the other so fast that it occupies every location in the shell or cloud at once, and it is only by being struck by a photon that its position or velocity can be measured. When this happens its orbital plane becomes clear and can be deduced mathematically. When this is done it is classed as second dimensional, following the first dimension of the singular charge itself with the polarity of that charge.

The standard, or Bohr, model of the atom is one of possibly multiple electrons sharing the same distance orbital shell from the nucleus, and this means that the multiple electrons must never collide, and that their orbital paths cannot all be the same at the same time. In other words, even though the different electrons can be the same distance from the nucleus, supported on a framework of microwavelengths emanating from the nucleus, and even in this way share magnetic polarity as the definition of their orbital trajectory, they must either be moving at different speeds, or be located in different places along the orbital track of the electron cloud. Of course, the only way to measure this is to strike them with a photon, and then we see that doing so only restructures the electrons in their orbital shells and absorbs or releases more or less energy, in other words, changing the very same environmental conditions for the electrons that we would be seeking to measure.

Because of the reliance on the photon, or real particle, radiation interference to study the nature of virtual particles, not much more than this is known about the electron, and therefore more profound questions, such as is the rotational vector of the electron cloud differential, like that of a gaseous body in space are left just out of the reach of the long arm of the law of physics.

b. orbital planes of galaxies, planets

It seems odd that, living on a spherical, three dimensional planet, in a fourth dimensional timespace, that many of the larger arrangements of heavenly bodies, between the third and fourth dimensions in scale, occur in the second dimension. The orbital planes of the planets around the sun are flat ellipses, and the path the sun takes around the center of our Milky Way galaxy is a more or less flat spiral on one of the galaxy's flat spiral arms.

Stars are usually found in nebulae, clusters, and galaxies, which can have many different shapes. The most common shape of galaxies is a flat, pi spiral. When stars are first formed in the gaseous nebulae by random spontaneous combustion — that is, nuclear fusion triggered by a concentration in gravity — they immediately take on differential electromagnetic rotation. This winds their electromagnetic field lines up around their surface, creating shorter, faster rotation around the poles than in the center. Thus, as they are gravitationally pulled toward one another and attract or create planetary debris, they orient themselves according to this planar polarity.

At the center of a spiral galaxy is a black hole. All of these black holes were once stars, and they continue to electromagnetically rotate differentially. Thus their accretion disks, which can comprise vast galaxies many light years across, orient themselves along the plane of the black hole's equator. Older black holes have another feature: they eject exactly straight electromagnetic field lines from their poles into space, and this radiation is the faster than light, gravitational superstring microwavelength of a tachyon.

3. third dimensional

Let me describe spacetime in terms of sound. Don't think of soundwaves as merely the disruption of the air particles, but as a layer in and of itself underlying them. Sound waves, like any other form of wavelength, generate resonant frequencies. Each one of these has its own geometry and these are what we call the exponential dimensions. According to the laws of physics, in the universe governed by entropy where dimensionality is compressed to the minimum possible number we see that only three basic tonalities in a harmonic accord permeate the entire universe.

Whether or not our universe is governed by the law of entropy is governed by whether or not there is restriction on the number of possible dimensions — that is, whether they are expressed as an open or closed set. Space is comprised of the three given dimensions, however time, the fourth dimension, is separate from these, and is in itself more complex, since it contains the three spatial dimensions within it. This is not a hard concept to grasp — simply remember how cavemen counted with stones: one for one, two for two, three for three and four for four, such that there would be ten stones altogether.

a. three dimensional branes

Having just described how the realms of space and time may be thought of as separate, it is important to remember that they are really one and the same. At the center of every intersection of the spatial and temporal dimensions is a fluctuation in potential towards probability — either a temporally singular virtual particle or a real particle that will ultimately end up as a gravitational singularity.

Therefore every one of these intersections is potentially two dimensional, with spacetime on one side and timespace on the other, containing space within them, with the fourth dimension of time outside. However, since each of these was a temporal or gravitational singularity, their ability to be expressed as upon a continuous surface also expands itself into what we know as height, depth, and span. Thus that which is inside the surface of the intersection of the three spatial and fourth temporal dimensions is the history of singularities flowing backward on a gradient from the more ideal temporal end of the spectrum of entropy into material form, and then dwindling into nothingness and returning again to the ideal realm.

b. six tachyonic flatline potentials

When these gravitational singularities become black holes they emit extremely straight electromagnetic field lines from their electromagnetic poles. This radiation travels faster than light, and is made up of tachyons. Since tachyons are above the limit of the speed of light, they are even more sunk into the temporal dimension of pure potential than the potential energy of electrons. Therefore it is easy enough to conceptualize of another wave of tachyons at an exact right angle to one occurring naturally in the gas jet of a black hole, and quite right to say that, in potential, it exists. Then it would be only natural to expand this to include as many more tangents at right angles as possible, and here we see that entropy has already limited us to only one, which gives us a total of three. Tachyons may not be able to provide a constant velocity any more stable than that of photons, but they are nature's straight edge rule.

c. spin occurs distributed between these fixed coordinates

Because the tachyons themselves are so very small, their measure of dimensionality can be applied even on the scale of the smallest known quantum particles, and in this way, by being mapped onto the surface of a particle, be used to plot the vector of its spin in Hilbert space, the space between quantum particles, as well as its particle spin. One of the most common of these types of vectors is the phi spiral, where the very dimensions of a particle itself are drawn by microwave gravity toward those of another.

B. Particle spin vs. force spin

There are two types of spin that a particle can have. Particle spin, or the intrinsic spin of a particle itself, can be measured in polar rotation, polar precession, and polar displacement. Each of these is assigned dimensionality for the measure of its contributing vector, however their cumulative effect can be compressed down to a vector expressed in even only one of these dimensions. The other sort of spin is projected trajectory, and is measured as a wavelength only, a two dimensional history describing a trajectory in three dimensional space compressed into a single dimension. This is the only way available for the measurement of virtual particles in states such as the electron cloud, since it is only when it is struck by a photon that the otherwise unknown dimensions of its charge and its polarity can be observed. It is also the dimension given for real particles over time, and in this way shows the link between the potential energy of virtual particles and the temporal dimension.

1. particle spin

There are no such things as real particles, as real particles are defined. All existing matter in the universe was created from energy, and will return to energy again in the end. Random quantum fluctuations in potential allowed probability through from outside the universe and filled up quantum units with the information of spin. These will carry on as long as their spin continues, and the life line of most real and virtual particles in our universe is destined to end in the gravitational singularity at the center of a black hole. Black holes will continue eating our universe until all that is left is the background tachyonic radiation. The evidence given us in our school text books also reinforces the truth of the fact that this has already begun to happen in the furthest reaches of our universe, and is sweeping in a giant wave right toward us in all directions. In fact, most of what we can see of the visible universe is the information carried to our eyes by photons emitted from very distant stars that died out long ago, and therefore it is not even safe to believe our very own eyes, for photons themselves are lies, and do not represent instantaneous information transport. The mask of photonic light worn by the universe is an illusion, concealing the fact that most of the universe is already dead. Such is the ultimate gnosis: that nothing exists, and that nothing is real.

a. vector

The most real thing that remains is mathematics, and it is therefore only because vector can be represented according to a mathematical model that modern scientists rely on its reality. The mathematical model of the vector is the matrix, a quantum mechanical, or wave mathematical, set theory. In this the number of potential dimensions in space are given perpendicular to the number of potential dimensions in time, such that they form a cross reference system or number lattice similar to a magic number square. Vector itself is represented as the transformation of the spin statistics in a diagonal across these interfacing dimensions, usually expressed in either whole number or as fractions, while the other probabilities are expressed in reflexive binary on either side of the diagonal. Matrices can also be produced using fewer spatial or temporal dimensions than the other, such as in only one spatial or one temporal dimension as a singularity.

To understand this concept it might be easy to look at the machinations of popular culture fads and trends and slower social agendas through the mass media. Since the media is not only one person, but very many persons, and not one machine, but very many machines, it does not present a picture of a single individual, but of a unified mass, working altogether for a common, ever-changing goal. Imagine, therefore, all the components in the media as if they were like pixels on a screen. Whose face seems to appear? Whose should? What shapes seem to appear? What should? This is the history of the changing of the human mind.

b. trajectory

Although there are an infinite number of possible trajectories that can occur in the inter-atomic space, a few seem to occur more regularly than others. These are the microcosmic representations of entropy, since they also arrange themselves as fractal, or infinitely self-terminating, and gnomonic, or infinitely self-replicating geometric patterns.

The fractal patterns occur at points of symmetry breaking, when the prior trajectory assumes a less ordered, more chaotic path. This occurs when energy is liberated. These patterns do not last long because the liberated energy and free particles tend to re-accumulate themselves towards one another due to microwave tachyonic gravity.

At these points of spontaneous symmetry formation, we see stable patterns propagating ordered change according to even cause and effect. Two such, negentropic patterns are the phi and pi spirals.

i. phi spirals

Phi spirals occur when symmetry is broken and immediately resets itself. This is pure dimensionality in the form of spin acting on the principle of nature abhorring a vacuum, and immediately finding pattern as soon as a previous pattern has been lost. The phi spiral pattern is taken by electrons changing energy level shells, for example. Phi spirals extrapolate out from the diagonal of cube roots.

ii. pi spirals

Pi spirals are even more prevalent and abundant in the quantum world than phi spirals. Pi spirals comprise the orbital path of the electrical charge in the electron cloud, as well as the orbital vector of all polar real particles. Pi is also the pattern that governs the rotation of our planets around the sun, and stars such as our sun in spiral galaxies around black holes. Pi spirals obey the law of inverse squares.

2. force spin

Before there were spherical, real particles of solid matter formed by raw probability there were microwave fluctuations in the subspace continuum comprised of pure potential. These generated the spinning real particles by swirling in upon themselves in a vortex that inverts itself and is filled with dimensionality by the vector information units that are the aggregate sum over histories of tachyons. So, when any real particle finally disperses, or is obliterated in the center of a black hole, it resorts itself into the ambient energy of these wavelengths.

Almost all of the known universe is comprised of electromagnetic background radiation. This includes the real photons that we see, as well as the entire rest of the otherwise invisible to the naked eye spectrum of radiation, from infrared to ultraviolet, not to mention microwave tachyonic.

The study of wave mechanics in general is therefore a rigorously mathematical discipline, since mathematics express the equations and axioms that govern the variable world of quantum uncertainties, and can describe elegantly concepts which if expressed using solid structure models would only be needlessly cumbersome and bulky. However, in the same way that the sums, variables and equations can express the quanta themselves, still geometric lattices may be employed representing such fourth dimensional holognomonic metaforms as the octaholohedron to express certain quantum relationships.

a. expansion/contraction

It is speculated that, in the same way that we can perceive the formation of alpha wave forms inside the brain through the rapid eye movement of dreaming sleep, so to our entire universe itself might be fluctuating between the moments, being obliterated into nothingness and instantaneously reassembled perpetually. This is the premise of simultaneous expansion and contraction upon which the tapestry of dimensions in the multiverse on the surface of timespace unfolds.

The ancients perceived this measure of time as the continuum's inversion between existence and nothingness, or at the very least, units of information and light. They identified the implosion of contraction with Draco, the constellation surrounding eleven of the twelve houses of the zodiac in the very northernmost of the celestial hemisphere and containing a star that was once the pole star.

This cycle was described by the Hindu as the snake of manvantara, that was turned by Vishnu, and it was commemorated in the architecture of Angkor Wat, the buildings of which are designed to align with the stars of the constellation of Draco. It is also identified in Hebrew literature as tzimzum, the auroborous, or snake that eats its own tail. The inversion of this was associated with the tao of the QBLH, and the lifting of the flaming sword from its setting in stone.

b. trajectory/transference

As the electrical charge of an electron travels through Hilbert space it oscillates its course around a series of probability wells that it forms from the raw potential. Because it only illuminates one at any point along the time measured by the wavelength, the electron cloud itself is said to ride along on only one well, however in reality it is the weaving around first one side of one probability well, then around the opposite side of the next, that produces the familiar shape of the sinusoidal wave.

c. tunneling and emission/absorption

The quantum tunneling of an electrical charge through Hilbert space, or of tachyons through timespace, is very different from the emission and absorption that occurs between real particles and virtual particles. In fact, the tunneling propagation of tachyons is unique even from that of the relatively slow electron, which only follows one wavelength. The tachyon, being a particle of a higher dimension, follows three wavelengths at once. Two opposite vectors trace around the outside of a probability well, doubling the effect of the electron. The other path, however, passes straight through the probability well of potential event, just as might an electrical charge from one well to another if a series of electrons were aligned in a line along their poles. However, whereas the electron would still follow a path outside the wells along these poles, the tachyon goes through the poles. And it does this not just in a straight line. It disappears into one side of the probability and reappears out the other instantaneously. The additional velocity is conserved by the field lines that surround the well the tachyon is tunneling through. The longer these surrounding field lines, the straighter the internal trajectory of the tachyon as it tunnels.

This is very different from the emission/absorption between real photons and virtual electrons. Photons emit tachyons and electrons orbital energy level shells are sustained as holognomonic resonances upon tachyons. Thus, there is a similarity with which photons and electrons can communicate their spin and energy to one another. The reason that the energy shell level of the electron expands when it absorbs a photon, and contracts when one is emitted from it, is that its quantity of tachyonic potential has expanded on the opposite side of the surface of spacetime from the energy of the electrical charge itself. It is for this reason that subspace matter traveling through hyperspace in a tachyonic wormhole is said to cast a subspace shadow of indeterminate probability, just like the electrical charge on the other side of the tachyonic energy.

C. spin vs. counterspin

Whenever there is any movement in the universe at all, its inversion is reflected in a higher dimension or dimensions. This is the principle of the conservation of counterspin, or reverse spin, in potential, or, on multiple higher dimensional levels. In this way the same mechanism of conservation that counterbalances entropy only serves to propagate it further in the hyperdimensional multiverse sum over histories of our universe. Therefore the law of the conservation of counterspin may as well amount to the rule of universal irony.

According to this law, we are moving forwards and backwards through time. As we move forward in time we are also moving backwards in time, through ourselves, all the time.

D. Matrices

E. Manifolds

The mathematical study behind the concept of motion, supported by the naturally observed theorems, axioms and formulae is based on the study of three separate fields, each of which can itself be read as a dimension. First there is the field of topology, which is based on the surface mapping of an object. The second is set theory, or the theory that all objects whose surfaces can be mapped can be graphed as a set of numbers. The third is matrix mathematics, the purely number field that theorizes the digital representation of the object's motions according to the set of numbers of its topological graph. Since manifolds exist in the fourth dimension it would be fair to say that a manifold is even one more further step in this long sequence of events in transformation. It is the graphing of a matrix on a theoretical surface such that the effects to the number set may be predicted for any given vector. While it can be said of the human state of consciousness as it is passed through this many in a series of transformations that it might be changed and wander, an objective measuring device such as represented by the manifold does not change between transformations of form, and establishes a fixed pattern of spin symmetry from one mode to the next.

1. sets and topology

To describe what function a manifold performs geometrically it would be right to say that it measures. It measures as numbers that can be graphed, plotted and coordinated, and it measures as a surface area which is comprised upon that number field. It does this in the ideal realm of mathematics well enough to be able to make predictions for movement to one neighborhood set on its surface relative to another when the values of the sums are known that define, delimit and determine its motion. What can be known of an idealized manifold in itself, therefore, is only that which may be perceived as upon the surface of the intervening field of math. Because math itself can be reduced in this way down to a flat dimensional brane separating the real and the ideal, it should be noted that the fourth dimensional manifold shows through by interpenetrating multiple fields of mathematics. For example, within the study of the mathematics of manifolds, known as Kirby calculus, the fields of set theory and topology are unified.

a. sets

Sets are something like number lines. The original concept of the number line was that at each equally marked unit there would be the next in the sequence of units where the center of six such number lines, the origin of a cartesian graph such as used to measure three dimensional space, was marked by zero. Imagine a three dimensional graph that has a portion of it divided against another portion, say, as in a hyperbola, such that what is on one side of it is measured as against what is on the other side of it. This is the basic theory of the set. What is in each half of the so divided graph is its own set, with its own definitions, diameters and parameters, all of which add up to figures. Therefore, a set is usually thought of as a set of these such numbers, and therefore even the two number points of any single point within a second dimensional fixed coordinate system are considered a number set.

The only difference between a set and a number line is that of depth of dimension. A number line is bound to expression in only two dimensions, establishing either up/down, right/left, or forward/backward and a distance. A set, on the other hand, can be a quantity with mass, weight, substance and volume. For example, the bagel I eat in the morning can be thought of as a separate set of matter molecules from the plate I eat it from, or the glass of water I drink with it. This seems about comparable to introducing what will be discussed in further detail later on, the shape of the torus as a 4-dimensional manifold. The torus, just like the bagel, though an idealized form, is a self-enclosed set. Another name for this is self imbedding or self-referentiality. These are fourth spatial dimensional traits equivalent to third spatial, fourth temporal self-enclosed sets. The torus, as we will see later, is an exemplary 4-manifold, a certain kind of a set that possesses many numerically expressible traits.

b. topology

As I have already explained how number sets can be derived from topological graphs in order to introduce the theory of sets, perhaps it would be best to also come backward at the idea of topology, and begin with the smallest number set, the single point of coordinate pairs. This is clearly a set of two numbers. Taking its absolute value we see that it can only have referentiality to a greater, more inclusive set such as a number line, plane or some other graphical function, and has no meaning on its own. Imagine these numbers as being representative of dimensions. Thus, the coordinate pair is also the smallest form of a matrix. While the coordinate pair's numbers refer it to placement in a numerical graph of number lines in two or three dimensions, the smallest form of matrix is binary code numbers representing the presence or absence of spin along three separate vectors each in its own dimension. So we see that the only difference is that one represents a stationary point in multidimensional space, and the other a moving spot in multidimensional space. To determine the difference between the two we must introduce topology.

Topology is simply the expansion, either arithmetically by coordinate pairs, or exponentially by spin, of a number set to derive constituent shape. There are several methods of doing this. One such, for graphing singularities in four dimensions, is the blow up method of manifolds. This consists in adding extra dimensions until the surface is hyperbolically or asymptotically curved enough to yield a measurable manifold neighborhood, or numbered subset, around a centralized singularity. A more common method is to simply further extrapolate the formulae through which the original coordinate pair or number matrix derived. It is usually the behavior of the other variables than the numbered pairs or matrices in formulaic equations that governs the full shape and motion of the object being observed, however, it is possible to expand upon just the numbers alone. The coordinate pairs can be expressed as angles in degrees or pi radians, imaginary numbers or fractions, the matrices can be of multiple vectors in multiple dimensions, posses more or less symmetry, be more or less in binary, and also can use fractions.

c. manifolds as sets with topology

The set of all sets that posses topological characteristics is called the set of all manifolds. A manifold is acting as both a number set and a topology simultaneously. While the example of the blown up singularity pertains here, there are an infinite variety of manifold types and forms. As I said earlier, also, the bagel is like a manifold. The difference between the bagel manifold and the blown up singularity manifold is that the bagel is real, while the singularity is ideal. This is not to say that the bagel manifold is real in the same way as the set of all real numbers manifold is real, or that the set of all real numbers manifold is in any way less ideal, or even less than identical, to the graph of a singularity itself. In fact, the graph in three dimensions of all real numbers would have to curl over and under to connect to itself, as would the physical dimensions if they were compressed around an actual singularity, whether that of the charged electron, the temporal torus or the gravitational singularity. This shape is also a well known manifold, and it is called the trefoil, or triple bound knot. A trefoil in one fewer dimension is a torus. A torus is a hypersphere at antipode. There are manifolds for all the regular solid polyhedra's metaforms that exist in the ideal realm of potential as well.

A set in itself, such as a matrix or coordinate pair, cannot posses spin, only describe it. A topology can spin, but without measuring out vectors on it and plotting them numerically into a system such as coordinate pairs or a matrix, the exact nature of this spin cannot be deduced. A manifold is there to bridge this gap. A manifold can spin, and by careful deduction of the formulae governing its behavior according to the laws of classical physics, the measurements of such spin can be entered in as vectors into matrices. Therefore, while it is comprised of sets and topology, it is not bounded by either one of them. Unlike them, a manifold can have measurable spin.

2. sets within manifolds

One of the main ways to know that spin is measurable on a manifold is to establish a matrix for a set within the manifold. This process involves first the establishment of a given manifold, and then the establishment of a neighborhood on the manifold. The number set for a given neighborhood is expressed as a sheet, which is defined graphically as a field and as a function in a physical formula. Once a sheet has been constructed a matrix can be formed for it based on the fourier transform of the sheet, an equation the processing of which factors in spin or motion of the sheet relative to the remainder of the manifold. Doing so produces a vector matrix for the sheet, which digitally represents its spin, and therefore, the spin of the manifold as well.

a. plane sets

Sheets are usually given only second dimensional depth, or assigned variable coordinate pairs. This expression of them, therefore, renders them flat planes on the manifold neighborhood, and their relative number sets bound to binary. This means that their spin must be expressed as a fraction, between the binary numbers one and zero, and the graph of their matrix plotted accordingly. This makes them equivalent to the electron's orbital plane between the binary dimension of it's poles, or like the flat accretion disk of the standard spiral galaxy orbiting around a central black hole.

This does not necessarily mean that the shape of the manifold to which the set is bound automatically complies to the torus shape of these two types of physical occurrences. While it can easily be seen to comply thus, the actual manifold shape can be in any form, because it is essentially describing a plastic, malleable consistency. The second dimensional sheet is only used as a matter of convenience for certain types of mathematical procedure, such as transformations that we will get into more deeply later, and is, by no means, an axiom in and of itself. Just as the dimensionality of a matrix, a number line, a function, etc. can all be expanded and contracted to fit the requirements of convenience and elegance, so to can the dimensionality of the sheet and that of the manifold.

b. vector sets

The other common kind of sheet is the vector set, where the only information comprising the sheet is made up a fourier transform formulaic equation, or a matrix. For such a sheet as this there is little additional information governing its composure or construction than what is being measured of the set in terms of its motion. It may be that the given matrix occurs within a larger equation for the determinant of the neighborhood set, or including the variable for the manifold itself. Usually the operative components in such a larger equation that are determinant of the spin of the manifold are given.

In such occasions we see the relationships arise between the different sets as one of them can be derived from another through spin. When a manifold is mapped as the topology of a number set, such as the set of all real numbers, and the element of spin is introduced, certain other number sets can be produced, such as the set of all real quaternions, etc. Just as with a second dimensional sheet set it is easy to determine spin, so with a vector set, or matrix, is it is easy to determine the effects of spin.

3. lorentz transformations on sets (manifolds in motion)

While a fourier transform can render a vector, or wave function, for any given particle or manifold, a lorentz transformation generalizes this onto the rest of the surface of the particle, or the topology set of the manifold. A lorentz transform is not harder to calculate than a fourier transform, but the two are very different from one another. A fourier transform creates a wave matrix for any given vector, representing this numerically as being relative to binary symmetry. A lorentz transformation is more like a linear equation through which an extrapolation can be factored.

a. the real number set

One type of lorentz transformation produces the set of all real numbers. In this transform a torus shape is usually used, although a trefoil could work just as well. The idea is that the ends of the number lines arc around parabolically until they all connect with one another. Doing thus is called in topology attaching a handle to a handle, and it forms a torus shape. When this is done with three handles it is a trefoil. Another name for such a kind of torus is a mobius band, or mobius strip. The Dutch artist M.C. Escher, who was famous for his depictions of optical illusions, was proficient in the use of such a mobius band, which he frequently portrayed with a half twist in the common symbolic description of infinity. A trefoil is a mobius strip with three full twists. There are many such depictions for the sets whose dimensionality expresses self referentiality, and these are classed under the heading of knots.

b. the set of quaternions

When speaking of topological knots we must begin with the set of quaternions. The plot of this graph has little, if any, definite form to begin with, and yet has strong aspects of spin. Thus it is better known what the lorentz transformation of the set is than the topology of the original manifold itself. I have described quaternions elsewhere in this exposition, however it would do well to go into them here again in brief. They can be extracted from the dark spaces of fractals when those fractals are given spin, and their appearance, thus, is only weakly symmetric, and usually is like a cloud of spun sugar that has been set in rotation. Thus, the plot of the lorentz transformation for a quaternion would only produce a globulous cloud cluster, rotating around itself, however due to the integrity of its internal composure it would be just as appropriate to refer to the manifold of its matrix as a knot or knotted.

c. other sets with symmetry

The number of manifold topologies that can have symmetrical number sets as matrices is not limited to the study of handles, tori, trefoils, or even twists in knots in general. There are many different geometrical shapes that posses an axis of symmetry around which they can be rotated with the bulk of their weight evenly distributed, and many of these with multiple such axes of symmetry. For example, the zonehedral of a cube is one such geometrical shape, and so is the tetrahedron, as is the isoceholohedron in higher dimension. There is no limit to the number of sets with symmetry that can be expressed as manifolds possessing spin. They are truly infinite.

4. 4-manifolds

One of the most useful of modern day manifolds is perhaps the 4-manifold. This has been written about extensively in the last recent half century and will undoubtedly only become more useful. The most basic types of four manifolds are the basic polyhedral metaforms. These include the hypersphere, hypercube, hypertetrahedron, and the other four regular solids in three dimensions. These also include, however, all of the other metaform polyhedra that might exist in four dimensions, as well as all the other rotational axes that each may also posses that would create alternate three dimensional lattice metaforms as three dimensional shadows of the shapes. These are only the rudimentary 4-manifolds, of which class there are an exponential amount from those which exist in less dimensions.

These sort of shapes have proved useful in many different fields. They have been used as lattices in the sciences for constructing flow-charts of relationships between variables, and they are also similar to the form and nature of buckyballs, extremely large macromolecules of carbon that are contained everywhere in nature serving various different kinds of functions, including smaller molecule carbon transport in the blood. Considering that life as we know it is carbon based, these macromolecules perform a very important function.

Beyond this there are even more, unknown four manifolds. There are as many different kinds of manifolds in four dimensional space as there are shapes and objects in three dimensional space over time. All of the natural patterns we perceive in all everyday objects are only shadows of these forms which occur in higher dimensions. Most of the examples I have used for descriptive purposes in this book have been four manifolds.

a. the torus (or hypersphere)
Primary among these is the torus. The torus is derived from the hypersphere in antipode position, from where we can see that the hypersphere appears to be a tube wrapped around in a circle and connected to itself. From apogee it appears to be a single circle, and from perigee it appears to be a sphere nested within a sphere. All of these positions describe the same shape, however each one is uniquely different as a shadow in three dimensions. The shape they are describing is a four manifold. It can be depicted two dimensionally as a diagram in which two chords wrap around each other in knots to form the outline of the shape, or in which it can be thought that two, two turn cylinder sets (handles) have been attached to each other by the intertwining and knotting of their topologies.

i. the phi/pi topology
Phi/pi is a very important concept in measuring the torus shape, because it is the description of a vector on its topology. As a point on the torus shape's surface would follow a phi spiral path as it wound down into the interior of the torus, so too would it follow a pi spiral path around in a circle out again as well as perpendicular to this as it spirals around the rotating circumference of the torus.

b. the tesseract (or hypercube)
Just as the torus had different appearances at antipode, apogee and perigee, so too does the hypercube. At apogee it appears as a regular cube. At perigee it appears

as a cube nested within another cube. At antipode it appears as a tesseract, or two cubes of the same size joined to one another at every corner by forty five degree angles. This shape is called a tesseract. While the torus shape is prevalent in nature as a recursive harmonic resonance, the hypercube is expressed repeatedly in the square and cube exponent functions of physics. Prime among these is e=mc squared.

i. the phi/pi topology

Just as with the torus, phi/pi is the measurement of a tesseract. It is the measure of the difference between the internal and the outer cube over the measure of the internal cube in the nested perigee form, and it is the measure of the cube over itself for the singular apogee form. This relativism between the sphere and the cube in higher dimensions, because it implies that, perhaps, in an even higher dimension, the two shapes are entirely unified.

c. the rest of the regular polyhedral metaforms

Lattices exist, as I have said, to depict the other fourth dimensional metaforms. These include the hypertetrahedron, hyperisocahedron, hyperoctahedron and hyperdodecahedron. Each of these would have at least one figure to represent them each at apogee, perigee and antipode. For example, the hyperoctahedron at antipode, or the three dimensional shadow as seen from above one fourth dimensional corner, is known as the hyperoctaholohedron, another name for the tesseract.

5. QBLH as manifold

The QBLH can be expressed as a manifold. Doing so means that it functions at multiple levels and in multiple dimensions. This means that it passes through different phases and different forms as the different parts of it pass through the portions in space that are governed by the different dimensional geometries. Insofar as it underlies and interpenetrates reality, the manifold of QBLH is equivalent to the will of God. There are many forms that such a complex idea can take, and they have usually been studied piecemeal by academics and mystery scholars in disparate parts of the world. As 4-manifolds are becoming more popular among mathematicians of today in expressing topologically complex histories, such as quaternions, so too can they be incorporated to the historical flow of ideas associated with QBLH. QBLH in and of itself is not quite rightly confined alone to the expression of topologically complex histories, however, it is expressed in the realm of ideal, that is, the one as ideation and the other as mathematics, such that we can see how, topologically, the two can be intertwined in a knot.

a. Dee's Enochian vectors on a cartesian matrix

The skrying of John Dee and Edward Kelly on the tyling boards of the great watchtowers of the four cardinal directions that has come down to us of this day in document form somewhat resembles the plot of spin in a number matrix. In a number matrix for multiple vectors in multiple dimensions, this is usually represented as a diagonal across from corner to corner of numbers around either side of which there is binary spin symmetry. While we cannot directly deduce spin from the gematria of letter number equivalencies in the tyling boards, we can only conclude that the skrying of the angel names by the connection of the letters resembles vectors on a cartesian coordinate system.

b. Enochian, Abremelinian, 231 gate magic number squares

For a long time magic number squares have been associated with the planets and with some of the stars and constellations. Some of the earliest calendrical astronomical observations were made by the Ethiopian Hebrew Enoch, and these were handed down through the centuries to John Dee and Edward Kelly, who further classified the sets and theories by graphing them on a letter tyling board, which is itself, much like a magic number square. Some time in between the observations of Enoch and the skrying of Dee and Kelly fell the Abremelinian magic squares, which are now largely incomplete. These gave letters, supposedly whose number equivalencies formed some sort of magic correspondence. Elsewhere I have demonstrated that the 231 gates of the sefer yetzirah, as given by Alexander Rockeach of Wormes, can also be looked at as a set of magic number squares that together form the vector matrix of a larger manifold shape.

c. the eight trigrams of the I Ching and set theory

It is probable that the king wen sequence of 64 hexagrams is meant to represent pairing in chromosomes during mitosis. There are 64 codons in each unit strand of DNA when it is wrapped around nucleosome cores to form the chromotid fibers that duplicate and split during mitosis and meiosis. The 64 hexagrams are numerically commutative with the 22 letters of the Hebrew alphabet, or 22 visible faces of a tetrahedronal band of DNA. They are also commutative with the 72 letter name of God, the shemhamforash, by the addition of the eight double hexagrams formed from the eight trigrams. The eight trigrams form the basic oscillation of a fluctuating resonance field expressed as the 64 hexagrams. As such they can be likened to a symmetrical matrix sheet for the manifold of the 64 hexagrams.

313

VII. Blackholes

Imagine you are outside the universe. All around you is utter darkness and far away in the distance is a dim light. As you approach it the light grows brighter and closer. This is the clear light of tachyons emanating from the universe. As you penetrate the surface of the fourth dimension the dim glow of the tachyons outside in the heavy black inverts to the third dimensional spacetime continuum of filaments, walls and voids of nebulae galaxies and clusters. As we approach one of the spiral galaxies we soar through the stellar mist of its arms toward the purple glow in the center. This Ultraviolet light is being pulled very rapidly into the black hole at the center, and as we pass through the event horizon we enter a tachyonic temporal singularity quantum tunnel into utter darkness. Far away in the distance is the dim tachyonic glow of the gravitational singularity.

One very all encompassing theory is that our universe is inside of another universe, even larger than our own. There are several possible facets to be considered about this theory. One would be the nature of the connection between our local universe and an outside universe, that is in particular: would it be a continuous connection, or would it be interrupted and in some way different outside from within. One particular end of this argument I like to consider is the idea that our universe is the singularity inside a larger black hole. Of course, in the same way there is no evidence nor any way to collect any for either of the former contentions, so is there no way to test the latter. This would require direct experimentation upon tachyons, and this would require they be effected by some larger particle than themselves, in the third dimension, while they are in the temporal fourth, existing as pure potential.

On the other end of the spectrum from the parental black hole is the baby universe inside one of the black holes in our own local universe. It is thought, even though they are the same size as the electrical charge of an electron when it is struck by a photon, that it contains as much information as has been taken in by the black hole surrounding it, which makes it equivalent to the smallest conceptually possible miniature universe in itself. There are probably billions of these inside black holes all over the known universe, and science predicts these are where all of the atoms comprising us now will eventually end up.

Between these two concepts at opposite ends in scope and scale, it is possible to consider the outer surface of the universe as a singularity covered in billions of singularities, inside a much larger black hole. This little model is cute as a button.

The truth, however, is less convenient. We have no way of knowing what it is really like outside our universe, aside from the fact that there are black holes in its surface and we think they go beyond. What we know of inside of this veil of the abyss is even more glum, since to all four corners the room is tyled and turned by time, which seems to be an invisible substance that we always desperately need more of than we have, and can never put a finger on without it changing.

Time is only the fourth dimension, another one of the individual exponents comprised of a sum of exponents that is included as part of the total of universal dimensional exponents. With, height, depth and length it is not alone either, according to experimental physicists. There may be a number of dimensions n, according to quantum mathematics done on computers, that define a hypothetical model of a universe similar to our own, and this number, as I have stated, is greater than infinity itself.

Therefore, if there is anything "outside" of our spacetime universe, it might be so different that we would have no way of comprehending it in such a way that we could, for example, explain it to a child or program it into a machine to compute. It might involve extrapolations upon the ideal realms of algebra and geometry just as black holes distort the more material laws of physics. If this were the case, we would have no more way of comprehending them than we do our emotions.

Long ago there was a fat, bald, dark skinned man who lived in the desert outside of Ethiopia. This man was one of recorded history's earliest wizards, and was a world traveler who had lived in Sumeria, Egypt and Israel before returning to the lands of his birth. He had the benefit of a long lifetime of magical experiences, and had mastered the learning of most of the ancient shamen by an early age. He read with rocks, he wrote with stones. He used crystal telescopes.

It was this man who has come to be associated with the anthropomorphication of the concept of a black hole's corpulence. He was the first recorded keeper of time for the masses of the people. History records many names for him, and perhaps he lived many lifetimes, or really was merely alike a great many people, however none of that is important. It was most likely he who discovered the concept of the vacuum of space as opposed to the sheet of the sky itself, and taught the people in the art of tracing lines between stars to remember them as constellations. Thus the comparison of him to a black hole is, though probably based on bias, actually quite in tune with his wisdom, knowledge, learning and skills.

However imagine this man holding a grain of sand and you will know how small the singularity is compared to the outer shell of the black hole.

In a black hole the infinitely dense singularity is housed within an outer sphere of unsurpassable gravity called the Schwarzschild radius. Where the singularity is an absolute probability one event, the surrounding, lightless, asymptotically infinite mass of the Schwarzschild radius that earns black holes their name serves as the absolute probability zero spherical well inside a boundary known as the event horizon.

If the mind is seen as a singularity, then the brain itself is the Schwarzschild probability well; and, just as inside the event horizon the laws of physics, especially the gravitational force, are bent to the extent of inversion of the uncertainty principle underlying the parity of particle and wave, so dreams, within the unconscious functioning of the brain act as a vast network of symbolic wormholes uniquely and fundamentally randomly organized for individuals, but bound by universal laws of probability to effect necessary similarity.

There are two categories structuring dream experience, which find their parity in the extra-singularity, intra-event horizon content of black holes. The first of these is governed by consumed substance and the innate characteristic patterns of organization thereof. Stimuli determined by physiological pursuit-escape functions during the waking hours accumulates along lines of behaviorally reinforced memory recurrence. So, for black holes, are the molecular, atomic, subatomic and virtual particles consumed the materials whose properties cause the patterns which arise therein. For example, an astronaut falling into a black hole does not immediately turn into a bunny rabbit, as he may if his particles penetrate a so-ordered singularity, but may be seen instead to undergo a prolonged series of dimensionally un-delimited transformations of morphology, perhaps allowing the *appearance* of a bunny rabbit, but insubstantial to the astronaut.

The second category is the more metaphysical, and deals with the mystical accessing of information in dream states, the equivalent of which in black holes is the crossing of paths of the worm holes that arise, built of the matter-energy consumed. The evidence of the revelatory nature of reverie is the foundation of all human creation, but this revelation occurs only within a preexistent physically ordered universe, rendering many of the most beautiful of human works to be merely discoveries of this fact. The experience of empathy itself derives from the sudden awareness that two lines of reasoning may share a common point. In a dream state this is alike the perception of what Jung identified as the collective unconscious, or the awareness of symbolic ideas beyond the individual's accustomed scope of a more broadly applicable potential. The reflection of this in black holes is the overlapping of multiple strings of consumed substance forming sub-singular, similar, quantum mechanically atypical events, or — if you like — event bridges between worm holes of a fixed aleph-null dimensionality. In the example given above it would be the encountering of one of these associative intersections that would cause the perception of a bunny rabbit for an astronaut who would have found time to be speeding up relative to his portable atomic clock in the gravitational distortion.

A possible hypothesis based on these similarities is that consciousness inside a black hole is worm holes, and unconsciousness entry of the singularity, or, for Schrodinger's cat, consciousness = -1 unconsciousness, and unconsciousness = pure -1 consciousness, where subconsciousness = (0 consciousness, 0 unconsciousness)

In the ten dimensional consciousness model this means that inversion occurs between the internal and the projected realms of consciousness, or between active consciousness and, by default, passive consciousness. This causes different changes to occur at different times in the functioning of the model. When the separation between active consciousness and true consciousness is as between a point on a sphere and the sphere itself, then the inversion is the second dimension. So here we see active consciousness acting as a carrier wave for true consciousness, and thus consciousness in the projective state is like the wave of probability within the formal system, and this is what surrounds us like an aura. We must also remember that what we are seeing is not unlike the implosions of active consciousness in the fifth and ninth dimensions. Therefore, when it passes through a wormhole of dream in a black hole, consciousness inverts, or turns inside out, as it enters the singularity. It then must invert back before it can emerge into a waking state again, as it would do in the mental black hole model, but not necessarily in a physical black hole.

The black hole model itself, however, holds up nicely to the dimensional mental model. The inversion that occurs for consciousness in a black hole can be directly likened to the inversion which occurs at the full extent of the functioning of the active consciousness and the lowest extent of functioning for true consciousness, through which the system can connect to itself and be cyclical. The positioning of these polarities relative to one another is indicative of the measuring of transverse waves, such as comprise consciousness, rather than regular waves, as comprise meaning. This is all thought to include itself in a basic suspension of ylem, and it is also ylem which is hypothesized to be exposed in the singularity of black holes.

A. input/output

Data passing through the surface of the event horizon inverts from below to above the speed of light. When it does this it does not simply become tachyons with flat line vectors because of its proximity to the gravitational singularity at the center of the black hole. Instead this additional gravitational force opens the tachyons up into wormholes by flattening the histories of their trajectories perpendicularly to themselves. However the tachyons still turn on their sides and tunnel through this, and therefore the ones that are not flattened into wormholes are pulled through these wormholes toward the singularity, and in this way continually divide, repeat and replace themselves. The time space which they measure, therefore, collapses into infinitely repeating halves, creating a quantum asymptote towards the clear light of the singularity.

It is only those tachyons at the electromagnetic poles of the black hole's event horizon that escape when they are warped through wormholes. The number of tachyons that escape is thought to be as many in the straight electromagnetic field lines emitted from the poles in the gas jets as are absorbed on the surface of the black hole. This does not mean this is a great many number, because there are only as many particles being consumed at one time on the surface of the black hole's event horizon as there are wormholes that form in its electromagnetic field, and these are essentially identical in cause to the sunspots that appear on the sun.

1. input

Whereas sunspots all represent electromagnetic bending outward away from the surface of the sun, the formation of wormholes on the event horizon surface of a black hole all bend inward. Inside everything is infinitely divided, as I have described, however all is perpetually passing through tachyonic wormholes toward the central singularity.

Just outside the Schwarzschild radius there are only whirling photons shifted far into the ultraviolet range of the spectrum as they are pulled as fast as possible toward the surface of the black hole. Further out than this is the accretion disk.

All matter is converted into photons before reaching the surface of the black hole. When it penetrates this surface it inverts to faster than the speed of light and becomes tachyonic wormholes.

a. the accretion disk

Surrounding the electromagnetic equator of the black hole is the accretion disk of dust and gas clouds, stellar debris and planetary matter. These can reach such sizes that they are visible for billions of light years and contain many billions and billions of stars that may shine for millions and millions of years. We know such massive clusters as flat or saucer galaxies that can be elliptical, spiral or new and yet nebulous. We live in one such spiral galaxy, and because we can see the circumference of it outlined through the arc of the heavens on a clear night like a giant river of stars in the sky, we call it the Milky Way.

The matter closer to the accretion disk orbits it faster than the matter further away, and this is what draws even the distant stars into the spiral path ever progressing towards their doom in a super massive black hole at galactic core. Why galaxies form as flat, saucer shaped circular planes is as much of an anomaly as why black holes, which consume all matter-energy, would retain an electromagnetic field to cause this to occur. The only explanation is that here we find the only necessary proof for the existence of tachyons as faster than photon electromagnetic radiation.

2. output

Indeed, we see that it is tachyons that are produced by the electromagnetic field of the black hole, in the perfectly flat history of their polar field lines, emitted out into space along the iron deposits in the gas jets of the black hole. By utilizing the technique of quantum tunneling through electron probability wells in the ionized iron, the microwave tachyons are actually traveling faster than light, and are therefore flowing outward from the poles of the black hole faster than entropy itself, and therefore measure out to a distance the inside polar axis of a galactic electromagnetic field as large as the circumference of the entire body of stars, with field lines connected to every star's electromagnetic field, guiding them in their courses and gradually pulling them in.

a. the gas jets

The gas jets of the black hole do not extend as far as the polar electromagnetic field lines project. They eventually dissipate into proto-stellar nebulae rich in iron, other heavy metals, and complex carbon compounds ripe for the spark of nuclear fusion or the gravitational seeding of both solid and gaseous planets. The only problem with them providing all the necessities for creation in the universe is that they do not condense into a greater nebula cloud or belt, and because there is an insufficient amount of them over too great a quantity of time to provide a steady enough source for the duration necessary for nuclear fusion to randomly occur and trigger star birth.

b. real tachyons (supergravity)

About the only useful things the gas jets of black holes do for the universe is replenish the background electromagnetic radiation with faster than light tachyons in the form of flat line history field lines. These are, themselves, superstrings.

The only reason that the quantum physicist researchers of superstring theory have failed to produce quantifiable results is that they have been looking for a universal phenomenon, instead of one that only appears visibly in certain rare cases. The gas jets of black holes is one such place.

Here we see the tachyons tunneling from pole to pole of electron probability wells at such a rate that their repulsive force is inverted relative to particles greater in size and moving slower than it, thus causing them to be attractive gravitationally.

This is not regular gravity such as is derived from the tachyons emitted by the photons and the electromagnetic fields of stars and planets. This is the much stronger gravity of the pure tachyon, accelerated to an asymptotically flatline wavelength.

B. components

The black hole is the son of the sun, and has many similarities to the structure of all the stars in the known universe. It has a corona of ultraviolet radiation, a photosphere of tachyons and wormholes in which the differentially rotating, coiled up electromagnetic field lines pull radiation in instead of ejecting it, and where the sun has a plasma surface on an ignited gas cloud undergoing nuclear fusion, the black hole contains infinite potential wormholes leading to a central singularity, a point of peace and clear light at the eye of a gravitational storm.

All of these are as fundamental to the black hole as are their equivalents in an existing sun. As I have already described, the wormholes on the surface of the event horizon of the black hole occur for the same reason sunspots occur on the surface of the sun — due to the winding up of the electromagnetic field through differential rotation and the band jumping between points on two separate field lines that occurs, however the wormholes of a black hole do not have counterpart pairs as do sunspots.

This is described as the weather of the black hole, or of the sun, just as the changes in pressure, temperature and condensation in the atmosphere of the earth are called its weather. Each of these forces causes the next. The pressure centers of the earth begin by concentrations of radiation on the earth deriving from the sun. The sunspot cycle of the sun is a function of its rotational position relative to galactic core. Therefore, each of these is only a kind of weather system within a larger weather system, or even, like weather patterns in this system.

This is an informational system that the ancients called the Akashic records. They believe it underlies the realm accessed by dreams, subconscious and unconscious states, as well as altered, induced or artificial states of consciousness. As I described before, it forms a massive electromagnetic torus around every galaxy containing a black hole at its core, and all of these are oriented relative to one another in the filaments, walls and voids of the greater universe according to harmonic frequencies.

It is believed that the quantum information units inside these orbs comprise the majority of the mass of the universe, and therefore determine whether it will perpetually continue expanding or eventually collapse, however this does not in itself explain how they would be interacting with one another to do so, since quantum information units would have to be being exchanged between them for them to be gravitationally drawn together. Because these enormous bubbles are made of invisible tachyons, it is most likely also invisible tachyons that are reaching out between them, in the same way that within the atmosphere of the earth it is pressure, in that of the sun it is electromagnetivity, and within a black hole that of gravity, all of which are lesser dimensional reflections of the same force and substance as tachyons.

1. the Schwarzschild radius and event horizon

The perimeter of the black hole that could be defined as its surface if it existed in any solid sense is known as the event horizon. The measure of the distance from the singularity in the center of the black hole to the event horizon is known as the Schwarzschild radius. This measure can be determined by calculating the age of the singularity and the size of the black hole, which are directly related, where the speed of spin of the black hole is a function of its size or mass, and the age of the singularity is a measure of spin determined as a function of the speed of spin of the black hole. The only practical problem with measuring the Schwarzschild radius is that we do not have the speed of the photon to act as a fixed constant within the black hole, and therefore quantum physicists predict that spacetime is warped asymptotically towards the singularity, and, as I have described, one effect is the infinitely repeating quantum halving of the tachyonic wormholes, such that any measure that could be made, even using the only possible existent constant inside the black hole, would be infinitely divided.

Therefore the equations may very well be able to provide predictions for the age of the black hole and therefore a possible measure as to how distorted outside into timespace the spacetime inside the known mass of the black hole is, but these are only conjecture that cannot be properly tested.

In truth, each of these baby universe singularities extends out into the tachyonic halo surrounding our universe to form spiral history wormholes that pull all the further tachyonic background radiation into them. These are the true time machines, for they penetrate the veil of photons upon which our picture of the universe rests, to the present, current and immediate moment, where the universe is a null void of random tachyons and super massive black holes.

a. null radiation

Because the tachyon wormholes feeding the singularity inside the black hole and the tachyon radiation and singularity wormholes outside the universe are faster than the fastest visible form of electromagnetic radiation, the real particle photon, they may be considered null, or negative energy. Not in the sense that they are negatively charged, such as the negative pole of a magnet. In the sense that, rather than give energy and contribute to entropy, they take energy out of the surrounding environment and create greater order.

The optical illusion we often witness of any repeating cycle of moving items in a loop, that they seem to begin moving backwards, is actually caused by this effect occurring for tachyons. Their wavelength is so rapid that, along a string of photons they would actually be moving in the opposite direction than the photons. Thus, when we perceive the objects moving backwards in a cycle, it is because of the microwave, tachyonic processing inside our brains of our thoughts that we seem to see these images flow together and then move faster than themselves such that their direction actually reverses.

b. manifest counterspin

In the same way that the tachyons are traveling backwards relative to the projection of photons in the continuum of entropy, they are also spinning in a way opposite to all other real and virtual particles. They involute, which not even the electron in its torus cloud shell does. The charge simply oscillates between poles in a differential orbit similar to the field of the sun, whereas the tachyon turns fully inward upon itself at one pole and warps through itself to emerge out from its own other pole instantaneously. It follows a phi/pi spiral as it does this, making it very like DNA. It follows the same pattern around other particle fields or wells at the same time that it tunnels through them, and it does all of this with spin that is opposite to that of those particle fields or wells, thus conserving their counterspin perfectly in a higher dimension.

2. quantum foam, quantum tunneling

Inside the event horizon of the black hole these tachyons are even further distorted by the presence of the rotating gravitational singularity, itself nothing more than a very old tachyon. The tachyons that are sliced and shuffled towards it into infinite halves as wormholes comprise a quantum foam very similar to that discovered by scientists to exist even in Hilbert space between the quantum information units, where it is thought to be too small for any other such particle to exist. Their experiment was between two metal plates compressed together very tightly so that there was less than an atom's space between them, and their findings were that the dimensions themselves seemed to fracture in a waveform geometry similar to the structures created by masses of bubbles. Since it was the compressed spin comprising the surface of these quantum bubbles that could be measured it was only their intersections that could be properly observed, however it was not then known what smaller type of substance could comprise these vectors, since they were far too small to be electromagnetic and still fall into the known spectrum. The only explanation I can think of is, of course, tachyons.

I have described the mechanism of tachyonic tunneling before, however I will do so again, since it can also be related to the tunneling of other particles that has hitherto remained a mystery. Tachyons do not oscillate exclusively like electrons. They have two external, opposite vectors that serve as dipolarity, just like the electromagnetic wavelength signature of an electron includes both the electrical wavelength and, perpendicularly, the magnetic wavelength. These two vectors encircle the surface of a probability well. When they reach one pole, they tunnel through to the other. Since these vectors are not isolated, as with the electrical charge of the electron in the three dimensions limited by entropy, information is moving through the tachyon at the same time as on the outside surface of the tachyon, and a smaller area is defined as the inside polar axis than that of the outside circumference. Thus the tachyon tunnels through itself faster than it moves, and thus propagates itself in perpetual motion.

Some quantum particles have been known to tunnel through others, or to exhibit tunneling type behaviors, however there is no known explanation. One such example is the photon in the double slit experiment. Even when projected one at a time at a pair of slits cut in a solid material, photons will appear through both of them, and cast light in a pattern as though coming through both slits. When particles are accelerated toward the speed of light they tend to disintegrate into their smallest possible components, which tend to end their histories when they come in contact with other particles of equivalent sizes. These particles are called antiparticles because of this quality, however their other attribute of coming into reality through essentially temporal singularity tachyonic wormholes is often overlooked.

Here we see that the information, the spin measured in vector, holding these particles together becomes increasingly inverted as it approaches the speed of light, until finally it reaches a point where it disintegrates under the existing laws of physics, and the particle breaks off into the constituent mutations. The factor holding tachyons together is that they are cohesively inverted from beginning, as wormhole, to end, as gravitational singularity.

a. wormholes and temporal singularities

Wormholes are formed on the surface of black holes and subdivide down into the central singularity, distorting spacetime in an asymptotical well that jets out like a solar flare from the surface of the universe into the dimension of pure time and the dim glow of tachyons. The precession of the poles of the black hole traces an orbit for the singularity projected out into timespace, and this trajectory forms a supermassive wormhole, feeding all the other escaping tachyonic radiation back into the universe through the black hole's poles. It is these tachyons outside the universe in timespace only that are the temporal singularities connecting wormholes inside one black hole to another, as they constitute a continuum between the asymptotical projections of the singularities out into timespace from all the different black holes in the surface of spacetime. This is known as the multiverse.

There are temporal singularities in the third dimensional universe as well. Insofar as an event constitutes a temporal singularity, that is, such as if it is observed and remembered, then the entire universe can rightly be counted as being a huge, temporal singularity comprised of infinite units of temporal singularity. There are also wormholes postulated by the special and general theories of relativity, however none have been as of yet observed. It is true, however, that when a space shuttle carrying an atomic clock is orbiting the earth, the atomic clock is moving slower than it would on earth, and therefore one can predict that this effect could be increased asymptotically to the nth degree until time stopped and one entered a temporal singularity, or wormhole, and immediately found themselves displaced in the greater continuum.

3. the singularity

Look deep within yourself and you will find the singularity. It reflects the light that you let in, and shines the brighter the more you look upon it. Some have called this inner flame the spark of life, some the ajna turned inward upon itself or the non-clinging lotus blossom floating on pure consciousness. To know it is to know thy self. To know it is to know God. It is only a single facet of our concentration, the knowledge of that which we share with the universe, however it is above and beyond all other radiant, calming, tranquil and ideal.

This entire universe is only one singularity. At the time of the big bang, the original singularity began to expand, forming our universe, which continues to expand to this day. The truth is that this singularity never expanded, and that we are nonetheless living inside of it right now. It is known that, in this universe, even the gravitational singularities inside black holes that are thought to contain baby universes are still the smallest quantum units in the entire known universe. Each one is nonetheless admitted to probably be about the same age as our universe. Therefore it is only these very small things that comprise the size of the universe, and therefore the size of the entire universe is very small, only the size itself of one singularity. This is not because its inside is small, because here we are. This only means that its outside surface area is only comprised of the gravitational singularities inside it that project outward into time as far as anything that exists. Therefore the difference between the inner area of three dimensional space, and the outer area of the fourth spatial dimension inside pure time is similar to an optical illusion, or even a play on words. Let me try to explain it this way: the further out in space you go, the less space there is around you. The less space there is around you, the smaller space itself becomes. Finally, when you are free in time, the whole of space itself is as foreign and difficult to comprehend a concept as time is inside the three dimensions of space. Here all you will find is a clear tachyonic light.

a. the singularity as surfacelessness

The fundamental geometry of the singularity is that it has no surface. If I haven't already obliterated enough beliefs, let me describe it this way. Imagine spacetime as like a stained glass window, perhaps with an intricate pattern on it or something. Time would be the tachyonic light passing through the window by which we can see the pattern on its surface. The singularity would be the surface of this light. Like the sun in this metaphor, the singularity would actually have more dimension than the depiction on the surface of the stained glass, however, since we are living in the continuum of that surface, this would remain relatively ineffable.

A more accurate description might be to say that the singularity has infinite number of surfaces, or even transfinite, or even n number of dimensions, since we have already established that the multiverse of wormholes can potentially have n dimensions. However, if the multiverse were the unified field continuum of n dimensions, then it would only equal one, and thus the singularity would only equal zero. This would be perfectly inverse to their fields of probability. Thus it equates to the spin of a gravitational singularity which occurs around it rather than upon its surface, because it is surfaceless, and comprised of the information of its spin alone.

b. the necessity of spin

As I have stated all singularities originate as temporal singularities. As such they are without manifestation: transient, warping tachyons. It is only when they accumulate excess internal spin and begin slowing down and disrupting their internal polar axis spin that their external surface begins to produce phi trajectory offshoots of lesser wavelength tachyons as it slows down toward the inversion of matter-energy at the fixed speed of light and in this way they generate centralized wells of gravity rather than unified force pull, and become black holes.

On a greater scale you could think of the black holes in the surface of spacetime as also equivalent to the sunspots on the surface of the sun, except that, like the wormholes on their surfaces, they do not come in pairs as do sunspots. They do, however, lead outside the universe inverse to the way solar flares and prominences eject material away from the surface of the sun. Their placement in the universe conveys temporary evidence of a much greater and more aethyreal pattern of force at work. This greater fourth dimensional holognomonic metaform would be equivalent to the differential rotation of time.

c. the primary clear light

The singularity projected out through the multiverse of tachyons into pure time opens up at the end into the clear light of the tachyons, emitting more that flow backward, and in this way pulling more in, and projecting them from the black holes electromagnetic poles. It is an invisible brightness within a complete clarity. Such is the external surface of timespace. It is the motion of these that is called time. Together they perform an effect similar to that of the superstring filaments and walls in the universal voids, the wormholes on black holes, the sunspots on stars, or pressure centers on planets. Such is the weather in the dimension of pure information: always clear with highlights of ultra-luminosity. Laugh it up, Buddha.

d. the singularity as ylem

Ylem existed before the universe, and will exist after it has disappeared. If we are a dream in the mind of Brahma that Kali destroys and that Shiva restores, then ylem is the stuff that dreams are made of and the soul, mind and brain of God. For such as are the dimensions beyond those of the sixth and those of the seventh. Still all are one in the clear light of the singularity. It is as if no darkness could be imagined in the multiverse. Still the heaven of man is yet like the depths of hell to God. Not even is it to Him like the earth is home to man, for to descend into it he must lower himself a long way. It is literally as much a fog or dream to Him as would be for we the living tiredness or sharp pain in a dream. I say this because I can imagine a Heaven governed entirely by the will and law of the most perfect thoughts, ideas and feelings of every living being, and yet I can still imagine a God that can imagine more than even all of this.

Therefore, outside of the dimension of pure information, there is a darkness of which absolutely nothing is known. It is here, beyond even the light of tachyons and the wormhole histories of singularities that is the multiverse surface of timespace on the other side of spacetime, that we can say that YHVH exists. I have said that the movement of wormholes on the surface of timespace or black holes in spacetime constitutes the pattern of time, whose movement is time itself. This is an effect occurring upon the body of God. God is, beyond this, ineffable, as though our universe were his heart and he were only his perception of it. However I would prefer not to allude to such parables or anthropomorphications.

4. perpendicularity to the history of the singularity

While the surface of our universe might be a seething, writhing cauldron of bubbling baby universes burning off the tachyonic glow of only one singularity, if we depart the history of any one of the gravitational singularities inside this universe, we depart the outside surface of this universe, and therefore find time to know God. Still, this is no good, for we are merely without the universe. We are back to the obese old Ethiopian shaman holding a grain of sand in his hand.

a. the "parent" universe

It has been my personal experience that it was the parent universe that induced the motion of spin in the local universe that has led to the production of baby universes. This occurred due to a function similar to the parent universe slowing down and solidifying while the local universe sped up and became more aethyreal. This would have been happening throughout the part of the history of the universe during which the stars of most of the visible galaxies were burning out, which would have happened long before we would stop seeing their light. As the universe was swallowed up into lighter tachyonic radiation by the super massive black holes, the darkness outside the tachyonic multiverse surface of the universal singularity began to compress itself more around the little universe, which was eating itself down smaller and smaller while inside of it, through the inversion of the speed of photons, it appeared, and continues to appear, to be expanding. This effect is no more an illusion than time itself, for such as are the same: the movement of the singularity, its gravitational spin.

The Egyptian civilization, which marked the end of the age of the father and the beginning of that of the crowned and conquering child, had the most entertaining and spectacular anthropomorphications for these levels and events. The man that I described before played an integral role in the creation of these mythologies, as well as those of Sumer and Israel. His resurrection as his own son would rise in the east to become Buddha, whose philosophies would reach as far as to Jesus and Mohammed. Perhaps he would be explained better by saying that he only appeared to be aging backward in time because he was from the future and traveling backwards through it into the past.

My description of this man has surpassed the limitations of a human, so allow me to clarify myself. Choronzon is a body-jumping archetype who manifests the genetic resemblances of beings on this planet in an accordance with their function in the grand design of the universe. He is not a he, and she is not a she, because there is no he or she or it. Choronzon has no body, and is merely the charge of ophanic energy in the earth's electromagnetic field. He is the one who guards it, and keeps secret the keys to the kingdom of access to the electromagnetic field of the sun, of other planets, of other stars, of galactic core, or other galaxies and of the filaments, walls and voids. Just as Gaia is the soul of the planet, Choronzon is its spirit.

When I describe first one man, and then another, and say that they are the same man, or say that one represents the son of the other, or say that this man has "aged backwards" or "gotten younger," I mean that the archetype of Choronzon was upon them, and was known of to them, and was alike them and that they were like better reflections of it, and that it was this archetype itself that had "aged backward" by becoming known to these men at earlier ages in their lives.

When one achieves Choronzon consciousness, also known as Christ consciousness, one feels the scrutinizing consciousnesses of all the surrounding souls of the planet, and it is even possible to speak in an angelic tongue with the voice of a person by moving with the consciousness as it transcends the physical limitations of the body and flies across the international channels of the em field. I have found it makes a useful hearing aid for eaves dropping on other people's nonsensical conversations, and that, as Buddhism predicted, there is amusement in this. However the majority of souls are as much trapped in time and space as their attention spans allow and conditioning demands. While this is not inherently an unpleasant experience it has had a petro rites based dogma attached to it by every religion and state society on the surface of the planet, such that knowledge of it is forced to be made equivalent in thought to karma, and repented for and atoned for by the ritualizing of its exercise in perpetuity.

The trance of nibana, in which one enters a nirvana from which there is no return, as well as the alchemical symbolism of Christian Rozencreutz both describe further the nature of following in the path of the cosmic fool, the Microposopus, represented by the constellation of Orion. This attainment of consciousness, usually called transcendence or enlightenment, is dated by the mythology back to as early as the fall of man and original sin, but to have only become a formalized system much later, culminating in the belief in the past messiah.

Choronzon is seen as the Holy Ghost, or spirit of Jesus, that stays with this planet and lives with and within its faithful. With God, the invisible All Father, and Jesus the living archetypal son, to initiate Choronzon to this status, according to the dialectic of Holy Scripture and the Lineage of the Ancient of Days, Choronzon is the microcosm of Gaia, the fruit of the womb of mother nature, the pearl of great price. When her age passed at the end of the last ice age, the age of the father began, and so on into the age of the son has the She Goddess continued only in spirit, in essence. Jesus the Body, Choronzon the aura, and YHVH the anthropomorphic abyss.

Neither am I impressed with the offer of a furtherance of this same Enochian system of a hierarchy of guardian angels that extends throughout the universe. I would prefer all were ylem, and still the only, old greatest common factor, and new least common denominator, I can hope to be remembered for is the geometry of temporal QBLH.

C. the black hole model of consciousness

While the crown of thorns may have represented the pattern of ley-lines which ancient geomancers had laid upon the surface of the earth to mark off karmic grid lines, a crown better fit for a king is the black hole model of consciousness. It shows us how the mind may be used as an exit from the confines of reality. This would make it actually more similar to the opium soaked rag given to Christ on the cross to put him into a false state of death. While one is active and expanding, the other is passive and contracting.

1. the mind expands (evolution)

We all know the old saying about the more we learn, the more we know, and the more we know, the broader the mind is expanded. This is true for the brain, but the mind seems to be pushed and pulled and jostled about in all directions generally, and yet remain calm, centered and at rest within itself. So let us, instead of giving credence to the saying, examine how the brain itself has been expanding, and only then when we have done this might we know the mind as apart from it.

It is obvious to say that between an insect and a human there is a difference in size between bodies, let alone brains, and the insect is the older species and humanity the younger. Thus we could say that our brains had expanded quite a bit, however bigger brains aren't necessarily better, as evidenced with the dinosaurs, some of whose brains were about as big as a sumo wrestler, and about as plodding and single minded too. Humans have the benefit of a well developed cerebellum, which acts inside the human nervous system in much the same way as pressure centers in the earth's atmosphere, the sunspots in the sun's electromagnetic field, the gravitational wormholes of black holes, and the movements of singularities in timespace causing the spin of the universal singularity, also known as the point of the big bang, or the inversion between the original nothingness and the present somethingness.

This always hung over the dinosaurs heads. They thought with their thalami, and were most fixedly intent on finding food, foraging, scavenging, preying on the weak and the lame. Their best skill was simultaneity, today known as manifestation or synchronicity, and, still popularly, as coincidence. This consists merely in stalking, sneaking up on or happening upon something, such as a source of food. They practised this on this planet for millions of years, walked the lands right where we are today. The part of our brains that is similar to theirs retains the inherent capacity for this skill, and our ability to project inference is a large part of religion.

However the columns and pillars of neuro-cellular grey matter of the cerebrum or cerebellum are far more complex than the holographic reflexive tissue of the thalami or thalamus. Here is where tachyons occur inside the brain, producing cascades of neuralectric waveforms, thoughts and emotions. On the other hand, producing cascades of the electrical current carrier neurotransmitter chemicals does not automatically induce inspiration, only gives for it a fluid basis in which to arise potentially and through which to be propagated and subsist. Inspiration is that which comes from out of the blue, that is, from the ideal of free time.

Also inspiration comes from the right hemisphere, for it is inductive, as opposed to the functions of the left hemisphere, which are exclusively deductive. It is still possible for inspiration itself to become a distraction. It is represented as the third eye in many ancient cultures, including the Indian and Egyptian, where it can be turned either inward or outward. This polarity of inspiration/distraction is equivalent to active consciousness. Active consciousness expands.

2. the mind projects (quantum thought)

There are many theories that have developed since the time of Einstein's relativity equations attempting to unify the realms of mind and matter.

Studies during the twentieth century by Carl Gustav Jung and Wolfgang Pauli proposed that there may be two connecting principles acting in nature. One is causality, or the cause and effect principle, which operates in the fourth dimension of time. The other connecting principle is acausal, and they called it synchronicity, believing it occurred at right angles in space time to the other principle.

More recent theories have proposed that it is actually information itself which is being transferred by both consciousness and material reality. The evidence is the increasing of order in the results of tests done on subjects for esp when they were randomly resampled. Since information is the negative reciprocal of entropy it would even go so far as to account for the potential for telekinesis, let alone John Bell's theorem that particles that were once in contact continue to influence each other instantaneously no matter how far apart they move, which would only require a meager acceptance of a faster than light mechanism for the transferal of energy.

Experiments with remote viewing or long distance telepathy show that there is a rise in right on target hits at the local sidereal times when the plane of the earth is facing perpendicular to the center of the Milky Way galaxy down the outer arm, around two in the afternoon, and that it drops immediately to its lowest point when the earth is aligned with the galaxy's core.

To describe the human ability of precognition, Jack Sarfatti uses the equation: $h = (e^2/hc) (mc^2 / H N^2)$, where h is Planck's constant, e is the electric charge of the electron, c is the speed of light, e^2/hc is the Fine Structure Constant, m is the electron mass, H is the Hubble factor $1/T$ where T is the age of our universe, and N is a number that will turn out to be the number of electrons in a coherent state of conscious superposition. Since a de-coherence event effecting any one of the N coherent particles will de-cohere all of them, it is reasonable that Tu is inversely proportional to N. This justifies the formula used by Jack Sarfatti, $Tu = 1 / H N$, for which at present, $1/H = 10^{10}$ years, which is 10 billion years, at which time $N = 10^{18}$, so that , to rough order of magnitude $Tu = 1 / H N = 3 \times 10^{17}$ sec $/ 10^{18} = 0.3$ sec, which, as Jack Sarfatti notes, is close to the Crick brain frequency of 40 Hz based on his model in which $Tgrw = 1/H = 4 \times 10^{17}$ seconds.

Tony Smith uses the same formulaic approach to extrapolate the potential for a conscious universe that can be known before its manifestations occur. He gives the long form of his equation as: $N^2 = (e^2/hc) (mc^2 / H h) = (e^2/hc) (mc^2 / H$ Epl Tpl$) = (e^2/hc) [(mc^2/Epl) ((1/H)/Tpl)]$, where $(1/H)/Tpl = $ Age of universe as a number of Planck times $= 10^{(7+10+43)} = 10^{60}$ Planck times; $(mc^2/Epl) = $ Electron mass/energy as fraction of Planck energy $= 10^{(-22)}$; $(e^2/hc) = $ fine structure constant $= 10^{(-2)} = $ square of amplitude to emit photon probability to absorb a "linking photon." The way he differs from Sarfatti is to take as fundamental, not $(1/H)$, but GRW de-coherence. The total time for which N coherent particles can maintain superposition is $Tgrw/N$, so he uses $Tu = Tgrw / N$ and gets for the present time $Tu = 3 \times 10^{16} / 10^{18} = 0.03$ sec, which is shorter than the figure of Jack Sarfatti, but even closer to the Crick brain frequency of 40 Hz, so there may be resonant tuning of human consciousness with that of our present-day universe.

The ancients said that awareness of all these types of things in all these types of ways was being in kether. This is what I call the heaven of man that is hell to God.

3. the mind is and is not (a self aware conundrum)

It has been proposed by some modern logicians that the root center of human consciousness is an impossible loop that is perpetually feeding information back on itself. This has been compared by Douglas Hofstadter to the saying of Xeno of Elea for the two step impossible loop, "the following statement is true; the preceding statement is false," and the one step impossible loop, "this statement is false," which derived from an earlier Cretan saying, "all Cretans are liars; I am a Cretan."

Due to this fundamental foible of consciousness, information inverts when it enters the brain and it inverts again when it enters the mind. This is just like the flipping of the image by the lenses of the eye and its reordering in the visual cortex of the brain. It is typical of the level of redundancy with which we seem to cope. It is fair of the Jainists to believe our universe is being digested inside the many stomachs within the belly of a cosmic cow.

The knowledge of this two-sidedness of ontology is fundamental to taoist zen Buddhism as well as to latter day western existentialism. Where existentialism gropes and moans over the quandaries and quagmires of the substance and motives of being, the eastern schools' array of meditative trances were designed to quell all such concerns and liberate the mind into the pure wonder of nothingness. Which is the dark and perverted, which the enlightened path we exist to determine and to decide, however we must also remember that all things are different from person to person.

The best euphemism for the human condition that I've seen was actually given in a Hollywood movie called City Slickers. In it Jack Palance's character, Curly, an old gristled cattle driver, explains to Billy Crystal's character, a New York jew, that only one thing is important in everybody's life, and that it's what that one thing is that everybody is trying to figure out. This is the old yet still effective strategy of giving life the finger.

Existentialism and Buddhism had gotten it pretty well figured out as upon a gradiated scale between the mind and the body, until Hollywood came along. Now, philosophers such as Noam Chomsky and Jean Baudrillard debate the imposition of the representational media on the minds of its consumers, and Hakim Bey proposes a counterculture of immediatism in response. The question of what we are told in and about the information age we can believe becomes needlessly complexified ontologically by hackers and corporate megaloconglomerations seeking to establish the terrain parameters of virtual reality, and so when we ask, "what is real?" the question echoes in our mechanical ears and comes back in our artificial dreams to haunt us. The difference between an archetype and a stereotype is that an archetype is an archetype because they chose to be an archetype, whereas generic stereotypes are manufactured waste products of the pop culture simulacrum.

Although the unwise fools are alarmed by the precedence the media have taken in recent times to such an extent that they fear for their churches and religious icons, these sort of listening/watching devices only represent an increase in the interest in electromagnetic fields and the return of my old perennial pals the sunspots of shemhamforash.

I believe as do the eastern schools of tantra, bahkti and karma yoga instruct: that we, as free spirits, chose to incarnate into this existence each with an individual purpose, and that the reason we seek meaning is life is to better understand this purpose, and that we have meaning in our life only to the degree that we have such a purpose. Nor do I believe we owe anyone at all any explanation whatever of what that purpose is or might be. I believe this purpose is God's gift to the individual, and that free will is our guide when our eyes are open and morality is our guide when our eyes are shut. Both will lead us to the complete satisfactory fruition of our goal.

However virtual reality is only a simulacrum. The more manifest reality resembles the flexible parameters of virtual reality, the more it will be a simulacrum. This does not actually apply to alot more than we think. Afterall, most of what the media have been used for is tactical video games practising strategies to advance agendas. These manipulations of the representational system reflect popular viewpoints, and therefore walk about like archetypes, manifesting synchronous stereotypes in their artificial auras induced by the views of the aggregate consumer basis of the populous in their wakes, thus creating artificial karma which can be followed independently of any of the players or even of the roles they play. This amounts to social consciousness or civic spirit, however it remains entirely a construct of a representational system that constitutes safe pockets for the intersection of fact and fiction that are equivalent to a voting system based on a sort of karma that is, whether fact or fiction, in itself false in origin, since its delivery mechanism is a machine.

Let me specify again the difference between living gnomons and mechanical fractals. Imagine you were stacking alot of ice cream cones. You would find that one fit into the other, and that thus they take up less space packaging. This is like gnomonic replication. On the other hand, imagine you are stacking aluminum cans. These are cylindrical shaped, and all the same size as one another. Therefore none fit into any others and all must be stacked separately on top of one another. This is like fractal replication. Nice, crunchy whole wheat grains of the ice cream waffle cone. Cold, hard metal wrapped in a paper pop art logo of the anonymous aluminum can. See what I mean? The strange thing is that, while the cells of the brain are stacked in columns like aluminum cans, thoughts propagate one right on top of the other like so many ice cream cones. In this way also, the mind is and is not.

4. the mind is a singularity (a force unto itself)
Nothing permeates the mind. It is nothing that separates the mind from the highest knowledge, which, when it is perceived, is like nothing. The light of enlightenment is clear. It permeates the universe. It permeates our selves. It permeates our cells. It permeates our souls. It is our free spirit. It is our free will. It is impossible for these things to be ours. Nothing is impossible when nothing needs

done. All is a clear light inside of a darkness. We will all eventually let go. Time is the flow. Inwards and outwards until never it goes. The light that is what it is, is all there is, and it is the information of the spin of time.

The mind is the singularity inside the singularity. It forms when tachyons leap into existence in our brains, generating a pi spiral probability well electrical charge out of the phi spiral path of a temporal singularity. It exists for a moment, flash fades, and then goes. It arises like a voice pulled out of the nowhere to speak our minds to itself, with more or less passion, and then go on about whatever was its business in a relatively dreamlike way.

The easterners believe that the mind is more or less like the many different animals, and in various ways can therefore be put to work tilling over the great mill of the questions of reality, or even domesticated and made profoundly docile. In the west it has been popular to equate the emotions with the weather, though only for the sake of those frightened by both. As I have described, the attainment of the fully functioning atman awareness, also known as Christ consciousness or Choronzon in the west and simply as Buddhahood, enlightenment, nirvana, Samadhi, etc. in the east are associated with the eight circuit ennegram only to a degree determined by the free will. In the west interest in the will equates kundalini with a lightning strike. In the east interest in the chakras equates divine inspiration with temptation.

This is probably a sign of lunacy and madness, however I don't believe in such things. It seems more likely to me that we are all the dreaming dreams themselves, dreams of the mind of God, in the mind of a dreaming God.

D. consciousness entering a black hole

This has always been a favorite of publishing quantum theorists, though for what reason I do not know. Take this as purely a thought experiment, since, of course, nowhere else but on the media screen has this ever happened (and even then the plot was really lousy). Imagine an astronaut in a space craft orbiting a black hole who decides to go out and take a walk around it. He (or she) gets out of the craft and floats over towards it, let's say, with a clock. They toss the clock in and watch as it slowly winds down and stops as it is turned into spaghetti inside the black hole. This part the scientists are always sure to include: the same happens to the astronaut if they get too close to the black hole! Scary scientists. Trips are for kids.

Now come out of your pleasant little reverie and instead of imagining lugging your cumbersome physical body around, just imagine your soul doing the job by projecting the electromagnetic aura of your astral body, or, if you are too lazy even for an out of body experience, let your geometric spirit or holy guardian angel take care of it for you. 23 skidoo.

Imagine yourself a projected singularity, transported instantaneously to before a black hole. The sense of second sight alone is sufficient to allow you to view it in detail, however creating a pictorial replica on a computer is equally easy. In this way it is possible to enter into it without imagining the ridiculous consequences to your physical body, since they are karmically irrelevant to one another.

1. inside the event horizon

The only effects that I can describe are those that occur upon the mind, and as I have already said I do not believe in any form of madness. The mind may dissociate while the person is in altered states, however it, along with the six fundamentals of reasoning and the free will which are always there, locked within, trapped with the body, inside the brain, until there is no more neuralectric activity in the brain of that person. The spirit is only pained when it is tried to be tamed. The soul is always free to go whichever way it wants to go. Our karma is bound to the garden of forking paths, but our self is the singularity of tao. We may literally see all of this before our eye inside the tachyonic wormholes of a black hole. This is equivalent to the sensation of a near death experience, where we might see our life flash before our eyes and wonder, "what did it all mean?" and "was it all worth it?"

The only baggage we bring into a black hole is the fourth dimensional emotional metaforms of the karmic histories of the spin of our relative form, and since we are using as our reflection a temporal singularity itself, we may choose to see the passage as it would, rather than seeing what we think of ourselves reflected. Here we see that the perfectly straight rays of tachyons would be swiss cheesed by a nebulous ambiance somewhat like cheap Hollywood smoke and mirrors.

This is not exactly like a laser light show, nor is it necessarily like the gate to Heaven. Like everything else that exists in the universe it simply is what it is. Let me describe it a little further before it becomes too "glamorized."

a. the waking dream

We move about through situations that imitate our inner mentalities and ideations. Because we are accustomed to doing this, our alpha wave sleeping state gives us an ego in our imagination that reflects this attribute to an even greater degree. Sometimes I am forced to wonder: who is the real me — the me I am in my dreams, or the me that I am while awake?

This is what life is like in the wormholes. Each wormhole leads to a different output pole of a black hole somewhere else in the universe. As you are pulled into the event horizon of a black hole you will be sliced into infinite halves, each of which warping through a series of all the black holes throughout the universe, however your view will not change of a single, distant clear light in the invisibly undulating darkness. Time becomes discontinuous, and the ego and the singularity gradually identify until there is an inversion.

The inversion for consciousness inside a black hole is not waking up, as the dream like state of the wormholes might make it appear. Instead the inversion would be the equivalent of going to sleep in a dream.

b. effects of the distortion of time

As anyone who has ever gotten out of prison will tell you, "things aren't that different over here from over there." This is about equivalent to the effects on the psychology caused by the distortion of time. That is, it is only what you have seen of the universe that can lower yourself, for there is nothing more beyond. However entry into the multiverse, otherwise known as the heaven of man and the home of the angels, is only one more type of condemnation to a life blown by the winds of chance, destiny, faith and karma, and cold comfort to Kurt Cobain's corpse.

Entering into the multiverse of intra-blackhole wormholes does not give one control over time, however it does remove all concept of time. Thus, by liberating the soul from it, time is also being removed from its hands altogether. This is, according to legend, what differentiates between angels and the ascended dead, that is — the ability to continue to do good deeds towards the earth by reentering it through the hypercube of time, or QBLH.

2. entering the singularity

When one enters a sleep-like state inside the dreamlike multiverse of tachyonic wormholes, one's mind becomes lucid, and the will a clear light. If one is actually asleep and dreaming, the dream will answer more to the wishes of the dream self or dream ego. If you are imagining this while you are awake, such as perhaps, right now, then you should see the entire history of a baby universe contained in the spin of the gravitational singularity.

Here you are back again, in the form of a stranger. This experience of meeting one's own higher power is one that should be even more highly recommended. For here is one of the elder tachyons, whose story is long and cyclical, and to say that this is not YHVH, or a god of equivalent ontological concept, is to throw a stone at the stained glass window of spacetime, for here is the very fire itself that burned and was not consumed, in the same way Brown's gas (a dioxy compound) can be used to generate extreme degrees of temperature entirely self contained in a controlled flame such as the acetylene torches used for welding that currently run off propane. Afterall, all singularities are really one and the same singularity, only with a different face.

This enlightened state of consciousness is equivalent to perceiving yourself as a temporal singularity, the electromagnetic aura of your astral soul, as a geometric spirit comprised of luminous tachyons, which you have been doing all along, for this thought experiment. Remember, though that you are not God. You are just a portrait of God. As the Dalai Lama said, "like a reflection of the moon on water." You are a shadow. You are an echo.

a. the equation of sleep unlocks the door of death

"To sleep, perchance to dream," said Hamlet, "aye, there's the rub; for in death we know not what dreams may come." We may visit our own graves as often as we want to in the end, but we will still be living. So these black holes are, for everything that exists now will eventually end up in them. They are the end of all history that record in fragments, the great cnoptic jars of the pharoahnic mummies, or the earthen wear jars in which were kept the dead sea scrolls. The singularity is the dreamless sleep, for to stand at the central gravitational pull of a black hole and close one's eyes to the dream of the multiverse is to awaken anti-consciousness, or to see through the eye of otherness itself. What you see will be yourself, as seen through your own eyes. This is not a dream. It is the self you see.

Otherwise, following along the wall into which the door of this final inversion cuts one finds themselves out past the breath of the name of God and outside His

Holy altar. This is what you could call the throne of God, for it is only the back parts of that yet beyond us, and we are told that it is this part of God that we are made to see. Here we are neither dreaming, nor sleeping, for we have left the wormholes and the singularity behind. Here there is nothing. This is where the lonely QBLHists squat, eating Sumerian dust, spinning out their little spirals of gravitational black holes and weaving the multiverse of time into the fabric of the universe.

The QBLH is merely equivalent to a regular ennegram in a living brain.

VII.1. Wormholes

Just as our three dimensional continuum is the surface of spacetime, so are the tachyonic wormholes of fourth dimensional potential the surface of timespace. We see that there are three sizes of wormholes: the greater size that is a phi/pi pattern traced out in the tachyonic multiverse by the baby universe histories of gravitational singularities stretched out in asymptotes from the black holes in spacetime; next the medium size, within and upon the event horizon of these black holes, connecting the universal histories of alternate dimensions distorted geometrically by the gravitational singularity; smallest are the temporal singularities of superstring gravitational microwave tachyons which comprise the subspace shadows of the quantum foam of potential probability wells that unifies the multiple dimensional histories of the universe.

To say that all of these are the multiverse is to miss the point of what is the multiverse. It is only a series of wormhole links between tachyonic geometrical archetypal concentrations of karma permeating the universe forming a unified field for all of historical spin over time. The multiverse is the gravitational rotation of the universal singularity. This creates tachyons that emanate in an aura around the universe, and are immediately pulled back in by the spiral histories of the gravitational singularities in the centers of black holes where they are either reabsorbed through wormholes at right temporal angles or are re-projected as tachyons through their electromagnetic poles at right spatial angles. The wormholes that move perpendicular to the history of the universe that are the geometrically fractured multidimensional bubble universes on the inner surface of timespace cause the addition of information into the third dimensional continuum observed as universal inflation. This is the aspect of the multiverse in which the microwave gravitational electromagnetic links between the fields of galaxies occur. Therefore this is the most active aspect of the multiverse, and equivalent to the main sequence of the lives of stars for the history of tachyons.

This does not mean that the greater and lesser forms of wormholes comprising the unified tachyon field of time or the multiverse are in any way less wormholes in and of themselves, or play in any way a less active part in the multiverse. Consider the similarity of the phi/pi histories of gravitational singularities projecting out into the tachyonic multiverse outside of the spectrum of photonic light of spacetime and their similarity to the torus shaped pressure centers that move around upon the surface of earth following the convection currents. If we were so inclined, we could even find similarities in the work of our own physical bodies. All is in all.

A. inside out = through

It has been said that "the only way out is through," in reference to initiation. This is, of course, a trap. It is easy enough to walk through it blindfolded if one uses the force of insight to guide them, seeking only wisdom, and being open to the truth.

One perspective on ancient mystery cults is that they know such truths as one cannot imagine, having taken millennia to make all the right moves. What they claim to represent is QBLH itself, a vector of spin throughout the multiversal history of the universal singularity in the form of brotherhood.

All this means for an individual is being studious and focused on what you are paying attention to, which makes the new age look like a paperback novel about zen and the art of motor cycle maintenance. To say that there may be something more in what is is like saying that some particular agendas might actually be more important than others. Therefore what more to it can there be without the boat rocking?

A better way to look at it is geometrically, examining the phi/pi spiral as a ubiquitous gnomon.

1. phi/pi
Imagine a sphere with two poles. Now imagine the surface of the imaginary sphere bending inward around the poles until the two tips touch at the center. This is a torus. If you drew a spiral that followed its surface area it would be a pi spiral outside and a phi spiral inside. This is actually due to the fact that you can trace a phi spiral around the vortex at either pole and the inside surface is made of a circular spiral such as a telephone chord, however each of these is a measure of the opposite surface, and therefore they are expressed as phi/pi, which, in itself, looks like a written inversion, but means phi without and pi within as the measurement itself and implies the opposite surfaces.

To say that this torus shape, whose surface can be traced as a phi/pi spiral, is ubiquitous means two things: one particular kind of its particles, the smallest, occupies and comprises all forms of all things, while also, the phi/pi spiral itself occurs on many further different levels and scales of particles, wells, and fields, and behaviors on larger, more complexly structured dimensional levels and scales. The former it does in the form of tachyons. The latter because it is a gnomon.

2. the torus
The torus is the shape of the field of which phi/pi is the wavelength measure. This is the shape taken by a wormhole between two points in spacetime through timespace. Ironically, this bears striking similarities to the world in which people are always trying to get into each others' heads to greater or lesser extents and degrees, and this is called the innateness in humanity for extra sensory perception. This is represented in the pop culture simulacrum that comprises the weather of the mediated information system. If extrasensory perception, timespace wormholes, and the stock market all shared the same torus trait, then it would be possible to prove that they were related gnomonically.

B. infinite potential
Accessing the geometrically distorted n dimensionality of the multiverse is tapping into infinite potential. From this is derived infinite potential energy, infinite potential information, infinite potential referentials, infinite potential light, love and life. However, since the multiverse is accessed through timespace distortions, and

these are the obliteration by inversion of karma that occur as spiral histories, then we may as well be left in the Himalayas to meditate why.

1. energy

It was Nikola Tesla who developed alternating current, which is the inversion between magnetic polarities of an electrical wavelength. Imagine this inversion to the nth degree, expressed temporally, until there is neither pole of light nor dark, and all is clear. This is infinite potential energy. It is all around us, waiting only to be liberated. Gravity is free energy. Gravity is the entropic effect on free and bound particles of attraction by faster than entropy light called tachyons.

These can be used to create wormholes in space through temporal singularities. Tachyons emanate from the surface of our planet all the time, however during the day there are also those of the photons of the sun. At night they project out at right angles to the horizon into the flattened electromagnetic field lines in earth's plasma field.

2. dimension

Pure dimension is the liberation of the mind and will, such that the mind is like a ship, anchored in its surroundings only by the free will, capable of doing all in its scope that it is capable to wish. The history of its free will is memory, the unconscious is the depth of the sea, the subconscious its surface, and consciousness the air the free soul breathes. This is all merely occurring within the flat land of the three dimension continuum, and emotions permeate all of this like metaform shapes floating in the free time of the fourth dimension. This is the cycling of karma, through the dimensions, such that memories can be triggered by reminders, and the orientations of beings are continuous, despite general relativity.

Morality is one hundred and eighty degrees from immorality, sickness in denial of itself. Now which is sick and which is denial is irrelevant. One will lead you one direction, and one will lead you another. Asking questions you can find your way around to answers. What is unknown and what is known are inversions until their essence is understood. As long as one clings to what is known, one cannot comprehend the unknown. What you choose to do is only a matter of time.

Geometry is native to dimension in the same way that phi/pi traces the surface area of the torus. It is the measure of dimension, and without it there would be no vector, and thus no such thing as spin. Geometry is based on links. It measures the connections between singularities, or points, in the continuum of quantum information units. It always obeys algebraic axioms that can be calculated and understood, no matter how many distortions and permutations thereof it goes through. These occur in the vicinity of spacetime due to gravity. Further out, it is thought, the same concepts and rules apply in the dimension of pure time due to the greater wormholes on the surface of timespace, but the exact measures are unknown, lost or forgotten.

a. the multiverse

The manifestation of the multiverse is probably the eschaton most people are expecting. This is a change of perception that evens out the expansion and contraction of our wills and the influx and co-creation of reality with our mind. Once one has achieved such a state of manifestation, one can manifest themselves historical tunnel realities through timespace. The only difference between hell, earth and the heaven of man is degree of freedom. Of course, since all the angels are said to work for God, their will is bound, however the spirits of the dead are believed to roam about in heaven freely.

The eschaton is the heaven of man on earth, and the coming of the kingdom of God to the people. This is a slowly cycling process that occurs over time. People have been building up the ruins since the last eschaton, and many have even come to believe that this was like participating in its immanentization, or bringing about, and though they have lived in luxury they are made to wait like all the rest.

Some cults believe that such perception can be induced, and it is true that there is plenty of technology for doing so. However, relying on the use of machines, drugs, or even the body itself is only scratching the surface of the multiverse. The system of the mind is so clung to by the discontinuous thoughts that it is called ego. When our personality connects to itself in memory, it reinforces this. The ultimate extension of the ego, the idea of God as self-actualization, is perpendicular in extent to its surroundings, which is thought to be the heaven of man, or the multiverse.

i. how to know when you've entered

Whenever you aren't watching the measurement of time pass, or keeping exact, careful, accurate account of its regular passage, you are in the multiverse. This only means going through a temporal singularity, and the multiversal history of the gravitational spin of spacetime that is timespace can be measured as such itself. Therefore, whenever the passage of the moments is unobserved it is entropically discontinuous in dimension upon fractal and gnomonic geometric structures. This is the shape of timespace, and the nature of time. This is not a matter of, as DesCartes described, the room disappearing from existence after we have left it, or even Schrodinger's cat, but more a matter of the old saying, "the watched pot never boils." Therefore it is not a question of whether or not something exists when we are not observing it, for it is necessary that it does, and that we know of it, for by doing so we can measure it. It is really a matter of the potential entropy increasing while not being bound down together into increasing order by the concentration of consciousness.

According to J.I. Gurdjief and Edgar Cayce there is even more to the history of human consciousness than the genetic manifestations of physical bodies alone. Like the eight circuits of consciousness model of Timothy Leary, they propose that we may experience ourselves in ways that even we don't understand, which are both geometric and emotional. They have all had different expressions for it, but the crux of their difference in assertion from the churches of their day was in saying that this had had a past, which could be remembered by or taught to a being in the present, which denies the religious assertion that the soul is timeless, and exists forever from the beginning until the end of eternity, subscribing readily to the mathematicians expression for infinity, the mobius strip. The soul is the part of the spirit that has not yet transcended beyond time, and in this way is the measure of a line on the surface of the mobius strip of infinite time that is the free spirit. Even this is only an idea in the mind of God, and an angel of a thought flies upon such wings. Therefore are these geometries emotions felt even by the strings of our personality, even that of a clone. The deists, for example, felt that the universe was a giant machine, created by God. Such a conclusion as this allows even the possibility of AI equaling human perceptions of life.

ii. how to get back out the same way you came in

Because getting into the multiverse means going through a temporal singularity, getting out means going perpendicular to the wormhole tunnel reality time stream due to its distortion to the timespace history. This only gets you into either the three dimensional continuum of another dimensional warping of geometry (or the original), or outside the space time continuum singularity altogether. I describe the effects on active consciousness as it passes through such dimensional inversions as these elsewhere in this book. It is karmic archetypal spiraling centralizations of tachyons passing through our heads that is the flow of our consciousness and font of inspiration, and thus, we are capable, in truth, of reading their histories as they do so, and deducing their origins as well as inducing their future projected trajectories. Doing so moves our center of conscious perpendicular to their histories, and thus, outside of time. We are, thus, looking at our spirit rather than our soul. Since there is no physical equivalent of this, as our center of consciousness is not even necessarily manifesting a perpendicular tachyon, only following a path in potential, then this is the difficulty in explaining such fine points to a machine, or any blank slate. It is a tongue twister.

iii. possible side effects on personality

While the most enlightened mentalities are those of meditation, it is difficult to find suitable conditions to properly clear one's mind from all the distractions of the urban sprawl, the suburbs and farms, and even from the concerns of global commerce that are good and honest responsibilities of any concerned citizen. This is not the only way to maintain clear light. It is also possible, and in fact necessary, to go about the events of survival in one's day to day life with a clear consciousness, that is able to make correct choices about conscience. This has been the lot of man since the hunter-gatherers, and is the cause and effect principle that has effected all manifestations of karma since the beginning of the universe. This clear consciousness is even reduced in the minds of metaphysical legislators to include the histories of this motion in the form of memory and imagination. However this is only QBLH in the primary clear light. Would that it were just someone's dirty little secret, for it is as necessary as entropy in the universe, or biological digestion. In this way, we all participate in the immanatizing of the eschaton, by redistributing entropy.

3. information

The only thing that withholds infinite potential information from every living being is the martyrdom complex. The more we are self-conscious, the more we are aware of our own eventual death. This is what makes us human, and also what separates us from the worlds of the animals. It is also responsible for the human condition of melancholy that leads to depression and even possibly suicide, a behavior thought to be unique to humanity alone. All other forms of life, whether genetic or merely holognomonic, are devoid of this form of self-loathing, and are therefore privy to infinite potential information. Even the damned, we are told, have eternity to reconcile with their fate, and thus come to better terms with their duties in life for their best self-fulfillment than the average person stuck in a dead end job.

Other than the perception of death there is no need for a concept of time, and therefore all those of God's lovely creations that do not fear death may release their cares of time, and in this way live eternally in the multiverse, taking whatever forms suit them, their minds always wandering freely. The multiverse is constructed upon the framework of the Akashic records, which are themselves, infinite potential information. Therefore there is infinite potential for geometric distortion into infinite potential dimensions in the multiverse.

a. infinite potential referentials

In the same way that certain words, phrases or inflections can convey additional layers or levels of meaning, so may coincidences or synchronicities in the manifest realm. So with the cataloging of information that asymptotically or exponentially approaches infinity, does the set of sums and substances reach a supersaturation of meaning, where everything is relative more and more to everything else, such that an acausal connecting principle such as consciousness, synchronicity, Bell's interconnectedness theorem, or negentropic tachyons (which are all facets of the same fourth dimensional metaform), is infinitely reducible to a one step impossible feedback loop connecting or geometrically linking anything anywhere with anything else ever.

This type of wisdom seems like it would lead to sensory overload in any person not suited or prepared for it, however this does not mean that it is not always a fact of reality. It simply surrounds the periphery of our sensory aparati and neural processing, occurring for most people more or less unconsciously, and only representing itself to them in the memories that feed their dreams as asymptotically archetypal anthropomorphications, dream symbols, sets or settings. These then bleed through into the person's karmic aura by the right brain's emission of tachyons through the holographic thalamus manifesting synchronicities to call attention to what it cannot convey to the left brain through dreams or the subconscious alone. This, at least, is one modern theory that accounts for the capacity for three second conscious precognition and remote viewing in an electromagnetic spectral network.

b. infinite potential intelligence

Modern secret societies are little different from the three musketeers of Dumas or Hugo's time in monarchial France. Free masons trace their origins to jacobite social clubs actually formed by peasants, merchants and soldiers of such European monarchies, including the French, after the monarchies began to become socially corrupt. The three musketeers had a saying: "all for one and one for all," that has become essentially synonymous with the fraternal structure of masonic organizations.

Like Weishaupt's 1776 Illuminati, these modern men and women advocate, "liberty, equality, and fraternity or death," which has set them in opposition to both Catholics and dictatorial or nationalist Communists. This remains, however, only the publicly visible face of such types of order, while there is a secret, hidden side to them as well. The slogan of the novus ordo seclorum is "fortune favors," which is, ironically, printed on American money, derives from the ancient Greek saying "fortune favors the bold." There is a secret, esoteric doctrine among the inner order, which might be best expressed as "all in all is all one."

What this means, or at least what it amounts to, is the realization of the individual's capacity for intelligence. Whatever the masons might believe, it is an established fact that the human being only uses, on average, ten percent of the capacity for neuralectric kinetics of the brain. This would seem to really bring home the Socratic saying that, "all wisdom is knowing that you know nothing."

When one considers that this saying contains, like the zen koans and the true gnosis, at its perimeters, the full fractal branching of infinite potential information already, then one must only wonder at how little of what is known is known of.

4. spacetime

The limbs of the astral body extend out through the geometrical distortions to the physics of hyperspace, penetrating, still, only a finite number of surrounding potential interdimensional wormhole tunnel realities of possibility in the hyperdimensional multiverse. These merely constitute the karma in our aura from which we extrapolate choice and integrate our decisions through third dimensional interactivity in the spacetime continuum. There are only as many choices as there is karma, and only as much karma as there is ego. There is only as much ego as there is thought of death and therefore of the passage of time. The ego, as I have already demonstrated, is reducible to an absurdity — that is, the measure of time as an infinitely repeating, or looped, impossible singularity.

When Buddha, DesCartes and others have discussed consciousness as being the basis for reality, this is the essential model they have been describing. A singularity of clear light receded from the surface of spacetime infinitely distorted around the edges of consciousness as the subconscious and unconsciousness, outside of which is a world that we perceive through the senses, but without which we know not. Whether this singularity represents a thought, an idea, or a dream in the universal mind is just a drop in the bucket beside the concept that it does not, as one day it will not, even exist at all.

Despite the metamorphosis of forms and vessels of incarnation, these remain infinite in the pure potential of the dimension of time, and it is from this that the three dimensions of space arise. Therefore spacetime is full of as many forms as have been caused to come into creation from time. Herein is all of the life, love and light resplendent in its glory.

a. going inside spatially your own temporal mind

Consider that you are an ancient philosopher mathematician, and you have discovered by examination the exponential expansion of vertices in shapes per dimension, and wish to correlate observations regarding time with the fourth dimensional shape.

First you will notice that there are three ways to represent the three dimensional shape, and those are first dimensionally, second dimensionally, and third dimensionally. Remember that a hexagon is a second dimensional representation of the cube, but that there are in fact infinite ways to represent the cube two dimensionally, for, as soon as you had exhausted the number of angles from which it could be depicted, you could always represent it mathematically. Because of this rule there are three different three dimensional shapes that we know of for a hypercube, or the fourth dimensional equivalent of the cube. Mind you, these are only particular shapes that happen to align in three dimensions, like the hexagon did in two, when really there are infinite potential forms of the hypercube that cannot be visualized, and exist only as mathematically provable relationships, but then, you could always express it in words.

The three, third dimensional shapes the philosopher mathematician, you, would have had to work with to develop further insight into the fourth dimension, at least, for a cube, would have been the hypercube at antipode — two cubes together side by side, the nested hypercube — a cube within a cube, and the hypercube at standard position — which would have just been a regular cube.

Now you are up to date with the origin of the Khab-Allah, the Body of God, Hebrew for metaphysics.

In consideration of the Kabballah, the number of ten is very important, because it is the number of sides showing on the hypercube at antipode. Its significance is identified by the ten corners of the two cubes which become visible when they are depicted two dimensionally as a pair of hexagons.

The hypercube at antipode has one face which is internal to it, which makes eleven, but this face is the combination of the two cubes, and so it is secretively twelve. These numbers were also important, and this face was described in a number of other ways as well.

Twelve are the faces of the nested hypercube, and six of the standard cube. The first of these became Astrology, which was used in divination, and the measurement of the passage of time by observation of the movement of the heavens; the second became the I Ching, which was based on eight elemental, three-lined trigrams comprising sixty-four eventual six-lined hexagrams, which was used in divination, and the measurement of certain particular cycles of time in the heavens, such as the lunar year of thirteen months and the eleven year sunspot cycle.

There are seventy-two dekans by day and night in the zodiac and seventy-two divine names in the shemhamforash. When these are aligned... let it be said thus: when we take a step toward the Eternal, the Eternal takes two steps towards us.

C. the alignment of black hole poles between galaxies

As I have described, the tachyons flowing perpendicular to the outside of timespace are pulled upward in tori around the projected spiral histories of gravitational singularities, through these and projected in exactly flat electromagnetic field lines in the gas jets of the electromagnetic poles of black holes, and that these field lines extend out much farther, forming a larger electromagnetic torus around the entire galaxy, that connects to the pole of every single sun in its galaxy, and that these gradually differentially rotate as their ecliptic orbits around these phi/pi galactic electromagnetic field lines. I have also described that this same effect occurs for our own sun, and that this causes the gradual fluctuation of the ecliptic plane in a flat spiral around the sun, as the electromagnetic poles of the sun precess. This effect on our sun also causes the precession of earth's own electromagnetic pole, which chases its tail, the gravitational pole, around in the polar precession that causes the precession of the Equinoxes and the gradual reorientation of our seasons to the constellations in the heavens that measures the cycle of the ice ages.

Because the earth is seventy percent water, there will always be ice at both poles. No catastrophe, no matter how great, could ever change that. The only thing that changes from time to time as the sun orbits the center of the Milky Way, its equatorial ecliptic plane revolves, its electromagnetic field lines wind up, and earth's poles precess the great ages, is that the electromagnetic polarity of the heavenly bodies occasionally reverses. The result of this on the earth is that there is either an ice age in the southern or the northern hemisphere. When there is an ice age in the southern hemisphere, little is different. Antarctica provides a large land mass upon which the water can gradually slow molecularly and divest itself of salt, becoming the vast continental glaciers of ice. Antarctica also has many high altitude hot springs in its geographical terrain, and these contribute the flows to the ice sheets as they inch their way out to sea to break off into bergs over the millions and billions of years. When there is an ice age in the north, then one side of the earth is going to get covered in snow, ice and finally great sheets of glaciers. It is known that the gravitational pole, around which the mass of our planet rotates as it orbits the sun, has changed position three times over the past 80,000 years. This has caused either America or greater Asia, that is Russia and particularly Siberia, to become covered in hostile permanent winter conditions over the different ages. In America this resulted in the great glaciers of the last ice age that covered the face of the northern continent, and absorbed world wide sea levels so low as to expose the Beringian land bridge between Russia and America.

It is probable that the electromagnetic polarity of the heavenly bodies reverses when their electromagnetic ecliptics align with one another, and this implies that all the spheres and bodies of the heavens are an arrangement of wavelengths like music itself, with perfect, regular rhythms in relation to one another. This would be very angelic, indeed, if it were conveniently the case. However, there is more to it than that. Many of the orbital planes of the heavenly bodies around one another are elliptical, and therefore the time it takes the spherical bodies to move the elliptical orbits is greater in some places and less in others, because the mass of the planet relative to the loci of the ellipse is propelled faster at some points in its orbit and slower in others. Therefore this does not make them necessarily exactly synchronized with the passage of galactic electromagnetic field lines through their poles relative to their equatorial orientation to a body at the center of their elliptical orbit. This might amount to little more than greater or lesser mass spheres producing or conducting greater or lesser wavelengths of energy, however since these wavelengths are tachyonic, they are flatline at all points, curved into a torus shape by a distortion to the underlying dimensional geometry of spacetime itself, and are therefore irrelevant to planetary or stellar mass. Neither should the difference in densities of masses be taken by classical physicists as a differential scale between the sizes of the mass of stars or planets, since tachyons are ubiquitous in temporal potential, and are therefore the fundamental substance of all greater matter-energy.

Therefore it seems to me that the reversal in electromagnetic polarity of the spherical heavenly bodies occurs as a function of both the factors of the elliptical orbit and the polarity of the next greater scale of body exerting gravitational force. This occurs because the gravitational force is two fold: it must operate at right angles because there must be a force and a body against which that force is being exerted. Both of these are tachyonic, and at perfect, sub-quantum right angle flatline histories to one another. One has been bent into the electromagnetic torus, the other is the direct gravitational force. These are the dual characteristics of the tachyon in the same way that the electrical and magnetic are the dual functions of the electromagnetic force, or that the electromagnetic force is a lesser measure on the same scale as the weak nuclear force. In this way, each galaxy is a greater tachyonic gnomon of all the lesser tachyonic gnomons within it — a torus, which is, of course, invisible because it is moving faster than light in the hyperspace temporal singularity of ylem. This is the same clear light that is in the gravitational singularity at the very center of the black hole at galactic core. The galaxies themselves are the lower dimensional reflections of the temporal "pressure systems" comprising the "weather" of the potential time just above the surface of timespace.

1. the filaments and voids and the nerves

When the electromagnetic poles of the black holes or the gravitational equators of the galaxies align there is an effect that can only be described as equivalent to cathexis in the Freudian model of the nervous system. Geometrically what is happening is that the histories of the gravitational singularities that extend outward into the spiral histories of multiversal wormholes are overlapping when the electromagnetic poles of their core black holes align. While there may be no visible event exchanged between them in three dimensions, in four dimensions they are like the overlapping of two "pressure centers," generating a massive and complex storm. The result is greater production of tachyons. These combine both the electrical and fluid dynamic components of an actual earthly rain.

As I have said, there may be no visible effect, there is a difference in the temporal dimension that might best be explained in terms of the phi cathexis system of the human nervous system described by Freud. As electrochemical induction was passed through a nerve, most would be passed on to the next nerve where the nerve was connected to a net or series of other nerves. Some of this transduction, which Freud called phi, stayed behind in each nerve and built up. Freud termed this the accumulation of ego, the ability to think, "I am." When the ego operated upon the nervous system in which it resided as an ambient electrical charge, it caused hypercathexis between the nerves of the system, such that more electrochemical phi was transported from one nerve cell to the next than was transported to it from the last. Since nerve cells alone, no matter how much electrical activity they have built up, cannot even reproduce themselves as such as the lowest forms of bacteria, it is at least questionable whether Freud's definition for "ego" would match the qualifications for consciousness defined by observations of life.

Imagine the way you would move if you were a large tachyon made of smaller tachyons, or the way an axon connects to a dendrite between nerves, as like an ant hill. The hive sends out individual ants as scouts, until they find a food source, and then they form a line (more or less) back and forth between it. If they need to, such as in a disaster, the ants could transport their entire hive community this way. Tre binary. However in basically this way, dendrites secrete neurotransmitters to either be absorbed by a dendrite or be reabsorbed by the axon, and galaxies emit tachyons to reposition themselves relative to one another. One might say of the universe that galaxies are arranged within it like flowers of stars on tachyonic vines. Another beautiful metaphor from the terrestrial bestiary would be the jelly fish, whose body is like the flat disk of a galaxy, but trails long tendrils behind it just as galactic hubs radiate tachyons.

By the gravitational force galaxies are brought together until the black holes at the cores of their electromagnetic poles align. By the electromagnetic force the galaxies are made to turn by the other galaxies around them, as the gravitational force pulls them together, for it could only be the electromagnetic, temporal singularity (or expanded wormhole) that would pull the electromagnetic poles of the central galactic core black holes in its precession, and these can only occur between the poles of one galaxy with another as the distortion to the surface of spacetime that causes the geometric patterns of the filaments and voids.

Wormholes follow arcing field lines between the electromagnetic poles of black holes at the cores of galaxies, and these penetrate another dimension of space, occurring only upon the surface and within the event horizon of the black hole at the galaxies' cores itself. To say that they traveled through spacetime in any form, even as a quantum shadow such as an emotion or a metaform, would be inaccurate. They travel outside of and beyond the three dimensional universe above and parallel to the surface of timespace, comprising the tachyonic multiverse. However it is not that this isn't there, it's just that we can't see it. Tachyons are wavelengths so small they are clear, and so we see space in the voids between the filaments as dark and empty. Here would be the breath of the air in which the music of the galaxies is played.

2. the holognomonic (wave) temporal matrix in the neuronal cylinders

Before, I have discussed ennegrams in the mind such as memories or lines of inductive of deductive reasoning, but I have not yet described them. They are a form of QBLH, or metaform, such as is the hypercube, or tree of life of Moses. Thus, studying these types of reasoning can open up ennegrams in the mind that induce experiences in sequences, since they are temporal patterns. Ennegrams are also formed by any type of study, however it is ordinarily only in accordance to the free form of our interest in research that shapes these inquests. Thus a good compass is the formal system of reasoning based on the six fundamental questions of reasoning: How, when, where, what, who, and why? (deductive) or how, why, who, what, where and when? (inductive). Still, such as this can only guide you between poles, such as fact and fiction. These types of poles constitute the parameters of our tunnel realities.

Often we see that the pattern we are looking at is mimicked by the pattern of electromagnetic energy in the visual cortex of the brain. This is the basis for the theory of ennegrams. Each memory, dream experience, idea or comprehension of perception occurs at many places at once as electromagnetic energy inside the cerebrum. The way in which these are related, it is thought, is holographic, similar to the holographic nature of the function of the thalamus just below the cerebrum. The thalamus projects impulses from multiple nerves to multiple nerves through single nerves simultaneously, thus acting holographically by storing multiple encodings of information on the same amount of electromagnetic impulse as is usually only used to encode a single information encoding. The cerebrum itself, though divided by the hemispheres, the functional sections, the stacking of neuronal columns and the fissures, is also thought to act as a single, unified unit, in which the information in different nervous tissue occurring at the same time is related as a thought form, memory, dream experience, or idea.

What we are seeing here is a holognomonic resonance that also occurs on a lower vibrational, higher dimensional frequency, inside the axon-dendrite gaps between the neurons in the stacked pillars and columns of the gray matter of the cerebrum. Here, the electric potential in the chemical neurotransmitters of the brain is so infinitesimal on the physical scale that it actually catches tachyons in hyperdimension like an ultra fine net, filtering them through into greater electrical potential, and virtually opening up wormholes inside the mind. These translate into the genetic material of the neurotransmitters, that are passed down through reproduction over time, as well as to thoughts in the moment for the minds of the brains themselves. Since this effect is resonant holognomonically, we see the electromagnetic field as occurring at multiple points simultaneously in distant parts of the tissue of the brain.

The result of this is the pattern of the brain's electromagnetic waves over time. There are several different states of brain waves, including the most regular, alpha, beta and gamma states associated with sleep. These are produced in the thalamus, and are thought to have a strong temporally regulatory influence on the monthly hormone production cycle of the hypothalamus in women and the growth cycle hormone production by the pituitary gland of pubescents. Thus, the sum total of our minds over times is not only what we think, but even how our bodies grow.

The waveform we are seeing that constitutes the mind in the moment, a holognomonic ennegram comprised of multiple synchronous electrical events in the brain, is the same thing as the mathematical matrix for spin that I described earlier. The spin of the substance of the fourth dimensional emotional metaform is the information that is the mind, traced out in the brain as the electromagnetic holognomonic projection of the ennegram thought form in a neural net. As the spin changes over time so it traces out the pattern of the brainwaves.

D. the Akashic records

The akashic records are a relatively recently revived concept deriving originally from the ancient far east. Edgar Cayce, the twentieth century psychic said that he derived his ability to read people's past lives from the akashic records. According to Cayce, he had learned of the akashic records in the same way Madame Blavatsky, the nineteenth and early twentieth century theosophist had, on journeys made to the orient in the east, in particularly from tantric yogic sanskrit documents. Besides these two there is no writing on the akashic records known to western society, and there are no translations of the sanskrit documents they claimed to have read into English or any other language. The documents themselves have allegedly been lost. Madame Blavatsky was considered a dubious character in her time, as was Edgar Cayce in his, and it is possible that the documents both were referring to were actually nothing more than the works of nineteenth century Russian novelist J.I. Gurdjief. However even these contain no mention of the akashic records per se, and so one is led to wonder how these two independent authorities came to discover the same descriptive terminology for a concept that no one else seemed to share.

It may be better to begin the history of this concept a little further back, with Ezekial's vision of the ophanim wheels, or even the Old Testament apocrypha of Enoch, preserved to this day by the jews of Ethiopia. These are accounts of visions of the heavens, and in particular of the mechanisms of their cycles. They differ from the Sumerian books of Enki in that they are more calendrical in nature, establishing patterns and cycles for the seasons. With the accounts of these perceptions begins the true recording of the history of the akashic records, which were, at the time, only known as the cycles of the heavens. The Egyptians, and later the Christians, would have large scale descriptions of the components of the heavens and their cycles, in great detail, however both of their calendars were as though frozen in time: the Egyptians set by the alignments of their megaliths to 12,500 years ago, the Christian gregorian calendar pivoting around the year zero, some 2000 years ago.

Meanwhile all the faiths the world over have always promoted the idea of an afterlife, and we believe this belief to date back as far as ritualized burial, practised even by Australopithecines, the first hominids. Thus, the temporal cycles of the extraterrestrial heavens have become associated by most faiths with the heaven of man, where the spirit goes after the physical body dies.

Thus, this could be called the akashic records: the sum over histories that is the multiverse of all matter-energy of the universe over its entire duration, measured as dimensions by the vector of geometry. It amounts to the exact location of any point in history, and insofar as these can be linked together, was the basis for Cayce's past life readings.

It can therefore be subdivided into constituent sections from the universal singularity through the gravitational singularities, through the galactic bubbles, through all the stars of the galaxies, through any planets of any star and so on and so on, through into the categorizing of information units themselves right down to the smallest tachyon. Doing so creates a perception of the temporal pattern of the heavens such as described by Enoch.

E. the Enochian system

There are also two origins of the Enochian system. One is the calendrical cycle described by the Ethiopian Hebrew prophet Enoch. The other is the system of the ayres described by the fourteenth century Elizabethan England skryers John Dee and Edward Kelley. The cycle described by Enoch is simple enough. It establishes many of the same calendrical features still in use today, such as the twelve months of the year, with both name, sign and dekan, and the four seasons. One feature of the Dee and Kelley system is in complete agreement with the elder system in these regards, giving angelic banners to the months and assigning Godly names as angelic dekans to each. Beyond this the Dee and Kelley skryings provide even greater insights into the temporal workings of our heavens, incorporating extrapolations of the four elemental forces as the four cardinal directional watchtowers at the corners of the universe. The Dee and Kelley model unwinds the shemhamforash of the 72 dekans of ten days and ten nights each, three per each of the twelve signs of the zodiac, including three positions for each determined by astrological alignments over an elemental grid. It is a very complex system, giving the names of a host of angels as the arcing intersections of letters within placed upon the grid of this cycle and assigning them into the multiple levels of spheres of the 30 ayres. While neither the names of the months and signs given by Enoch, nor the importance of the letters skryed by Dee and Kelley seems to have held up, the systems with which they measured are still out there today, and can still be used to understand the cycles of the heavens and the place of our moment in this universe.

Think of the akashic records as the contraction of the same medium as the Enochian system is the expansion. While the akashic records provide information from without, the Enochian system derives information from within. Both are merely movement on the geometry of the QBLH, which is phi/pi. Therefore the akashic records and the Enochian system represent all the same things: the universal singularity, the gravitational singularities, the galactic bubbles, the coiling electromagnetic fields of stars, and so on and so on down to the sorting of information itself down to the smallest tachyon. Think of the akashic records as what is being accessed, and the Enochian system that is used to access them.

1. the satellite telecommunications system

Whatever demiurge or guiding principle there may be in the universe, the satellite telecommunications system is made by and for humanity. Like stonehenge it will stand as testimony to the greatness of the human mind. Its usage, on the other hand, seems to be unanimously agreed upon as contributing more often to human stupidity. This is through no fault of its own, for every ingenious component of these beautiful scientific marvels functions accurately to perform their goal. It is only because of politics between the people on the surface of the earth that these are used in the ways that they are.

One type of satellite is the military satellite. These observe and have very strong camera lenses, capable of reading license plates on cars. Some of these are the left over Star Wars satellites from the 1980's, that have laser guidance targeting systems for destroying intercontinental ballistic missiles. These were never used, and will probably eventually become flotsam. A popular theory among some citizens of America, the country that constructed these kind of satellites, is that they contain scalar wave technology. Some evidence for this, as well as its usage, derives from those seeking legal suit against the military and the state for secret projects involving the ongoing use of microwave frequency transmission from satellites for use in mind control.

The more popular kind of satellites are those used in telecommunications by large, international capitalist corporations such as television, telephone and internet service providers. These carry all the frequencies of mainstream culture in the air waves high above the heads of the secured and insulated masses, while outside the ridiculous garbage noise of our culture over the span of history since we first put satellites in space reaches out into the vacuum of the electromagnetic background radiation of the universe, screaming life on planet earth to all our surrounding neighbors. When you surf the net, when you channel surf, when you turn the radio dial, you are traveling through frequencies broadcast by these beacons.

There are also satellites sent up by the various space programs of the nations of the earth for the purpose of conducting different types of research project and conveying various different types of survey. Some of these look down and monitor such things as the weather, tectonic continental pressures, pollution from population densities, and exotic ecosystems. Others are aimed outward and make measurements on such things as background radiation levels of the universe, a survey of galaxies, or, like the hubble space telescope, send back direct observational data from extra-planetary objects. Some are simply internally motivated, containing biological experiments to be recollected later, or measurement equipment for telling the difference between the time inside the satellite from the time at the launch site for coordination of the alignments of windows. Some are simply time capsules sent up by lucky classes of children. Others contain plutonium.

The space shuttle is the vehicle used to deliver many of these satellites into space, while rockets are launched off containing others inside a breakaway shell. The space shuttle is an enormous airplane that is attached to a fuel tank and two jet boosters to propel it beyond earth's atmosphere faster than the pull of gravity. After the shuttle is outside of the atmosphere, the fuel tank and thrusters break off and become space flotsam. The use of solid fuel rocket propulsion systems such as are used in the space shuttle boosters has been common practise since the German blitzkrieg of England in the 1930s.

There is also an international space station in orbit around the earth, and there are astronauts living inside of it right now, at the very moment I am writing these words. This is merely the newer, collaborative replacement for the Russian space station, Mir, which had been in space at least twenty years before it was decommissioned. Currently the astronauts that live on the station are competing to break one another's records for longest time spent in the microgravity conditions of outer space.

a. the global communications system

349

Perhaps the primary purpose for the satellite technology in orbit around the earth is telecommunications. This is the goal to link all members of society to one another through mechanical media. Control of these media may, themselves, be able to fit in the palm of one's hand, however they are becoming increasingly reliant on the satellite telecommunication system, as people demand greater and greater coverage areas for their chosen connectivity to each other. This process is known as globalization, and this is a multicultural and societal pattern that is going on all over the world now. It is an agenda of the united nations, who encourage it be taught in schools, and that it incorporate equal rights for women and minorities. It is the cultural phenomenon of mediated press coverage of international events, more of which is offered with the more expensive services.

b. the pop culture simulacrum

Meanwhile, the cultural phenomenon equivalent to this is the grossing of the classes of the masses into the aggregate mainstream. This attempts to appeal to people from all walks of life, and to sell them further acculturations for their affectations. Its sum and substance is the pop-culture simulacrum.

In the 1930's there was a disagreement between the aryan Germans and the German Jews over the value of the human soul that led to the six billion slaughtered Jews of the holocaust. This led to the capitalist culture's cold war with the communist Soviet Union. The argument of both was the dollar as being a representative simulacrum or symbolic token exchange system of nothing greater than an addictive substance such as a drug. Nowadays, many people have caught onto this idea, and are forced to part with their foolish identification of themselves with their money, for the value of a person is not expressed by his or her worth to the economy. This does not mean that the same parties who promoted drugs and money as being relative have not continued to make their insinuations about the constitution of the soul of a person as being relative to some chemical substance or other, because there is little difference between the value of the human soul to the economy and the puff of smoke from a cigarette.

People still assert their personal desires on reality, few realizing what agendas they might be triggering the furtherance of, even fewer caring that these should benefit their fellow people. This is merely the pop culture simulacrum. It doesn't matter if you are real or not. You are only as important as a number, one out of an unknown amount. The system can be bent as much as it bends you, and to use it is like looking long into the face of the abyss, for it is looking just as long into you.

2. the innate perception of archetypes

The names skryed from the set of Dee and Kelley's Enochian system are assigned to angels, however insofar as the manner in which they are interconnected so are they arranged upon the ayres, such that attention should be called to their arcing curvature. Although Dee was not officially entitled to call these angels archangels, we are free to call them archetypes. They are not alone as such either, because the trait of their arcing interconnectedness is not a necessary characteristic of archetypes. The patterns described by Dee and Kelley do offer an additional level of insight into the ontology of archetypes however we see that the phi/pi spiral is archetypal in itself due to its geometrical ubiquity. Archetypes may not only be angels, or even archangels, but they are the geometrical equivalent of the infinite potential free will of the singularity of the spirit raised by enlightenment of the conscious ego.

This may be another name for the universal life form, that is, a singularity one with the universal singularity. This is not necessarily a biological life form, though by its inhabitation a physical body could be motivated. This would be the essence of the soul that is equal in heaven to the orders of angels before God.

Even the demons of the qliphotic shells are archetypal, though not in their given form. The hierarchy of Hell represents an ordered system in which every one that clings to the side of the pit is given duty and knowledge for its conduct. This is merely to say that, like the periodic table, even quanta may be classified according to referentiality and relativity.

In this way the most high and the lowest are called like the same, and this is essentially the consciousness of entropy. The rule is that the old greatest common factor perpetually becomes the new lowest common denominator, and this leads to the measure of time the degree of clinging to which determines the gradient between the ideal, temporal archetypal and the manifest probability quantum archetypal.

Our minds are always perceiving these things, around the periphery of consciousness. They come through in synchronicities of manifestation, in the dreams of the subconscious and as unconscious emotions themselves. Another way of expressing them is the metaform of the neural net of electrical activity in the brain.

a. Plato's ideal forms

Plato described the ideal archetypal realm as according to pure geometry, and gave as his example of perfect or ideal forms the five regular solids. These are the only five three dimensional polyhedra for each of which all the sides are the same shape. These are: the tetrahedron, octahedron and isocahedron of triangles, the cube of squares and the dodecahedron of pentagons. These he held to be physical forms that transcended the realm of the physical, as representing the same purity of concept as algebraic mathematics. On the other hand he saw all Athenian sculpture, some of the finest in all of history, as being merely representational of national trends, themes and agendas, without recognizing in it even the geometrical beauty divined by Leonardo DeVinci. Again, it is little wonder Socrates, Plato's own avatar in his writings, willingly ingested the state's hemlock.

Similar conclusions have been reached recently by Hebrew rabbis, who found that the shapes of the Hebrew letters might actually be derived from the shape of the horn of a ram. It has long been held that the letters of the Hebrew alphabet, when taken as equivalent to given number sums, are thought to be the structure for equivalencies in meaning upon which the Hebrew language is based. This, it is thought, would make Hebrew writings more ideal.

It is probably true that the layer of meaning the Hebrews encrypt in their writings is about equivalent in a rich, resplendent tapestry of beauty and awe as the Muslim illuminated arabic scrolls. However it would be a blessing from a bull indeed to call this necessarily valid, accurate, true or even, necessarily ideal, for these are all words that only apply to what can be known in this physical realm, and expressed in even these less than ideal ways. Similarly, to anyone who doesn't believe in geometry, Plato's forms are little more aesthetically pleasing than a good game of baseball, or the concept of an ineffable God.

b. the malleable manifest

Whether we can consider our realm ideal or not is a matter to be legislated exclusively by the free will, however every day we are forced by circumstance to consider it real, and for the good of our continued physical survival we do this in the ways that make this reality the most apprehensible possible. Perhaps it is only because of the scarcity of the ideal imagined due to relativity that it is even valued at all. The ancient Egyptians had an expression for this: the weighing of the soul of the individual at the time of their death against a feather by Maat. Here we see that there is no scarcity, because there is no relativity, because the feather itself is the soul. In this way man is the measure. The free will manifests its own karma. Time is the spin of karma.

We are continually digging our own grave, a quantum tunnel reality with its end forever in the back of our mind. Which is more ideal, the karma which we manifest in our aura, or the desired result of the end, is ultimately irrelevant to the fact that we are who we are and what we are, each individually.

The love of life is clinging, the love of light is clinging. The life of love is fleeting, the life of light is fleeting. The light of love is fading, the light of life is fading. We suckle upon sex, we suckle upon drugs. We die for all our hatred, we die for other peoples' ideals. Our hearts change like the seasons, everybody is always busy dying. These things that are meant to give life greater meaning are reduced and diminished themselves by their own use and application. King entropy wins.

The ideal of reality is not, as most people would expect from reading the Republic, a perfectly ordered, rigorously enforced governmental construct. It is the ability to manifest what we will that runs perpendicular to the history of our tunnel reality. Because this accesses the archetypes of the subconscious and unconscious this should not be thought of as an affront to the realm of the ideal of time locked in the potential energy of the right hemisphere of the brain. Remember that the archetypes represent the full neural net activity, not only that of the left or right hemisphere. This only seeks to include the contribution of the right hemisphere in the tunnel reality defined by the holognomonic archetypal brain wave metaform. This allows for the free use of the thalamus to reflect projection external to the self, and in this way effect probability on spin, and in this way manifest physical reality. The irony of this is that it does not constitute less entropy, as does the utopian governmental idea, but actually increases it by exponential amounts similar to the pattern of growth of the human population.

Manifestation is an affront to the ideal of time itself, however, since it obliterates the concept of it by relativity in the multiverse. Therefore, more potential energy does not mean more time. The liberation of additional energy from the mind could only contribute to the entropic aging of the universe. Therefore the angels of our better selves may as well be the thought police, since they are only trying to remind us of what we would only have forgotten by going to sleep. In this way time has its revenge by warping the geometries underlying the hyperdimensional multiverse. It is because of the supremacy of the ideal of time, king entropy, that manifestation is shunned.

VIII. Phi/Pi

Phi/pi is ubiquitous. It is most of what is, and the rest of what is is not that much different from it. It occurs in genetics and it occurs in physics. It occurs in architecture and the arts, as well as music. It is the pattern of black holes and it is the pattern of consciousness. It is the pattern of time itself.

Phi/Pi is light, love and life. The three are one. That which I am bows before phi/pi. YHVH is trapped within the time of Thoth. He lives, and the measure of this life is Thoth. He loves and the depth of this love is Thoth. He is a light, and the shadow of this light is Thoth. Yet there is no difference between them, and they are one and the same, the universe and the multiverse; the singular point of the tao of chi and the endless nothingness of the tao of chi are the one and only tao of chi. All these things are phi/pi, also called the QBLH. QBLH is the Hebrew gematria for the sum of phi/pi. In QBLH there is no over, under, around and through, nor any such thing as perspective. Everyone is a star of their own show, making movies, and nobody pays any attention. These are all things well worth forgetting, since they will only add contrivance to convention. Modern terminologies pale before the awesome grandeur of dreamscape. It is said that that which is outside of QBLH is also that which is within QBLH. In this way it is imbued that it may measure the forces that govern time.

I have shown how the torus shape pervades the scales of the cosmos, and how it describes everything from the universal singularity, through the gravitational singularities, through the galactic bubbles, through the coiling of the electromagnetic field lines of stars, and so on and so on down to the sorting of information itself to the smallest tachyon. I have described how phi/pi is the verbal expression for the mathematical equation expressing the geometrical spiral measuring the surface area of the torus.

Phi/pi is the shape of the electromagnetic polarity that occurs in the atomic nuclear fusion of the sun and the biological lifeform of a pretty flower growing with many others in an English style garden.

A. various shapes

Phi/pi can be expressed mathematically or geometrically, however in its geometry there are many forms. As I have said, some are organic matter, others are tachyonic time bubbles, however, as with the Platonic forms, there are a finite number of shapes of which phi/pi is the geometric measure. These are usually even more extrapolated from manifest reality than Plato's solids, existing only as fourth spatial dimensional metaforms in fourth temporal dimensional potential. Some of these are the hypercube, the hypersphere (or torus), and the hypertetrahedron (or cone). All of these can be expressed in three dimensions, however what is necessary to understand is that these are only third dimensional shadows of fourth spatial dimensional shapes known as metaforms, whose actual appearance is so complex that it cannot be expressed in three spatial dimensions alone. These third dimensional shadows are no different than the second dimensional shadows cast by third dimensional objects on flat, second dimensional surfaces, only cast from the fourth dimension. Therefore it is not these metaforms alone, but their involution, that is the measure of phi/pi, the mathematical metaform they describe.

1. the hypercube

Imagine a regular cube, with six square planes, each with four edges at ninety degree right angles to one another. Now imagine that this cube is inside of a larger cube in such a way that twenty four other cubes the same size as the smaller one could also fit inside of the larger cube all around the first cube, such that the smaller cube would be the exact center of the larger cube and its volume one third that of the larger cube. This shape is called the hypercube, and it is used as a measurement of a fourth spatial dimensional cube. In reality the hypercube is more complex than this, because it is not, in reality a fourth dimensional shape itself, only a third dimensional shadow.

Imagine this comparison from the novel Flatland: how would someone living in a second dimensional universe perceive or comprehend a three dimensional sphere? One answer would be to say as a circular second dimensional shadow. Another would be to imagine the three dimensional sphere passing through the second dimensional plane of Flatland, causing a circular second dimensional slice that would expand and contract as the sphere passed through. This goes a long way towards helping us understand the model of wave mechanics as electromagnetic tachyons wormhole through perpendicular microwave gravity tachyons, however what can it show us about the fourth spatial dimensional geometry itself?

Any of the known positions of the hypercube represent alignments of the fourth spatial dimensional metaform's shadow in three dimensions with a geometrically regular polygon in the same way that a three dimensional cube can cast a square or hexagonal shadow on a two dimensional surface. These only represent the regular shapes, however, and, just as we see that there are, in fact, a very high, but finite, number of ways to depict the shape of the shadow of a three dimensional cube on a two dimensional surface. Similarly there are further, regular and irregular, three dimensional shapes that are the shadows at different positions of the true metaform of the fourth spatial dimensional hypercube, and many of these represent the patterns of cycles as the history of spin.

The regular positions of the hypercube, those which can be expressed in three dimensions as well as those which can as of yet only be modeled in two, I will discuss in detail in the following paragraphs. The irregular forms of the hypercube can be thought of as its motion. This is because the regular shapes of the hypercube are like steady, unchanging shadows, while the irregular shapes form as slices due to the shadow's three dimensional movement as the fourth spatial dimension passes through the lower dimension over the fourth temporal dimension. It is such patterns as these that cause the cycling of emotions.

The hypercube moves in two ways. First, it "jitterbugs," like some of the geodesic shapes designed by Buckminster Fuller which rotate around themselves as the contract, similar to the old popular dance "the jitterbug" or later, "the twist."

The hypercube, however, jitterbugs in a different way. Rather than rotating around a polar axis or central point, it elongates along a precessional axis radiating out from one corner to the opposite corner, then to each of the other corners on the same upper square as the opposite corner. This it does from all eight corners so that each face is bent inward like a sail, and so that each corner is pivoting in a vortex around itself. The second way in which the hypercube moves is that it involutes. One external face becomes the corresponding internal face, while the opposite internal face becomes the external face. This happens, again, everywhere on the hypercube simultaneously, so that the hypercube is perpetually "shedding its skin" and then "eating its own tail," much like the auroborous of the aurora borealis.

It is difficult to imagine these motions. It is possible to depict them using computer graphics, however, even then, limited to moving second dimensional images that can only describe compute programs based on translation of multiple dimensionality into the three dimensional shape of the hypercube, and doing so constrains these images to represent only the rotation around one corner rather than all eight, or the fluctuation of two sides rather than all six, and not even both of these at once. Thus, it is possible to expand upon these graphics, and imagine the full motions of the hypercube, but the difficulty is in why they move this way.

As I have said, the hypercube is actually rightly expressed as several different regular shaped three dimensional polygons, and at least one, the octaholohedron, that remains as yet expressed only in two dimensions. Therefore the irregular slices only occur between the regular shadows, as the fourth spatial dimensional metaform moves through the third dimension over the fourth temporal dimension.

a. the cube over time

Imagine a cube as I have described. This is the simplest form of the hypercube, viewed from above one of its fourth spatial dimensional "edges" such that the inner cube and outer cube line up exactly, and there appears to be only one cube. According to the exponential expansion of vertices per dimension we can see that, even the shape of a solid, real third dimensional cube such as the mome of chemical physics or the cubit of the ancient world, has only eight three dimensional vertices because it is an illusion of the fourth temporal dimension, seeming to posses finitude, and therefore being governed by dimensionally limiting entropy. Because the fourth spatial dimension is even more native to the fourth temporal dimension than the lesser components of the three physical spatial dimensions, it is true that the three dimensional cube is, in a higher reality, actually a fourth dimensional hypercube.

Similarly, all the vertices, or corners, of the five platonic solids are the intersections of only three faces, each of which is a second dimensional plane. This is because they are regular three dimensional polygons. In the same way that the cube is actually concealing the exponential dimensionality of the hypercube, so too do the rest of the platonic solids conceal their hyperdimensionality. Each governs the geometry of the reactions of a different elemental force of physics. The cube governs earth, the strong nuclear force, responsible for nuclear fusion such as occurs in our sun and in all other stars. This force was associated with the earth because of its solidity and accumulative effect. It was associated with the cube because of its sturdy, unassailable construction, being the most stable design on a flat surface. In all these ways it was taken by the ancients as representative of the fundamental unit measurement of time, which has come to be called by modern physicists an event.

b. the hypercube at antipode

Now imagine a rectangle with short side of one unit and long side of two units. This is the golden division or golden ratio rectangle. Another measure for the same rectangle is short side two, long side three, and it is from this rectangle that a phi spiral is derived. This rectangle is comprised of two equal cubes, side by side, and it is the hypercube viewed from above the middle of one of its fourth spatial dimensional "corners," such that the inner cube and the outer cube appear to be next to each other. It is the case, however we look at it, that the two cubes that comprise the hypercube are equal to one another in area. As I have already said, if the hypercube were a regular three dimensional shadow such that one cube were nested as one twenty fifth portion of the other, the lesser volume would be one third that of the rest of the cube. However there is additional dimensionality that must be factored into the lattice, and therefore there is a relativity of measurements, such that the actual mathematical area of the interior cube is equal to that of the exterior cube. This is another one of those exotic characteristics of the hypercube that would make a modern physics student yawn, but which will doubtless end up being fundamental to the piloting controls of flying saucers in the eventual future.

The antipode point is a point midway between the two cubes, such that one appears even with the other. Although this point has only recently been discovered, the lattice of the tree of life of the QBLH has been based on the shape of the hypercube metaform since, according to rabbinic scholars of QBLH, the time of Moses, when it was supposedly delivered to him on Mt. Sinai as the first stone tablets, which he smashed upon finding the zadok priests worshipping the blazon apis, to subsequently deliver the ten commandments that have since become the basis of law.

There is some evidence now, such as passages from the Egyptian book of the dead only translated from the hieroglyphic in the last few centuries, that seem to confirm the ten commandments are borrowed from a list of negative declarations before Maat at the time of the weighing of the soul at the beginning of the soul's passage through the underworld of the Duat, the origin of our word for death. Some scholars also associate the tree of life with the Egyptian pantheon, however there is no archaeological evidence to validate this having been done by the ancient Egyptians. The tree of life lattice therefore appears to be an authentically Hebrew construct, attributed to their divine lord, YHVH.

YHVH was the name of the highest initiate in the Hyksos, or lower Egypt mystery cult, equivalent to Thoth in the Nubian, or upper Egyptian mystery cult. These two cults were the same at least until the time of Akhenaten in the fifteenth dynasty, long after the unification of upper and lower Egypt under a single pharoahnic crown. However the priestly advisors of Akhenaten attempted a coup against him, and he and his wife Nefertiti fled into the eastern wilderness. Since it is believed that this occurred after the time of the writings of Moses, it is not usually held by researchers that Akhenaten, who would have held the titles of both YHVH and Thoth, might have been Moses, leader of the Hyksos exodus out of Egypt. Further, because the Hebrew calendar was fundamentally different from the ancient Egyptian, there has been little way left for scholars seeking any connections between the histories of the two people to accurately cross reference the two.

i. the phi rectangle and the pi circle equal in area

As I have said, much of the Hebrew claim to divine authority derives from gematria, or the equivalent sums as equivalent meanings encryption theory. Although it was written in Greek, the New Testament is only supported in its claims to Hebrew lineage by this same gematria. Greek letters, when taken as equivalent to number sums, can also be shown to have, somewhat less direct and more intuitive, connections in meaning to their quantitative equivalencies. Using this system, Jesus, or at least the authors describing him, encoded many different types of important measurements into their vocabulary, just as the shemhamforash has encoded the precessional number 72 into the Old Testament. Some of these were measures for the heavenly bodies, including the sun, the moon and the earth itself. Others were golden rectangles and phi spirals. Sometimes the numbers described both.

One of the fundamental geometrical truths that these ancient number scholars seem to have grasped is that there is a phi spiral golden ratio rectangle that is equal in area to any circle. Thus, they used the golden division as a measure for latitudes on earth. This had been done in very ancient times by the builders of the pyramids and the aligners of the megaliths such as stonehenge, using the geometrical pattern of phi as well as the sum of the number itself, however the measure encrypted in the gospels is that for the latitude of a phi rectangle equal in area to the circumference of the earth. While the ancient ley lines plot sacred measurements from the previous gravitational poles and from other sacred karmic centers, the golden mean area of earth gives us the coordinates of Nazareth, home of the Nosrei-ha-brith, the keepers of the covenant, the essene sect to which Mary and Joseph may have belonged.

All of this esoterica aside, what is being described here is that the phi rectangle and the pi spiral circle can have the same area, and that they may be seen to overlap one another. This overlap is literally looking down upon phi/pi as from above one of its spiraling poles. Thus, imagine the pi spiral rotating inward and the phi spiral rotating outward as around.

c. the hypercube at apogee

This is the shape that I described as being the one cube inside the other. It has several odd properties, including its involuting motion, the discrepancy between ratio, volume and area, and it is the regular third dimensional shadow of the fourth spatial dimensional hypercube viewed from above one of its "sides." Thus this completes the phi/pi spiral around the surface of the fourth spatial dimensional metaform used to measure the fourth temporal dimension. This final regular position is referred to as apogee because it is at a point in an orbital ellipse around the metaform similar to that at which the point of observation is further away from one locus than the other, as opposed to perigee, when it is equidistant to both. As I have described, this occurs for the orbits of such heavenly bodies as planets around stars, despite the fact that the star is at the center of the orbit, and usually large enough that it incorporates both loci. As I have said, Newton's laws of gravity and astrophysics alone to do not account for this, and modern scientists believe that it may be due to the relativity of the mass of the sun to the planets. What we are seeing, however, is not only the planet's orbital path being spread into an ellipse, but the entire ecliptic of the orbital plane of the sun precessing with the rotation of its electromagnetic pole. As I have said, this is because of the passage of gradually fluctuating tachyonic field lines from the poles of the black hole at the center of the galaxy. The question only remains then, why should the spiral orbit through the passing of a fourth dimensional metaform like the hypercube be elliptical?

The answer is that it doesn't have to be. It is only the resultant perception of the phi/pi diagonal spiral measure of surface area as the shadow of a metaform becomes more complex, and thus changes up the gradient of its component lesser dimensional shapes, as well as between metaforms on a scale of complexity. A hypercube is more complex than a hypertetrahedron, which is more complex than a torus, in which the orbital path of observation is perfectly circular, and the degree of reorientation zero around the origin point. A hyperstelloctahedron would be more complex than a hypercube, and so on and so on, just as an octaholohedron is a more complex second dimensional rhombus than a square alone because it represents a higher dimensional shape. Therefore the hypercube is only a hypercube because of the elliptical distortion to the phi/pi geometry of the spiral spin measuring its dimensionality as information. The more distortion to the geometry, the more geometrically complex the metaform.

i. phi rectangle on cube surface over spherical diameter of cube

The measure of phi/pi in the hypercube at apogee is by tracing out a phi spiral between the surface of the inner cube and the outer cube on each face, and measuring the spherical circumference of the central cube by a pi spiral connecting these. The phi spiral on the surface of each surface of the inner cube begins with the centered phi spiral in the golden division rectangle that has the same area as the circular circumference of the face of the inner cube. It fills out into the pi circle circumference of the face of the outer cube. This can be done for all six faces such that each becomes a vortex. This is thought to create the temporal undulation of the hypercube, as well as to describe the shape of the warping of galactic electromagnetic/gravitational fields such as for the stars, and the trajectories of tachyons around probability wells creating real particle photons.

Another way to measure the same effect is using the square root of three diagonal of the cube and the square root of two diagonal of the square exterior faces, however this is third dimensional, and the same pattern is better expressed in four dimensions, spatial and/or temporal, as phi/pi.

d. octaholohedron

The octaholohedron is a second dimensional shadow of the fourth dimensional hypercube, however, insofar as it is also a regular polygon, we cannot deny that it also describes the exact same shape. This is the shape that expresses the geometry of a wormhole, where it represents the interconnection of two cubes, in the case of a wormhole each representing a separate event in spacetime, by a bridge of relativity. It has also been used as a lattice in quantum physics for expressing the relationship between coordinate pairs of vector probability in the expression of chemical atomic structure and quantum relativity. This is one of the better possible diagrams of the archetype of the mind, representing the union of the past and future universal sets.

Imagine one cube. Now imagine another cube offset along the diagonal. Imagine these two being connected at all corners by diagonals. This is the octaholohedron. It can only be expressed as second dimensional, however, such that you must really think of each of the cubes as only the depictions of two offset squares, also connected by diagonals. The reason for this is that, just as we know in the second dimensional representation of the cube that the diagonals, even though they do not appear the same length, are meant to represent the same measure as the squares. So it is also for the octaholohedron that the diagonals connecting the two cubes represent the same area as each of the cubes itself.

This does not mean that the octaholohedron is made of three cubes. It is merely a hypercube: a cube inside, the cube outside, and the difference between them, which can be represented mathematically as phi/pi.

2. the hypertetrahedron
The tetrahedron represents cosmos and the color green in the spectrum. The internal tetrahedron of the hypertetrahedron is directed inversely to the external tetrahedron. This should be noted because it applies also to the hypercube, which seems, because of its constituent shapes, to have no such inversion in its nested position. However, just as the cube can be derived through the steloctahedron from the tetrahedron, so can the hypercube be derived from the hypertetrahedron, and here we see that the inversion of the interior tetrahedron's direction from its exterior counterpart does carry through into the construction of the hypercube.

The hypertetrahedron's equivalent temporal measure is the cone, such as that of the precession of the polar axis. The zodiac, for example, is a measure of twelve because it is the circular circumference of a cone that reaches ahead and behind twelve wavelengths in the polarity's reversing cyclical frequency. Each time one of these changes to the next according to the gravitational alignment of the ecliptic, the earth's electromagnetic poles reverse. Every time the ecliptic aligns gravitationally with galactic core, the electromagnetic polarity of the sun reverses. While the earthlings have kept track using the zodiac only of their earth things, they have not developed a calendrical system to measure the solar cycle since the time of the Maya and, it is thought, perhaps the Chinese I Ching. Romulus and Rhemus were suckled by the bull.

Study of metaforms has only just become popular. The application of kabbalah to quantum mechanical lattices is so recent I own the first book about it ever. What the ancients left is obscured by sexualized anthropomorphication, and this dates back as far as the ancient Egyptians, who were this planets first metaphysicians. Even then, Thoth was seen as a doctor, equivalent to a tribal shaman, and the contemporary carvings of the Ica stones near Nasca in Peru depict open chest surgery being performed upon people by other, human, people. When all even gnostics care about is the body, why ever even study metaforms such as the hypertetrahedron, whose history is a cone? What good can it do?

Study of the hypertetrahedron is relatively recent, however study of the cone dates back to Euclid, who conducted extensive surveys on the nature of spirals and arcs derived from following different paths and courses around solid cones, and is credited with discovery and the naming of the phi spiral. Pythagoras was known to have used phi extensively in his own school, so much so that the pentagram, from which the phi ratio can be derived from any one of the legs of the star, became its mascot symbol. If you saw a pentagram inscribed above the archway of an inn, you would know Pythagoreans were drinking within.

a. phi as looking down the cone at an angle

The phi spiral is the only regular spiral derived from an irregular cone. An irregular cone is one whose base angle is not perpendicular to the cone's point. The phi spiral can be extrapolated from such a cone, and it is a regular pattern that can be expressed mathematically as a ratio of base angle to diameter. Another way to accomplish the same result without measuring the base, thus allowing you to simply use a ninety degree angle cone, is to think of phi as a regular, pi spiral around an equilateral cone, seen from a forty five degree angle. This is not exactly accurate but it is only meant as an illustration.

The ancients would have come to know of this by examination of nautilus shells on the beach. Here we see perfect phi spiral shells being lived in by mollusks, crabs, oysters and scallops, and further up the shore, on land, we find snails with the same kinds of shells. We also see phi in flowers, in the arrangements of their stamen and pistil, in the spiraling growth patterns of leaves around stems, branches around trunks and in the currents of the clouds and the waves of the sea, and all of these can be seen as being traced upon invisible cones, the history of spin over time.

i. the ratio that spirals

Spin over time indicates the phi spiral around the pi surface of the cone. Each of these is a ratio, and between them they represent a ratio. This only means that each one of them expands in an exponential gnomon, and that, when they come together, they orbit around one another, and when they do their combined history is a cone.

Stephen Hawking has represented the history of the universe as such a cone, and calls the projected future of this the light cone. This does not imply that the universe has contracted into the single moment at the center of the now and that it expands more over time because of this, although that is a convenient, and even fittingly esoteric, interpretation. Instead, as I have said, the history of our universe would follow a spiral path around this cone. Other modern researchers have speculated that time is comprised of a series of massive karma bubbles, or fluctuations of spheres of various sizes, and that the trajectory of our universe follows a spiral path around these. Spin permeates the universe on the informational level. It is the dimensionality that constitutes spacetime, and also timespace. To imagine homogenous spin for the histories of all particles for the entire universe is lacking in accurate data. The truth is always less convenient than the far fetched.

Temporally our universe involutes. Because its poles of involution are the finite number of gravitational black holes, it is incomplete. It will continue to involute until all that remains of the universal singularity are gravitational black holes whose histories are the wormholes in the last scattered remnants of tachyonic radiation. This is the clear light of YHVH being consumed by the darkness of Thoth. Although the involution of the universe follows the phi/pi pattern of the torus holognomonic shape, while it is within the universe it obeys the pi spiral more than the phi spiral. Entropy acts upon the pi shaped particles and spiral wave fields in the form of phi microwaves, and this as well contributes to the elliptical orbits of the heavenly spheres. Such as all of this is only the multiverse, never changing in being ever changing.

b. pi as looking straight down the cone

While we perceive time as spiraling in a circle around a central point, such as on a sun dial or clock, it often eludes us that the same shape is used by the hypnotist to put people into a waking sleep. It is doubtful that most of what I have described would have been deduced by Euclid, studying the cones, unless perhaps he had ergot. However the pi spiral was known of in the golden age of Greece, and is thought to date back originally to China, however there it was derived mathematically and using second dimensional geometry on grids drawn on rice paper with sumi-e brushes rather than with cones.

One saying is that "money makes the world go around." It is true that people's thoughts do seem wound up when they revolve around money. However, it is, if anything, time we all lose while pursuing the most in life, following the unfolding of our negentropic evolution as it flows around inward upon itself until our ends. Thus, if we look back from the moment of the now into time, it may very well be thought of as a light cone upon which is our own phi/pi spiral history, plotted hypercubically as the different ennegrams along the QBLH.

However this should in no way be thought to determine our future. Our ability to read the signs of the zodiac calendrically gives us not only a way of describing both the past and the present times, but also a way of predicting future cosmic events. This might go so far as to establish one's fate, yet still free will is infinitely more than this one particular tunnel reality. Free will is only bound to acting upon karma, that is, to change it, and, if it chooses to, reacting to karma as well. However the spirit of the free will is beyond even all universal karma, and therefore its future cannot be predicted. This is not only because it is outside of circular space measured by pi time. It is also outside of the phi light cone. It is outside of phi/pi and beyond the dimensions of both space and time. This is that which says I am. It falls upon the body of the universe that is YHVH, that it should turn into Thoth. For this there is a scapegoat complex. When I am becomes I was, time consumes the holy breath of all. For such is death, and death especially hurts God. Death is at the center of time.

3. the hypersphere

The simplest geometrical from of phi/pi is the one with the least geometrical distortion to vector, and that is the torus, or hypersphere. Because it is reducible to a point, the torus is holognomonically ubiquitous even within the material universe below the speed of light, although as clear light moving faster than photons, through the photons of the universe. The next smallest size particle is the electron, a point with an electrical charge that gives its probability well polarity. Then is the gravitational singularity inside a black hole, a baby universe comprised of aeons of infinitesimal spinning information. Then the temporal singularity, or microwave tachyon, expandable by scalar waves into a wormhole. After this are quarks, leptons and gluons of atomic nuclei, and then electron clouds, bosons, mesons, and baryons. Most of these particles contain electromagnetic polarity, and by their relatively (that is statistically averaged) phi/pi vectors we can see that the torus is the holognomonic geometry of all of these particles, as well as the pattern of propagation for all of the waves. As I have already explained, it also governs the greater tachyonic field lines of the galactic electromagnetic bubbles of the poles of the black holes at their cores. This is also the shape of our electromagnetic auras, or souls. It is because of this that the brain has evolved in the shape that it has over the countless millennia. Our nervous system does not extend out below us because it has grown behind the brain like a tail, although it has, but because it is acting as a tachyon collecting net held in gravity above the surface of a world the solidity of which produces the acceleration of microwave frequencies to tachyonic velocities.

The hypersphere itself is expressed in the coronas of suns and the atmospheres of planets. Like a hypercube it is a sphere surrounding another sphere. This is the perimeter between which tori form, such as the pressure centers in planetary atmospheres, and the sunspot forming loops and prominences in the corona of the sun. This effect also produces the wormholes on the surface of the event horizons of black holes, however here there are infinite possible hyperspheres, and therefore the wormholes are miniaturized by infinitely repeating halves.

b. the torus

The geometric shape of the torus is a sphere collapsed in at the two poles to a central point in the core. One hypothesis about the structure of timespace is that it is simply a big one of these made out of a hologram, or fractal, of smaller ones of these. These theorists never mention what the source of the holographic light is, though. Nonetheless, this is probably a true theory, since it would render our vector relative.

The torus, two-torus, or tube torus, is a circular shape whose surface is circular that wraps around once and connects to itself. Think of a circle drawn on a flat, two dimensional surface. Now imagine that the line of its circumference is a circular tube. These can be imagined as very open or very closed, according to their geometries, however for the unit torus, where the circumference is twice the circumference of each circle along the center, and the sum of the origin point is zero, the measure of its surface can be obtained exactly by a phi/pi spiral.

The unit torus is an ideal measurement, and not seen to occur in nature. The zero sum origin is essentially equivalent to the zero point energy of the singularity, whether electrical, gravitational or temporal, however, again, it is an ideal measurement not capable of being exactly expressed in material reality. Because of this there are not more exact examples of phi or pi measurements in the visible universe. The earth has an equatorial bulge, the sun is oblate, orbits are elliptical, black holes precess and their histories all differ. All of this, however, is due to the exactly phi/pi tachyons that pervade the universe, unifying timespace as a huge holognomonic torus called heaven by people and also called the multiverse. Still, there remains no explanation for the source of this tachyonic light, unless it is the singularity at the center of a black hole in a larger, parent universe, in the same way we believe gravitational singularities in the center of black holes in our universe may contain baby universes in the ancient histories of their spin.

i. phi/pi as area of vortex over spherical circumference

Phi/pi is given as the measurement of a spiral on the surface of a unit torus such that phi is the measure of the polar vortex and pi the measure of the spherical circumferences, of the surface of the tube torus itself, as well as around the outside of it from pole to pole. Thus the spiral measure of the torus is taken as the measure for the surface area of a torus.

This is an exact measure for the exact unit torus, and this is an ideal form which never occurs in material physical nature. Instead only derivations of it occur, and these include most of the forms I have already described. What is important to remember is that the torus being reducible to a singularity or zero origin means not only that it exists within our universe, but that it is simply the smallest of a series of geometrical extrapolations through distortions to geometry underlying dimension that produce the rest of the more complex hypershapes or metaforms, only one regular set of which would include the platonic solids.

Since the platonic solids are akin to colors of the rainbow, it is really something quite like seeing a rainbow, comprised of the prismatically refracted photons emitted from the sun cascading through a shower of millions and millions of rain drops comprised of millions of microscopic molecules of hydrogen-2 and oxygen atoms. The "rainbow," which can also be thought of as a solar loop or prominence, is simply occurring outside the singularity of the universe as the spiraling vectors of wormhole histories of gravitational singularities projected into the clear tachyonic light of hyperdimension surrounding the surface of timespace. In this way it is equivalent to QBLH, or the study of phi/pi, even though none of these exact histories perfectly follows that exact path.

Perhaps the best expression for phi/pi as a vortex over a spherical circumference would be a combination of all of the preceding views, measured using the different geometrical shapes. For the purpose of two dimensional depiction, however, it would be wise enough to draw a pi spiral circle with a phi spiral superimposed on top of it. As the pi spiral rotates inward, the phi spiral rotates outward and around. What you are seeing is the top of a torus. To see the torus all the way through, imagine the round spiral arms forming circles with one another, and infinite phi spirals surrounding the surface from pole to pole.

c. the hypersphere

As I have stated, the hypersphere is, ideally, a sphere within a sphere, however, in reality, here we see tori occur. These are the pressure centers in the atmosphere of planets, the loops and prominences in the corona of the sun, the wormholes on the surface of, and, exponentially halved, inside black holes. All of these lesser tori occur because they are in greater hyperspheres. Most tori do not.

The difference between a free or a bound torus is whether it exists within a hyperspherical system. For example, gravitational and electromagnetic tachyons inside galaxies are bound within a temporal hypersphere, while the wormholes of temporal singularities that float about in the voids between galactic filaments are free.

Because the hypersphere is the same thing as the torus, they are equally ubiquitous through spacetime and the multiverse, and because it is the beginning of a sequence of n dimensionality, the geometry of phi/pi that measures them extends out even further into the pure potential of time.

It is thought that these histories of the three dimensional gravitational singularities that project out from the surface of timespace feed spacetime into itself, absorbing the tachyons of the surrounding extra-universal glow, and in this way expand the universe. Therefore, time itself is the creation of space, while space itself amounts only to a finite history over time.

Time is the hypersphere of space. The shape of the geometry of timespace itself is unknown, and the spacetime continuum of the universe mostly uncharted. We do not know the actual relationship, or even the orientation of space to time, but it can be described by the geometry of the hypersphere, which we see is reducible to the universal singularity that is what is.

i. phi length area circle around a pi circle

The hypersphere is the shape of the phi spirals around the pi spiral surface of the torus, such that the part of the torus that it is really measuring is the diameter of only one circle of the two-torus, or only the radius of the torus. This is because phi without is with phi within. Now I have not mentioned pi within before. I have not fully described the phi vortex through the axis of the poles. I have only said that the surface of the two-torus was circular, and so was pi bound. To describe the hypersphere's mathematics, it is necessary to remember that the interior phi spiral is a measure of a pi surface, while the exterior is a pi measure of a phi surface, and that at no point do the phi and pi spirals measuring these surfaces cross over one another. Just like the hypercube's many different regular shadows and irregular slices, the torus and the hypersphere are only two regular shadows describing a shape that is involuting.

What this involution means mathematically is that the surface area of the torus flows inward upon itself at one pole and outward at the other pole. However there is also more involved than this. The external surface area of the torus can only be divided into exactly seven different areas such that the single line dividing them will spiral around the complete surface of the torus only after all seven of these areas, which can be color coded according to the rainbow for reasons I will explain momentarily, to pass through the central vortex of the torus. Remember that the torus is also a hypersphere, however, and so the internal surface area and the external surface area exchange places as the torus involutes. Moreover, the more the surface area of the central vortex of the torus is seen to depend upon the phi spiral measurement of it, the more distorted it becomes as this phi measurement is turned around the circumference of the torus as the two tubes of its radii revolve. Lastly, remember that the torus is reducible to a singular point, and therefore the different movements of vectors on its surface area cancel out one another's spin.

The surface area of the three dimensional shadow two torus of the fourth spatial dimensional hypersphere can be divided into the same number of areas as there are colors in the rainbow because it is the mathematical expression for the visible spectrum of photonic radiation whose speed comprises the boundary of our universe. If you split a beam of white photons with a prism and produce the seven color spectrum, what you are really doing is fracturing the uniform encoding of the tachyons emitted by the photons, such that their different frequencies of wavelength, and therefore the different color properties of the photons, emerge. It is not the photons that are traveling at uniform speed. It is the tachyons that pass through them that do so at a uniform speed. When they are passing through electrons, tachyons move at a different speed, and other particles, other speeds, such that it is really their wavelength vectors that determine the mass of the particle wells. The different wavelengths of photons, for example, each represent tachyons traveling at different velocities motivating the photons. All the photons of the visible spectrum move at a relatively uniform speed, however the speed of their tachyons differ. In the ultraviolet and infrared ends of the photon spectrum, the photons themselves are moving at relatively slower or faster rates, while the tachyons themselves are closer to moving at the same velocity. Light that is slower than the speed of light forms matter-energy, while light faster than photons forms time, and these two can be graphed perpendicularly to one another, with the visible spectrum bouncing off the hypotenuse between them just as in a prism, white light rising up from matter-energy and the seven color spectrum measuring time according to a torus, as I have described.

B. the ratios

While phi and pi, being transfinite numbers, can be used to measure many of the variety of fourth dimensional metaforms, it is important to remember that they only amount to numerical sums, and are, each individually, and in ratio to one another, ultimately mathematical. Though they both describe geometrical spirals, each is given the name of a Greek letter because they represent decimal sums that stretch out forever. The sums themselves are first dimensional, though the actual measure of decimal places between whole real integers into which the mathematics of their measurement drops implies the infinite dimensionality expressed by both n exponentiality and the relativity of figures of measure.

These can be derived in a number of ways, the quickest of which is the mathematical, where each can be rendered as either an inexact sum or an exact ratio, and these can be plotted geometrically as the internal angle of two different kinds of spiral, which are, as I have said elsewhere, really only two different views of the same spiral. The difference between them can be thought of as the measure of the angle of movement of this spiral from one position to the other, and it is expressed mathematically as a ratio of two ratios, and this is called phi/pi. Here we see that these letters are also a mathematical way of expressing the numbers, since they are each the given names of variables used to express them in short form for equations.

A good way to express them is by first order of difference. This is that in their set of multiples and divisors that would form a pattern upon which counting would constitute an expansion and contraction. Here we see the tables of the eighth and the ninth factors become relevant, as they represent this same pattern, though not for the same numerical sets as for the phi and pi. Here we see that pi forms a numerically repetitious pattern, and phi a more open and exponentially arcing one.

Elsewhere in this work I describe the tables of the eighth and the ninth in greater detail, and this description also includes a greater expansion on some of the fields in which phi and pi can be found. About the tables of the eight and the ninth, I believe these to be those same tables of the eighth and the ninth that Hermes Trismegestus is attributed to writing a divine dictation to his son upon. This text is attributed to ancient Sumeria, and it is thought, to before even the time of Zoroaster. It was possibly the knowledge received from these tables, perhaps originally belonging to Enkidu and Enlil, as indicated in the epic of Gilgamesh, before being passed on to Hermes, who would become Thoth of Egypt, later called King of Israel, Zoroaster, Mithra, and so on, until the time of Christ, and the institution of the title.

In the tables of Hermes he advises his son to pray by "singing a song in silence," which eludes to the beginnings of the initiatory occult, who, according to the sepher yetzirah, swear to the breath of conspiracy and the oath of secrecy according to the "circumcision of the tongue and the circumcision of the membrum." Such practises as these have lead to the secret societies of today. The theft of the tables by Abraham was played out by Imhotep from Sumeria, Akhenaten from Egypt, Moses in the desert, Menelik, son of King Solomon, the Knights Templar with the Dead Sea scrolls from under the second temple, and, according to modern Masons, his assassins from Hiram of Tyre. The archetype of the grand thief or great jester was associated with Loki by the Nordics of northern Europe, and his Greek equivalent was Dionysus, later known to the Romans as pan and to medieval witches as Satan. He was associated with Set, God of serpents in Egypt and with Shiva in India. In America, where a copy of the Decalogue carved in stone has recently been found older than 1000 years, he was known as Quetzalcoatl, the feathered serpent, caught in the midst of molting.

The only probable explanation for this being described as a divine pymander by Hermes is to further associate the archetype with the form of the dragon, the constellation we know today as Draco, which surrounds all the constellations around the current north star Polaris. The reason that the archetype is described as more reptilian in the Southern latitudes, and Loki, as a jester, as more human in his role, can only be because such are the climatological conditions in which lizards, snakes and reptiles live, whereas they do not live in the fjords of Norway. This is an ancient attribution to the stars of the fixed constellations, dating back to the very birth of civilization in the cradle of the fertile crescent between the Tigris and Euphrates river. This, archaeologists have discovered, was the source of many zodiacal finds and artifacts whose histories can be traced throughout the world trade routes during all of interceding history. The origin of its attribution, although northerly from Sumer, as a dragon, lizard or reptile, probably refers to the movement of the position of the pole star in the heavens relative to the climatologically temperate zones, where the descendants of the dinosaurs still live. Had the ancients any knowledge of the thunder lizards, as the engraved stones of Ica, Peru, near Nasca, indicate they did, then it would certainly account for the cultivation of their belief in dragons. The movement of the pole star relative to the seasonal latitudes of the earth is an effect of earth's precession, whereas the movement of the seasonal latitudes of the earth relative to the pole star is caused by the position of the sun relative to the center of the galaxy. Both of these are ratios on the same proportion of first order of difference as phi/pi, and can be thought of as tonalities on a frequency scale.

1. phi: (1 + the square root of 5) / 2 = 1.61803

This is a very widely known and truly universally ubiquitous pattern. It occurs as a statistical average of many measurements in the natural sciences, and it is a variable used to represent a relatively constant kind of vector in physics — that taken by an electron changing energy shells. As the measure of a light cone at a forty five degree angle, it derives directly from the ideal realm of pure Time. It is one of the patterns governing psychology, which Freud implied by his choice of this letter to stand for the transducted electrochemical substance in the nervous system.

It can be expressed using mathematics, the language of the ideal realm as either an exact ratio or an inexact, transfinite decimal, integer. The exact ratio is given as (1 + the square root of 5) / 2, where the square root of 5 = 2.23607, therefore 1 + 2.23607 = 3.23607, and where 3.23607 / 2 = 1.61803. However, this sum is not exact, because the equation includes the division of a square root, and this results in a sequence of infinitely repeating halves that can be graphed as either a fractal or a gnomon. In this case, phi, a numeral that continues on infinitely after its decimal point because of such information sorting using the relativity of measurements, is one such gnomon, or regularly repeating pattern. The Julia set of fractals is based on the Fibonacci division of multiples sequence, similar to the subtraction of sums of Blaise Pascal's triangle, and this differs from the phi gnomon by only a minute amount, such that they share the same expansion, but that the sum of phi itself is not necessarily contained upon the number set of the Fibonacci sequence except as the product of any of the given ratios. This is only because such a number set can only describe finite sums, such that phi could be a fraction of itself in a larger ratio, and since phi itself is a transfinite number it can only be expressed in a series of other transfinite numbers, which is the proper exponential expansion of phi. The difference between this true gnomon of phi and the fibonacci sequence, which produces the Julia set fractal, is similar to the difference between the actual sum and number set of the phi spiral and the number set of sums given by the tables of the eighth and the ninth. Their is a virtual similarity, however phi remains ideal beyond even the spiral fractal of the Julia set, the basis for this described by the Fibonacci sequence, the number set of Pascal's triangle or the tables of the eight and the ninth, and, to my knowledge, the actual number set of transfinites in which phi belongs, and which also includes pi, has never even been deduced. It is possible that this pattern underlies the gematria of the Torah, and even of all social systems such as crime rates, economics, and the entertainment industry. It is only one kind of such naturally recursive cyclical patterns that may be observed to occur as multidimensional vector sums over histories, and therefore is not all pervading on the observational level. However, in the unseen realms that connect the galaxies, this measure is integral. It is this cycle I refer to when I shorten the ratio to phi/pi.

a. phi in nature (leaf patterns)

The spiral concourse of branches growing around trunks, stems growing from limbs, and leaves growing from sticks is, in some plants, particularly many of the smaller, flowering plants, a phi spiral sequence. This is a product of statistical averaging, and there is usually little attempt made to explain or account for axioms believed to be universal arrived at by statistical averaging. Take for example Planck's second constant, given as η, which represents $1.05457 * 10^{-34}$ Joules per second, believed to be the average speed of energy of an electron. As I have repeatedly described, the only way any measurements can be made about electrons is by interaction with photons, and therefore Planck's second constant is the average difference of the measured effect from the constant speed of light. Because as I have shown the only fixed speed of light is visible photons, while the invisible spectrum is comprised of different frequency pulses of particles, and is determined more by the fixed speed of even smaller, non-electric, energy. Therefore, not only does Planck's second constant derive from interaction with photons, it only describes a spectrum of velocity that can occur within the range of the realm of the speed of photons known as the continuum of the local universe. There probably are, however, electrons even in the voids between galactic filaments, at least enough to give charge to the random electromagnetic temporal singularities that expand into wormholes measured as gamma ray bursts.

Because Gamma rays are bursts of energy moving at a different velocity than light such that they constitute the sudden appearance of solid particle radiation like photons, this alone disputes the concept of the speed of photons as being representative of the full spectrum of velocity of radiation for particle energy. Also, because the angle of spiral an electron takes when it changes orbital energy shells, such as when it encounters a photon, is phi, thus we see that even these blossoming designs seen as though telescopically through the photonic lens of the visible universe, their sums over history reaching infinitely out into the voids and their origins dimensional perpendicularity points are only heaven's mirror of the tiniest flower.

b. phi in music (theme)

One of the most important elements in fluent communication strategies, such as music, speech, or narrative story telling, is the element of rhythm inherent in timing. This contributes a meta-anomonopea connecting disparate elements of idea, such as character, set or setting, conceptual constructs (all of which occur also in dreams), as well as the perpendicularity in music of theme and motif.

Here we see that a motif is a recurring variable and the pattern of its recurrence constitutes a theme. This is also true the other way around. These are thought to be steps in logic along the same ladder as the combinations of notes to form chords, the combination of chords to form rhythms of harmonies called songs, and even the division of half and quarter notes, minors and sharps (which are the same thing).

Here we see that exponential expansion is a fractal that sounds rather flat, and that the phi spiral gnomon evokes emotion more than the pi spiral gnomon. Thus it may be found to underlie a great many of the musical themes that flow from the stream of consciousness through the instruments of the muse.

It would be a wrong belief however to think that such implements of stylistic element are in any way necessary, though they are both useful and beneficial. Far too frequently people believe ideas that seem equivalent on a different level of implication than can their corroborative implications support. People labor under the misconception that the better the one thing follows from the other, no matter how related or unrelated the two may be, if the two ideas rhyme enough to inspire or evoke instinctive response, then there must be a direct causal connecting principle to link them, imputing to this negentropic statistically averageable "universal constant" the role of consciousness, and calling it "God." In due time people will discover that the same effects can be accomplished using ESP and telekinesis, however this will only be following a higher path of karma through multidimensional reality than the paths commonly followed now.

c. phi in architecture (the alignment of the pyramids)

The builders of the ancient Egyptian pyramids, it is thought, did so that they might meditate upon their own grave. Thus, this pattern they aligned into the tops of the three pyramids, that are also aligned by the bases with the stars of the belt of Orion. The constellation of Orion, known then as Osiris, was believed to represent the king of the underworld beyond the river of the Duat, or the heavenly sphere divided by the Milky Way. This belief derived from the Dogon people of the Eastern savannah of Africa, near Ethiopia, where the Nubian Egyptians of southern, or Upper Egypt were from, having followed the northward flowing northern hemisphere Nile river from its source in modern day Aswan towards the Mediterranean delta. According to Edgar Cayce, and later confirmed by satellite imaging of Africa's geographical topography, the Nile river once flowed from the same source westward, emptying out into the Atlantic. It is possible that the Dogon people, who studied the star Sirius extensively, and even made depictions of its binary system sidereal orbital periods as paintings on the walls of their caves, only moved Northeastward into Nubia and began building pyramidal graves when the direction of the flow of the Nile changed, although it would be unlikely that this happened in such a short period of time that these people could be thought of as having been swept there, as in an Exodus or upon an ark. Sirius is known as the dog star, and its helialical rising at sunset marks the beginning of the "dog days," or hottest days of summer in the northern hemisphere. It follows at the heel of the constellation of Osiris or Orion, who was associated with Hercules and depicted as a stone geoglyph in England similar to the astrological Nasca lines. Thus it may have been that the ancient Egyptian architects were attempting to express Sirius as a phi superimposition over the Giza pyramids dedicated to Osiris. In either event, at the core of this super spiral karmic center is an underground chamber tunneled into the hard granite rock of the Giza plateau, presumably using high pressure water and precision water pressure powered diamond, stone and crystal drills.

2. Pi: 22/7 = 3.14159

Pi is a transcendental number determined as the area of a circle derived from the circumference divided by its radius and measured as a circular spiral around a central point. Pi is transcendental, as opposed to phi being transfinite, in the same way that the point of the pi spiral is central while the point of the phi spiral is offset from center, and these all really mean the same thing in that the pi and phi spirals both describe the same measure of the expanding circumference of a cone seen from different angles, and the movement of this precession of the central point of the spiral, or the sum over history of the spiral comprising the cone, can be represented by the number set of phi/pi that includes phi and pi, as well as by the fibonacci set, Pascal's triangle, and all the other real vectors that exist in the material universe.

A good example of a pi spiral in nature is the coiling up of the sun's electromagnetic field bands due to differential rotation. Here we see that, like phi, pi is expressed in pure form in both the reality of the physical dimensions of nature, and beyond these in the abstract realm of ideal as a mathematical equality. The expression for this is derived from only one possible numerical ratio representing the circumference of a circle over its radius that happens to fall both upon whole real numbers, and it is from this that the endless decimal integer factor is derived. It is thought that the ancient Egyptians were the first to calculate pi as 22/7, and to use this exact measurement to square the base of their pyramids, however the pi spiral can be found depicted on rocks and in caves all over America, Europe and even Australia dating back as far the Cromagnons. The quotient of 22/7 has only recently been delved into deeply by mathematicians using computers to figure the number out towards the nth decimal point. 3.14 is the jumbled number equivalent of the infinitely repeating pattern of exact fraction decimals, such as $1/3 = 0.3333...$ where the threes after the decimal place continue on forever according to mathematicians.

This property of exact fraction decimals was known of in Ancient Greece, where they studied the square root of two diagonal of the square and the square root of three diagonal of the cube and understood the fractal geometric implications on these measurements of fraction square roots.

a. pi in nature (trunk rings)

Trunk rings aren't exactly pi spirals, but they aren't exactly individual circles either. They mark the pattern of the growth of the bark on the external surface of the trunk of the tree, and this is more in some places than in others over the different years. For example, more chlorophyl forms in the surfaces of smaller, flowering plants on the side facing the sun, and therefore plants grow more cells on this surface, and cause the flower to bow down towards it in a spiral. The same happens for tree bark, though, because trees live much longer, they may be used as calendars to measure precession.

371

The precession of the equinoxes happens according to the interaction of many various pi spirals, such as the movement of the electromagnetic poles of the earth and the sun, that are smaller measurements within larger phi spirals, such as the tachyonic torus of the electromagnetic poles of the black hole at the center of the galaxy. Phi/pi is the measure of all of these galactic bubbles, however pi is the basic measure of their circular circumference, and this determines the accretion disk upon which spiral galaxies such as our own Milky Way occur and revolve.

b. pi in music (motif)

A motif is the musical equivalent of an event in time, and, due to the relativity of measurements described between the phi/pi gnomon, the fibonacci Julia fractal, Pascal's triangle, the tables of the eighth and the ninth and the impossible loop of consciousness itself, as well as exponents, decimals and phi and pi individually, all of these are of generalizable duration, in synchronicity with the universal singularity.

Most often in music a motif is a small, evocative phrase of notes, or a harmony, that is expanded upon throughout a piece in a theme. Doing this, it is believed, is equivalent to the painter putting paint upon a canvas with a brush, such that a scene of set or setting is evoked in the mind of the listener, however more recent bands, such as the Beatles, have introduced Eastern trance composition stylings into the classical sensibilities of western music, and the other base systems for harmonic cording than the octave, such that their interpolation and juxtaposition creates even more intricate "sound shapes" occupying the mental projective space of the listener.

Many of these forms of contra-posed modality describe the various angles of Platonic metaforms as they expand geometrically out into hyperspace. These are described best in the tables of the eighth and the ninth, and ideally by phi/pi. Each of these sound shapes individually is represented by a pi spiral sephira, just like grapes along the vine of the QBLH. This is the formula of theme over motif.

c. pi in architecture (the measurements of the pyramids)

Not only did the Egyptians probably use 22/7 in the calculations for the square base of the pyramids at Giza, it is thought that pi was also used in the reckoning of their height. This would mean that the pyramids were each the stellations of half the diagonals of unit cubes, each relative to the other. This would mean that the vast sum of specific numbers used in the construction of the Great pyramid could overlap holographically the lesser two pyramids. This would seem to refer back to my metaphor earlier about stacking cones, however it may be thought of as more relevant to the stellar activity in Orion's belt.

Just like Sirius, the smallest star, offset from the alignment of the other two, is a binary, pulsar system. Along this belt, and between these stars, are the vast gas pillars of the Orion nebula, where we can see stars being born, coming into existence as supernovae, even right now as I write these words.

If you would like to think of the ice cream cone metaphor, also remember that the Egyptians aligned the air shafts of the great pyramid's King's and Queen's chambers, as well as the pyramids themselves with the sphinx, to a date when the constellations of Isis (Sirius A and B) and Leo aligned with them, which was six precessional cycles of reversal of the earth's electromagnetic polarity ago, 12,500 years past. There is also the greatest concentration of galaxies in the perimeters of the constellation Leo in the visible sky.

3. phi/pi

As I have described phi/pi is a universally ubiquitous statistical average, however rarely occurs in its ideal form as an exact measure in nature. Imagine all the arrayed petals of the lotus blossom, or all the convex seeds of the sunflower in the center of its bloom, and imagine all the electromagnetic polarity field lines of the stars of spiral galaxies that arc above and below to be bent to the poles of the central black hole, for such as are the myriad of the faces and the forms of phi/pi.

Perhaps the best example of phi/pi in nature is the vector of precession. As a sphere, such as our globe, a star such as our sun, or even a particle such as a photon, rotates around a pole, so too does this pole orbit around another central point such that the sphere itself is a pi spiral, and the cone from the center or core of the sphere to the circular precession of the pole can be measured by phi, thus rendering the surface area of the tachyonic torus equivalent, or potential event in time.

This can be done for the earth, and yields the earth's electromagnetic field, distorted on the side of the sun into the Van Allen belts, and behind in earth's shadow the plasma sheet of flatline history tachyons. The same process can also be performed for the sun, and yields the same measure, its electromagnetic field, but even more distorted towards phi/pi as pi is the winding up of the sun's electromagnetic field lines due to differential rotation and the precession of its electromagnetic poles can be traced according to a cone drawn up from its polar circle to a measure above (in the "future" light cone) or below (in the "past" cone) as the phi measurement of the difference between the circumference of this circle and the radius of the tachyonic field line wavelength connecting the sun, or the star, to the black hole at the center of the galaxy. Phi/pi is represented even more homogeneously by the enormous electromagnetic torus of galactic bubbles, comprised of millions and billions of such conjoined and connected star systems. The most exact measure of phi/pi in nature however is the tachyonic emission field of the photon that makes it glow. It is because of this that a single photon viewed through a convex lens will shine in the pattern of the unfolding of a hypercube, and also why the type of photons emitted by the sun refract prismatically into hexagons at an angle through a convex lens, and it is also the reason for the seven color spectrum of rainbows.

As I have described, phi/pi is not only the measure of the holognomonic metaforms of reality, but a number set expressable in idealized form as a series of sums. This would represent the essential signature of the universal code that is the average harmony of all statistical probabilities over the history of potential. This number set contains both pi and phi, as well as phi/pi, which is a decimal or fraction smaller than one, as well as pi/phi, which is an integer greater than one, and is a fraction of two decimals that almost counterbalances to the whole number three. The dichotomy between phi/pi and its reciprocal pi/phi is similar to that between the cube exponent and the cube root, where the former is a whole number and the former is the irrational square root of three in a unit cube.

This number set is theoretically comprised entirely of transcendental and transfinite fractions, and all of its relationships between these numbers are identical transcendental and transfinite ratios of a higher order, such that the infinite number line of the set as well as the exponential expansion of it are both reducible to the singularity of one, and this one invertible to nothing. Since only a few of these sort of numbers are known, such as phi, pi, the square roots of two and three and 1/3, most of the set remains obscured within the realm of imaginary numbers, such as the square roots of negative numbers (sums representing absences which, because nature abhors a vacuum, cannot be proven to exist in the the multidimensional reality described by the square function), which are so complex a concept that mathematicians using them usually simply annotate them by affixing a variable and leave them further uncalculated, except relative to contextual equations.

Because of the nature of these numbers, they more await discovery passively than present themselves as common variables. There are, in reality, an infinite amount of them however. It is to this entire number set that I am referring when I use the term phi/pi.

a. electron shell, magnetic torus

As I have described, the rainbow is the same effect of electromagnetic banding of the earth's electromagnetic field as the coiling up of the electromagnetic field of the sun, and these are both the same as the orbital path taken by the electrical charge of the electron as it orbits an atomic nucleus in its energy shell. This forms a pi spiral spherical probability well so rapidly that the charge is distributed simultaneously to all points within the field, and this effect is known as the electron cloud of potential. The electrical charge is held in orbital distance in the energy shell as static by microwave tachyonic gravity, faster than photonic entropy. Differential rotation occurs for the electrical charge because precession is occurring for the polar axis. Thus the actual shape of the electromagnetic field or energy shell is, holognomonically, a torus, although such can never be found expressed exactly in nature, and most are found surrounding solid spheres anyway, and so are not visibly observable as passing through the core.

When the poles of these electrons are aligned, a tachyonic microwave current is formed between them that can accelerate energy to beyond the speed of light, and can quantum tunnel through solid particle wells such as photons.

b. tachyons — from microwaves to wormholes

When such a concentration of energy as this occurs, as can happen between oscillated frequencies harmonizing in a scalar wave field, the tachyons that form the potential substance from which the size and shape of probability wells arises can be expanded into hyperspace field lines that connect disparate points through the multiverse in a temporal singularity. The moving shadows of these in subspace are the fourth spatial dimensional metaforms of the third and second dimensional representational hyper shapes and the platonic, archimedian and stellated solids and regular polygons, fractals, gnomons and tesselations. Since as long as we exist, we exist relative to time, the ultimate understanding of time is ultimate understanding of the letting go of our existing self, and this leads to the font of emotions, which are these fourth temporal dimensional geometries passing through our clear concentration of consciousness. It is the history of these archetypes which we are, more like some than others, and so is our reality changed, and so it is, as well, for the greater wormholes that are the sum over histories of the gravitational singularities, or baby universes, inside the black holes at the centers of all spiral galaxies that distort the continuum of three dimensional subspace into the wormholes inside the event horizons surrounding these singularities, which are inverted at the surface of the black hole's electromagnetic poles into the perfectly straight tachyon field lines that arc out to connect to the precessing differential rotation of the surrounding galaxy's stars that regularly reverses their electromagnetic polarity, inverting the electromagnetic polarity of any planetary bodies, and thus causing a direct effect on the electromagnetic fields of any and all sentient souls with nervous systems that generate a karmic aura. Thus we are alike within and without, as above so below.

c. the involution of evolution over time

The sum of phi/pi expressed as the equation of its ratio is given as light and love and life, that it should be known that all these three are one. There is no greater purpose to wit than what is given by these three in one. There is no trinity more holy, more divine or more sublime than that which is the truth, and the highest truth of the most high can only amount to scaling this mountain that is the triangle of Light and Love and Life. Here we see the triangle hexes, and that the hex stellates into a steloctahedron, and that this form bears all the burdens of our reasoning. Thus it is the metaform given to time, that it should be free of the imminent eschaton of the Great Because, loose the shackles of the hunchback whose burden is questions, and be forgotten of that curse upon it that it Should, so that it always was and always will be, and so it is.

This translates into the forward and backward flow of entropy, the inversion of the temporal chi that represents the progression of the universe of visible light and matter energy into greater disordered states of chaos, and the opposite effect to this in the evolution of living beings and the clear light of tachyons. As entropy inverts, the great torus of the hypercube of time involutes, and the multiverse breathes with light, love and life. Such is the involution of evolution over time: only good karma, or else no good or bad karma, and never bad karma. This is also known as QBLH.

IX. QBLH

The soul is only the accumulation of karma of an individual free spirit, manifest in its aura. It was this that was believed by the ancient Egyptians to have been weighed against a feather by Maat at the time of an individual's death. Now there have been studies on the metaphysics of ethics since the beginning of recorded history. The idea of karma originates from the Indus river valley, and the idea of ethical legislation derives from the peoples of the Tigris and Euphrates river valleys who brought the Nubians out of cannibalism according to Egyptian texts. In the east the school of karma arose to develop its own ideas of right and wrong behaviors based on trial and error with natural consequences. In the west the school of ethical legislation adopted as its scales the Manichean dualism of Good and Evil.

The occult school on the underworld or the afterlife is based on viewing archetypes as like business establishments known by their names, and all things, including their own souls and other people, as mere collages of karma, so much property to be occupied by advertisements for information. According to this gnostic belief, the soul decays much like the body will eventually decay, such that the physical body itself is equivalent to the microbacterial organisms, maggots and worms that digest the divested corporeal remains. The longer we live, the more damage it does to our souls, however these are merely vessels, meant to carry us so far before we move on into another form.

QBLH extends upward and outward above and beyond the regular surface of timespace that conforms to whatever geometry (open, flat or closed) that governs the universe, causing the observed expansion of spacetime. These are the histories of the gravitational singularities that, from distortion towards them of multidimensional wormholes through hyperspace, project out as vast spirals beyond the speed of all photons. They expand through dimension as the torus of timespace gradually involutes all its continuous substance through itself, leaving it behind and measuring first a vast cone inside another torus, which can be measured as a hypertetrahedron. From this can be extrapolated the next level of geometry, occupied exclusively by the fourth temporal dimension, and this complexity of structure can be represented as a hypercube metaform. These continue on through the rest of the platonic solids as hypershapes and measures a ubiquitous phi/pi gnomonic spiral throughout them all, that continues even beyond them into yet unknown geometries.

Thus it is not only this itself that we are, for here is recorded the history of the timespace from which our immortal souls come, but through this that we must pass once again when we leave the vestments of incarnation within this universe behind. Our spirit is like a free floating temporal singularity that passes perpendicular into the multiverse or through a gravitational singularity outside into ylem. From here we seek our own origins by applying our understanding of the geometric rules of dimension just as we do our inherent physical understanding of the rules of physics in the local universe. This is the path trodden by YHVH, for in that these things might be, they may be of and for Him, for they are truly the rule of QBLH, and this has long been thought the rule of YHVH.

A. history of the QBLH

The official story of the origin of the QBLH is that it was the message engraved on the first table of testimony brought down by Moses from Mt. Sinai and smashed upon his finding his people worshipping a brazen idol. According to the story Moses went back up to the top of Mt. Sinai and came back down again later with two different stone tablets, these containing the ten commandments, the first of which guarded against the keeping by the people of any brazen idols.

As with many sites in the middle east, which has adhered as much as possible to the economic status and social structures of Biblical times, a site for Mt. Sinai has been selected by archaeologists, geographers and historians both internal to the sovereign nation wherein is the chosen mountain as well as international experts. This was done by Muslim Sultans and Catholic Templars to increase tourism, however it occurred well before the scientific method was adopted to keep record of the reasons for such things. Similarly Mt. Ararat in modern Turkey is supposed to house the last of the wooden planks and rusted iron nails of the real Noah's ark, covering a vast, flat plain on one side of the mountain's otherwise sheer drops and craggy faces. Most of the archaeology done in the middle east is as much based on superstition as the prophecies written by those ancient scribes that used to gaze at the fires of Miggido. Other than this no archaeological evidence whatsoever exists to substantiate this biblical story. According to the jews of Ethiopia, they are in possession of the stone of testimony, and it is a flat rock which they parade around annually under a sheet. Any collection of evidence to substantiate this however is barred by the seemingly perpetual governmental insanities of east African politics.

When considering the history of QBLH it is important to consider several separate things as related to one another. This means following the history not only of the stones upon which QBLH is said to have been engraved, but also the keepers and the meaning of the oral tradition of its heritage. These can be thought to include the scribes and rabbis who have written on this subject, however the details of their lives are meant to be sorted out as much as possible, when possible, since it is the belief that when a person is studying the QBLH, it is really God studying God.

1. origins in stone

Trance channelers, remote viewers, astral travelers, psychics, precognitives and prophets have difficulty seeing ahead in time beyond a date in the near future, thought to begin after December 12, 2012 on the gregorian calendar, as if there were some giant wall of ice or gravity well reflecting only what they project from their present upon the future. There is a similar difficulty in examining the past, since history is always changing, and each generation knows more about the world than the previous generation. In truth we are each given the same infinite potential understanding, and it is only in what we reflect back on our realities that makes our mark on this dustbin that is history. However, the soul is only free of stupidity in the same way that the mindless body is free of intelligence, and so we see that, while all minds might well think alike the greatest, it remains actually only a few people that can contribute themselves in this regard to history, as having freely known beside infinite nothing.

Therefore at the time of the writing of the rabbinic scholars of Muslim Spain regarding the origins of the QBLH in the biblical story of Moses and the ten commandments given at Mt. Sinai it is not thought to have been known that Cromagnon man in the Lascaux caves in nearby France had, hundreds of millions of years ago, used various ground minerals, herbs and berries applied to the walls of the cave to depict the same figure of a bull, and that, concurrently in America cromagnon men had left carvings of pi spirals on rocks similar to those thought to have been made by Australopithecine hominids in the Australian savannah. Even if the rabbinic scholars were being allegorical, it is still correct to address the actual evidence of ancient learning rather than subscribe to those beliefs that comprised it.

a. prehistory (petroglyphs and geoglyphs)

The history of humanity is recorded in stones, from the date of the earliest hieroglyphic tags on small stones representing the oldest known writing, the linear a and b cuneiform alphabets of the Sumerians impressed with river reeds on clay tablets, to even earlier, in the building of stonehenge and the ancient depictions of game animals and shamanic hunters on the walls of the earliest human cave communities, the history humanity has kept to describe itself, to communicate its beliefs to itself, to preserve its metaphysics, has always been recorded for all time. There is much human life that floats by in the intervening millennia like so much water under the bridge, whose works for the survival of and contributions to the communal societies goes mostly forgotten by their own descendants, and it is true that even the date of such monumentally important inventions as the wheel cannot be known for this reason.

What little is known of this ancient civilization remains only in its ruins. Here we find not only carved stones such as those from Ica in Peru, depicting knowledge not only of the skeletal structure, but also the visceral physiognomy of a pterodactyl, and cave paintings, such as those at Lascaux depicting early human animism, but also the vast Megalithic pyramids of China and Merubecka, South America, which have largely worn away to only very large burial mounds, as well as the vast, ancient canals of South America and the number of sunken cities off the coasts of modern shorelines. This period in history is marked by such great edifices as the Bimini road as much as by the flint chipped arrowhead or the atlatl. It is by the presence of the ancient petroglyphs, those edifices that remain in stone, such as America's stonehenge (recently called mystery hill), the stonehenge of England and the ancient temple mounds of China, Europe and Oceania as well as chipped arrow and spear points, carved stone pipes and the carved stones of Australia and America, as well as that of the ancient geoglyphs, such as the Nasca lines, the giant hunter and serpent mounds of England and the South American canals that we can identify the presence of global, stone age civilization. Thus here is where we must look for the origins of QBLH, for, if QBLH is a truth to be known beyond mankind, then it would have existed already before it could have been earliest discovered, like a flower awaiting cross pollination.

It is true that all of these features of ancient history are only now so much dust blowing in the whirlwind of time, however this whirlwind itself is phi/pi. The meaning of the symbols and relationships we can observe represented in ancient petroglyphs and geoglyphs may have been as mysterious to the people who crafted them as it is to the majority of people today, however we can see that there was clearly an astronomical predilection, as many of the megaliths are either maps of the heavens according to animist or anthropomorphic constellations or markers to measure the place relative to the horizon at different times of year of different heavenly bodies, and there are pi and phi spirals depicted the world over.

Of the first type we can conclusively include the nasca lines as representing the Peruvian constellations attributed to the stars in the southern hemisphere, as well as the great hunter and serpent mounds of England. Stonehenge in England, Carnac in France and mystery hill in America, as well as literally millions of other neolithic sites, represent alignments of stones to measure the place in the heavenly hemisphere of various celestial bodies and to mark the time of certain seasonal events. The later, great pyramid of Giza also contains numerous such alignments in its architecture, however it is unknown if other, more ancient pyramids, such as those in China and Merubecka, which have deteriorated into mounds, served the purpose of making similar such celestial measurements. Pi and phi spirals can be found carved into and painted on rocks and caves in Australia, Asia, Europe, Africa and the Americas. Then there are some that may only be cultural, such as the meglithic heads of Easter island.

It is thought that, because some of this civilization was lost when the coastline changed at the end of the last ice age, many of its original records were destroyed. Most civilizations, from the level of the most superstitious tribespeople of Africa to the most refined religiosity of Europeans, share a common myth about a flood and a savior who bore the seeds of the new civilization aloft the waves on a boat. Thor Hierdahl has proven that even the modern Peruvians, who live much the same way their most ancient ancestors did, can construct a sea faring ship of reeds and therefore could have crossed either the Atlantic or Pacific ocean. Another modern theory derives from recent translations of the lost books of Enki, and describes how the earth has been nurtured by extraterrestrial entities in space ships. In either event the moral of the story is best expressed by Pontius Pilate washing his hands of moral responsibility for Jesus.

While most of the world's myths of the flood and the great civilizer remain insubstantiable, the history of events described in the lost books of Enki, as well as other ancient Sumerian clay tablets, do survive to describe the history of a particular stone, called the Ram, or what would become the table of testimony, upon which they set down a record of the history of all of civilization. This was described as being two stone pillars secreted away in an underground chamber accessed by a river through cavernous tunnels in the Labyrinth beneath the Giza plateau by the Greek scribe Solon, who was told that they were from Atlantis. It is thought that it was this same stone, or stones, held by the hands of the historical Moses.

b. the legends of the sacred stones

All modern religions started with the early practise of savage, petro rites. This consisted largely, as testified to by Moses, of the worshipping of graven idols and the making of blood sacrifices to them. These stone deities were largely thought to represent the weather, the chief, or some other desired person or object. Human sacrifices were gradually replaced in the cultures of oldest ethical legislation, however the practise of voodoo animal and even human blood and murder rituals perpetuates to this day. These times, now steeped in mystery due to the subsequent repression of dogma by the Judeo-Christian descendants of the earliest ethical legislators, have been called by modern occultists the age of the Mother or the Moon, because of the insanity associated with the moon and because of the blood of monthly menstruation. The records of history of most of these types of civilization have been lost, and must be pieced together again by their sheepish conquerors in order to decipher their original meanings to the original ancient practitioners.

Perhaps the best way to understand the impact that stone age civilization had on ancient humanity is expressed in the I Ching, where we see even the simple act of counting yarrow rods transcended by a complex hierarchical referential system based on number sequences that include the 384 and 1/4th night lunar calendar. Understanding QBLH is only understanding the ancient hominid who conceptualized counting using externalized materially manifest memory referentials. One equals one, two equals two, three equals three, four equals four, and thus are there ten. Even this simple counting game can be used to explain the exponential expansion of vertices in regular shapes per dimension, such that the more intersecting planes come together at the point or corner of a regular three dimensional shape, such as the five platonic solids, the more faces they will have, and the more edges these faces will have; this axiom continues through all dimensions geometrically, though in the fourth spatial dimension, for example, it is possible for four plane faces to intersect at each corner point, rather than only three, as with the three dimensional platonic solids, and so on in direct relationship. Here we see that, as with the caveman, a line drawn between the fourth dimensional connection of four sides, the third dimensional connection of three sides, the second dimensional connection of two sides and the first dimensional connection of all the sides in a singular point represented by the line itself measures a ten sided shape.

i. as energy source

Some of the earliest records of history describe the sacred stone of the Ram being used as an energy source, and a clay jar containing an iron rod in a citrus alkaline suspension has been found in Sumeria possibly dating back 4000 years that is still capable of generating a one volt electrical charge.

The mythology is unclear and incomprehensive, however a liberal summation of events might be to say that, at some time before the building of the great pyramid of Giza the stone of Ram was brought out of Sumeria and the Tigris and Euphrates river valley area and into Egypt and the Nile river valley area. It has been theorized that such a stone as this was used in a way now unknown inside the chambers of the pyramid to power a massive hydroelectric pump built into the design of the pyramid.

Kirilian photographs of stones from the Giza pyramids show a strong aura, with even more bolts of charged energy being emitted invisibly from them than the average human hand. There are high amounts of radioactivity deposited around the Dead Sea area, northeast of Egypt, that trace back according to radiocarbon dating to at least 4000 years ago as well. Also it has long been known to English dowsers that some of the ancient petroglyphs are charged with ley line energy fields that overlap underground rivers, and the Irish still kiss the Blarney stone for good luck and love.

ii. as cutting device

The legends of the Ram being used as a power source also describe the shamir, or stone that cuts stone. While many modern theorists can provide possible explanation using Brown's gas for high temperature heating, or possibly water pressure to account for high speed drilling, none of these can fit the account of the shamir, which was described as being "capable of cutting through the toughest of stones, even the hardest of diamonds, without friction, heat, or noise." According to legend the shamir, "could not be stored in any iron vessel or metal container because it would burst through such an enclosure." This sounds similar also to the effects described by Moses as being possessed by the ark of the tabernacle housing the stone of the ten commandments, that it could stop armies, or to the power imbued to the trumpets of the army of Joshua when they brought down the Jericho city wall.

It is true that the stone bricks of the Giza pyramids are not only measured to within one millionth of an inch, they are cut this exactly as well, and show not even a hair's breadth deviation from their alignments. Considering that many of them weigh as much as ten tons, and considering that there are more than a few thousand stones used in the construction of the pyramids, one must wonder at the precision tools that must have been used that could have made such infinitesimal intricacies of ideality that even modern laser and diamond drills cannot match. At Machu Pichu in South America there are also stones quarried to this precision fit, so that there is not even a micron of space between them, and here we see that the stones are irregular shapes beveled down to fit together, and that some of them weigh even more than ten tons.

iii. as meditative object

One can imagine the masses of workers that constructed the great pyramid at Giza and who are buried in the smaller neighborhood of pyramids in its shadow willing to have built the great monument if only for want to be able to meditate upon their own grave. Imagine Imhotep, seated in lotus position, on the flat stone lid of the sarcophagus in the king's chamber, solemnly fixed on contemplation of himself being lowered into the pit, and his third eye being freed into the heavens. Such confirmation as this is given by Hermes to his son over the tables of the eighth and the ninth, and this would progress Mithra worshipping bull-jumping Rennes into Zarathustrian prodigal sun, baptist — coptic and gnostic, sects. The dilemma is general: to live though dead; to be beyond and yet still be around; to be remembered well and to forget the self; yet none of us is born in control of these traits of our fate.

The bending of our circumstances into accord with our will is called the work of karma, and it is this that amounts to an individual's control over their destiny. In the east, where this model comes from, it has long been believed that the best type of behavior for clearing one's mind of the consciousness of karma is meditative trance. There they believed that karma carried over from one life to the next, and that meditation prepared us for the moment of our death. The Egyptians believed that karma was weighed at the time of death to determine if a spirit was free of having to reincarnate again. Later, the Christians believed that knowledge of and faith in their Lord alone would forgive all their sins and gain them entry to Heaven. The free and associated Masons use as symbols the craft tools and implements of stone masonry to represent the moral work that transforms the karma of one's soul.

While not all of these derive from the stone of Ram, all have become associated with the mystical study of QBLH. QBLH has become identified with karma itself, as well as the tools with which to work upon it and the desired goal thereof. QBLH is the object of concentration for much trance meditation, and has been equated by occultists as the western version of the eastern tao of chi, or way of all energy. Because of this QBLH has even become associated with meditation upon one's own mortality. All of this is described in the first three initiatory degrees of masonry.

iv. as law

It is true to say that the financial token exchange reward based trade economy is as old as civilization itself. In Sumeria, many of the oldest written records describe financial transactions for goods or services, such as bills and receipts, inventories and prices of merchandise for sale, and financial holdings accounts, usually for the purpose of kingly taxation on the properties of the people in exchange for their protection by the guards of the city and armies of the sovereign lands. At the same time the ethical legislations levied out by the king or the court began to be recorded also, making them a truly open society in terms of exposing corruption. Their laws, mostly revolving around protection of and accumulation of material properties, even including wives and slaves, were extremely stringent, and it is from one of the most ancient books of ethical proceedings that derives the saying, "an eye for an eye." Perhaps the longest lasting of these harsh, 'people are less important than property' sort of laws was the practise of cutting off the hand of a thief, and this practise is made mention of in masonic dogma.

It is not true to say, however, that all of civilization has been based on the token exchange economy. Concurrent to ancient Sumeria's rise in the Tigris and Euphrates river valleys, the Hindu caste system arose as the social code of civilization in the Indus river valley. This system was based on classes, much like those described by Plato in his ideal Republic. There was a slave class, a merchant class, a soldier class, a priest class and a ruling class. Each of these classes served the next class up, and within each there was a social hierarchy or chain of command as well.

Both of these only represent the earliest modalities of ethical legislation, and, as I have made mention before, both began as petro rite, sacrificial cultures. As surely as Christ said, "Let you who is without sin cast the first stone," so has ethical legislation progressed much since ancient times. The subject of the Law was meditated upon by Confucius, Moses and Aristotle, each who began different philosophical schools on the ethical social legislation of karma. Just as Moses contributed the QBLH, so did Confucius the Yin and Yang of the I Ching, and Aristotle recapitulate the idea of the five regular ideal solids. Insofar as all of these meta-objects integrate as models describing the right proper ethic of karma, so too has this been grasped and understood by all those who have worked with and studied it.

v. as perpetually stolen

Anyone who has perceived ylem as the tao of chi, opened the third eye of the ajna, or extrapolated karma and integrated QBLH, understands that this forever flows, ineffable to the apprehension. As Thales said, "you cannot even step in the same river once." To anyone who has studied karma as information collage, this is the saturation effect of Samsara, or sorrow, the symptom of wisdom. Since we have defined QBLH as the meditation of ethics upon one's own death, we can see now why it can come at first only in small and fleeting doses, as the concerns of surviving take precedence instead, and why it can only come through meditation that brings these concerns over the ongoing flow of one's karma to an end. QBLH, similar to the Buddhist Samadhi or Hindu nirvana, is death to karma. As much as concerns over karma are one's ethics of life, QBLH is the trance of ego death that opens one to awareness of the higher ethics of the free spirit. Just as karma is perpetually attracted to the static field of our electromagnetic auras, so is the true self above and beyond this function of incarnate existence, and flies from it just as karma flows.

This has manifested itself in the history of the actual stone of Ram repeatedly, and again the first three initiatory degrees of free masonry describe the event archetypally. According to the oldest traditions, the stone of Ram contained a written account of history that dated back to the times of civilization before the last ice age. It might have contained the descriptions given in the lost books of Enki, which describe events in the heavens that would also be mirrored by the events described by Utnapishtim in the epic of Gilgamesh. Whether either of these is factually accurate, representationally allegorical, or fictional contrivance has been lost to history. These were given by Enki to the recording scribe to carry out of Ur.

We are told that the name given to Abraham means, "he who has Ram," or the stone of Ram, also known as the tablet of testimony. The bible accounts how he left Ur in ancient Mesopotamia where he had lived after meditating upon an upright sacrificial stone in the wilderness where he tended his herds and experiencing transcendence in the presence of the lord he called El, whom he made his Elohim, or chosen God.

Around this same time in Egypt, the vizier, court magician and grand architect for the first pharoahnic dynasty was a man named Imhotep. We are told about him that he designed both the earliest step pyramids of Djoser and the great pyramid of

Giza for two different kings, and that his successor, the man who designed and oversaw the building of the second, lesser pyramid at Giza, was named Ptahotep. Since the style of pyramid building he used combined architectural styles both of Nubian burial pyramids (as we can see corrected in the so called "bent" pyramid) and of Sumerian palace ziggurats (as we can see with the above ground interior rooms of the larger, later pyramids), it is speculated that Imhotep had traveled to both these lands.

According to Plato, writing as Socrates, he had heard by word of mouth from an old scribe named Solon who had been initiated into the ancient Egyptian mystery cult in his youth, there were two pillars carved of Orichalc on a small island in an underground grotto connected by tunnels of rivers in a labyrinth under the Giza plateau that contained written on them in a language he could not understand a history of the world and of civilization. The Egyptian who accompanied him told him that they were from Atlantis.

We are told that the name given to Moses means, "saved from water," and that it was by parting the red sea that he led the Hebrews out of Egypt. This event probably describes the Hyksos rebellion under Akhenaten following that Pharaoh's attempt to institute solar monotheism when he and his people were forced out of Egypt by the priests and viziers of the mystery cult. The bible recounts that Moses received the tablet of testimony on Mt. Sinai, brought down the original QBLH and smashed it when he found his followers worshipping a golden bull.

The story does not end there, however. According to legend the stones of the ten commandments were carried around in the ark of the tabernacle until the Hebrews entered Canaan and slaughtered the native peoples. After establishing a royal bloodline as King, the Hebrews began work on the first temple to be dedicated and consecrated to the Elohim of Abraham. The ark of the covenant was kept in the holy of holies at the center of the temple, and various other Hebrew treasures were placed in the hollow columns called Jachin and Boaz that marked the entry.

At the time of the Babylonian captivity the first temple was destroyed down to its foundations, and most of the contents of Jachin and Boaz, though eventually returned, were looted by the rulers of Assyria. At this time, according to the kebra negast, a holy book of Ethiopia believed to be a genuine apocrypha of the old testament, Menelik, the mulatto son of Solomon with the Queen of Sheba, took the holy stones from their enclosure in the Holy of Holies before the destruction of the temple and returned with them to his home of Ethiopia. The Jews of Ethiopia claim that the stones reside there to this day, although they were recently moved from a lone monastery on the Aswan lake at the source of the Nile into a larger city.

vi. the holy stones and the holy skulls

There is no question that, just as the stone of record was revered by ancient civilizations, so were the stone megaliths revered by ancient peoples as long ago as the last ice age, when modern homo sapiens lived side by side with communities of Cromagnons in Europe, the middle east and America. It is even possible that construction on the carving of the stone that would become the head of the Egyptian sphinx was begun much earlier, by a different species of human beings altogether, and that the explanation for its forward sloping lower jaw is not meant to represent either a bearded man, nor the face of a cat or any other kind of animal, but the long, sloping face of the Australopithecine and Neanderthal peoples, who lived millions of years before modern humans.

There can be no doubt that, whoever were their neolithic builders, such sites as stonehenge have long remained sacred to the indigenous peoples, and were used as ancient sites for worship and prayer, as I have described was true for Abraham, as well as for communal meetings such as the formation of social circles, where people congregated from distant regions to share news and trade goods with one another.

Thus, the message of the Law encoded on the Ram, as well as the astral alignments of such sites as stone henge, may have been largely unknown to ancient peoples, and still have been used as the basis for their civilization and trade centers. In this way the stones gradually became less holy than the minds of those meeting at them, until finally the meaning of the stones could again be understood by many.

A particularly good example of the difference between the mentality of ancient people around the time of the end of the last ice age as opposed to that of people in the following generations up to today comes from Ica, near Nasca, in Peru, western South America. Here we find a collection of engraved stones depicting all manner of knowledge and technology of many advanced modern civilizations dating back, it is thought, some 10,000 years. The original carvers of these stones are unknown, however it is believed that they belonged to the same culture from which three, elongated skulls have been discovered. These skulls had been lengthened since birth by wearing tightly coiled bands, similar to the neck stretching rings of Nairobi, Africa, or the lip and ear hoops of the Dnenge culture of Uganda, Africa. Nothing at all is known about these people, however it is clear from the evidence that they must have possessed an incredibly high degree of culture. They may also have been responsible for creating the Nasca lines, enormous pictographic geoglyphs that can only be seen in their full size and scope from in the air high above the dry plains.

c. the living archetype

In the ancient eastern mysteries they taught of seven chakra points inside the body, and of the Kundalini serpent force that rises upwards through them, calling this model of the human form the atman, or the self aware self. In the middle east the myth of the first man, Adam Homo began from the same lands where the oldest fossils of homo sapiens have been found. Kabballists call the biblical Adam the second, or lesser Adam, and the son of Set, and call the archetypal Adam Kadmon the holy guardian angel or the soul of man. It was Adam Kadmon, according to the ancient Hebrew legends, whose wife was Lilith, who became the demoness of the desert, an evil spirit associated with infant mortality, and warded off by magical charms. According to the kabballists, the biblical story of the expulsion from the Garden of Eden is only an allegory for the expulsion of Lucifer, the morning star thought to be the thief of first light, from heaven, despite the fact that the existence of an actual place called Eden is confirmed by independent contemporary Sumerian records.

According to some modern kabballist scholars, the entire old testament is an allegory for the same passage of a single soul through the underworld described by the Egyptian book of the dead. This code, called the Nefesh, may have been discovered by the historical Jesus, and account for his gnosis of himself as the son of God. The belief in a messiah began much earlier, though, at least at the time of Moses, who recorded all the requisite prophecies describing the coming savior into the Torah. Nor is the belief in a messiah as archetypal savior of humanity unique to Judaism. The Inca and Aztecs greeted the Spanish conquistadors with open arms, thinking them their returning civilizer, Quetzalcoatl, who they remembered as being fair skinned. The Buddhists, too, continue the practice of seeking out Dalai Lamas as reincarnations of the original Buddha, and await the coming time of Matrieya, the last terrestrial incarnation of the Buddha.

More recently the belief has become popular that we can all better ourselves, and thus become more enlightened, by individual karmic work on the soul, even simply by creating a list of goals for our self interest and then meeting them, and in this way become greater self actualized. One popular modern social movement states that we cannot love another until we love ourselves, and that we cannot know another any better than we know ourselves. This may be true, but the can of worms it opens up is the existential identity crisis of faith that has shaped mankind since birth.

To know the self is impossible, for it is the self that is knowing. Instead we say that all the self knows is itself, and this accounts for great amounts of peoples' personal karmic baggage, as it tends to attract reaction from the aeyther. In truth the self can know many things, and in many ways. The self best knows itself as through the body, though we know that loss of the sense organ that is the body would not necessarily diminish the self any more than loss of any of its five senses. Some say that there is a sixth, intuitive sense as well, however there is not any reason to believe that loss of the sense organ of the body even necessarily means the loss of the senses, either. To describe the self as archetypal, therefore, is impossible, since it is impossible for the self to comprehend that part of itself which is comprehending.

However, since the self cannot be divided against the self without its knowing that it is being so, neither can calling the self an archetype harm, damage or change the self in any way. It can only augment the definition of the self, which also has a long list of names throughout history, including the soul, the spirit, the ego, superego and id, the conscious, the subconscious and the unconscious, the aura, the atman, Choronzon, not to mention its notorious reputation for being ultimately reducible to only a set of memories and behaviors, brain tissue, or even only DNA.

By living better lives people throughout history have hoped to greater actualize the archetype of their self. According to the highest doctrines of most modern religions, in the deepest sense and on the highest level, we are all the same self, and this human, mammal, animal, terrestrial, elemental, universal self is only buried beneath the surface of our individual personalities, which we have only adopted for the convenience of satisfying our own lifelong survival agendas. Thus, the archetype of the self is in truth so transcendent that it is, in the end, right to think that all the self is is consciousness knowing consciousness, self knowing itself.

This archetype seems more ideal than to say that God is an asymptote and human population growth created in His image.

i. the ancient occult leaders

History has many different descriptions of the early civilizers. In most cultures the world over they were revered as Gods. The Sumerian culture describes them as the Nefilim, who descended to earth in flying ships. The Egyptians called them the neteru, who were all thought to be aspects of the sun god Ra. The bible describes them as a generation of half-breed children of angels with the wives of men and as "leaders and men of renown." Buddhism describes them as the "ascended masters," or Boddhisattvas. In the Mayan Popul Vuh they are named Jaguar Quitze, Jaguar Night, Not Right Now and Dark Jaguar, and the Aztecs recalled Quetzalcoatl as a bearded pale skinned traveler from across the eastern ocean.

With most of their cultural centers, the archetypal civilizers established a class hierarchy, free trade, and profession specialization. They are usually credited for bringing either writing or math to the indigenous people, or both. In the Egyptian book of the dead, amongst the negative declarations of the soul made before Maat at the last judgment which also include the ten commandments, there is a reference to cannibalism, indicating that there had been some problem with that in the Nile valley before the unification of upper and lower Egypt, when the book of the dead was composed.

This was likely accomplished by Sargon the Great, called the Scorpion King of Sumeria, and he was one of only a few pre-dynastic rulers of Egypt. He is credited with the first writing, having drawn a pictogram of a scorpion representing his signature beside a battle plan carved on a rock near the battlefield where, it is now thought, the decisive victory unifying upper and lower Egypt was fought and won. This pictogram was also later found on a set of small, flat stone tags containing engravings of small hieroglyphic figures that represented the names of different ancient cities, some in Egypt, some in Mesopotamia. These are believed to be the oldest writing in existence. The similarities between Sargon and Imhotep are something like the similarities between the assassinations of Abraham Lincoln and John F. Kennedy.

Imhotep was probably the person whom Sumerian religion would record as Enki, Egyptian religion would record as Thoth, and Hebrew religion as YHVH. In addition to these titles, he may also have been known as Sargon, Abraham, Pharaohs Djoser, Khufu, Khefren and Akhenaten, Moses, Utnapishtim or Enoch, Uhurumazda, Mithra, Zarathustra, Lao Tzu, Lao Tse, David, Solomon, Confucius, Siddhartha, Yeshua Ben Padiah, Jacques De Molay, John Dee, Comte St. Germain, Rasputin, S.L. MacGregor Mathers, Aleister Crowley, etc. since what we are dealing with is a body jumping, reincarnating holognomonic archetype. The story of Osiris may be read as prophecy fulfilled by Khufu. Similarly each of these people have only lived their lives fulfilling the prophecies described by their predecessors, because each of them was only another to hold the same titles and position in the mystery cult held by Imhotep.

There ought to be little doubt that all we know about the QBLH comes to us originally from Imhotep, however the mythology makes it quite clear that this can only be dated back, even at the time it was first recorded, as far as the last ice age.

ii. sun symbolism

Most of the mythologies, religions and philosophies of these individuals shared one thing in common. Each of them displaced a lunar, feminine cult with a solar, masculine cult. Worship of Enki predominated over worship of his brother Enlil, worship of Thoth, a lunar deity, was replaced by worship first of Osiris, representing resurrection, then of Horus, the solar hawk, then of Ra, the sun itself. Worship of YHVH, Uhurumazda, Mithra and Zoroaster drew people away from the worship of Inanna, Ishtar or Astarte. In India, petro cult worship of Kali was replaced first by cults of Shiva, then of Vishnu and Ganesha. Zen Buddhism displaced Shinto across the face of much of the orient. Later, Catholicism would burn witches as pagans, just as early Christians had been made to fight gladiators and lions to the death in imperial Roman amphitheaters, and Spanish Catholic Conquistadors would lay waste to the animist Aztec and Incan cultures of South America.

There is a very racist, Mormon joke about Quetzalcoatl that goes, "as Jesus said to the Mexicans, don't do anything until I get back." This seems especially indecent in observance of the fact that the peoples of pre-Columbian Mexico city, a monumental architectural feat being built up over a marshy lake and housing more people per capita than any other city on the planet at that time, as well as up to the twentieth century, have always been known as "the people of the sun." This is typical of the senseless macho contempt these solar cultures have developed for one another as their campaigns of bloodshed and violence against one another continue. They appear to be in a sort of competition with each other to breed a Messiah or a master race, and thereby come to rule the entire world, as well as collecting tithes.

Perhaps the ancient followers of Baal or the Mayan ball court players had the best expression for this kind of competition over the sun sphere. The Enochian system also reveals it as like a ratings system of karmic points that collects over different geographical areas, people or peoples. When it is centered upon an area it is called a karmic center; about a person, Christ consciousness or Choronzon, the Holy Ghost; for a group of people, mass hallucination or game reality. The crown of kether is passed around the round table of the globe. Despite humanity's chase of it being called civilization, like Louis the XIVth in his palace, the sun rolls over our heads every day, from dawn to dusk, disappearing only to reappear again.

It can only have been the pounding heat of the sun that would drive the workers into the frenzy that must have been needed to have built the great pyramid in as short a period of time as most historians of the time accredit, less than one generation, and it can be seen that the Nubian culture of southern Upper Egypt began to dominate more and more heavily over the Hyksos culture of northern Lower Egypt in the art of the three kingdoms of dynastic Egypt, as the characters are depicted as more and more African in appearance. Similarly in ancient Ethiopia, where they were said to cut shanks from live cattle, eat the flesh raw, and patch the animal's wound with mud. At these early stages, beside doctrine, the Solar cults and the Lunar cults displayed little difference in their low level of rational humanism.

By the time of Jesus, who represented the sun amongst the twelve zodiacal houses of his disciples, the message had become one of peace, love, and understanding, and by the time of Jacques De Molay, a Christian crucified by the Christian church, one of liberty, fraternity and equality. One must wonder at what such gushing openness on behalf of the sun fails to reveal about its true nature, since beside the Mayan tzolkin, a very complex calendar based on many different cyclical variables observed in the movements of the planets, and the works of Ottoman astronomers, none of these have even hinted at any scientific understanding of the nuclear fusion gas cloud of our solar system's sun, preferring instead to mask whatever actual data they had collected in anthropomorphic allegory and conceal this in a hierarchical system of social order.

iii. death and resurrection rituals

The central paradigm of social order is the death complex, and so I will use the words occult and Order of Death (or Daath) interchangeably. Here we see that the human species has attempted to cope with its apparently unique perception of the end of life by idealizing death, and making it holy by enshrouding it in mysterious rituals, magickal legislative dogmas, and often hallucinatory beliefs.

The oldest known conception of what became of the essence of an individual after their physical body had died is the djinn, or nature spirits, the origins of which can probably be traced back as far as tribal animism. Around the time of the building of the pyramids, however, the middle easterners invented the process of brewing grain barley with fermented hops and formulated the recipe for beer, and at this time the Shamanic trances that had only been practised by the tribal medicine man far out in the wilderness began to become popular amongst the common citizens in the early agrarian city state civilizations.

Around this time there were several attempts made to formalize or categorize the djinn. These attempts can be traced along with the history of the Ram stone.

Supposedly the oldest formalized conception of the djinn is given in the Sumerian Necronomicon, and describes the djinn as having once been the Annunaki, lessers of the Nefilim, and as able to control various certain natural processes, such as the gathering together of great numbers of people, the working and refinement of natural ores and minerals to produce metals, and control of the weather of the firmament of the atmosphere. Though only fifty names are listed, there were supposedly more, and though only their usefulness to making war is described, they supposedly possessed as much personality as people. According to the history of the Sumerian people, these fifty were actual people who had lived and died during a "war in the heavens," and according to the Necronomicon they had then become ghosts who could be summoned using ancient magical sigils.

Next, the Egyptians described the Annunaki as the dekans, giving them a place in their Sumerian twelve sign zodiac based calendar, though not attributing constellations to them. The Egyptians divided their calendar first according to the twelve Sumerian constellations, then assigned three dekans to each sign, calling each one the watcher over a week of ten days, such that there were thirty days in each Egyptian month. Since this did not correspond correctly to the actual 365 & 1/4th day long solar year measured by the earth's orbit, at some point in Egyptian history, perhaps when Akhenaten attempted to establish solar monotheism, the calendar also began to include five intercalary days at the end of each year, and these were called the holidays of the thirty seventh dekan.

According to the occult throughout history, the shemhamforash, or the spell that Moses used to part the red sea when the Hebrews fled Egypt, is a seventy two long letter code triply encoded into the verse describing his doing so. Many modern Jewish researchers today have proven that, along with such gnomons as phi/pi, there are many different types of number or sequence coded geometric lattices used in the writing of the old testament. The number seventy two is important because it is the doubling of the 36 standard Egyptian dekans, and because it is also the number that is the measurement of years over which the gravitational pole of the earth precesses one degree, or 1/360.

Buddha, who describes Samadhi as the transcendence of Samsara, or the cycle of reincarnation, is said to have slain all of his inner demons of doubt in one night while meditating beneath a bo, or bodhi, tree, and it is from this that the Boddhisattvas take their name. Buddhism teaches that, by following the path of right thought, right speech and right action, we can attain higher states of conscious enlightenment. In this way, Buddhism defines the afterlife as the attempt to escape from the yugas, or cyclical ages of growth, destruction and renewal described by the Hindi, which may be thought to represent the annual seasons as well as the 41,000 year ice age cycle.

The shemhamforash was also used by Solomon, according to rabbinic scholars and kabballists, to construct the first temple. Supposedly, also, the author of the angel scroll, a list of angels associated with the old testament apocryphal prophet Enoch, the essene Yeshua Ben Padiah, also knew of the shemhamforash. Mohammed also describes the wars in heaven, and gives the names of some of the key players. The Enochian system describing the Annunaki was later elaborated on by the skrying of John Dee and Edward Kelley, who expanded upon the day and night duality of the Shemhamforash's expansion on the Egyptian dekans by factoring in the planetary astrological alignments and associated elements, and who gave a fuller, more complete set of archangel names as tessellated lattices.

Most recently involved in the definition of the hierarchical initiatory levels of the death experience are the Free Masons, who trace their origin as far back as to the building of the temple of Solomon, but who may date even further back to the Sumerian workers who immigrated from Sumeria to Egypt, bringing with them beer, writing and civilization, before the building of the great pyramid. The free masons, like the gnostic essenes and the Buddhists, attempt to demonstrate to people that death and reincarnation are perpetual parts of life itself, and can be ended in one's death.

Perhaps the irony of the occult leaders teaching people freedom from reincarnation coming back to haunt the leaders by being accused of reincarnation themselves is best expressed in the passion of Jesus Christ, although the same trend is carried on to this day by the seeking out of reincarnated dalai lamas and lamas by

Buddhists, the election of popes, cardinals and bishops by Catholicism, the initiatory rituals of free and associated masonry, and the Jewish belief in the generational prophets and the coming Messiah. As I have said, this only leads to competition among the sovereign cultures over the essence of Choronzon, the archetypal ghost.

iv. Poimandres: the dragon of animism

Along with the twelve signs of the zodiac given by early Sumerian astrologer astronomers, the ancients also defined an entire host of other constellations, each culture recognizing the same stars and symbolizing them in different ways. For example, the Egyptians and Greeks identified Osiris or Orion as a human hunter, while the Peruvians and Mayans identified the same constellation as a spider. The Egyptians identified the North pole star constellation as an ape and a plow, the Russians as a Great Bear, and Europeans as the Big Dipper. One constellation for which most ancient and modern cultures share in their description of, however, is Draco, the Dragon, that circles around the pole star through all twelve longitudes of the heavenly hemisphere divided along the ecliptic by the twelve signs of the zodiac.

Draco was significant to many ancient cultures, and is depicted in the snake mound of England, the Nasca lines, and the alignment of the temple of Angkor Wat in Cambodia. It was associated with Kundalini in India, Quetzalcoatl in South America, Set in Egypt, and Poimandres by Hermes. Insofar as the north polar star was in Draco during the time of the last ice age, as well as much longer ago, during the time of the dinosaurs, it has been speculated that it might represent not only the ancients' knowledge of precession, as demonstrated in the shemhamforash, but also their knowledge of dinosaurs, as depicted on the Ica stones of Peru.

It should also be considered significant that the majority of constellations described by the ancients are depicted as animals, much like the majority of cave paintings by stone age homo sapiens and cromagnons. Perhaps the best representations for this irony, which did not go unrecognized by subsequent Greek golden age temple astronomers or later Ottomon Muslim astronomers, is in the depiction of the zodiacal sign of Capricorn — half goat and half fish — marking the middle of a northern hemisphere ice age, or Sagittarius — the centaur, half man and half horse — marking the middle of a southern hemisphere ice age.

v. mendicant moralities and aescetic martyrs

The result of the conflict between the solar cults and the lunar cults over the precession of Poimandres has led to the mass delusion of the populous. The ones who win feel obligated to lose, all that is gained is ultimately sacrificed, nothing remains sacred, all victories are pyrrhic, all desires fleeting, and "thank you" is never said without "I'm sorry." The ego is associated with the archetype of suffering and this is raised unto the Most High. Still, God is moving across the face of the deep.

Zen Buddhism, Taoism and QBLH offer ways through the garden of forking paths that is the neural net of decision making for the pursuit of karma. It is thought that, by identifying and transcending our goals, we can become at greater peace and harmony with our inner self and with the rest of the universe as well.

Most of the rest of the world religions are transfixed by the passion of Christ, such that so much of the world's population are Christians that it gets in the way of the anti-Semitic teachings of Mohammed and the Marxist Zionism of reform Judaism. While the new testament has changed the entire philosophies of the people of the Holy Land, it has left them in financial ruin and economic poverty; at the same time it has done the exact opposite for the western corporate world, making industrial developed nations rich beyond their wildest dreams and leaving their minds stuck in biblical times, reenacting the last supper.

Thus we are either given mendicant roles cobbled from the Hindu caste system or made to identify ourselves with a martyr. Neither of these seems sufficient to meeting the goals of simple survival, and it is obvious the way in which both of these schools of philosophy evolved from originally aescetic petro cults.

In the case of the east these were elderly Vedic wanderers who had given up all but a cloth, a pair of sandals, a walking stick and a bowl. It was their place in the Hindu caste system, having lived long, socially productive lives, to be exiled into the wilderness in order to contemplate what meaning this had had, if any, before dying. It was to these old men that young Siddhartha came, though they would eventually learn from him.

In the case of the west these were the Baptist Essenes, a group of Jews who had been exiled from all the surrounding communities for being even more religious than the anti-Roman terroristic zealots. They were called the Nazorenes, or the poor, for being the Nosrei-ha-brith, the "keepers of the covenant" of the commune of Qumran. It was to these well worshippers of Damascus that Yeshua Ben Padiah would come, though his time there is best described by Jesus's final visit to the temple.

It is possible that these teachers of wisdom only briefly encountered the existing occult, and it was by the establishment's cursing of them upon their leaving from its folds that the subsequent religions were born as their mirror image. For Buddhism this consists in the slaying of all one's inner demons, doubts, turmoils, etc. For Christianity this consists in the pleasant gaze of Jesus, hanging on the cross. Perhaps one could be thought to represent the external visage of the divine countenance, and the other the internal mental state of serenity at the conclusion of all cognitive dissonance.

As I have said, all this leaves little left over for the common man, who does not necessarily care for the mendicant lifestyle, or the archetype of the martyr, as much as for his own salvation, and for bread and water. It requires understanding on another level of reasoning, and as with the formal system of metaphysics, this implies the depth of dimension of Had and Not, and the potential dualism of fact and fiction. The ornate rituals of religion and the grotesque spectacles of the media do not change the economy, the politics, or any of the rest of the natural functions of the real world, they only reflect, represent, and attempt to elaborate upon them.

d. philosophies of the most high

It is clear that the youngest of the great religious teachers have all been more in rebellion to the existing philosophies than those of their elder counterparts, such that, whenever one seeks initiation into the mysteries, one will react to the gnosis, or the knowledge of God, in a way congruent with one's age. We see that the young, slender Siddhartha would become the jolly, round Ho Ti; that Sargon (Hamurabi) would become Imhotep (Abraham) would become Thoth (Enoch) would become YHVH (Moses). We see that Jesus was missing during most of his youth, and known only during the three years of his preachings before being crucified. Just as Imhotep and Ptahotep were also Khufu and Khefren, so were they also Osiris and Horus, as Shiva of Sheba was Isis of MeruBecka, and Akhenaten was Siddhartha was Yeshua were all reincarnations of Khefren, the Ra of the free spirit soaring through the set souls.

Recently, the British occultist Aleister Crowley described the situation this way: "the Khabs is in the Khu and the Khu is in the Khabs," meaning that "the greater is within the lesser, and the lesser within the greater," or, "the above is as the below and the below is as the above." Another way of thinking of it is given by Lewis Carrol in Through the Looking Glass, where the white queen describes memory as being potentially backwards and forwards, but never in the moment, as the law of, "jam tomorrow, jam yesterday, but never, ever jam today." Unfortunately, that story ended with Alice awakening from a dream, and the truth is that such axioms as she perceived in archetypal forms throughout the plotline persist more in waking life.

One of the types of litmus test for agelessness that can be applied to the teaching of any ancient sage is its translatability into modern terminology, although this will as much reflect the age of the translator as the original did the age of the writer. Another is based on the use of chemical substances to encourage, induce or enhance enlightenment, such as soma for the Buddhists, peyotl for the Cherokee, blood for the Black Mass, and sacramental wine for the Christians.

Perhaps the best is through the use of systems and their degree of complexity, for such as can be used to measure in terms of wisdom rather than in terms of age. The ultimate extent of systems, however, is arbitrary, since it is an ideal, equivalent in essence to a dream object, something that can exist only in the mind, and which transcends the mundane reality of such character traits as identity.

The modern information systems theory hinges on inversion, and this is for astrological reasons, as the Montauk - Sedona time tunnel opened in the sign of Libra, which is really Virgo according to the social concealment of precession. This is thought to represent the same event as predicted astrologically by the encrypted prophecies of the new testament in the parable of Jesus overturning the scales in the temple. Such time tunnels occur in accordance with the equivalent in earth's ozone layer and ionosphere of the coil crossing, band jumping prominences and flares of sunspots in the sun's electromagnetic field. These cause karmic centers to occur below, and such interaction may also be stimulated by the direct solar radiation that stimulates the movement of atmospheric pressure centers.

At night, the air molecules move more slowly and drift further apart, as the solar ultraviolet radiation is not disturbing them, and everything assumes a more regular form as the flatline histories of tachyons penetrate through the ground and trail out into the plasma sheet in earth's shadow cone like a solar windsock. The slight distortion of them in the atmosphere makes the stars sparkle and the planets twinkle. They form spiral vectors on the rods and cones inside our eyes. People who have driven long distances know this effect. Children are exposed to it by watching too much tv. It is the air in which we communicate.

During the day there is only more of it, as tachyons are being emitted by photons, and here we see that the regular matrix that these particles assume over night becomes a chaotic field of static. People who have been soldiers in a war know this effect. It is from the accustomed perception to this that the Muslim religion gradually derived its geometric tessellation art over the many ages to represent the face of Allah. It is also this which is associated with Kether, and with the primary clear light of ylem, although it should be specified that there is still clear light even beyond the tachyonic multiverse comprising the border of the universe bound by the speed of photons (zimzum), even as Ain Soph Aur (limitless light) in the Ain Soph (limitlessness), in which the I Am is the Ain (limitless nothingness).

This has led the media into the millennial wild goose chase for an explanation for their existence. What should have just been a good Picasso painting now has to represent Taurus. What should have been a family photo is now a crucifix.

i. sacrifice

The beginning and ending of self-aware existence is sacrifice. In the beginning we sacrifice the womb life, comfortable and warm, in a state of miraculous suspended animation, only to become people, and cover the face of the planet. In the end, we sacrifice all we have gained from this life, and give up all advantages.

This is the way it has always worked, and this is the way it always shall.

There is very little choice on the part of a star, for example, as to whether it wants to have begun from random quantum fluctuation. Nor can it hope to change its fate if it is not destined to become a black hole. Therefore the Law that every man and woman is a star is wrong, because humans have free will and can change their own fates and destinies. This is because the spirit is free, though even the soul ultimately is not. The soul, insofar as it is electromagnetic, can only exist within the natural universe, below the speed of photons. Only the free spirit, in essence, tachyons, could travel into the gates of heaven, outside of this universe. Stars may be like us, but we are not much like stars.

We give up the solace of the womb for the control of our own destinies.

Outside our parent organisms is a reciprocal interchange system that has been artificially created and must be supervised and maintained that supports and gratifies our desires, and we understand this force however we choose. It may be seen to represent God the father, karma, memory, money, civilization, other people, the identity. It is what gets us home safe at night. The replica, or pop-culture simulacrum, hovering in the Enochian system is the ghost of our self control, the social consciousness or Choronzon that we are perpetually letting go. The rest remains the memory of the involved individuals, like a toy for them to take home.

We give up the control over our destinies whenever it's our time.

Since long ago it has been this that has been the substance of sacrifice, while we have kept hold of material goods only slightly longer than they have had value as referentials, or representative meaning. Since the sensory stimuli increases in velocity, its meaning is asymptotically forgotten and gradually redistributed to other referentials. When the sensory stimuli has achieved a certain point on this vector graph, it simply ceases to exist. Here is the death of sense perception, here is ego death, and here is the death of the body.

I am not that which I am. I am not the substance of my sacrifice, though it may be for me my most exalted replication of karma. I have detached myself from its umbilical chord and let it go beyond. Such is the devotion and dedication necessary of a mendicant.

ii. asceticism

On the island of Japan, Buddhist monks discovered a way to mummify themselves. They ate tree bark from indigenous pine trees for one year, taking daily, rigorous constitutionals across the mountainous terrain. Then they would drink ink containing lead, and water from a local well containing arsenic. One of them buried himself in a small cubical hole beneath the ground with only a string leading out to a bell in the outside world, and a cup of the lead based tea. He rang the bell for three days, and on the fourth day he was dead.

The remains of his corpse are still preserved to this day, however, almost 1000 years later. The tree bark toughened and thinned his skin and muscle tissue. The lead and arsenic killed the microbacterial organisms in his intestines that, in life, had helped him to digest his food, and that, in death, would digest him from the inside out. There are three such mummies in a Buddhist monastery in Japan.

This represents the height of the mendicant monk's lifestyle, and the culmination of reward for it in the form of preservation of the body after death on par with the Egyptian Pharaohs. So, forever, may it be seen that both the greatest of kings and the simplest of monks may be equivalent before the ages. It is possible to extract extremely accurate genetic samples from any, once living, artifact, so long as it has not become completely petrified. On the other hand, it remains possible to extrapolate the genetic patterns even of petrified particles using holographic resonance imaging to map the projected material into a computer program.

Asceticism is as useful to the common man as a bicycle to a fish. You can sell a fish a bicycle or you can teach one how to ride. You can lead a horse to drink tequila but you can't make one swallow the worm. Why should we all beat ourselves up all the time inside over the fact we will all one day die? Good times are always passing by.

iii. martyrdom

Consciousness researchers believe that our first impression of death comes from the birth experience, however that our beliefs about the nature of life after death derive primarily from our self-transcendent experiences, or what I have called elsewhere our personal transcendental experiences. Various researches have shown that these are related in various ways to various events in our lives, such that the most archetypal levels tend to filter down statistically into the manifest realm of event. Some experience these from orgasm, and it is likely that this is the most primal form. Some experience these from use of chemical substances, however this often creates a clinging to pre-disocciative reality as a central point. Some experience these as profound religious experiences, and these come in too many different forms to list. Some may leave you with good memories of them, others might vex, trouble or worry you. Some may make you see yourself differently, others might even change who you think you are. Researchers have tried to categorize these according to various different systems, such as the Annunaki and Nefilim, the Essenes and Nazorenes, and the Enochian Ayres.

This is where we see the similarities and differences between the lifestyle of the monk and the welfare of the king, and their relationship to the common man. Since everyone has personal, self-transcendental experiences then it becomes necessary for the group to have shared personal transcendental experiences, and this is where the social simulacrum comes into play.

It has long been a superstition that the more people involved in an event the more energy there is in it, and that it is both healthy and well that this energy should be guided towards the implied desired ends. Passive energy, such as the holy mass given by the Pope, requires fewer leaders to orchestrate. Active energy, such as an American football game, requires, other than strategically, mindless players, and hundreds of screaming, mostly drunken fans.

People share personal, self-transcendental experiences almost everywhere they go and in every situation they are in, to either greater or lesser extents and degrees. Similarly, if only one person tells another their thought, and that thought seems worth it and catchy enough, it will eventually travel memetically throughout the entire community unconsciously. This effect has been observed with moths living in nineteenth and early twentieth century industrial Europe, where the moths that lived inside the heavily polluted cities were turned black by coal soot, and the moths of the same species that lived out in the surrounding countryside also naturally changed their color to black as well. This also occurred to the tree bark of the indigenous trees.

In this same way do people adopt personalities from the aeyther, and one that is particularly camouflaged is the martyr complex. This is based on the Greek Oedipal or Elektra complex, which, according to the original story, goes recognized too late, and Oedipus puts out his eyes. While before this time the practice of animal sacrifice was standard operating procedure at the temples, not until Greek drama had human self-martyrdom been portrayed on stage. Thus it was as much, if not in fact, to the Roman sense of irony, much more, this theatrical flare that was behind the practice of crucifixion, than it was intended to be, as it was held by the first Jewish Christians, equivalent to the passover sacrifice.

iv. inversion

At the pinnacle of self sacrifice is the inversion of self. This means not only the sacrifice of what is known of the self, but the self that is known, as well as the self that is knowing. This seems like alot of inversions, but according to people since the time of Christ, it can all be done in one, if, of course, you believe in Christ. Perhaps you might wonder how, since the inversion between all of those inversions and only one inversion only seems to be another inversion.

Only imagine there is no you. Then it will all become clear.

It is this that it is the crux of the argument among many of those who consider themselves the most high, since on this level, or from this perspective, however you want to look at it, there is no difference between life and death, according to all the recorded legislation on the heaven of mankind that is the free and accepted multiverse. Once we have begun to see in this way, which is called the trance of Samadhi by Buddhists, and the tao of chi in China, as well as Christ consciousness or being in the presence of the Holy Ghost in the west, then one can begin to freely and truly question whether they even exist at all. It appears possible that we might simply divest ourselves of karma and dissipate whenever we wish.

Doing so, however, creates inversions of karma. An expansion in one dimension creates a contraction in the next, and another expansion in the one after that, and so on, on all scales. This even includes the projection of our thoughts in mental projective space, and this effects reality when it involves referentials. We create attachments between things that remain within our memories and act associatively upon referentials, creating geometric lattices. These are ennegrams in the mind that overlap three dimensional reality in a mental projective map. Even fish do this. It is thought this is what makes the grass grow.

What does it mean for karma to be inverted? Only that it is a digital part of a holognomonic referential infrastructure that is at once expanding and contracting, at once extrapolating and integrating, at once an influx and output. This means the karma is coming and going, and, if one will remember that it is, on the most fundamental level, all tachyons, it is doing so simultaneously backwards and forwards in time, and this seems to create the fourth dimensional temporal cycling of the three dimensional heavenly bodies.

Karma can be inverted in as many ways as it can be seen to be relative to any other object, material substance or essence of force in the entire universe, since these all comprise the actual geometric relativities that comprise the acausal connecting principle. None of this has any more need for meaning to you than in that it can be useful, and moreover no more meaning than what you give unto it.

The ancients knew of this by the name Shekinah, and it was only after the same ideas became associated with Choronzon that the Christian trinity of God the Father, the Holy Ghost and Jesus the son of God could come to replace the maternal occult dialectic of the aeons of the Father, the Mother and the Crowned and Conquering child. In the inversion system, the old greatest common factor becomes the new lowest common denominator.

v. nihilism

Nihilism is the belief that nothing exists. Nothing is real, nothing is true, and all that we perceive with the senses, the mind, the ego, et al. is a false illusion called Maya by the ancient Indians, and Mana by the Mayans and Olmec. There is no need in nihilism for religion, because it allows for atheism as well as agnosticism and anthropomorphism, so long as the applicant realizes that these are essentially false and illusory, and can only contribute to a further delusion of the senses resulting in greater clinging to the ego. The goal of nihilism could be called clarity, but nihilism is iconoclastic, and therefore would only smash such a symbol. There is no moral purpose for nihilism, yet it is the grounded philosophy of countless people on some level worldwide. There are many moments where the self is transcended purely out of frustration, and since this amounts only to distraction, it cannot accomplish anything, and is therefore considered ultimately irrelevant. Nonetheless, this is embraced by the rebelliousness of youth culture as the finale of the haunted house ride that is reality. As well as the existing American media, the immediatist collage philosophers of tomorrow seem anxious to rush headlong through their roles in manifestation of social adjustment karma and assert their thalami on the karmic cutup machine, claiming to be the inheritors of Burroughs and Nietzsche.

Nihilism began conceptually as the inversion of tao, the concept of the anti-ylem. It is described as the Ayin that is above and outside of the universe, where even the tachyons of the multiverse do not reach. It is thought, therefore, to be the philosophy of the darkness, and associated with yang inside yin, as opposed to yin inside yang, associated with ylem. It is therefore very deeply rooted in socialized, acculturated philosophies, as well as to the psychology of the self-cognizant bicameral mind. It is not known what purpose it serves. It motivates the id, which actuates the superego. It differentiates between consciousness, subconsciousness and unconsciousness. It is the everywhere and nowhereness of the self. Yet it is inverted from all of these things, and is therefore none of them in itself. You cannot put your finger on nihilism without it hiding inside of you. Whenever you go looking to find the source of all your nihilism, you will not find it, it will not be there, because there is no such thing.

Nihilism perpetuates itself invisibly like a computer virus, decomposing the moral fiber of all of history's grandeur. This splendour is a mirage, say the nihilists, and, like Satan's temptations of Jesus in the desert, must be let go.

Perhaps this was best described in the children's book, the Never-ending Story, where there is a great nothingness that threatens to consume a magical kingdom. If you feel threatened by this description, as well you should, for it is the Jabberwocky, then you will remember that all that I am describing is what would make you ever forget all that I have been describing. That is what nihilism is. Forget forgetting.

e. schools of the light

Our bodies are made of material that has already been on this planet for millions of years, our genetic sequences gradually evolving, our bodies reproducing themselves cellularly, the materials the same since the first bacteria. This is one source of memory. Our souls are electromagnetic energy that has traveled throughout the known universe, and this is another type of memory. These are the neuro-genetic and neuro-atomic circuits described by Leary, and represent Binah and Chockmah, where the eight circuits are arranged on the tree of life excluding Malkuth and Kether, and where Malkuth represents the peripheral nervous system, and kether represents the source of memory of the free spirit, comprised of tachyons, whose home is as a gravitational singularity in the heavenly multiverse.

Because there are many forms that appear to catch the light of ylem, there are many different systems and models that have been cobbled over time, and therefore many different mystery schools have arisen throughout the land. However, remember, it is as I have said of kether, that it is passed around, for the Rosicrucians described the true inner order as "the traveling lodge." The most ancient peoples used to have ceremonial fires. This was probably the beginning of the craft.

The philosophies of these schools have always reflected the philosophies of the most high. It is a rather difficult operation to identify the most high, unless it is you. It is genuinely unfortunate that more people do not do so, because then there would be even more, and therefore, perhaps, better philosophy and metaphysics. Some of these philosophies actually say exactly the opposite of some of the other philosophies that supposedly originate from the same school. This is usually because even the most high do not always agree with one another. This can result for any of the various reasons I described in an earlier section as the different causes of personal self-transcendental experiences, as well as the different sources of memory. When such things as these are different, as Nietzsche claimed they were for men and women due to frequency, then disagreements in the recorded doctrines may occur. Usually these will either have been corrected and replaced, or corrected by a third, separate factor, and remain in the doctrine alongside their resolution. In the cases where this has not occurred due to the natural process of information resorting over the countless ages for which the Order of Daath has existed, it may be considered a doctrinal mutation equivalent in essence to a genetic mutation, which occur usually due to gene hopping transposons that replicate genetic coding out of sequence. Thus, the error is not really in the doctrine itself, but in the time at which it is being legislated, such that the doctrine at no point disagrees with itself from different perspectives of timing. This occurs because the doctrine itself is a holognomon. It is only spin being put on it that causes it to appear ever changing. In truth it is only the mantra, "om," reverberating throughout the continuum.

Think of the lotus blossom that floats upon the surface of water, without the water droplets clinging to the underside of the broad, thick, smooth petals and leaves. See how the water bends under where it collects in pools on the flower, and how it cannot run off, nor be absorbed. The Lotus blossom is like the doctrine of the QBLH, and the pools of water upon its surface are like the schools of the dharmas. It floats in the infinite continuum that is not what is, and it does not sink into it and it does not drink from it. See the Lotus blossom in the daylight as the soul warms to see it. See the Lotus blossom by the cool light of the mourning soul by moonlight. See the Lotus blossom in the rains and sleet and snows, and the temperate breezes and warm glow of the seasons, and see how these same ravagers and destroyers of the apocalypse renew, regenerate and restore the Lotus blossom. See the million petaled Lotus blossom with the wonder with which it would look upon you. See your way through the Lotus blossom and you will see it through and through.

Such is the mind of God, and beyond this ineffable even to the mystery schools.

i. gnosis: self abnegation (there is no thou)

The highest state of enlightenment, according to western mystics, is gnosis, or the secret and revealed knowledge of the nature of the universe, according to

QBLH, the ancient oral tradition of the wandering Hebrews, that is written down in such scrolls as found at Qumran and at Nag Hammadi. These are usually considered apocryphal literature, since most of them were editorially excluded from the canonized Latin Vulgate, as well, mostly, from the subsequent Lutheran Gutenberg translations, and comprise the doctrines only of the eastern Orthodox branch of the church that included the Greek and Russian churches, as well as held sway over the Holy Roman empire of Prussia. The oldest gnostic documents have been discovered alongside Coptic documents, some of which predate Christianity. The religion of the Coptics was similar to the religion of the gnostics, which differed little from that of the Jews. The Jews believed that there was a coming messiah, and believed he would be born among the Hebrews, because of their covenant with God. The gnostics believed this as well, however they also believed that the common man could learn to become a Messiah. The coptics were a very late dynastic Egyptian cult similar to the Mediaeval alchemists, who used Egyptian symbolism to promote gnostic messiahism.

In their own time all these sects were very unpopular with the surrounding townships, because they were associated with the poor and homeless mendicants, or beggars, as well as with the wandering lepers. The coptics were viewed by Egyptians and Hebrews alike somewhat similarly to the way a modern Southern Baptist might view their cousin who attends a snake handling church. They were thought to worship Set, or Satan, and shunned for the devil. The gnostics faired little better, being associated with eschatologists and proselytizers, and were shunned by Zealots as well as Pharisees. Similarly, the Hebrew people were, at the time, an enslaved people of Rome, which was a pantheist culture that saw little value in an intangible, all knowing God.

The early gnostic and coptic apocrypha share much with Egyptian and Buddhist ideas of the soul and the spirit. They identify the ruach, or breath of the soul, with the seven Ba of Re, and the Nefesh, or holy spirit, with the Akh of light. Above this, and animating it, are Chiah and Jechidah, or the will and the way. While much of the late Egyptian metaphysics remains untranslated, the apocryphal Hebrew texts are usually the same, allegorical style as the rest of the bible, some including names of bloodlines, some recording calendars, some apocalyptic visions, some simple parables of Hebrew law in Jewish daily life. One by an Essene named Yeshuah Ben Padiah, gives a list of angel names similar to the Book of Enoch, and thought to be an elaboration thereon.

Some of the works found at Qumran and Nag Hammadi are copies of works that were included in the canonized bible, however there are many more that were not. The first council of the Roman Catholic church that met to establish the contents of the book of the bible chose for the old testament from the Torah, the talmudic works, including many psalms, and the other contents of the Tanakh, and for the new testament from the Essenes gospels the works of the few selected disciples, apostles and a few epistles later written by Paul of Tsarsis, who had established himself as the voice of the pre-Constantine Christian church. This excluded a vast amount of additional books in the old testament, and much later the wealth of rabbinic commentary on the message and the law that had been produced by the Jews since the Babylonian captivity. It also excluded a wealth of works written by other disciples of Jesus, as well as epistles supposedly written by him to Pontius Pilate. There are a wealth of revelations from Adam and Seth, the son of Cain, Abraham and Moses, to Mary and Joseph, John and James.

Although these were all excluded from the latin vulgate, the various sects and orders of the church who had possession of the few then known copies of some of these such works kept them stored along with dozens of other scrolls, many dating back into the same ages of antiquity from various other cultures around the world. When the Roman Catholic Universal Christian church was invaded by Muslim Turks the order was divided into two branches. One was comprised of the people of south western Asia and eastern Europe, living around the Urals and the Caucasian mountains and as far east as the black sea, and this was called the Orthodox church. The other was only the refuge of the Roman Catholic church, however, when the papal seat moved back to Rome, the Prussian Holy Roman Empire remained ordained. When the pope was returned to the Vatican, the seat of the Orthodox church was moved from Constantinople to Prussia, modern Germany after the end of the Crusades. This brought a swarm of visigoths and ostrigoths from the Cossacks simultaneously to the Mongol hordes invading the Muslim Ottoman Empire. The goths brought with them ideas of trade masonry and architecture that were new to Mediaeval Europe, and which contributed to the Renaissance and to the age of Reason.

I will continue the history with the age of enlightenment momentarily, however first let me describe gnosticism. The highest gnosis is that there is no self.

This is based on the gnostic documents which contradict the official biblical version of the history of Christ's life. According to some he was administered a tranquilizing barbiturate, such as opium, in a soaked rag on the cross, in order to go into a death like sleep, and thereby stage his resurrection. According to these documents, the actual historical Jesus lived a very long life, married Mary Magdalene, and moved with her to France, where further gnostic documents exist describing their life together at the abbey of Rennes Le Chateau. According to some, documents were discovered under the abbey's altar tracing the dead sea scrolls back to Qumran before they were found. This all contradicts the official heresy of Paul, whom the Essenes describe as the Wicked Priest, and who may have only been another name for Pliny or Peso, two Roman families that may have actually written the new testament gospels, fragments of copies of which were found at Qumran.

The reasoning behind the gnosis is this: Jesus was a fictional character representing a real person. The Catholic trinity of Jesus as the son of man born the son of God is juxtaposed by the Orthodox belief in the ascended man, who achieved the highest gnosis and transcended mundane reality. I am not that which I am.

ii. enlightenment: nothing is true (there is no not)

At the time of the Renaissance, great painters, sculptors and architects, such as Fillipo Brunelleschi, Donatello, Masaccio, Fra Fillipo Lippi, Boticelli, Leonardo Da Vinci, Michaelangelo Buonarroti and Raphael Sanzio filled the lands of Europe. This was largely due to the Muslim occupation of Spain and the introduction of older cultures along the eastern trade route, traveled by Marco Polo, reaching as far as the Silk Road in China. Europe imported silk from the orient, spices from India, rugs from Persia, and along with this came a wave of eastern philosophical schools, such as Chinese and Muslim astronomy, Greek idealism and gnosticism, and Hindu concepts of cyclical time. These were all dutifully anthropomorphised by the Christians, however the study of the natural sciences would ultimately break away from the Universal Church, and establish its own sets of rules and parameters by which to define the universe. The early scientists, such as Leonardo DeVinci, who understood geometry and applied physics, would eventually lead to the later scientists, such as Isaac Newton, who applied geometry to physics and understood classical gravity. Little more than what the men and women of this day brought to light would be equaled by physicists until the twentieth century. The rest of the intervening years were spent trying to mature Mediaeval alchemy into chemistry and medicine.

Here we see more being accomplished intellectually in a shorter span of time than any other pattern of growth throughout human history other than monumental works projects such as the pyramids of Giza or the Great Wall of China, or the asymptotic exponential growth of the human population over recorded time. The enlightenment occurred entirely as a natural, social process, due to various concurrent international factors, however it was seen as a time of rapid change by the people of the day, some of whom lost and some of whom gained. Those who were older or stood to gain more from the existing cultural trends began to see the natural progress of the times as rebellious against established institutions, and this culminated in the conflict that progenated Protestantism.

With the invention of the movable type printing press towards the end of the enlightenment, Martin Luther, a jesuit priest of Bavaria, nailed a list of demands of changes in Catholic dogma and doctrine to the wooden doors of the church of Prague. One of these was that the bible be printed on the Gutenberg movable type face printing press in translations into European national vernaculars from the Latin vulgate. These demands were roundly rejected by the Papal seat, and Martin Luther turned around and created his own church, beginning a wave of similar churches, called Protestant, some of whom adopted his same views, called Lutheran. The scandal over the selling of indulgences, Luther's primary issue with the church, so badly hurt the Catholic church's reputation that, when it became inconvenient under Catholic doctrine for the King of England to be annulled from his wife, he simply turned and created his own protestant church, known as the Anglican church. Of course, the Catholic reaction to all of this was the inquisition, where roving bands of religiously mandated executioners roamed the lands holding impromptu witch trials in a pogrom against Jews, gnostics, goths and pagan protestants.

This period of history cast the church in such a bad light that its fundamental authority was almost eradicated altogether. This led to doubt and depression across Europe, as all the once proud members, now victims, of the Church suffered a crisis of faith. If the church that told them God was real was lying about its own practises, as well as other of its own dogmatic doctrines, then perhaps even the idea of God was to be doubted. Philosophically this led to the radical humanism of DesCartes,

Hume, Kant, Hobbes and Locke. DesCartes, though a professed devout Christian, began his philosophy by working backwards, deductively rather than inductively, and assumed that nothing was real, and all perceptions are false. Hume proposed the dialectical method of Greek rhetoric be applied to ontology, and Kant did so in an attempt to create a humanist ethic. Hobbes elaborated upon this humanist utopia naturalistically, defining all the mental life of precognitive man as being savage, and Locke proposed the exact opposite, that mankind's natural state was more in equilibrium with St. Augustine's City on a Hill representing the ideal kingdom of God, which itself was little more than a warmed over, Christianized version of Plato's Republic. Locke applied the liberty, equality and fraternity of the Knights Templar to the class hierarchy problems of the Republic, to compensate for the absence of God. The result was illuminism.

iii. illuminism: all things are permitted (there is no shalt)

The knights Templar had probably been the first westerners to encounter illuminism, a fatalist form of gnosticism, in the Saracen occult of the Hashishins. This Ishmaeli sect of Mohammedism were extremists, who believed their leader, Hassan I Sabba, to be a direct descendent of Ishmael, son of Hagar, who they claimed was the real son of Abraham that he was prepared to sacrifice before God, rather than Isaac, son of Sarah, in the Jewish biblical version of the story. The members of this sect were given opium and hashish in the mountain fortress that earned their leader the name, "Old Man of the Mountain." They were trained in murder and sent to infiltrate administrative positions in the governments and religions of the neighboring nations, and were willing to kill or even die themselves at the word of Sabba. Their philosophy was, "when nothing is true, everything is permitted."

These people should not be looked down upon for their trance like state of belief, which no more robs them of their independence and individuality than any other form of religion robs its respective members. They were almost never used in their capacity as assassins, and actually accomplished a bloodless cultural coup, instituting many popular humanist social reforms.

This politic of illuminism — a cultural coup de ta, humanist social reforms, and enslaved soldiers from mind controlled secret agents — seemed to ring especially true to the ears of the Knights Templar, and upon their return to Europe they instituted an international banking and commerce system and were accused by the Pope and the French aristocracy of worshipping Baphomet, now believed by modern scholars to be a derivation of the name of Mohammed, the prophet of Allah.

Illuminism derives its name from the illuminated manuscripts of the Ottoman Turks and the scholastics of enlightenment era Europe. Many fine reproductions of the Bible, the Koran, and various religious and mythological texts from both religions exist dating from this time, and the nouveau bourgeois of the mercantile class of masonic trade guild unions bourgeoning after the Crusades was awash with them, as well as the spices, silks and Persian rugs of trade. The tripartite hat marked jacobin and jacobite social clubs that predominated under the wig wearing aristocracies of late enlightenment era Europe, and which were often funded by the elder, Scotch Rite Masonic Lodges, which the Knights Templar had become, and which were even more zionist, anti-papacy and gnostic enemies of the Catholic dogma than the Lutherans, Protestants or Anglicans. The tea sipping commoners took eagerly to the tactics of political upheaval represented by Illuminism.

The most prominent of these was the former Jesuit monk, Adam Weishaupt. He began an order called the Illuminati in Bavaria that had some unscrupulous ties to the Scottish and then young York Rites of Free Masonry through some various shared colleagues and Lodge members in western Europe. This he had done in an attempt to protect him from the Jesuit order to which he had belonged, and from whom, it has been speculated, some of his gnosis was stolen. The Illuminati became infamous throughout Europe many years later, after Weishaupt's death, when the also new media of the printed newspapers uncovered and circulated copies of documents supposed to have originated from Weishaupt's sect plotting a plan for the world wide overthrow of all existing governments and religions. There is actually less evidence linking these documents to the Illuminati, or any Masonic organization, than there is that Adam Weishaupt faked his own death and moved to America where he became known as George Washington. Ironically, around the time that the Illuminati manifestoes surfaced in the press, the issue of the first national bank, now known as the federal reserve, (from which America loans money to third world countries as well as still to European nations just as under the Marshall Plan, and against which we charge them interest that causes our interest rates to rise, the value of the dollar to decrease relative to the consumer cost basis, measured usually by the scale of minimum wage over inflation, and the international exchange rate to become imbalanced) was being argued between James Madison and Thomas Jefferson in the Federalist papers in America, and introducing the press to the yellow journalism and the control of the issues covered by the news media by interested, independent, rich capitalist parties.

It was also around the time of the illuminated manuscripts that fed the imagination of young Jesuits and Jacobins that the religious philosophy of Christ as the grand architect and the universe as essentially mechanical and beneath the removed deity became popular. This stemmed from the fresco depictions of Christ seated with a globe, an open book and a compass on His lap that were common and popular in eastern Orthodox Christian churches, and it became known as deism.

iv. deism: the structured universe is not a living organism

Imagine the pole of the world held in the hand of its mass like the parallel bar of an acrobat, and see that it is not really the pole that moves, but the mass of the world around it, spinning like a circus gymnast. But the bar of this pole too, is held in the grip of the gravity of the sun, and spun circularly around it in its orbit, so that the fixed directions of the cylindrical pole do move relative to the fixed background. So too, however, is the body of the sun twirling up around the bar of its poles, and this bar held in sway within the discus of the galaxy, spiraling around the core. So are all the galaxies of the universe engaged in a vast cosmic ballet, and our own Milky Way will one day collide with the galaxy called Andromeda. All of this preexisted life, though since much of it only exists now as recorded information in rays of photons, it is more alike the essence than is mundane human existence.

With the heliocentric model of the solar system proposed by Galileo Galilei, which occurred around the same time as the beginning of Lutheranism and immediately followed the Spanish resurgence of QBLH, more and more people began to believe in the universe as being like an enormous clockwork machine, created by God and maintained by God and/or the angels of the spheres of the heavens. This began with representations of the spheres of the solar system that functioned gyroscopically, however the idea would go on to flourish into an age of industrialized machines, culminating in modern space age computers.

According to deism, the only universal laws are the laws of physics, and God may have created the universe, however there is no evidence that can be gained using the laws of physics to deduce His continued presence or absence. In this way deism was an attempt to reconcile the Catholic concept of the Holy Ghost with the gnostic concept of Christ as the ascended man. This is similar to the shemhamforash of Solomon, which demi-deified the workers on the first temple, or to the Necronomicon of the Sumerians, which demi-deified the workers on the Great Pyramid. Just like gnosticism, deism can be traced back to the times of ancient Egypt, where arose the practise of the anthropomorphic representation of cosmological processes that continues within the traveling Lodge of Rosicrucianism to this day. This is also when representing the universe on earth by the leader of the occult began, that continued on through the Buddha and was passed to Jesus Christ. Here is where the Christian argument over the universal generalizability of the ghost king Choronzon from kabbalistic mythology became associated with the concept of the Last Sacrifice by God of His living son, Jesus the anointed Christ.

This was an argument between early scientists and the institutions of the Christian church only because after the time of Christ the occult became subject to the decline and fall of the Roman empire, and exoterically became the Christian church, which had a duly elected representative head, the pope, and made no claims to divine right by blood line as had the Kings of Israel and as was attributed by the writers of the New Testament gospels to Jesus, known as the "first pope." Meanwhile the early scientists sought for God in natural consequences, and turned further and further from the path laid out for them by the representationalist church. Martin Luther encouraged these scientists to "sin, and sin boldly." The gnostic occult became the esoteric aspect, the inner order of the sects, cults and orders of the greater social institutions, whose secret doctrine was the agenda controlling the world.

Much of what influenced deism and the mechanical revolution was the abundance of south American codices brought back into Spain by the Conquistadors, containing beautiful pictographs describing the heavenly cycles, that were attempted to be translated into QBLH. This began with a rethinking of early Hebrew mythology, but would ultimately lead to John Dee's Enochian system as well. Most of these Mayan and Aztec codices were, like the Dead Sea Scrolls of today, remanded over to the Catholic church for inventory and translation.

f. the holy concepts

The oldest form known of the QBLH is the tree of life diagram, and it is supposedly in this form that it was delivered to Moses on Mt. Sinai. This depicts the ten sephira, or attributes of God, which are equivalent to the ten renunciations of sin that constitute the Law of Man. Perhaps the QBLH is ultimately best meditated upon as upon a mountain, for it is a height of its own, and a mass of its own, and a measure that must already be known.

The QBLH is much more than these concepts alone, and thus it is known of psychologically, just as two entities know one another. It is, like humanity, more than the sum of its parts, and its knowledge only represents a reflection of its light. Understanding it rightly, we are clear in the presence of this light, and perceive it as wisdom. So do we know that wisdom is perceiving us, and yet that we are clear light, and that this is, moreover, nothing.

It has been said about the study of QBLH that "fools rush in where angels fear to tread," a warning similar to the description of the weather of the month of March, "in like a lion, out like a lamb." This is simply to warn brave fools that QBLH might make them into messaliengels. God may giveth one eye and taketh away another.

QBLH in itself cannot kill, but it can be used to kill, and much of the black arts of Magick, thelemic Satanism and voodoo describe how such techniques as astral projection and remote viewing can also be used to induce experience at target sites. These techniques are not necessarily harmful, however during the Cold War both the Soviet and United States militaries experimented at psycho-kinetic assassination. The experiences at the projected locations, that can induce sensation in observers, are merely lower level reflections of the mental conditioning of the remote viewer or astral projection's physical source in the mind of the body. Therefore the effects that can be turned to harm are the same that must be overcome to acquire skill with the mental aspects of the QBLH. This is similar to the radiation induced hallucinations caused by exposure to the stones of the Ram that eventually, with conditioning, heighten mental sensitivity and increase perceptual awareness.

One of the predominant energy cycles identified by the QBLH is the double helix of mind and matter, and here we again find the qliphoth, as juxtaposed to the QBLH, considered to be negative mental aspects, or equivalent to negative numbers on the same number line of QBLH upon which mental thoughts are positive numbers. Since all physical materiality in the continuum of the universe is made up of the quantum probability wells of the qliphotic shells, which themselves represent the phi/pi spiral of geometry through the n dimensional multiverse, this is considered the exoteric aspect of the QBLH, while the true nature of the way of QBLH remains concealed, and has only been described anthropomorphically as the oral tradition of human history.

i. the QBLH: phi/pi

The naturally occurring form of the QBLH is phi/pi, where this is given as the measure of a wavelength that approaches, circumnavigates and escapes from the opposite side of a sphere, represented by a hypersphere, or as the measure of a torus. This metaform, or fourth spatial dimensional hypershape, is considered ideal, along with the hypercube and the platonic solids, which also all occur along the same line of mathematical extrapolation as the torus. Numerically this is encoded into the multiplication tables of eight and nine, and revealed when the numbers of their factors are summed. Since this is an inherent gnomon in the mathematical structure, it exists irrelevant of its discovery, reflected more or less in all the rest of its parts holographically.

While the meaning such an abstracted concept as a transcendental or transfinite number might have to the average person is probably very small, the fact that such concepts exist at all, and do actually occur in the measurements of natural phenomenon, particularly evident according to statistical averaging, has universal applicability. In this way the secrets of Heaven can lay about directly beneath the noses of us all, and many times we might go without noticing them, particularly since the fuzzy logic processing systems of our neural nets perceive natural patterns as digital static.

The comparison between phi/pi, as I have described, has largely been in the form of national and personal histories, as well as the records of cyclical events in the heavens. This has led to the animism of the zodiac and the anthropomorphication of religions. Because phi/pi describes negative entropy it is justified in its associations with living forms, since according to our survival instinct, we evolve opposite the odds that define the rest of the natural world. However, since phi/pi is a mathematically ideal archetype in the fourth dimension, it also represents the revolutions of entropy, the essence of time.

Thus, phi/pi is as the inversion between the perpetual, active survival of life and the gradual, passive encroachment of death.

— numerical equivalency

As I have said before, the number set of the gnomon containing pi also contains phi, and thus, on one level, these numbers may be thought of as equivalent. However, more than this, by ancient reckoning, the gematriacal sum of the Hebrew letters of QBLH (Qoph, Bet, Lamed and Heh) was equal to the denominator of phi/pi. In this way, moreover, did phi/pi as the greatest common factor of the factor sums of the multiplication tables of eight and nine become QBLH, the lowest common denominator of all mystery school esotericism.

— geometric spiral

As I have also said before, the graph of phi/pi is a geometrical spiral in three dimensions that looks very similar to the coils of DNA inside human genes or the banding of electromagnetic field lines around the sun. This can be depicted as the seven color fields banded around the surface of a tube torus. It is also thought to be the measure of the histories of gravitational singularities that are baby universes.

ii. the Kabballah: the body of God

Insofar as phi/pi is universally generalizable, that is, ubiquitous, QBLH was thought by ancient peoples to represent the body of the living universe. In the east this was expressed using harmonic vibrational energy fields, called the atman or auras, and interconnecting, multidimensionally faceted and leveled relationships, called the chi of feng shui. In the west these concepts were both humanized into the soul and the spirit, called by Egyptians the Ba and the Akh and by the Hebrews

Ruach and Nefesh, and QBLH associated with the measurement of the geometries underlying and governing these forces. It was attempted to humanize QBLH, as had been the auric field into the astral soul, and the chi of feng shui into the free spirit, and the result of this was the myth of Adam Kadmon, or Kabballistic representation of the universe.

According to this mythology Adam Kadmon, whom the Sumerians called Enki, was the twin aspect of Adam Homo, whom the Sumerians called Enlil. In the epic of Gilgamesh we see Gilgamesh, representing Enlil, trying to restore the life of his friend Enkidu, representing Enki. Similarly in the bible the story of Cain killing Abel immediately follows that of Adam and Eve, and Cain's son was named after Set. Here we see that it was Lilith that was the wife of Adam Kadmon. According to the mythology Lilith died and was buried and Adam Homo was made from the adama, or red clay, of the earth of her tomb. Eve was made from Adam Homo's rib, which follows a phi spiral, and was tempted by Kadmon in the form of a snake. According to Christian mythology, it was Adam Kadmon who would be reincarnated as Jesus Christ, and therefore his death would be taken to atone for all the intervening sins of humanity since the original sin of Eve's temptation and the banishment from Eden. Enki was Sargon and Enlil was Imhotep, and they were all one. All of this was actually occurring at that time in Egyptian pharoahnic succession and was described by the indigenous mystery cult religion in the myth of the death of Osiris at the hands of Set and resurrection at the hands of Isis, his wife. Eden was an actual ancient town in the Gobi peninsula, and it was from here that the Hyksos emigrated into Lower Egypt until the reign of Akhenaten, after which they were expelled and entered Canaan. Other than the archetypes of the slender Nazarene King in the west and the corpulent, opulent recumbent Ho Ti in the east, no more attempts have been made to anthropomorphise the primary archetype of solar deity mystery schools since they first began supplanting feminine, lunar cycle cults of the late neolithic age.

Subsequently to this the Kabballah has taken on increasingly geometrical forms in the hands of Ottoman Muslims and the Jews of Spain they left behind after their occupation. The geometrical lattices that can be extrapolated from the tree of life have been expanded upon as tesselations, or tiled patterns, and these regular patterns produce regular frequencies of brainwaves, and induce regular emotions. Thus they have come to replace anthropomorphication of deity for Muslims as the true visage of God. It is true that this lattice can be extrapolated into many different natural and ideal patterns.

— the tree of life diagram as a lattice

The placement of the ten sephira as corners on the geometrical lattice that can be extrapolated between them depicts a shape that may be thought of as representing a hypercube at antipode. This references all the other forms and shapes along the gnomon measured by the number set of phi/pi. In this regard it may be thought of as similar to the lattices of quantum mechanics, which depict quantum behavior predictability as occurring in relationships between coordinate pairs of variables.

— the ten sephira and attributes

In ascending order the ten sephira are Malkuth (the kingdom), Yesod (foundation), Hod (splendour), Netzach (severity), Tiphereth (beauty), Gevurah (victory), Chesed (mercy), Binah (understanding), Chakmah (wisdom) and Kether (crown). There are also the non-sephiroth of Shekinah (the bride of God) and Daath (the veil of the abyss), as well as the spheres beyond the tree of life, Ain Soph Aur (Limitless Light), Ain Soph (Limitless Nothing) and Ain (Limitlessness). These are arranged as upon three pillars, called passive, active and neutral, such that, on the passive column, understanding is above victory is above splendour, in the active column, wisdom is above mercy is above severity, and in the middle, neutral column, the crown is above beauty is above foundation is above the kingdom.

iii. the Qabala: the art of precession

Have you ever bumped into a cup of liquid, and observed how the surface might wobble as it settles down, or ripples flow inward from the sides of its volume? Why does it turn around in a circle as it fluctuates up and down? Where does the current of the inward flowing ripples go after it disappears from the surface at the center of the cup?

Such is all occurring for our galaxy, right now, so slowly it would take many modern lifetimes for there to be a change in it that we could see. It is slowly pivoting around in a spiral wobble around the central black hole, and the gravity of each of the stars is feeding back the gravity of the black hole just enough to slow their fall into galactic core to where they are in orbit around it. This leads to the sunspot cycle of stars and the wormhole cycle of galactic black holes.

It has long been believed that the study of Qabala is associated with the study of sorcery and Magick, and it is the case that much of occult esoterica is steeped in obscure terminologies and the attributions to various characters of different kinds of above average skills. However much of the campaign to occlude the occult in such mythology and ritual has occurred only among kabballistic scholars of the twentieth century, such as Frazer's "Golden Bough," the writings of Aleister Crowley and Jungian psychology. Before this time, at which also occurred the publication of the Necronomicon, little association was made between the dead and the djinn, as priests were thought to be the earthly representatives of God.

Supposedly, many of these esoteric occult documents are translated from very ancient source texts, such as Sumerian scrolls and tablets, Egyptian papyri and stele, Hebrew, Coptic, Gnostic, and early Christian apocryphal scrolls, although many of them the history of which can be traced through the scholasticism of the Medieval ages, when monks transcribed and translated many various older documents, some very badly indeed, are more dubious than those archaeologists have only recently discovered, which can be carbon dated, or those which historians have the originals.

What these modern occult writers describe of the ancient pagan rituals is their knowledge of a cause and effect correlation existing as an ambient energy field, and the use of referentials to represent interactivity within this field.

This has been catalogued and categorized by various different scholars, beginning with J. G. Frazer, who traced the origin of this belief to ancient neolithic seasonal rites at sacred sites beginning at the time of the agrarian revolution. Crowley discussed the use of a wand, a weapon, a chalice and a discus as being the components of magic ritual representing the four elements, and the altar as representing the QBLH, particularly that of the four directional Enochian watchtowers, the four elements and cosmos. He also draws relativity between the sacraments of Abramelin and the penances of Buddhism.

According to this system, which is preferably practised out in the wilderness such as the desert, or in a closed room which can afterwards be tokenly cleansed, the magician draws a magical circle around themselves to seal themselves off in the astral, magical world. Inside of this they draw a pentagram, which seals their minds from all other minds (so it is thought). Outside of the circle they trace out a triangle. They stand within the circle and summon Choronzon into the triangle. Here they have discourse with the spirit realm. In the end, the triangle of Choronzon, the circle of the aura and the pentacle of man are all swept away and the winds cleanse.

It is claimed by modern scholars that this is the oldest form of ritual system, and that it predates the beginning of the belief in the djinn in the middle east. The difference between this and skrying is about the same as the difference between a Ouija board and a séance. Suffice it to say that skrying tends toward gematria, while ceremonial ritual tends more toward astrology.

— astrology

Sumerian Astrology may date as far back as to the time of a great flood of the Tigris and Euphrates river and Nile river valleys as long ago as some ten to 8,000 years, that may also have been the time when the Nile river changed the direction of its flow from east - west, emptying into the Atlantic, to south - north emptying into the Mediterranean. It is possible the nasca lines date back even farther. Much worship of bulls predominated in the middle east at that time, such as the apis of Egypt and the bull-hoppers of Minoan Crete. The zodiac could be used to measure the precession of the spring equinoxes, or the constellation along the ecliptic at sunrise when the earth was at northern hemisphere perigee to the sun. According to this cycle, the sun rises in a new sign about every 2,000 years. We are currently entering the age of Gemini. The age of Taurus is coming to a close. The time of Christ ended the age of Aries about 2,000 years ago. King Solomon's temple marked the ending of the age of Pisces about 4,000 years ago. The Great Pyramids marked the ending of Aquarius about 6,000 years ago. The birth of civilization began with the ending of the age of Capricorn about 8,000 years ago. This marked the end of the last northern ice age, and it was probably during this time that there had been flooding in the middle east.

— gematria

Gematria is the ancient practice of assigning number sum numerals to letters. This was done to contribute to a language another layer of meaning. The number sums of words could be related to one another, and this would usually serve to accentuate the complimentarity of the different words' meanings. As number sums they could be arranged relative to one another in ways that could produce any type of mathematical relationship, and the use of magic number squares as token spells to ward off evil and accomplish one's goals dates back to the time even of Abraham. The form that this gives to the evolution of the Hebrew language is similar to the root words of latin in the Romance languages, but on a more basic, calculable level.

iv. the Cabala: the psychic sanhedrin of Messai

During the twentieth century many esoteric documents were submitted to the French biblioteque nationale. Many of these, all submitted by the same man, are purportedly documents of the modern esoteric priory of the Crusades era Catholic Order of Zion that may have backed the Knights Templar. These give the names of many famous, less famous, moderately obscure or completely unheard of people throughout history as the succession of heads of the order. The claim these documents themselves make to authority derives from several cypher documents containing crude geometric coding, allusions to disposed royalty and citing of a few sacred locations, one of which was supposed to have contained buried treasure.

The Knights Templar probably discovered the dead sea scrolls buried under the ruins of the second temple in Jerusalem, and, under the auspices of the King of Israel, transported them to the cave near Qumran, which would have been under the water level of the Dead Sea at the time they had actually been written. It is possible that some of them may have been copied down and brought back with them to France, and that their persecution at the hands of the French monarchy and the Papacy was the result of a search for any such copied documents. The Knights Templar that escaped fled to Scotland and established the thirty three degrees of Scotch Rite Masonry, which would be shortened in the Anglican York rite to exclude the degrees between seven and seventeen, and between eighteen and thirty three. It is thought that these initiatory rituals derive directly from the gnosis brought back from the Holy Land by the Knights Templar.

Some of the locations of sacred sites listed in the priory documents include churches in northern Europe which were early Masonic meeting sites, and these churches align with one another, often, in Masonic ley line geodesic geometry similar to that crudely represented in the cypher documents. The story of these documents describes the deposition of the Merovingian bloodline from the throne of aristocracy in France, and the list of heads of the order includes these deposed kings, and dates even further back than them, to the time and person of Christ. The documents do not make it clear if the list of heads of the order adheres to a bloodline based, committee selected, chosen or initiated manner of succession.

The root of the word Messiah, the same as that of Moses, means both, "heir" and "saved from water." This probably dates back to the earliest tribes of Africa and the first post-deluvian civilizations of Mesopotamia and India. The Priory of Zion is, just as Judaism itself, messianic. While in the west this influence has been forced to remain veiled, esoteric, obscure, and occluded, it was embraced early on in the east with the creation of the social caste of the Vedics, or Brahmans and Lamas. In the west the priests do not claim to posses enlightenment, only to be seeking it.

The Order of the Golden Dawn, an esoteric organization that arose early in the twentieth century, was the occult celebration of the ending of the dark night of civilization's youth and its coming of age ritual, the Equinox of the Gods. It was they who resurrected popular, public interest in the arcane arts such as ESP. It should be worth noting to what lengths the world has gone subsequently to get itself electromagnetically interconnected.

— the gospels as history

Recent archaeological discoveries in Chenoboskion, of the Nag Hammadi library, and near Qumran, on the shores of the dead sea, reveal a wealth of newly discovered apocrypha left out of the bible, from both the old testament and the new testament. These include the treatises and stele of Seth, apocalypses of Adam, Peter and Paul, apocrypha of James and John, gospels of Marcion, Niccodemus, Matthew, Philip, Peter, Thomas and Bartholomew, as well as Arab narratives of the nativity, a sophia of Jesus Christ and the letters of Pontius Pilate. There have also been countless discoveries of ancient Sumerian tablets, including hymns to and stories about Inanna and the lost books of Enki, which describe the same times as the old testament of the bible and its apocrypha, as well as the other existing written descriptions of the times of the new testament such as the writings of Pliny, Josephus and the Muslim Koran.

g. students over time

According to history the oldest mystery school students were the Vedics of the Indus river valley caste system civilization. These were all of the elderly, who had devoted their entire lives to the community, and were then exiled into the wilderness to seek the meaning of life. We see in tribal communities that continue to exist to this day in Africa and Australia that it is usually the elderly of the tribe that are the medicine men and healers, while the younger men and women are hunter gatherers. The vedic caste of the class system merely codified this, and banished the wizened to outside of the populated towns and countrysides. They established small communes, often out in the wilderness far from society, and here the art and practise of meditation was born. It is believed that the idea of the third eye originates from this time, and it is an idea that continues to be propagated to this day.

Supposedly the first Vedics were a pale or blue skinned race called the Aryans, who wrote the Rig Vedas and then migrated north of India, into China, where several anglo-saxon, celtic looking bodies have been found ritually mummified. Thus it is believed that long ago, perhaps at the same time stone henge was built, the Gualic, gaelic, germanic, anglo-saxon, slavonic and cossack peoples immigrated westward across the Urals and Aryan mountains into Europe from across the Siberian steppes and displaced the last of the cromagnon culture of the late last ice age.

At around the same time there was a wave of Hyksos migration out of Sumeria into Egypt. Their series of leaders (Enki, Sargon, Imhotep, Khufu, Khefren, Akhenaten) are complimented by a series of Sumerian names (Enkidu, Scorpion-man, Humbaba or Hamurabi, Gilgamesh, Inanna and Utnapishtim), a series of Hyksos names (Kadmon, Satan, Abraham, David, Solomon, Jesus) as well as a series of Nubian names (Thoth under Shu and Tefnut, Set, Thoth under Isis and Nepthys, Osiris, Horus and Re). Later the Hyksos would leave Egypt and settle in Canaan on the Gaza strip.

It is to these people that we owe all early metaphysical legislation. The vedic philosophy is known to have influenced early Egypt, where the belief in the third eye became associated with the ajna serpent of the crown of upper Egypt. To understand the significance of this sixth sense of three hundred and sixty degree mental conceptual projective space is to realize that the thalamus in which mental holography occurs, connecting multiple neurons simultaneously through individual neurons, is what separates mammals from reptiles. Remember also the attribution by Hermes Trismegestus of Poimandres to Draco, and that constellation's position as pole star above the heads of the dinosaurs.

The pale races had obviously come from the continental shelf of India through the exposed lands of Oceania, known in mythology as Lemuria, while the Nubian

413

pyramid builders of Africa had come out of South America, where their presence has been recorded in some sixty ton stone carved Olmec heads as well as the various Jaguars of the Popul Vuh, along with later depictions of Quetzalcoatl, a caucasoid, bearded civilizer, who also departs to the east, promising to return. It seems that the first race of homo sapiens, who lived alongside Cromagnons, emigrated out of Africa, along the coast line of Asia and Lemuria before the continental shelf of India had joined with Asia, into Australia, then upwards through and populating the mongolian Orient up to the Beringian land bridge, where some crossed over into the Americas to settle, probably by simultaneous land crossing, becoming the sparsely populated north americans, and following the shore line in reed boats, becoming the vastly overpopulated south americans, all then known as Atlanteans.

i. the Saqqara of Ethiopia

Hominids have been living in Africa longer than anywhere else in the world, followed closely by Australia. The indigenous lifestyles of these two continents have remained much the same over the long aeons of the millennia. They are tribal people who hunt and fish and trade between villages the game that they catch and the fruits and vegetables the women and children gather. The elders are star gazing shamen. In Australia they are known as the Watchers over or the Keepers of Dreamland.

In Africa the dogon tribes people, who are thought to be history's eldest surviving astronomers, still preserve a ritual with a may pole thought to represent the sidereal rotation of the binary stars that comprise the Sirius system at the heel of Orion the hunter. It is not known how their observations can be so exact, however some of their ancient cave paintings preserve depictions of ellipses that seem to represent Sirius, as well as spirals and labyrinths as are seen in many other paleolithic cave painting and rock carving sites around the world.

It is in Ethiopia wherein is the Aswan lake that is the source of the Nile. Ethiopia at the time of the end of the last ice age, as well as the entire sahara, were fertile grassland and savannah, with thick, dark, rich soil, that received ample seasonal floods that turned the grasslands gradually more and more awash. By the time of the building of the pyramids, the Nile had stopped flooding regularly, and the Sahara and the Giza plateau dried up and turned into the sands of the desert. Now, more than 4,000 years later, Ethiopia is no longer receiving the amount of precipitation it once did, and the Aswan lake has been dammed in order to produce electricity and form a water reserve to protect the people against their crops dying out, the soil turning to mud or sand or clay, and the children starving.

In ancient Africa dwelt he who was known as Thoth, Enoch, or Utnapishtim. It is believed to be to this man that the art and craft, the tools and instruments of magic as they still exist to this day, can be credited as his creations. It was also, later, the biblical era home of the Queen of Sheba, believed to be a reincarnation of the holognomonic archetype of Shiva of India, Inanna, Ishtar or Astarte of Sumer, and Isis of Egypt. According to the kebra negast, a sacred Ethiopian old testament apocrypha, she had a son with Solomon, the king of Israel and chief architect of the first temple to YHVH, and when the Babylonians came to burn the temple, this son, named Menelik, rescued the ark of the covenant from the Holy of Holies and brought the sacred stone tablet of testimony to Ethiopia. This stone was stored in a secluded monastery near Lake Aswan until the end of the twentieth century, when it was moved into a church within the surroundings of Axum, a larger city.

ii. the I Ching of the Chinese

The ancient Vedics all shaved their heads and wore their skulls bald to the skin. The hair follicles are simply columns of dead skin cells that contain better traces of genetic material than fingernails, which contain keratin, though both continue to grow after death. Both pubic hair and fingernails grow flat, and faster on one side than the other, such that they spiral and kink. Some people in ancient China and modern south America have grown their fingernails as long as almost half a mile, and such that they wrap together, by never cutting them. It was also thought that the hair follicles falling before the eyes were like the filaments of galaxies, only a material distraction visible as an illusion of light. This practise seems to have been initiated with Siddhartha, just as the compulsive Muslim cleansing rituals and the ritual Baptism were introduced among the Essenes around the time of Yeshua Ben Padiah. Before these two teachers, who may been reincarnations of the same holognomonic archetype, the Hebrews and the Vedics wore their hair and beards long, and such marked the faces of Abraham and Moses, Lao Tse and Confucius.

In the days when men wore beards, in the oriental lands, a token game arose similar to the game from Sumeria that would come to be called chess. Both were based on the sum sixty-four. The chess board has sixty four squares and the I Ching has sixty four hexagrams. These both represent the sixty four codons of the genetic code. Sixty four times six gives the number of tao rods or chi sticks, represented originally by yarrow spliffs of bamboo, in the complete, sixty four hexagram King Wen sequence of the I Ching, and this number, 384, corresponds to the number of nights in a lunar cycle, the time it takes for the moon to orbit around the earth and to rotate about itself. The lunar cycle corresponds to the sunspot cycle, as the moon goes around the earth while the earth goes around the sun, such that the combined gravity of the earth, the moon and the other planets generates more or less sunspots, prominences and solar flares in the electromagnetic coiling on the surface of the sun. This was marked along with certain other events, such as the pentacular sidereal rotation of Venus, in accordance with various cycles including the precession of the equinoxes.

iii. the safhardic semites of Muslim spain

Many of the semitic Muslims who had moved into Ottoman occupied Spain during the Late Middle Ages did not want to leave when the Christians re-conquered Spain and re-instituted the Spanish monarchy. These people were forced to become Jews by the conquering Christians, since the two terms were interchangeable pagan concepts in their doctrinally indoctrinated minds. However, after the Christians retook Rome and the rest of Italy, they began waging pogroms on these anathema peoples who had chosen to remain behind, and who were, by then, also being joined by gnostic goths from across the Urals.

During the short period between the end of Muslim occupation and the beginning of Christian pogroms, however, more study of QBLH was generated by a single group of people than any group of people since the ancient Egyptian mystery cult some 4,500 years before. Here we see the works of the various Rabbis of Biblical commentary compiled in the Zohar, or book of splendour. It is also thought to have been at this time that the sefer yetzirah, and commentary thereupon, was first written down, and this text had only been preserved as an oral tradition since before the time of Moses. These two books are considered to be the two key texts on the Kabbalah, and their understanding is great wisdom. It is thought to be this time period also that produced the work of Abramelin. This period in history led to the beginning of the Catholic Order of Zion and the Renaissance, as well as the beginning of banking and finance rate monopolies by non Church officials or royal bloodlines.

At the same time as the Knights Templar transported the dead sea scrolls, in Europe they established an international banking and interest bearing finance system similar to the taxation system of the governmental institutions and tithing system of the edifice of the church. This was handled by the order of zion under the auspices of instituting the gnosis of the Ishmailis, and it rigorously recruited the former Muslims living in Spain, who had converted to Judaism, the Jews of France and the Orthodox of Italy. Thus, when the Templars were persecuted by the Roman Catholic Church and the French aristocracy, they were able to funnel their channels of currency through other countries, and elude the full seizure of all of their assets. Then the remaining Templars of France sailed to Scotland and established Masonry at Roslyn chapel.

Much of what constituted Templar riches and wealth was the gnostic documents they had copied and translated from Qumran. These formed much of the basis for the symbolism of speculative, initiatory masonry. This is the beginning of the additional, now lost degrees of regular craft masonry. In these rituals it states very specifically that the Lodge must first be opened to conduct business of any of these levels of initiated degree before it can begin to do so, as well as that anyone who is caught discussing the content of these degrees with non Masons should be killed.

iv. the Tarot of the Visigoths and Ostrigoths

Simultaneously to the persecution of the Templars at the end of the Crusades and the success of the anti-royalist and antichurch banking system came a wave of migration into Europe from the east of fair skinned, dark haired goths. These were usually gypsies, traveling merchants and traders who brought with them various souvenir items from the Silk Road, Persia, Turkey and Greece. It was thought to have been they that invented the card game and deck of playing cards of the Tarot. Although this was originally a seventy two card set, the twenty two trumps and the Princess or knave cards have been omitted from the playing card decks of today. At the time this game was known to the French aristocracy and was popular in Italy. Characteristic attributes were assigned to the trumps and they were used by gypsy sooth sayers in fortune telling, and endless varieties of games derived from the endless shuffling of the numbered cards of the deck. Such packs and decks would continue to be produced throughout the late middle ages, the Renaissance, the age of reason, and the enlightenment. Playing cards and prophets are still popular today. The time of the Tarot's earliest appearance begins with the gnostic documents and ends with the bringing back from America of the south American codices.

It is unknown if it was thought at the time to be associated with QBLH, however subsequent researches, such as the Rosicrucian Order of the Golden Dawn or the thelemic Order of Oriental Templars, have associated the Tarot trumps with the twenty two paths on the tree of life diagram, and with the attributions of letters and states of consciousness to these given in the sefer yetzirah. The seventy two cards of the tarot, when each is assigned an attribute, form all of the dekans by day and by night over the three periods of ten that make up each of the twelve signs of the zodiac. This number can be extrapolated from the ancient Egyptian annual calendar and represents the number of years in which the earth's pole precesses one degree of three hundred and sixty, since there are exactly five times two times three of these per sign of the zodiac.

v. the Mayan tzolkin

Around the same time Christ was being crucified in the middle east, the pyramids of teotihuacan in the Yucatan peninsula were being erected to the sun and moon, aligned with Osiris, and the way of the dead between them flooded with water. From this culture, which also produced Pacal Votan of Palenque, comes the great wheel of time called the tzolkin, that is aligned to calculate for sidereal rotations of venus, the precession of the earth's pole as seen at the equinoxes, the lunar monthly calendar and the solar eleven year sunspot cycle. This seems like an awful lot for a neolithic culture and only agrarian civilization to have accomplished in such a short amount of time, however remember that their equivalent of the dead sea scrolls were the Ica stones of Peru, the equivalent to the stone of Ram was the perfectly carved crystal skulls ubiquitous to south america and elongated human skulls of Nasca. Remember, also, that they followed the Olmec culture, who, it is thought, constructed the Nasca lines.

There is little reason to doubt that Plato's Atlantis described the Altiplano in south America — the only place to find orichalc in the known world, that lay far inland, but which was a grid of circular canals and square lots of ground with a channel leading all the way out into the Atlantic ocean. Similarly it is possible that this occurred at the same time as the flooding in the Tigris and Euphrates, Nile and Indus river valleys and the submersion of cities off the coast of Japan, Spain and Cuba at the end of the last ice age, and that the mythology of the subsequent cultures may reflect an earlier learning involving knowledge of precession. It is clear that some of the existing neolithic sites with astral alignments were built before a change in the world's ocean levels, since they are only partially submerged.

In any event, at the same time the Muslims were beginning sufism, the Mayans were engaged in a rubber ball sport.

2. present modes

Philosophy killed God and Magick consumed Him. Beyond this there is no trace of God. The heavenly spheres all function deistically, while mankind struggles with gnosticism, attempting to obtain mathematical ideals. These are all only metaphors.

In modern times there is a great need to perpetuate karma, allowing it to flow through one without hindrance, clinging or obstruction. This may only amount for some to the exchange of consumer goods and services, but it amounts in Ethiopia to starvation. The QBLH of the heavenly spheres proceeds as if indifferently.

This is the reenactment on the deified archetypal level of the death and resurrection of the central solar deity, reflected in the rise and fall of human civilization. This occurred first between Sargon and Enki, then between Imhotep and Ptahotep, then between Khufu and Khefren, then between David and Solomon and then between Buddha and Christ. After this, if it continued, it did so esoterically, as since this time Christian calendars have not reflected the precession of the zodiac.

In modern times the multiverse has become so congested with interdimensional traffic of information that it is difficult to discern the distracting from the inspiring. This is because the octagonal and orthogonal leg of this polarity, representing Thoth, is missing from the hexagonal representation of the Ychthos Christos cross with the rex ioerem crown.

In the present mode there is concentrated activity to construct and improve upon electromagnetic communications, and these extend as far as research involving synchronicity as an acausal connecting principle that can be manifested disparately electromagnetically, such as the simultaneity of a phone call or a live broadcast on a television set. This permeates the airwaves of earth, and spreads far off into the distant cosmos. Beyond this rays of photons communicate electromagnetic signals from distant, ancient stars. Just as the clock of earth's precession is measured by the movement of the entire galaxy of the Milky Way, so it is thought there are connecting fields between the electromagnetic bubbles of the galaxies in the filaments, walls and voids.

Traveling in a temporal singularity is as simple as measuring the time of a radioactive decay atomic clock on the ground or in orbit outside the atmosphere, for when these are compared they will be seen to have become different. Where waves of electromagnetic information overlap, they form temporal wormholes, and some information travels backward in time, even through the mediated communication system. While some wait for the coming of wormwood as one, big event, earth has already become riddled with spiraling karmic centers, ley lines and commercials.

a. QBLH as matrix lattice for vector of entropy

As I have described, phi/pi is the measure of the histories of gravitational singularities unwinding outside of the universe into baby universes, the measure of the perpendicular wormholes through the multiverse inside the event horizons of black holes, the measure of the statistical average of electromagnetic to gravitational relationship between stars in spiral galaxies and the black holes at their cores, the measure of a star's electromagnetic polar field lines coiling around its differentially rotating gas surface, as well as the measure of the earth's polar precession. These comprise most of the processes we would define as the causes for entropic matter-energy interchange in the universe. It is reasoned that, because the hypercube is one measure of phi/pi, this is why the ancients depicted QBLH as the hypercubic tree of life.

QBLH as the hypercube of time can be thought of as a geometric lattice depicting the mathematical matrix for phi/pi, the vector of entropy. As I have described, the tree of life diagram is far from the only representation of this shape. The hypercube can be depicted as nested, at perigee (as the tree of life, at antipode), at apogee (as the octaholohedron), as well as as a regular cube. The same shape is represented by the singularity, line, plane, sphere, hypersphere and tube torus, as well as by the cone, the tetrahedron, the hypertetrahedron, the tetrisociholohedron (hypertetrahedron at antipode) as well as all the other Platonic solids and their hyperspatial equivalents. Mathematically it is represented by the sums of the digits of the factors of the multiple tables of eight and nine.

i. shems as time traveling transport ships

The ancient name for neolithic petroglyphic sacred sites was shems. Their builders were associated with the earliest civilizers, called the Nefilim in Sumeria and the Annunaki by Enoch of Ethiopia. The beatific visions of the celestial cosmos of the earliest cultures rival the pictures of the Hubble space telescope today, and their understanding of the intricate workings of temporal relativity often superior even to modern science. Their source of inspiration was only sacred stones, they say.

Sites such as stone henge do lock in a temporal vector as the world turns and as it orbits around the sun. The amount of visitors to such sites, generating greater amounts of karmic energy, might determine how strong the signal is for that particular site at a particular position in its cycle, and, it is speculated, this might also bear some relationship to the position of the other planets as well as the stellar background. Such sacred stone shems often align with celestial events in the heavens at certain times, as if to mark the cycles of this karmic energy flow, and link the lives of their visitors to the stars in the heavens.

While the stonehenges in England and America, Carnac in France, the pyramids in Carral and Merubecka and the Nasca lines in south America are all believed to be prehistoric sites, for which no histories of their creators still exists, the remainder of the world's monumental petroglyphs post date this, and a history of their construction is recorded. While no mention is made of the construction of the great pyramids in the Hebrew bible, the earliest reference made to such monumental petroglyphic construction is to the Tower of Babylon. While the Sumerians themselves record a history of the palatial hanging gardens of Babylon, they do not make reference to a tower, and so the tower is probably a metaphor for the ziggurat style that was adopted for the later pyramids of Egypt. The Sumerian history does record a war in the heavens that is not described in the old testament, though.

The nefilim, who had come through the shems, "descending to earth from above in flying ships," had brought with them and implemented all the elements of early Sumerian civilization, from agriculture and metallurgy to free trade and the law. They produced a ruling lineage that would also establish pharoahnic kingship over unified Egypt and the far distant Indus river valley. However the Hyksos and Nubian Egyptians turned on the Sumerian Nefilim, saying, "let us build a shem of our own, that we may go up and look into the heavens," and with the building of the pyramids broke from Sumerian and Akkadian influence, and this also ended all communication with the east until the time of Alexander the Great. Since this time control of the stele of the shems has been in the possession of the sanhedrin.

c. thelemic esp and tunnel realities

We are perpetually bombarded by subliminal suggestions, and these guide us through the definitions of what choices we have to make. It is thought to work essentially the same way for the universe as a whole, bombarded by gravity, which creates the curvature of spacetime. The vector of entropy can be thought of as a tunnel reality, a wormhole through the multiverse.

Our perception is primarily controlled by one thing: brainwaves. The memory ennegrams we access, the fourth dimensional metaforms that trigger emotional cascades of neurotransmitters and the difference between waking consciousness and unconscious dreaming are all functions of the electromagnetic waveform state of the neuralkinetic energy field native to our brains. This single factor controls how we see the entire world, and it is itself entirely regulated by the thalamus.

The thalamus sits just ahead of the hippocampus, and it is in the hippocampus that the electrical biorhythms that regulate the autonomic and peripheral nervous functions originate. However, the constant projection of energy into the hippocampus, where it is then pulsed, derives from the thalamus, which sits just before it in the interior most portions of the brain. This system is self regulating, such that the biorhythms of the hippocampus control the nerves and the heart, and these send signals through the thalamus which contribute to the projection of energy into the hippocampus. Note that the flow of energy into the thalamus, out of the thalamus into other parts of the brain, and out of the hippocampus are all pulse regulated, however the flow of energy from the thalamus to the hippocampus is a steady steam. It is from this that our sense of time derives, and by light of which in relation to time that we define ourselves.

The will power also derives from the thalamus, as it is the holographic relay depot that mentally projects outside of the body. In this respect it serves the function of the ajna or the third eye of ancient mythology. It is by holographic projection through the thalamus that karma arises in our aura, and it is from this karma that we extrapolate to make choices. Thus, the history of karma comprising our aura can be thought of as such a tunnel reality, or wormhole through the multiverse.

i. memory gaps, missing time and cognitive hypnosis (waking trance)

Two different groups of people describe the same conditions for two separate forms of experience. One is ufo abductees. The other are victims of mind control abuse experiments.

It is widely known that the memory is selective. We pick and choose what we want to remember when we want to remember it, in as much as we are able to. It has also long been known that memory can, like the perceptual senses, fade with age, and one particular disease of the brain that can cause this that has been identified is Alzheimers. Beyond all of these natural conditions or ailments, the memory occurs much like sense perception in the brain, with neuralectric activity triggering neurotransmitter cascades in different areas of the cerebrum, often coordinated through the thalamus. The only physical effect of memory is that neurotransmitters trigger re-uptake inhibitors, opening gates for increased cathexis through axon dendrite gaps which exchange more neuralectric activity, and this causes neurons to grow together into the clusters that comprise the columns or pillars of cells in the cerebrum. The nerves function holographically, actually reproducing a small image from electrical impulses in the many millions of intricate neural connections.

It is thought that cognitive dissonance triggers mental activity, by the juxtaposing of two holographically projected ideas in the mind occurring simultaneously. Thus it is also thought that the sudden resolution of two contradictory views can trigger a euphoric state of transcendence and subsequent loss of focal concentration. If forced this effect can trigger a sudden moment of memory loss. This is the concept behind brainwashing and the practise of memory recovery and erasure.

By maintaining a state of placid, detached, suspension of disbelief, one can enter a tranquilly flowing tunnel reality, and drift away into obscurity and a life of moderation. This concept is based on the brain wave state of day dreaming, or waking trance, usually associated with cognitive dissonance produced by dividing one's attention between the external reality and the internal realm of imagination, dream and memory.

ii. remote viewing, precognition and telekinesis

As I have described, the history of our auras can be thought of as tunnel realities, or wormholes through the multiverse, and our thalami as interactive choice making mental projectors. I have also described how, in cognitive dissonance, internal and external stimuli can be processed by the brain simultaneously.

In remote viewing the mental projective space is attempted to be transported to a target site while the aura and the body remain in a fixed location. This can be simple, short distance esp of correspondences and acausal synchronicities, telepathy upon or between people, or it can be long distance remote viewing of objects, people or places. Just as it is possible for an autistic to consciously do large computations it is possible we all do unconsciously, so some part in esp and remote viewing might only be awareness of recurrent patterns, and telepathy between people only attention to context.

This leads to the field of precognition, or the ability to foresee events before they occur. This often happens to people at random, and in these cases, which can come in the form of a dream, a recurring theme in their lives or personal observations, or even simply a deep sense of foreboding, the preminisced subject matter is usually a disaster. This has also been accomplished intentionally, as the result of guided practise to focus the concentration. This sort of waveform of perception has a strong interactivity with the field of its awareness, and can often produce negative emotions in those who have randomly foreseen disasters. Thus it is thought to be the synchronicity of emotional energies between the receiver and the victims that causes the distortion of time in hyperspace connecting them.

Telekinesis is the control of spacetime matter-energy by the conscious bending or warping of this thalamic mental projective temporal wormhole through the aura on the karma in the multiverse. The thalamic mental projective holographic space is a two way circuit, as information flows in and out through the thalamus. This perceptual space sees in three dimensions and three hundred and sixty degrees in all directions. It has long been associated with the ajna, or the all seeing eye.

f. the messianic breeding program

The concepts of right thought and right action have been held as virtues by Buddhists since the times of the ancient Vedics. As I have described, their goal was to create an entire class of holy and enlightened seers and seekers of truth. In the ancient middle east, the Hebrews believed that a prophet came among each generation to interpret the teachings of the Lord. As time progressed these two beliefs grew closer and closer together, until the time of Christ, when messiahism was at it historical height. There were countless members of the generations claiming to be prophets, visionaries, illuminated seers, cult leaders and messiahs. Few of these people are remembered, and besides the name Jesus, the name Marcion means next to nothing to the common person.

The belief in the messiah is the belief in a chosen one, an anointed representative of the age of man. The goal of producing one has existed among the people since the first bloodlines arose as conquerers. The elect is the king of the masses, and held by them as a king over all the other kings. This practise predominates even though the entity may only exist as an ideal archetype.

The Order of Zion, who had funded the Templar expedition to and archaeological excavations at the ruins of the second temple, may have had some interest in the messianic breeding program of the Hebrews. Much work had been done during the middle ages by Alchemists attempting to recreate the conditions of conception for various different kinds of homunculi, from the metal and the mineral to the vegetable and animal. The order may have adopted a position of sympathy for the Hebrew conception agenda as part of a campaign to preserve Mediaeval alchemical manuscripts. While the Catholic order of zion would eventually become, according to the documents, the priory of zion, a more detached body of the Catholic church, or even a separate sect altogether, they are thought to have remained the secret chiefs of the inner order of the free masons after the first Lodge was formed by the last Templars.

This has allowed the agenda of the messianic breeding program to continue without the Catholic church's direct assent or even, necessarily, awareness. It has been one of the primary occult currents throughout masonry as it has grown internationally, and has influenced many new religions which have arisen since then, including the polygamy of Mormonism and the descriptions of ufo abductees.

The result of this has been the upswing on the asymptote of population growth that has occurred in the past couple centuries. Many people emigrated from old world countries to the new world, redistributing the world's population, and since then many third world nations' populations have been expanding due to lack of contraceptive technologies. Here we see that populations grow where the food is, but follow where the wealth goes, and that a lifestyle of leisure and philosophical repose is considered by the first world bourgeoisie, the class created by masonic jacobins, to be more ideal, and this is in keeping with the ideals described throughout history.

3. future potential extrapolations

Many people in the western world follow the gregorian calendar, which is based on the estimated birth date of Christ as around the year zero, and divided such that the dates before Christ are counted backwards from year zero, and those anno domini after death counted forwards. This calendar, unlike the Buddhist, Hebrew, Muslim, Ethiopian and Mayan calendars marks a specific event in the cosmic cycle of time for which the other calendars give dates of the beginning and intermittent endings. The Buddhist calendar is based on the Hindu conception of the Yugas, or ages, that are the cosmic cycle of the universe's beginning and ending, given by the beginning of Buddhism. The Hebrew calendar is based on the date of the fall of man and beginning of the generations recorded in the Torah. The Muslim calendar's dates revolve around the battle of Massada, recorded as being prophesied at an earlier date as the fall of the rebel angels. These enochian angels actually describe the divisions of the precessional cycle of earth's poles, and this the Ethiopian calendar measures by the twin star system's revolution in Sirius. The Mayan calendar measures the precession of the electromagnetic poles of the sun by calculating concurrently the sidereal rotations of Venus, the thirteen lunar monthly phases of the moon over 384 days, and the eleven year sunspot cycle. The gregorian calendar only calculates figures for the cyclical patterns of orbital alignments that occurred during the variable year zero.

The calendars based on the beginning of man, the beginning of Vedic Buddhism, or a battle in the Jewish revolt, do not make predictions about the future, while those based on astronomical events, such as planetary alignments, sidereal rotations, or electromagnetic cycles of the sun or of the planets, do. The path of man is phi bound, while the path of the heavenly spheres is bound to pi. Phi is tangential, pi is regular. In this way humanity is the measure of the heavens, and the heavens the measure of humanity. Those systems that do account for predictions of humanity's future according to the heavens can only do so as a superimposition of both types of calendrical system.

15,000 years ago mammoths were flash frozen in Siberia while eating tropical vegetation, 12,000 years ago, Egypt was a fertile savanna, 6,000 years ago Sumeria and India were lush gardens, 500 years ago Ethiopia was fed by the flooding of the Aswan lake and Nile river, and it is due to the gradual precession of the electromagnetic and gravitational poles that environmental conditions in these regions have changed. The fossil record indicates that there are no lands on earth that were not settled by homo sapiens shortly after the end of the last ice age, as well as populations of cromagnons in America and Europe, and, even earlier than the last ice age, as much by paleolithic australopithecine and neanderthal hominids. The shems stand testimony to the fact that earlier humanity made measure of the celestial events in the heavens, and the existing calendars of the world remain today produced from many such observations. There is no reason to doubt that such calendars can be used to predict future celestial events in the heavens.

These calendars predict that, as polar precession continues one hemisphere of the earth will gradually be moved more toward the sun at one time of earth's annual orbit, such as at perihelion or aphelion. This is the effect that causes the seasons. As the hemispheres of the earth are moved around the precessing pole as it tilts toward or away from the sun per half earth's orbit, the seasonal conditions of the two polar hemispheres gradually reverse. Thus, when it is summer in one hemisphere it is winter in the other. As the pole gradually precesses around, a different hemisphere will eventually come to be exposed to the sun at the same place in its annual orbit. This process takes about 26,000 years, and is measured by sunrise in the spring equinoctical sign of the ecliptic zodiac. For about 12,500 years or so, summer and winter conditions gradually increase in each of the two polar hemispheres, such that, as the summer of one lasts longer, the winter of the other lasts longer. This is thought to be in accord with the sunspot cycle. For about one third of this time there are ice age conditions in one hemisphere or the other. We are currently in the middle of northern hemisphere summer and the beginning of autumn, while the southern hemisphere is at the end of its winter, entering spring. As spring becomes summer in the southern hemisphere, autumn will become winter in the northern hemisphere.

a. the immanent eschaton

Since the end of the last ice age people have believed that the world will end. Since this means different things to different people, just as we are perpetually passing through predicted dates for the world to end, so is the root emotion of these beliefs the fear that the world is already ending. The ancient hominids practised ritual burial, perhaps with the foresight that their buried remains might one day be discovered and brought back to life. This practise would eventually lead also to ancient mummification, which may have been the discovery of the peoples of the late last ice age, when freezing conditions preserved bodies.

The belief that the world will end dates back to the rig vedas, in which Vishnu is described as controlling the tug of war over a giant serpent, as depicted in the temple of angkor wat, which is aligned with Draco, the constellation that surrounds the north polar star. The Vedic calendar gives the ages of the creator, the maintainer and the destroyer, and, like the later Mayan calendar, records the destruction of the world by the different elements at various times before.

Astrology, or the recording of the alignments of the planets with the twelve signs of the zodiac, began concurrently in ancient Sumeria, and was thoroughly recorded openly by Muslims and in secret by Catholics before being brought to the European nobility during the crusades and subsequently strongly influencing masonic jacobins as a motif for design elements in architecture.

Some apocalypses are actually included in the canonized bible, though they are usually misinterpreted as literal descriptions, rather than as depictions of heavenly cycles or, later, holognomonic archetypes associated with them. The descendants and inheritors of the same people that included these revelations and not others, even more orderly and detailed, have based their apocalyptic dates on various gematriacal interpretations of the text, relative other biblical prophecies, and on the gregorian calendar. Different apocalyptic cults have formed over the intervening years since Christ that have given different dates, all of which have come and gone. Some of these cults have disbanded, but others were willing to give up their lives to their belief. The most recent such cult was the ufo cult Heaven's Gate, who sacrificed themselves to the comet Hale Bop.

It is probably because of the need to preserve law and order that the truth of precession is not known more commonly. It has been thought, based on spartacus alone, since the time of Marx that the masses would rise to revolution if they discovered that religion was only a concealment of a global environmental fact. More than being a revelation, and vehicle for the transmission, of this fact, religion offers one a glimpse at the gateway to escaping precession forever. This, it alleges, only appears physically to be a humble life of moderation and contemplation.

As I have described, churches are centers of karma. They have existed longer than sports arenas, and are considered archetypal rather than stereotypical. The worship in a church differs little from independent worship on one's own, except that it allows for fellowship among the parishioners. In this way it differs from the theatrical experience of the stage or film. All of these things are merely the displacement in mental projective space of superfluous karma, or a cognitive dissonance triggering a mental electrical ennegram, and the free will to invert this into mental clarity.

The shems were the original karmic centers, and they are often positioned near or above underground rivers. At some intermittent point the distances between them were measured as geodesics on the surface of the earth, and these lines of connection called ley lines. The alignment of later abbeys, churches and cathedrals in Europe all describe exact geodesics relative to one another also. Some of these are even based on latitudes and longitudes that existed during previous positions of the poles.

i. the rapture

Since 1,000 ad gregorian people have believed in the immanent return of Christ as foretold in the apocalyptic scripture. At this time, the devout followers believe that they will vanish from the surface of the earth, and be taken directly up to heaven. This belief, called the "rapture," was started by John Smith, and is a central precept of his religion, Mormonism. After his predicted date(s) for the rapture had elapsed, some members of his cult still remained to establish the Church of Latter Day Saints of today.

This corresponds to the belief in wormwood, described by the prophet John of Patmos as a portion of the heavens falling to earth. This is simply good apocalyptic literature, since it can be interpreted in a million different ways. For the dinosaurs it had a very literal meaning, since we think it to have been an asteroid that resulted in their extinction; for the later cavemen and early Sumerians this may have meant the correspondence between the representational zodiac and the constellations in the ecliptic; for Enoch this meant the fall of the rebel angels; for Solomon the summoning of the goetia to raise the temple; for Buddha, Christ and Mohammed their teachings; for all of these have followed in the messianic current of coming to stand for or represent the meaning of their time. The return of Christ is based on the zodiacal predictions for the astrological events that occurred during Jesus's life, from around zero to 33 ad. These are simply part of a recurring cycle that marks the spring equinox's entry into a new constellation in the ecliptic zodiac. At the time of Christ the age of Aries was just ending and the age of Taurus just beginning.

Insofar as shems are karmic centers, they can serve as calibrating vectors for timespace teleportation, such as can be accomplished by expanding open a wormhole. Doing so accesses the multiverse, the n dimensional gravitational warping of geometry comprising hyperspace, or the sum over histories of all possible dimensional extrapolations upon our universe that comprises the exterior surface of timespace, the interior of which is the surface of spacetime comprised by the speed of photons, that occurs as the infinite potential for random quantum fluctuation producing probability wells that is the essence of the universal continuum. In hyperspace the laws of physics are distorted such that mental projection, such as occurs for the thalamus, instantaneously manifests. There is no other force to induce cause and effect, and therefore no such thing as real karma, since the manifestations are only holognomonic reflections of the light of the self, extensions of the free will, and therefore none of these projected reflections continue existing beyond their sight, and cannot therefore correspond with one another to form a webwork of synchronous acausal consequence. This experience is equivalent to the dream reality one would experience inside the event horizon of a black hole as one asymptotically approached the singularity.

ii. the multiverse

The stele of the shem have come to be seen as keys, unlocking the gateway of the knowledge of precession. What lies beyond is the multiverse. As I have described, the only difference between perception of the universe and perception of the multiverse is the perception of time. In the universe, entropy measures time linearly from the past through the present to the future. In the multiverse, time is relative to measurements of space, and insofar as the one can be distorted, so may the other. This means that zero, one and infinity are all simultaneously measures of the sum over histories, and that space, in substance and essence, is only one polarized concept about which we do not need to revolve our entire lives. Here we find QBLH extending upwards as the spiraling histories of gravitational singularities that unwind wormholes between one another comprising the dimensions of the multiverse contained within the event horizon of black holes. Here we see that the wormholes open perpendicularly to one another, and that therefore the only way to get through the singularity outside of the universe is to go perpendicularly through them. This produces the phi/pi spiral helix lattice of QBLH.

Here, the projection of what is called the mind, the soul or the spirit only exists in the form of a baby universe, the body of which is wormholes. This is equivalent to evolution of life on earth, since when the photons now reaching us left the most ancient galaxies our solar system was only a nebulous gas cloud. As these galaxies were consumed by black holes a supernova fused the nebulae and formed the sun, the planets and the asteroids. As the wave of the informational ending of photons by super massive black holes approached us, life formed on earth, and precessed over the ages, the sun falling into tempo with the rhythms and patterns of the surrounding spiral galaxy, the Milky Way, as it slowly approaches the Andromeda galaxy.

The sort of entity such life describes might best be termed a reincarnating holognomonic archetype.

b. reincarnating holognomonic archetypes

When describing various types of more advanced, evolved or enlightened entities there have been widely varying experiences recorded throughout history. The first of these to be recorded were those of Enoch, describing the wars in heaven and the fall of the rebel angels. After this are the Sumerian accounts of the history of humanity in relation to the djinn, or the discorporeal form of these entities, which they thought remained bound to earth, and associated with the shem. The Egyptians structured hierarchically various levels of Gods, demigods, natural forces and calendrical influences as associated with the spirit realm that the disincarnate soul passes through in the underworld of the dead. Shintoists of early Asia and modern Japan associate the djinn with nature spirits. Siddhartha associated the karma in his aura with negative attributes of the self, and slew these in his mind. Hebrew gnostics described the soul in kabbalistic terms, and the times as apocalyptic. The Catholic church structured various orders of angels and seraphim, as well as of demons and the damned, all of which ruled the realms of the afterlife that a soul could enter at death. Muslims reiterate the descriptions of the wars in heaven and the fall of the rebel angels. Masons reiterate the djinn in their elemental and calendrical form.

As I have described, the baby universe sum over histories of gravitational singularities in super massive black holes beyond the projection of photons at the speed of light unwind in a phi/pi spiral through the wormholes connecting the black holes in the multiverse, and this connection underlies the quantum information exchange between photons within the universe, such that the same information projected by the photons has been consumed by the black holes and is contained in the wormholes expressed as the history of the gravitational singularity. This is the same pattern of life described by the double helix of DNA.

i. the freeing of the spirit from the karma of the aura

The knowing of the higher self as above the influence of the distracting karma that accumulates in one's aura has long been considered a more ideal state of achievement associated with enlightenment. Whether one knows the higher self as God, a higher power, their own spirit or guardian angel, as their soul or astral body, or even only the temple of the physique, these have all been encouraged by Vedic and gnostic doctrines. One technique for this is the inversion of distraction into inspiration, such that one is drawing superfluous energy off of their surrounding environment. Insofar as this energy is more often redirected and channeled into creative expression, it creates an alternating entropic circuit generating regular brainwaves. This does not cause the negative consequential effects of distraction that perpetuate both internal suffering of the individual and external suffering of the environment. Thus, the entropic circuit is considered more stable than the simple out flux of entropic waves into the environment. It is associated with time, and considered an anchor in a trance or meditative environment.

As I have said, this manifests as a stable energy flow between the thalamus and the hippocampus and functions as the internal perception of time. As one gradually releases the clinging to this as the center of the self, one will become more and more free of the karma in their auras. If one expands upon this, as the teachings encourage earlier aspirants, one will cleanse the aura of its karma, the free will permeating all depths of mental projective space, clear light projected out all around. However for later aspirants it is necessary to remember that the thalamus, the root of the spiritual sense of free will in the biological organism, is not the only center of free will in the universe, and that one can, by extending their free will, travel in mental projective space through all the dimensions of the multiverse at all the centers of the universe.

The path of the mendicant is to seek enlightenment. The path of the enlightened seer is to watch over the sacred. For the monk this first condition may mean tithing alms, while for the master the latter condition often involves the stele of the shem or the djinn of precession. As I have described, the soul and the spirit have been described in various ways, but it is thought that the mind of the mendicant monk is more like the soul, and that of the enlightened master more like the spirit.

It is also thought that the soul is more like the body, while the spirit is more like the mind. Thus, the freeing of the aethyreal spirit from the karma of the soul is like the freeing of the mind from the body, and is a matter of the relativity of time as much as perception makes use of the external senses, dependent on the vessel's death only insofar as the circuit of entropy is centralized and associated with the ego. As the soul is thought to inhabit the body, so the aura tends to create a tunnel reality of socialized routine. This is for the soul what the steady stream of brain waves in the thalamus is to the brain. Insofar as this circuit is associated with ego, the measure of its activity in the brain is considered the biological organism's lifeline or history. So, similarly is the tunnel reality of the aura the lifeline of the soul. The soul would see its tunnel reality — as the seat of the spark of the free will or the spirit, as the body would see the thalamus — as the seat of the spark of life or the soul. Therefore, just as in order for the mind to be free of the body it must unseat itself from the association of the ego with the neuralectric activity in the thalamus, so for the will of the spirit to be free of karma it must unseat itself from the association of the soul with the routine tunnel reality of mundane existence.

ii. ramifications of uncentered ontology

The primary entropic circuit of the aura functions on binary inspiration / distraction, where distraction creates cognitive dissonance, burning off external entropy, and where inspiration creates a solution in a closed inner circuit. Notice that this occurs in a phi/pi spiral, where phi is the trajectory of interiorization and exteriorization of karma through mental projection, and pi is the circuit of the stable state exterior referentials and the spherical probability well of active consciousness. Leaving behind the interactive karma of the external environment of the body and the mental projective tunnel reality aura of the soul means identifying the self with this geometry, for this is the form of the free spirit. At this point in meditative trance one has awareness of all other beings and objective things as gnomonic reflections of this.

The highest gnosis, then, can be applied to the most fundamental natural gnomon. The epoché of "I am not that which I am," can be interpolated upon the underlying fundamental geometric pattern of the structure of our DNA and the filaments, walls and voids. There is no self. There is no not. There is no will but that which we project. Phi/pi can stand for all of these things and still more. To say that it is these and yet is not anything at all in substance or in essence is to say that there is no phi/pi, that none of the ratios in nature are exactly or perfectly precise, and that it would only be by such an exact measure that we could think the universe capable of conscious communication within itself, such as between or among people.

This is what it means to say there is no center or no self. Many researchers of consciousness have associated trance states with ego death or with dissociation. This is not a necessary requisite any more than any of the ways of inducing trance consciousness exclusively cause the nature of personal transcendental experiences, which can also be seen to occur at random. Dissociation ultimately requires the incapacitation of the physique. The more one enters another realm of thought than their perceptions are accustomed to, they often must readjust themselves in relative proportion to it, in the same way the living distinguish between themselves and the new. This is in no way a condition or necessary symptom of trance states of consciousness. Neither is delusion to the sense perceptions, for such karma is not relevant to the trance consciousness, which is the entropic impossible loop that is and is not at the center of ego or self, a gnomonic singularity that is the seat of the soul and the root of the free will. As I have said, ego death and dissociation are not conditional components of trance consciousness any more than the auric soul or the most elegant geometry necessary components of the free will.

There are a great deal of zen koans written about ylem, chi, the tao, and the binary yin and yang. All of these agree, ultimately, with nihilism. That is because the primary clear light, the energy forces of the universe, the similarities between manifestations and the duality of matter-energy are all invisible characteristics of nature, thus many common people simply choose not to believe in them.

c. actual timespace machines for physical transport

From the 1930s until the 1950s, the U.S. military experimented with scalar wave technology weather balloons. These balloons, elliptical cylinders with technological equipment attached to the undercarriages, were supposedly used only to test atmospheric conditions, however could also allegedly control the weather by orientation of orgone energy to the ionosphere. They are thought to have been equipt with technology similar to that developed by Nikola Tesla involving regular interference patterns of pulsed field currents. These balloons match the description of late nineteenth century lighter than air ships that were the first reported ufos.

It is possible that the zero point energy null field created by the pulsed scalar waves inside the devices, which can also be thought of as the first, inner atmospherical satellites, caused the zero time orientation field they generated around the machines to become de-resonated from the earth's 40 mega hertz electromagnetic field, and thus for the balloons to slip out of phase with the fourth dimensional entropic flow of time, yet remain in the earth's electromagnetic gravitational well.

Nikola Tesla, who also invented broadcast radio, claimed to have made contact with off world sources. Much of later stealth technology is suspected to have been reverse engineered from flying saucers crashed by using the early weather balloon technology and later satellite and radio technology as a net disruption effect to the electromagnetic field between the ionosphere and the van allen belts and plasma sheet.

This is the age of space travel in solid fuel rocket propulsion technology such as the space shuttle and time travel in the form of faster than sound flight at varying atmospheric altitudes that would actually measure different times on atomic clocks from those that remained below on the ground. It is possible that outside of gravity wells such as that of the earth, the dimension of time is asymptotically distorted toward a null field.

i. places to go (universal tourism)

The church has long legislated the roles of the heavenly bodies in the daily lives of the people in opposition to those astrologers and gypsies that have claimed the movements of the celestial cycles govern the daily behaviors of the mind. It is evident that such processes as the precession of the poles effect such cycles as the seasons, however beyond this it is unlikely that the positions in the planets relative to the zodiac bear any ultimate relevance to the meaning of a person's life.

It is possible to warp spacetime into hyperspace through wormholes and travel freely through the multiverse using null field scalar wave zero point energy, and this kind of technology can be contained in even hand held sized devices, such as the orgone gun of Wilhelm Reich, similar in appearance to the implements held by stone carved Olmec and Mayan figures in south america, or which may have been Brown's gas welding torches capable of producing the over 6,000 degrees celsius heat needed to fuse the metal alloys of the "I" shaped clamps that held their massive stone masonry in place. It is thought to be this type of null-time technology possessed by ufos that allows them to bend the fabric of the continuum and travel relatively instantaneously across vast distances of space.

ii. people to see (entropy inverts work and vacation)

When you are working, you can only think back or ahead to being on vacation. When you are on vacation, you can only think back or ahead to being at work. The natural process of entropy inverts binary, opposite states such as these in its regular functioning. When you are at work, you know you will eventually be on vacation.

When you are on vacation you know that you will eventually be back at work. The longer these conditions continue to invert, the more faded away their divisions become, and they become blended together. This is only the tunnel reality. In order to find your way from here you must go perpendicularly.

B. the most holy all of the most high

There are not two selves. The ghost of the second self is Choronzon, an ancient name for consciousness. All the while the body survives this is the self. In sleep and in some meditative trance states this inverts and becomes the dreams of the subconscious while the consciousness is unconscious. When the self is not the consciousness of the body, it is believed to be consciousness still. This consciousness, it is thought, either finds refuge in an incarnation, or enters the heavens of the multiverse. Just as when it is cloaked in the flesh it knows itself by the flesh, so when it is brought before the lord God must it shed its highest definition of self to be one with the creator and that is only the consciousness. There are not two selves, because you do not stand beside your own consciousness. There is not one self, because there is you besides consciousness. There is no self because there is no consciousness.

Choronzon was first, in the age of the ancient Egyptians, associated with the djinn of the heavens and the dekans of the zodiac. He came to represent the disincarnate, discorporeal underworld summoned by magicians, and the concept he represented came to be suppressed along with the other arcane arts of metaphysics by the subsequent solar central masculine cults that have become modern religions. The deified ideal that replaced him in mainstream terminology was the Holy Ghost. When people lift their spirits up with the Holy Ghost they are expanding their consciousnesses. Just as the Holy Ghost is only part of the deified trinity of God the father, Jesus the sun, and the maternal Holy Ghost, so was Choronzon called the disincarnate spirit of the original head of the Egyptian occult, who first held the title YHVH. It is thought to have been Choronzon that Siddhartha conquered in the form of his inner demons, clearing his consciousness to teach the way to the light. The Holy Ghost represents the idea of the conscious universe, derived from the Hindu belief that we are all the dream in the mind of God. In this capacity then the angel scroll of Yeshuah Ben Padiah may be thought of as a list of mentations in the mind of God. This has been part of the cyclical process of precession, all of mankind's contribution, as much as the ritual of sacrifice that makes a martyr out of the scape goat. The occult orders have always spoken for the precessional cycle. It has only been since the time of Christ that the prodigal sun cults have had younger and more public representatives, thus ending the age of the father and initiating the age of the crowned and conquering child. This age is marked by the fool, or jester, archetype of Choronzon, representing Orion in the tarot deck, because it has been marked by rapid cultural change and technological progress.

It has long been theorized that to understand the Most High you must think of the inversion of the most low, and as I have described, in QBLH, the former greatest common factor inverts to become the new least common denominator. Thus whatever God was or has been, Satan becomes. This works similarly to the river of Thales representing the flow of consciousness or of time. That which God was,

Jon Gee

Satan is. God is within the moment, and what is past for Him is Satan. God, also, is seen as the Light of humanity's future. Thus, what is the future for man is the present of God and the past for Satan. In this way, the lowest is that which is shed by the Most High that is its sense of self.

Bibliography

The world of the written word is rather like the ocean of politics described in ancient apocalypses, or the modern concept of the seas of finance, in its mercurial viscosity, turpid temporalism and Buddhist impermanence. Few works survive their author's lifetime to remain known to the gross masses as classic. Most works go unrecognized even during the author's life. There are those books which serve the purpose only of fueling the popular flow of the mainstream of cultural idiom, and it is, presently, these works that are, by majority, lauded as soon as they see print, and, for better or worse, forgotten about just as fickly. It is usually those books that are on the fringes of this movement — in the etty currents, as it were, of the mainstream — that receive longer lasting recognition for their stylistic contributions; however no matter how unique they are, they generally manage to spawn imitation as if by dint of mechanical replication, and create subcultural or countercultural movements of their own, which are eventually, that is — that the walls of society not erode — as rapidly as possible, reintegrated into the great gorge of popular opinion just enough so as to alter the direction of the current as little as possible. Of those works that survive their author's lives, some do so only for a short while before being recalled from booksellers' shelves and warehoused, awaiting recycling and gathering dust, already beginning to become completely forgotten. Some of these remain afloat in the market, circulating the rounds of used bookstores and private collections, where they do not stand out, but where they might be cherished close to the hearts of their prospective owners, some of the few people remaining alive on earth to have read them. Some books go out of print only to resurface again years later in recomissioned editions, shedding new light on their forgotten postulates. The life of a book is truly a harrowing and beleaguered plight, considering that the vast majority of people's writings do not even see print, and those few of these such that are copyrighted remain as unbound, loose-leaf manifestoes in federal libraries, to survive the centuries, barring trial by fire, and perhaps be discovered hundreds of years later by an arduous and ambitious chronicler or translator of the subject and, if they be understood by this true magician, to finally come into circulation at a much later time. This so rarely happens that it should not even be hoped for as an exception to the popular rule.

So what quality is it about a work that makes it a classic? There are hundreds of theories about this in the realm of literary criticism; there are almost as many theories as there are classics. Some works, considered essential by some but which go wholly unknown to others, are considered classics of their field. Indeed, for each field or genre of writing there are a select grouping of works considered requisite for any serious consideration of the subject. This is not the matter I intend to address under the heading of the classics, though I will return to this topic of specialized manuals shortly, since most of the works I will cite for this present book fall into this category.

My theory about what a book a classic comes from the intersection of two things: 1) rhymed ideas and 2) new ideas or expressions. The first concept goes to the construction of the work. It has to follow the one idea from the last in such a way as to appear cohesive to the reader, that the reader not get lost along the way for wont of sufficient transitional material. The antithesis has to correspond to the thesis, and the synthesis has to be able to partner with either equitably. As to the second case it is self explanatory, however of such rare occasion when it actually occurs the work is automatically destined for reconsideration throughout history. This pertains to style as well as to theory, and there are certain schools that, in my opinion quite rightly, argue that these two are inseparable.

Now, as I have stated, and it bears reiterating, very few books are considered classics. Most books lack technical eloquence and originality, and are doomed to the dustbin of mediocrity. To this extent there has arisen an institution to teach people how to be prepared for this course of events. Scholasticism, like an assembly line in a great factory, produces works of referentiality rather than originality, of formula rather than formulae. This process has only just begun to be recognized since the birth of the industrial era, when nontechnical colleges came under the auspices of state funding. However it is not the, admittedly, prison-like surroundings of the average college campus that has alerted a rogue few scholars within the hierarchy to the dehumanization of the Prussian educational system. It is the representationalist curriculum at which these teachers raise their eyebrows.

At first it was merely a political issue. History books weren't including certain minority groups to the full extent of their actual contributions. This issue took precious time and deliberation to amend. At first it was thought that setting aside a single month per year to teach more about the minority group in question would suffice to quench their vocalization. But then the question came from the leftists who had long associated themselves with the minority groups: what about the other ethnicities? What, for that matter, about the entire southern and eastern hemispheres? This was turning into a very vexing problem for the staunchly conservative deans and faculties. Afterall, what about all the peoples whose histories ended with their being conquered, or began with their being enslaved? And then again, what about all those strange little social variables who had chosen an alternative lifestyle? Would not they want equal representation next?

In about the mid sixties scholasticism began to break down. Rather than try to amend itself to please everybody, it simply declared bankruptcy, like any other morally bereft corporation, and sold itself off piecemeal to its detractors, who proved entirely unprepared, as the conservatives had expected, to inherit the lie factory that had been created to suppress their past and mold their present for the purpose of controlling the future of society. The liberals that flooded colleges thought that they could simply make rapid changes to the system that been fixed in its ways for over a hundred years. Within twenty years they had realized how difficult it would be to go against the grain, and after forty years most would come to wonder if their very presence as harbingers hadn't been intentionally manipulated to evoke a stronger push towards the sciences by the conspiratorial and secretive boards of trustees.

As regards writing there are two rules scholasticism cherishes above all. 1) quote. The essential idea behind this is that there is nothing new under the sun. If you think you have a new idea, you simply haven't done enough research, because if you had you would be able to find at least one other person, preferably with some scholastic name recognition, who had the same or a similar idea sometime ago in the past. As long as your thesis statement is a quote, your work is merely revision, and this is non-threatening to an establishment that is firmly rooted in tradition. The quote one must revolve all other instances of quoting around is that which says that all great men throughout history have themselves been "standing on the shoulders of giants" (meaning other great men). The end result of this diagram, if one actually dwells upon it long enough to follow it through to its conclusion, gives one the origin of that often overheard whisper among the student dorms on the campuses of scholastic institutions, "brown-nosing,"

The second requisite for all writers who expect to be taken seriously by their peers in the self-respecting scholastic ant farm is: 2) cite. Whenever you quote someone, no matter how egregiously out of context you are taking their words, make sure you leave a footnote in the margin directing the inquisitive subsequent researcher attempting to follow your paper trail to the actual source of your reference. The reason for this is, allegedly, to discourage people from taking quotes out of context and presenting them in an unfavorable allegory — the idea being that no one but a professional clown would make a joke that could only be gotten in on by referential scholastic research, and there do indeed exist laws protecting against this sort of event, so serious is it considered a stylistic breech of the social contract of good taste. Nevertheless few students of Democracy know that the words carved in the Jefferson memorial regarding tyranny are actually referring to multinational banking.

Now metaphysically these two guidelines create a very specific, though subtle state of mind. As long as one is eminently quotable, one can, in a rather Kafkaesque sense, go on "living" indefinitely. Also, as long as one cites thoroughly, than every work, no matter how utterly pointless and socially unredeeming of the time wasted on its research, emulates a machine, where all the circuits are autocorrelated, and all the loose ends are tied up into references to other, identically self-referential, mechanical works. Now it doesn't matter what the thesis says the purpose of these term paper machines is supposed to be, the only real social function they serve is to express what amounts to no more than opinion, in a preset scholarly style, and insofar as they rigorously agree with the terms of that style, to be leant an air of credibility which is, and this is essential to understand, entirely artificial — entirely false. All papers written according to any existing style only further serve to underline the imposed reality of that style, and to add to its authority for future reference. This process is what Chomsky understands as "propaganda" and what Baudrillard understands as a "simulacrum."

Just as the ocean of politics is the nemesis in ancient apocalypses, and the sea of finance the source of all strife to modern conspiracy theorists, so is the liquidity of the written word a shadow land, from which grotesque monstrosities emerge. They emerge in the form of clichés, aphorisms, euphemism and bureaucratic neologisms. The rules of style allow for all of these to crop up; they, in fact, encourage them instead of new and uniquely original ideas, as long as they obey the rules of style.

The two primary demons of scholastic style are sic and ibid. Sic rears its ugly head in quotations, as a sort of tongue in cheek apology for replication with error. Because of the scholastic rule of quoting, one is forced to make reference due to its proximity to tried and true factual material to certain other material which has, since the time of the quoted writing, fallen out of favor with the dominant belief system by being disproven using scientific method. Sic is usually used to denote grammatical errors, such as occur between eras in one language (for example, started written in old English would be ftarted), or in translation from one language to another (for example Satan translates as "adversary," however it is most recognizable when misused as a proper noun), but it can just as easily be applied as editorialization to highlight anything the quoting author disagrees with about the quote (for example, "God is Dead" *sic.* — Nietzsche). The denotation of sic is that, "this is not the fault of the editor. Ibid is the counterpart of sic in the other rule of serious scholastic writing, citation. Ibid is used only when multiple citations follow sequentially from a single source, and it appears in the foot notes or end notes. The connotation of ibid is that large quantities of data are being reproduced from one work to another, with or without the original author's consent, and in lieu of an original idea.

What sic and ibid both reveal about scholasticism is its thin veil of armor to protect its adherents against prosecution for the crime it then turns them around and sends them out craven to commit: that of plagiarism. While quoting is not considered plagiarism, it meets the exact definition thereof. It is the replication of the work of another under the auspices that making reference to it constitutes work accomplished by the author making the reference. Quoting is a short cut to thinking. Why come up with the words to prove your own point when you can overawe the belittled audience by simply saying someone else whose name they should know, if they are part of the educated elite, already said it for you. Equally citation is propped up as the antithesis of plagiarism, because if you can say where the idea came from, then you cannot possibly be recriminated for having stolen it. Nevertheless, that is exactly what you have done, and citing the source of your stolen information is no better than thumbing your nose about that fact. The unspoken consequence of citation being considered necessary to protect against plagiarism is that it isn't safe to have any new ideas that are unique and original to yourself; it is only safe to restate and revise other people's ideas. This is rigorously institutionalized blasphemy against the eternal light of the individual intellect.

Anybody who subscribes religiously to scholasticism will eventually be hoodwinked into thinking that their efficacy in pompous preset style transubstantiates their opinions into facts. I am not going to say whether this actually occurs or not, because in my opinion reality is what you make of it. The flow of the cultural mainstream is painstakingly controlled by a carefully constructed canal system of intellectually elitist institutionalized scholasticism. Therefore, the majority of people believe that if you prove using scientific method that the moon is made out of swiss cheese then the moon, by God, must indeed be made out of swiss cheese. Most people believe that, if something is accepted by scholars, then it is their responsibility to accept it as well, right or wrong. What most people forget is that scholars are just like them; they're just ordinary people, conditioned to believe in the final accuracy of the scientific method as the formalization of their empirical senses. But, just as people can make mistakes, neither is even the sacrosanct scientific method ultimately infallible. It can only make predictions of the known and measurable universe based on generalizations of past observed laws. There is another world beyond what is known by measurements and observations. And that is the domain of metaphysics.

After having said all this, I would still like to be kind to my readers by providing for them a list of other books that have acted as sources or inspiration for this one. The best way of doing this that I can think of is by listing all the books in my library by category, some with brief descriptions of their content, so that the curious reader can pursue some titles that have interested me. I have neither the patience nor the desire to list all the publishing houses and dates of publication, as I have not made reference to any of their content except by paraphrasing nor quoted anything exactly that should be attributed to a certain page number of a certain volume or edition. Many of these books are considered classics in their fields, which simply means that they have met one or both of the requirements of a classic and can be categorized by genre other than fiction.

Politics
This is the section with the greatest number of books I haven't read. All the books about communism, I mean.

Class. Paul Fussell.
About the class structure in America at the time of its writing.

The Prince. Machiavelli.
The rules of ettiquette for acquiring and maintaining power in hierarchies.

Levithan. Thomas Hobbes.
Proposes the idea that the ruler is a sort of demiurge instinct within the multi-headed hydra of the state.

Steal This Book. Abbe Hoffman.
But the book told me to steal it. More timely in the sixties, with references to radical groups and happenings long outdated. A worn out version of the anarchist's cookbook.

Capitalism: the Unknown Ideal. Ayn Rand.
Capitalism supports independence by competition. I never understood this.

the Federalist Papers. Alexander Hamilton, James Madison and John Jay.
The first yellow journalism in American history. Hamilton and Madison go back and forth with Jay as the voice of moderation, discussing the concept of a national bank. The fact that the rich controlled the media even at this early time proves what DeTocqueville forewarned.

the World in the Twentieth Century. Louis L. Snyder.
Written in the sixties.

Report of the Warren Comission on the Assassination of President John F. Kennedy.
According to Jimmy Carter this is not a fabrication.

the Kennedy Government. Stan Opatowsky.
A list of heads of state under Kennedy before the CIA firings.

the Words of Martin Luther King Jr.
Quotes from the king. This book can only be found at the memorial museum in Atlanta, Georgia.

the World as I see it. Albert Einstein.
Thoughts ranging from atomic potential to antisemitism.

the Abolition of Work and other essays. Bob Black.
An essential for all generation x-ers and echo-boomers.

Let's Abolish War. Tom Hudgens.
A well intentioned but not well reasoned treatise on one of the most important issues facing the united planet.

Civil Disobedience and other essays. Henry David Thoreau.
More about retreatism and escapism, boycotting and banning than Ghandi's applications of the idea in mass protests.

Finite and Infinite Games. James P. Carse.
A finite game has a goal and a winner. An infinite game has no winner, though it may have many goals.

Today's Isms. Ebbenstein.
Fascism, Socialism, Communism, Liberalism. Written in the fifties.

The Imperial Animal. Lionel Tiger and Robin Fox.
Human biology and behavior analysed in the realm of politics.

Social and Political Philosophy. John Somerville and Robert Santoni.
A book of excerpts of the writing of famous political philosophers, from Locke and Hobbes to Marx and Hitler. An invaluable resource to anyone studying the original expressions of the ideas that actually shaped the political battlefield for the past four hundred years.

International Security Systems. Gray.

NATO, the IMF, OPEC, the UN and security council. Predated NAFTA and the EU.

The Civil War in the United States.
and
The Communist Manifesto. Karl Marx and Frederik Engels.
For the Marxist view on how banks orchestrated the civil war, and what the unions can do to stop them.

Das Kapital. Karl Marx.
The dialectical view of history reveals the real motivating factor behind all politics: money.

Essential Works of Lenin.
and
Imperialism. Vladimir Illich Lenin.
Lenin's nationalised communist apology. Hard edged.

The Communist Party of the Soviet Union. Leonard Schapro.
About the structure of the duma and the history of the Bolshevik and White Russian parties.

History of Russia. Walter Kirchner.
From tsarism to Stalinism by way of revolution and castrated Trotskyism.

Colonial New England: the Cashless Society. Landmark Series Publishing.
How the colonies of New England survived exclusively on funding for the governors by the Queen's royal account. A working welfare state and cashless utopia.

the Wretched of the Earth. Franz Fanon.
A scathing description of living conditions on par with Studs Terkel.

Community/Peasant Society. Redfield.
How the other half really live. Only partially propagandistic.

German Social Democracy. Bertrand Russell.
The Bertrand Russell, political philosopher, writing at the time immediately prior to the rise of what would become the Nazi party.

Varieties of Fascism. Weber.
Totalitarian dictatorship to Kafkaesque bureaucracy.

The Global Public Management Revolution. Kettl.

Modern. Based on changes in information processing dynamics and the quickening pace of the rate of business exchange.

Propaganda and the Public Mind.
and
Language and Mind.
and
On Power and Ideology.
and
Language and Responsibility.
and
Secrets, Lies and Democracy.
and
What Uncle Sam Really Wants.
and
The Prosperous Few and the Restless Many.
and
the Umbrella of U.S. Power.
and
Media Control.
and
Acts of Aggression.
and
9-11. Noam Chomsky.
One of the best modern writers on propaganda and the media. A professor of language, his writing is often befuddled and at other times crisp and witty, but always scathing and accurate. He is prone to quoting without citing from his encyclopedic repetoire of research.

The Spirit of Terrorism. Jean Baudrillard.
How terrorism in effect cuts both ways by forcing the terrorised regime to raise the stakes of its security and in essence terrorize its own citizens by enforcing patriotism.

You Are Being Lied To.
and
Everything You Know Is Wrong. Russ Kick. editor.
Collections of essays by modern thinkers about pop-culture propaganda and American social ills.

Psychology/Philosophy
Most of these titles I read in high school, when I decided I wanted to be a philosopher. Metaphysics is just a specialisation of that career, which, if one thinks about it, is not only the only one idealized by the greatest thinkers from the Classical world through to modern times, but also the only one that goes without payment or recompense of any kind.

Civilization and it's Discontents.

and

Psychoanalytic Psychology. Sigmund Freud.

In this work, which I think is a posthumously re-edited compilation of several disperate essays, Freud introduces his theory of the nervous system. It is from this I have borrowed the term "cathexis," meaning the amount of residual electrochemical substance that passes through a nerve which remains contained therein, and "hypercathexis," which means the sudden release of this surplus electrochemical substance. Freud thought that cathexis was equivalent physiologically to the psyche, and that the ego, or choice making faculty, only came into play when hypercathexis occured. Interestingly, Freud called the electrochemical substance in his model of the nervous system "phi."

The Silva Mind Control Method. José Silva.

In this book, Silva, supposed by some conspiracy theory authors to have been connected to the CIA, proposes that the best angle for rolling back the eyes while meditating is about fourteen degrees. He describes different optical effects which occur. Experimentation with this method will produce comparable results at varying angles.

Psychoanalysis and Religion. Erich Fromm.

the Edinburgh and Doré Lectures on Mental Science. Thomas Troward.

Interesting lectures from two scholars, the first from the early twentieth century, the second from the mid to late, on the nature of mind as it relates to creation and to spiritual matters.

About Behaviorism. Burrhus Frederic Skinner.

"A scientific analysis of behavior must, I believe, assume that a person's behavior is controlled by genetic and environmental histories rather than by the person himself as an initiating, creative agent; but no part of the behavioristic position has raised more objections. We cannot prove, of course, that human behavior as a whole is fully determined, but the proposition becomes more plausible as facts accumulate, and I believe that a point is reached at which its implications must be seriously considered."

— B. F. Skinner, chapter 12 "the question of control"

Man and His Symbols.

and

Synchronicity. Carl Gustave Jung.

Jung proposes an acausal connecting principle that underlies otherwise seemingly unrelated events. His definitions leave something to be desired, as his concept of a synchronicity is broad enough as to include events that occur at disperate points in time, and are only familiar to one another in trivial ways. For the purpose of this text I have endeavored to make a synchronicity a more direct link between two simultaneous, or nearly simultaneous, events, rather than allowing, as does Jung, the definition to border imaginatively on the, in my opinion unrelated, phenomenon known as dejá vu.

The Hero Within.
and
Awakenning the Heroes Within. Pearson.

Essentially just personality tests that use mythological archetypes instead of psychological disorders. I was a "wanderer" when I took the test many years ago. Not alot of method to it, but very romantic, and definately uplifting and inspirational.

The Hero With a Thousand Faces.
and
Myths to Live By. Joseph Campbell.

Looking at Philosophy. Palmer.

A cartoon book from which I learned the history of philosophy. Particulalrly good for studying the pre-Socratics, who have no writings of their own, that I have ever found, to study.

Dialogues of Plato
and
Timaeus and Critias. attributed to Socrates.

I read the section on Atlantis. It was very short. I recommend it for anybody studying Atlantis, since there never was such a thing by that name before Plato called the global coastal civilizations wiped away at the end of the last ice age that.

Selections. Aristotle.

The father of Metaphysics. For someone who wrote when writing was young, Aristotle's style certainly is voluptuous. He knows how to draw the reader in, and how to change the subject just before things get truthfully enlightenning.

On the Nature of the Universe. Lucretius.

It astounds me how much classical philosophers knew about quantum mechanics without having any sort of aparati by which to make measurements thereof. Truly an inspiring book.

Discourse on the Method and the Meditations. René Descartes.

The discourse on the method was the inspiration for my rant in the introduction to the style section about the methodology of philosophy. Descartes essentially sets out to question reason by denying everything that appears empirically true. All he is left with is an abyssmal void of his interior self and the still small shining spark of his ego, or soul, and even this, he argues, might be only the dream in the mind of a sleeping monster.

Philosophy in the Age of the Greeks.
and
The Birth of Tragedy and the Case of Wagner.
and
Genealogy of Morals and Ecce Homo.
and
Beyond Good and Evil.
and
Thus Spoke Zarathustra.
and
Twilight of the Idols and the Antichrist. Friedrich Wilhelm Nietzsche.

Philosophy in the age of the Greeks deals with lots of philsophers I've never heard of in any other book. Nietzsche was an epistemologist who finally concluded that there is "no original text." For a while there was Gilgamesh, but that was after his time. In Philosophy in the Age of the Greeks, he searches for truth by dissecting the minutia of classical reasoning. It is diffucult to say if he was really crazier at the beginning of his career or the end. Nietzsche is one of those philosophers who is eminantly quotable. My personal favorite is: "the same effects in man and woman do not cease to differ in tempo."

Existentialism from Dostoevsky to Sartre. Walter Kauffman.

Walter Kauffman was the translator of most of my editions of Nietzsche, so when I saw this book I immediately picked it up. It is excerpts from the different writers of the late nineteenth to middle twentieth centuries, including Dostoevsky's "the Grand Inquisitor."

Irrational Man. William Barrett.

Barrett is a Christian apologist for Nietzsche.

A Kierkegaard Anthology. Bretall.

Kierkegaard is a Christian apologist for Kierkegaard.

Being and Time. Heidegger.

Heidegger is a Christian apologist for Christ.

Origin of Geomtry. Edmund Husserl.

Introduction by Jacques Derrida; my favorite topic by Sartre's mentor and creator of the epoché.

Being and Nothingness.
and
Truth and Existence.
and
The Psychology of Imagination.

and
The Emotions.
and
The Words. Jean Paul Sartre.

While most of Sartre's shorter writings are almost as unapproachable as my own writings, with sentences in length paralleled only by the paragraphless writing of Franz Kafka, his longer works are quite the opposite: engaging and entertaining. Like Jack Kerouac, if one can decipher the rhythm of Being and Nothingness, one will find themselves transported even more directly into Sartre's world than in Nausea. Most of his metaphors for existence are lifted directly from the coffee shop set in which he was sitting and writing, so the sense of immediatism is strong, and one wonders if it was only necessary to argue the humanism of existentialism to those people who had never read any.

The Second Sex. Simone de Beauvoir.

Existentialism as a humanism from a feminist's standpoint.

The Myth of Sisyphus and other essays.
and
Notebooks 1942 - 1951. Albert Camus.

The myth of Sisyphus is considered required reading in some scholastic settings, both preperatory and collegiate. Seeing as how it only describes the torture of a damned soul, one wonders at what it would be like if Danté, or even story of O, were taught in place of Dr. Seuss or Dick and Jane?

For the New Intellectual. Ayn Rand.

Rand advocates smoking cigarettes as a liberty. If she were alive today and were informed about both sides of the issue in modern times: cigarette additives being addictive and causing cancer and the clean-living fascist no-public-smoking laws, I wonder what she would think.

Madness and Civilization.
and
This is Not a Pipe. Michel Foucault.
and
The Foucault Reader. Rabinow.

In Madness Foucault, as only a philosopher could get away with, concludes by comparing living conditions in victorian era mental asylums with the paintings of Goya. On the other hand, This is Not a Pipe, about the painting of the same name by Magritte, manages to examine the piece from every angel except, as it makes note, the purely aesthetic.

Illuminations. Walter Benjamin.

Parabolic wisdom, soothing to read.

The Pleasure of the Text. Roland Barthes.

An essay for its own sake about writing for its own sake.

Theory of Reigion.
and
Literature and Evil. Georges Batailles.

One of my favorite authors of fiction in terms of style, bordering on expressionistic surrealism. Here he analyzes a handful of authors throughout history who have dealt with the subject of, or been accused of being, evil. What stands out in my mind is the section on the Marquis De Sade, since I was reading him at the time, and I found that I disagreed with almost every point Batailles made. De Sade is just puncturing the aristocracy. On that point, alone, I think we agreed.

America.
and
Cool Memories.
and
the Ecstacy of Communication and Forget Foucault.
and
Simulacra and Simulacrum. Jean Baudrillard.

I don't care much for America, which reads like a family slide show after a cross country winnebego trip, but the ecstasy of communication and simulacra / simulacrum are to live and die for.

Immediatism.
and
Temporary Autonomous Zone. Hakim Bey.

Hakim Bey, an anomolous modern Muslim philosopher, has some new ideas, such as immediatism, which I make repeated reference to in the MPDR, as well as the T.A.Z. or Permanent Autonomous Zone, where anarchy reigns, as well as an excellent literary style, but his works are rare and hard to find.

the Electronic Disturbance. Critical Art Ensemble.

Interesting critique of modern dub and sampling media as derived from the cut-up method of Burroughs and collages of Picasso. Goes on and on about "the Body Without Organs."

The Matrix and Philosophy. William Erwin. ed.

Some essays by modern philosophers, mostly professors or grad students, about the metaphysics of the movie the Matrix and its impact on the media of today.

Science

This is my favorite subject, but I probably understand it less than any other. I was never that good at it in school, so I decided to teach myself from what books I could find on the shelves at book stores. The criteria for books I have included here, as the first serval will no doubt strike the average reader as belonging more to the psychology section, is the rigorous application of scientific method, which depends on a lab setting unavailable to most psychologists, or to a sample test group kept under complete control and absolute observation, a feat beyond most philosophers.

Info-Psychology.
and
Self-Determination.
and
the Psychedelic Experience. Timothy Leary.

Despite his reputation as being far out I find much of Leary's writing to be very conventional from a scientific standpoint. He analyzes every event and experience using arduous and rigorous methodology, and formulates systems based on his data. Info-psychology introduces the eight circuits of consciousness model I make reference to in this work. Self-Determination is a must read for anybody who is opposed to behaviorist determinism. The Psychedelic Experience is where I first read about ylem and the primary clear light, although these are real Buddhist concepts.

Altered States. James Hughes.

A picture gallery history of drug scenes from the beginning of the counter-culture to the present.

Human Survival and Consciousness Evolution.
and
The Cosmic Game.
and
The Holotropic Mind. Stanislav Grof.

Stan Grof, founder of the breathwork movement, in which I am certified to teach, was a disciple of Tim Leary during his Haight Ashbury days in the late sixties.

Coincidance. Robert Anton Wilson.

A more serious and in-depth look at synchronicity.

Your Psychic Powers and How to Develop Them. Carrington.

A serious look at ESP as being an innate instinct which, with practise, can be honed into a fully functioning sensory skill.

Psychic Phenomena. Bradley and Bradley.

A somewhat less scientific approach to the same subject matter.

The Truth About Psychic Self-Defense. Keith Randolph.

Although it gets a little side-tracked talking about strange voodoo vampires, this pocket-sized book does demonstrate how to visualize armor and other defenses to protect oneself from other peoples' ill intentions.

447

Drugs and the Mind. Robert S. De Ropp.
This book came out in the early to mid seventies, when drug experimentation was highly prevalent. It contains sections on all the drugs that were popular before crack cocain was introduced, and the history sections on marijuana and psychedelics are particularly enlightening.

The Doors of Perception and Heaven and Hell. Aldous Huxley.
"When the Doors of Percept are cleansed, the world will appear as it truly is: infinite." — William Blake. Early experimentation with psychadelic substances, probably hallucinogens, led Huxley to make the first inquests into the fine line between psychology and religion in the 1940's.

Mysticism and the New Physics. Michael Talbot.
Explores links between ancient shamanic techniques for transcendence, eastern cosmologies, and modern quantum mechanics and relativity.

The Way Things Are: Basic Readings in Metaphysics. W. R. Carter. ed.
A compilation of essays on the mind/matter dilemma, including one by Rene Descartes I haven't yet found elsewhere, as well as a particularly bemusing one by Richard Swineburne entitled "Body and Soul."

Dream Symbolism. Manly P. Hall.
Included in science rather than speculation becuase of Hall's application of the hueristics to trace the history of dream prophecies.

Mathematical Magic. Simon.
Topology, statistics, number squares, card tricks.

The Geometry of Art and Life. Ghyka.
A book of some of the finest porportions in art, architecture and nature, dealing at length with phi, the Golden Ratio or Divine Porportion.

The Divine Porportion. Huntley.
A history of phi and tens of beautiful examples from nature, incl. formula.

Platonic and Archimedian Solids. Sutton.
Beautiful art depicting some of the mathematical nuances of the only regular solids.

Geometry, Relativity and the Fourth Dimension. Rucker.
Proposes the interesting notion, diagrammed elsewhere as lightcones by Hawking, that space and time might actually run perpendicular to one another, and only appear to us to be parallel because we are only looking at their intersection along the line of the plane's edge; stops short before going too deeply into the toroidal model of spacetime by saying that the only thing that could distort

spacetime to the singularity at the centerpoint of a torus is matter, but that then only energy could pass through it, essentially taking the model far too literally.

Relativity
and
Sidelights on Relativity. Albert Einstein.
Presented first as a visualization, followed by several thought experiments, this is the source of what is probably the most elegant formula for the matter-energy exchange of entropy ever devised. A must read for anybody who is interested in either physics or why the world is the way it seems to be.

General Thoery of Relativity. Dirac.
One of the masters of quantum mechanics picks over the formulae of general relativity. The section about scalar waves boils down to zero point energy, though it avoids saying so, and the next logical leap from there is the spin of the tachyon.

Theory of Relativity. Pauli.
Another master of quantum mechanics who gave us statistics for waves goes over the special (entropic) and general (gravitational) theories of relativity with formulae.

Einstein's Theory of Relativity. Born.
Another contributer of statistics for subatomic particles rhapsodizes on the formula of special and general theories of relativity.

About Vectors. Hoffmann.
A book that explains what I have herein called the conservation of dimension for matrices as the parallelogram law of vectors.

Thermodynamics. Fermi.
A book about very slight temperature variations in subatomic perturbation systems by the man who discovered quanta, or wave-packets, of energy. Essentially, a book about Fermions by the man after whom they were named.

The Physical Principles of the Quantum Theory. Heisenberg.
One of the pioneers of atomic energy discusses with minimal formulae the essential factors of quantum theory such as uncertainty and exclusivity.

Elementary Particles and the Laws of Physics. Feynman and Weinberg.
A good book, it essentially just talks about fermions (waves of energy) and bosons (particles of matter), listing all known of each kind, and giving their properties, etc.

The Nature of Space and Time. Stephen Hawking and Roger Penrose.
A very good book about the geometry of spacetime and the possible histories of world lines depending thereon.

Quest for a Thoery of Everything. Stephen Hawking and Kitty Ferguson.

Hawking's answer to string theory before he began speculating with M-theory and branes.

the Holographic Universe. Michael Talbot.

A very good book that talks about the hologrpahic nature of the brain and speculates about the holographic nature of the light upon which we base our view of the universe.

Faster than Light. Nick Herbert.

This book came out long after I had completed writing the MPDR. It's the first book I've seen that deals seriously with tachyons, and I even had to go back and double check every time I mentioned the word in the MPDR to make sure it didn't disagree with the definition he had given. This is also where I learned the term "transcendent" tachyon.

Hyperspace.
and
Visions. Michio Kaku.

The primary modern spokesperson for superstrings talks about them in no uncertain terms as being tightly coiled histories of multiple dimensions in Hyperspace. In Visions he talks about the possible breakthrough discoveries that can made in the next fifty or a hundred years in science.

Millennium. Vivololdo and Dychtwald.

A book from the late seventies essentially the same as Visions by Kaku. An edited compilation of essays by various different scientists and speculative scholars about what scientific breakthroughs they might expect by the year 2000. Many are the same as in Visions.

Reasoning with Statistics. Williams.

Teaches exactly nothing about quantum mechanics.

Understanding the New Mathematics. Ray Kurzwell.

This book is from the sixties, and so is probably a little dated.

Future Shock. Alvin Toffler.

A book after my own heart, whose style immitates its subject matter: the ever-changing face of ultramodernity.

the Naked Ape. Desmond Morris.

From the author of the Human Zoo, in this, his original work, Desmond Morris explores the connections between man and his primate ancestors in such areas as gesture, personal space, facial expressions, etc.

Origin of Species. Charles Darwin.

The premier book on evolution. Contrary to the popular opinion prevalent since the Scopes monkey trial, Darwin's evoltuon does not contradict creationism. He doesn't deal with fossils at all, only with already existing species, mostly exotic, and traces how they adapt from one generation to the next as they struggle for survival.

The Double Helix. James D. Watson.

The true story of the discovery of the DNA double helix by one of the two scientists responsible. A fascinating plot, made dull by being written by a scientist, is almost amazing in spite of it due to its brevity and quick pace. A much more approachable work than Crick's grandiosities.

Human Origins. Abrams.

Probably not available. A mini text book about evolution and palentology.

Apes, Language, and the Human Mind. Taylor, et al.

An Oxford thesis work on overcoming verbal communication barriers.

The Cosmic Connection.
and
Dragons of Eden.
and
Cosmos. Carl Sagan.

One of the only very few serious scientists to speculate about extraterrestrial life, interstellar germination, the levels of the human brain, and other "far out" concepts.

Colorado Springs Notes.
and
The Inventions, Researches and Writings of.
and
the Complete Inventions (patents)
and
My Inventions. Nikola Tesla.

This represents the greater part of the collection of works by the foremost proponent of free energy broadcasted wirelessly as well as extraterrestrial communications systems, subterranean death rays and torus shaped energy coils.

Stephen Hawking's Universe. Filkin.

A picture book too beautiful for them to keep on stands, this will probably not still be around after a few years, but it has great pictures of clusters as well as several research projects being conducted at super colliders.

The Force of Symmetry. Vincent Icke.

This is about the force that seems to govern the dualities we see in nature, from the greatest galaxies to the smallest vectors on quanta. In particular of interest are the concepts of spontaneous symmetry breaking, which pertains to quantum level jumping by electrons, and supersymmetry, the concept that when symmetry is broken it is only to conform to a greater or deeper symmetry we cannot perceive.

4-Manifolds and Kirby Calculus. Gompf and Stipsicz.
4-manifolds, discussed in the MPDR, are fourth dimensional metaforms whose topography can be measured by vector sets in neighborhoods, a process known as Kirby calculus.

Dreaming the Future.
and
the Zen of Magic Squares, Circles and Stars. Clifford A. Pickover.
It is from this book that I give the formula for finding the sum of any magic number square with sequential constituent numbered cells given only the number of cells and first number in sequence. This book has some really beautiful forms of magic shapes, and I would highly recommend the study of this to any would-be mage.

the Essential John Nash. Kuhn and Nasar.
This book gives Nash's full post collegiate work on 4-manifolds, and is truly essential for a study thereof, as he demonstrates how symmetry can be gleaned from topography and how topography can be gleaned by the fewest possible vectors possible.

Black Holes and Time Warps. Kip S. Thorne.
A work by the co-discoverer of the exact mathematical proof for the existence of the previously theoretical black holes, long before one had ever been observed or it had been speculated they occupied the centers of spiral galaxies, Kip Thorne gives a fascinating look into the worlds of no escape gravity wells and the conservation of information.

Cellular Automata and Complexity. Wolfram.
An interesting book, though dauntingly thick, about self-replicating micro-mini shapes, studying how they propagate to fill different given spaces.

Chaos. James Gleick.
A fascinating look at fractals such as the snowflake, Mandelbrot and Julia sets, as well as "sponges" with infinite volume and zero mass which I compare in the MPDR to magic cubes.

Physics of Waves. Elmore and Heald.
This is where I got most of the information about perturbation theory in the section on entropy.

The Big Bang. Joseph Silk.
　　A qualified look at the few minutes around the first quantum fluctuation.

The Structure and Interpretation of Quantum Mechanics. Hughes.
　　Formulae in text book form.

Princeton Guide to Advanced Physics. Tribble.
　　Formulae and diagrams in text book form.

the Secret Life of Quanta. Han.
　　A very lay-person friendly look into some of the mysteries of quanta such as tunneling, superposition and the double slit experiment.

Classical Mechanics. Corben and Stehle.
　　Formulae and diagrams in text book form.

Atom. Isaac Asimov.
　　A detailed history of the discovery of each different aspect of the atom and of quanta.

The Universe in a Nutshell.
and
A Brief History of Time.
and
Black Holes and Baby Universes and other essays. Stephen Hawking.
　　The Universe in a Nutshell, Hawking's latest on physics, introduces M-theory, or the theory of branes, two dimensional superstrings. A brief history of time describes the four elemental forces, and Black Holes and Baby Universes describes the way in which singularities accumulate more information than they lose to entropy, and may therefore be negentropic, forming bubble universes.

Knotted Doughnuts. Garder.
　　A book about tori and 4-manifolds that approaches them from the angle of topology. A better book for the lay-person.

Tapping the Zero-Point Energy. Moray B. King.
　　An Adventures Unlimited Press book, so a fringe publication. This describes different electronic devices that can create null fields. Some use scalar wave dispersion fields to create an interference pattern where zero point energy is achieved. Others use counter-rotating electromagnetic fields around differentially coiled wires.

A History of Pi. Beckman.
　　Just what it says, this book traces the number from its earliest applications in Egypt and China through the Age of Reason to modern supercomputers.

The Golden Ratio. Mario Livio.

The same as the history of Pi, only for Phi. A more recent book, as certain west coast controversies have rekindled new age interest in this transcendental shape.

Godel, Escher, Bach.
and
Fluid Concepts and Creative Analogies.
and
Metamagical Themas. Douglas R. Hofstadter.

Very large books which tend to hop from subject to subject in a way that might at first appear random but which is meticulously preplanned.

The Reflexive Universe. Arthur M. Young.

Most cited author by the Philosophical Research Foundation, established by Manly P. Hall, 33rd degree Scotch Rite Masonry, Arthur Young introduces his concept of an arch in the structural complexity of organisms that peaks in molecules between quarks and spirit.

Dreams of a Final Theory. Weinberg.

A history of scientific reductionism over the past one hundred years.

Masters of Time. Boslough.

Working on a cyclotron.

Handbook of Current Science and Technology. Gale.

A little dated, but a thick and thorough reference work.

A Little Book of Coincidence. John Martineau.

With beautiful diagrams, plots the retrograde sysles of the planets and explains their eliptical orbits.

Text Books

This is the only section of books you can be reasonably sure that I've read all or most of, even if it was only because these ephemeral ideas called "grades" dependended on it. You see, we pay good money to participate in our own institutionalization, in the form of education and in the name of self-betterment, whether the institution succeeds in educating us or not. The point is simple: an "F" costs the same amount of actual cash as an "A." I guess the moral of this fact is that you don't have to believe I've read any of these books. The truth is even worse: I did read them, cover to cover, got excellent grades on the tests about the material they contain, and then, after the pressure to remember them for testing purposes, I have forgotten everything I learned in every subject. Go figure.

Psychology. Wade and Travis.

Psychology and Adjustment. Cohen.

Sociology. Applebaum and Chambliss.

History of the Theatre. Brockett.

American Government. Wilson.

Invitation to social psychology. Philip Chalk.
    A new branch at the time I studied it combining sociology and psychology in crowd and varying size group dynamics.

A Short History of Western Civilization. Harrison, Sullivan and Sherman.

United States Diplomatic History in two volumes. Clarfield.

The Enduring Vision (American History). Boyer, Clark, Kett, et al.

Street Law. McMahon, Abbetman and O'Brien.

Art History. Stokstad.

Western Civilization. Perry.

A History of Narrative Film. Cook.

Global Politics. Ray.

Understanding Politics. Magstadt and Schotten.

International Law and the Society of Nations. NYSBA.

Arts and Ideas. Fleming.

The Visual Arts: A History. Honour and Fleming.

Algebra and Trigonometry. Swonkowski and Cole.

A Survey of Mathematics. Angel and Porter.

Economics Today. Roger LeRoy Miller.

Anatomy. Gray.

Prentice Hall Literature. World Masterpieces.

College Algebra and Trigonometry. Dugopolski

Perspectives on Arguement. Nancy Wood.

Advertising: Principles and Practice. Wells, Burnett, Moriarty.

Biological Science. Heath.

Biological Psychology. Kalat.
    Neuropsychology. Also a new science when I studied it.

Persuasion. Larson.

Social and Political Philosophy. Sterba.

Introduction to Quantum Mechanics. Griffiths.

Astronomy. Zeilik.

Horizons (Astronomy). Seeds.

Geometry: the Easy Way. Barron's

History
    I have a sufficient number of books in this genre that I thought I should make a separate category for them, even though most of them classify as what I would call text books.

World History. Heath.

The Lost Civilizations of the Stone Age. Rudgley.
    The inventions and explorations of neanderthal and clovis man.

The Global Past. Fields, Barber and Riggs.
    One of the first text books to reflect the new, corrected dating for the pharaohs based on comparison to the Sumerian records instead of to the Hebrew Bible.

Annual Editions: World History, prehistory - 1500. Fifth edition.
    A brief pamphelt of updates from archaeological finds.

Great People of the Bible and How They Lived.
and
Atlas of the Bible. Reader's Digest.
    Tourist guide-like text books.

Guide to the Bible. Isaac Asimov.
Quotes, cross references, contextual explantions and contemporary comparisons.

James the Brother of Jesus. Robert Eisenman.
This book was put out right before they released the story to the media that they had found a tomb bearing this inscription.

Handbook to the Old Testament.
and
Handbook to the New Testament. Westermann.
Cross references to text and location with maps.

The Ancient Near East. Prichard.
A cultural atlas from before the release of the Sumerian records to the public.

Old Testament Theology. Von Rad.
A clerical text book.

The World's Great Religions. Life.
An oversized picture book.

Mythology. Paul Hamlyn.
An invaluable resource of world mythology, particularly comparative flood mythology.

Cultural Atlas of Mesopatamia. Roaf.
Including the King's List and a description of Sargon. A new edition it isn't on most stands yet.

The Times Atlas of World History. Hammond.
The source for the location of the garden of Eden.

The Secret Architecture of Our Nation's Capital. David Ovason.
Much astrology was built into the monuments of Washington D.C. by master free masons.

The Outline of History. H.G. Wells.
Not a textbook, but it could be. Wells, author of many books about the conspiracy to instill public anarchy, has herein set down his entire chronicle of history. A valuable resource.

The Jewish War. Josephus.
A first hand account by a Hebrew trained Roman scribe on the Jewish revolt against Roman occupations and the battle of Massada.

The Pentecost Revolution. Schonfield.

A well researched description of the battles of the time of Massada.

The Jews in the Time of Jesus. Wylen.

A description of Jewish culture leading up to the historical span of the lifetime of the literary Jesus and for some years following his death.

The Secret Archives of the Vatican. Ambrosini.

An investigation of the depth of information locked away in the building beside the Vatican library.

The Meaning of the Dead Sea Scrolls. A. Powell Davies.

1956, barely ten years after they were discovered, this book describes the conflict of the Dead Sea Scrolls to existing Christian dogma. Since this time we have seen a massive relaxation and liberalisation of the papacy.

Secrets of the Dead Sea Scrolls. Price.

Historical context to their finding and an interpretation of their content.

Secret Books of the Egyptian Gnostics. Doresse.

A book describing the archaeological finding and an interpretation of the anthropological context of the Nag Hamadi library in Chenoboskion.

Speculation

These books are generally considered fringe thinking. It should be noted, however, that no truly dangerous idea to the establishment would ever be able to make it into print in the popular media, and even the writings of modern conspiracy researcher David Icke are so interwoven with "far out" extraterrestrial associations that his average reader would probably be led by kneejerk reaction to even disbelieve those aspects of his well researched writings that are, in fact, exactly accurate. Similarly, because Atlantis has an air of mythology surrounding it, most people would shy away from a serious and scholarly investigation of the subject even in spite of it dealing specifically with well documented evidence and factual data. While it is necessary to take the style of most of the speculative writing with a grain of salt, there are, at the same time, many nuggets of gold in every one of these types of books. It only requires enough interest in the subject matter to put up with the irrelevant positioning of opinions in between the interesting new information.

Atlantis.

and

The Destruction of Atlantis. Ignatius Donnelly.

The original resurgance of interest in the trend of the mystery of lost land mass was initiated in the mid nineteenth century by Ignatius Donnelly, and many of his ideas found their way directly into the movement of Theosophy.

The Lost Continent of Mu.
and
The Cosmic Forces of Mu.
and
The Second Book of the Csmic Forces of Mu.
and
The Sacred Symbols of Mu.
and
The Children of Mu. James Churchward.

Written around the time of Thor Heyerdahl, the Mu series represents an increasing trend in fringe archaeology towards the speculative. Mu, also known as Lemuria, is the counterpart of Atlantis connecting India to Australia where modern day Oceania and New Zealand are located. These books deserve much greater recognition in the world of speculation on lost land masses than they receive because it is now an accepted fact that such a land bridge actually existed during the last ice age.

Mu Revealed. Tony Earl.

Another in the Mu series.

Unexplained Mysteries of the Twentieth Century. Bord and Bord.

Mostly about frogs raining and other inexplicable incidents, some crop circles.

The Maldive Mystery.
and
Fattu Hiva. Thor Heyerdahl.

From the writer of Kon Tiki who sailed across both oceans on boats made of river reeds, thus proving intercontinental migrations might have occured for primitive peoples even only given the simplest means. The Maldive Mystery describes a find that appears to have been a storehouse for various religions throughout the world built all at the same time.

The Mysteries of Easter Island. Schwartz.

An excellent analysis of Rongo Rongo writing, the mysterious petroglyphs found throughout Easter Island.

Stonehenge Decoded.
and
Beyond Stonehenge. Gerald Hawkins.

In Stonehenge Decoded, Hawkins describes for the first time Stone Henge as an enormous astronomical clock. In Beyond Stonehenge he surveys other neolithic sites in the surrounding countryside and finds similar conclusions.

Uriel's Machine.
and

the Hiram Key.
and
the Second Messiah. Christopher Knight and Robert Lomas.

In Uriel's Machine these two thirty third degree free masons speculate that Enoch might have travelled to stonehenge, and postulate that there might have been other such neolithic sites that retained pilgrimage among the wise men of ancient antedeluvial societies. In the Hiram Key they follow a collection of scrolls from under the Temple of Solomon back with the Knights Templar that they believe revealed the linneage of the Merovingian bloodline from Jesus himself. In the Second Messiah they continue their fascinating revision of history by speculating that the shroud of turin was created by amino acids burned off the tortured body of Knight Templar Jacques De Molay before he was crucified by the Roman church.

Gateway to Atlantis. Andrew Collins.

Introduction by David Rohl. Collins proposes, similarly as I do, that the center of the great Atlantean civilization was in modern day South America, and, as do Christopher Knight and Robert Lomas, he proposes a comet struck the site. His date, however, is in agreement with that of Graham Hancock. He even goes so far as to speculate, in my opinion quite rightly, that the descriptions of angels such as the Annunaki as men with wings derived from the Atlantean tradition, preserved in South America to the time of the conquistadors, of wearing robes of feathers.

The Atlantis Blueprint. Colin Wilson and Rand Flem-ath.

The angles of ancient cities and monuments indicate alignment to previous locations of the north pole and the equator.

Atlantis. Andrews.

A purely speculative book about Atlantean society and culture.

Edgar Cayce on Atlantis.
and
Mysteries of Atlantis Revisited. Edgar Cayce.

Channeled teachings on Atlantis by the sleeping prophet regarding the Law of One and the Children of Belial.

Atlantis: the Eighth Continent. Charles Berlitz.

Atlantis in Antarctica. Probably true, among other places as well.

the Giza Death Star. Joseph P. Farrell.

The idea that the great pyramid really served as a power plant whose energy could be directed outward.

The Sirius Mystery. Temple.

Explaining some of the mysteries regarding the alignment between the sphinx and the dog star, representing Isis.

the Sign and the Seal.
and
Fingerprints of the Gods.
and
the Mars Mystery. Graham Hancock.

In the Sign and the Seal Hancock traces the theft of the ark of the covenant by Menelik, the son of Solomon and the Queen of Sheeba, to modern day Axum. In Fingerprints of the Gods, Hancock introduces the concept of precession to modern archaeoastronomy. This is a classic of the genre. In the Mars Mystery he ponders the similarities between the face on Mars and the sphinx and pyramids.

Message of the Sphinx. Hancock and Bauval.

Hancock and Bauval's take on the alignments of the sphinx 12,500 years ago with the helilacal rising of Sirius.

Genesis of the Grail Kings.
and
Lost Secrets of the Sacred Ark.
and
Bloodline of the Holy Grail.
and
Realm of the Ring Lords. Laurence Gardner.

The ultimate turn of the millennium revisionist/conspiracy theorist history series. Following from descriptions of the Annunaki and the Nefilim, shems as upward fire stones and shewbread made of monatomic gold to rex deus, to the mythology of dark ages Europe, written by a mason and derived from voluminous research over masonic and rosicrucian documents. I only found these after I had written my history section, but, upon seeing it agreed with my theory that Moses was Akhenaten, I had no problem making brief mention of the shewbread.

5/5/2000. Richard W. Noone.

A book speculating the world would end on this date with the grand cross alignment of the planets that occured then. It has some interesting factoids of speculation, like that the King's Chamber of the Great Pyramid might have been used as a mechanical pump.

The Orion Prophecy. Patrick Geryl and Gino Ratinckx.

Speculation based on pression and the Mayan calendar that the world will end in December, 2012 due to a solar storm.

the Time Travel Handbook.
and
Anti-Graviy and the World Grid.
and

Technology of the Gods. David Hatcher Childress.

More from adventures unlimited press. The time travel handbook has only a little about the nature of spacetime and is mostly dedicated to ufo ship designs. Anti-gravity and the world grid talks extensively about what I call the Enochian Communications System, connecting all ancient petraglyphic sites with the locations of all modern satellites in a giant electromagnetic net. In technology of the Gods Childress explores ancient electricity, Brown's gas, lasers, etc.

Cosmic Trigger (vol. 1).
and
the Illuminati Papers. Robert Anton Wilson.

This is where I originally learned about Leary's eight circuits of consciousness model. The Illuminati papers is just silly.

Philadelphia Experiment — UFO Conspiracies. Brad Steiger.

An actually fascinating book about multiple realities on intersecting dimensions and the US military versus the denizens of this realm.

Montauk Project.
and
Montauk Revisited.
and
Pyramids of Montauk. Preston Nichols and Peter Moon.

A little stretched in terms of characterization, but according to a strict interpretation of Einstein's theory of relativity, particularly that regarding wormholes, an entirely believable scenario. Proposes the interesting concept of temporal signatures, unique to our personal em fields, and temporal locks, the pulse phasing of tractor beams through wormholes to beam these signatures back and forth in time.

Nothing in this Book is True but it's Exactly How Things Are. Bob Frissell.

Purports that the flower of life mandala of Drunvalo Melchizedeck was taught as the Left Eye of Horus mystery cult in ancient Egypt. Tré new age wishy washy.

New World Order. Still.

You can't find this book on shelves under a republican president. Only under a democrat. It is masonic propaganda posing as religious right wing anti-masonic propaganda. Gives quite a good history of American masonry.

The Watchers. Raymond E. Fowler.

One of a series of knock off abduction scenarios that flooded the market after Whitley Streiber. This one stands out, however, as its description of the little gray aliens gives the name for them of the Watchers, the same as the word Annunaki translates to. Perhaps t is just coincidance.

Rule By Secrecy. Jim Marrs.

Traces the origins of modern conspiracies back to the originally alien bloodlines of the Anunnaki in ancient Sumeria.

....And the Truth Shall Set You Free.
and
the Biggest Secret.
and
Children of the Matrix.
and
Alice in Wonderland and the World Trade Center Disaster. David Icke.

Despite his conclusion that everyone he doesn't particularly care for is a reptile from the Plaidies, David Icke is probably, along with Jim Marrs, the best conspiracy theory researcher in the business right now.

Mind Control World Control. Jim Keith.

An indispensible guide for the devoutly paranoid.

Liquid Conspiracy: JFK, LSD, the CIA, Area 51 and UFOs. George Piccard.

particularly good for its speculative sections on underground bases, a subject I don't go into very deeply in the MPDR.

Channeled/Divine

These works are those of the new age movement that are written by people who claim to be other people. I have explained how this is possible in the MPDR, but it can be explained just as easily by saying that the prophets they channel are sentient programs that can access all the software connected to the mainframe of the computer program we call reality.

Starseed, the Third Millennium. Carey.

This is really the one that started all the others. Received some time in the mid seventies, I think, it supposedly originates from outer space. It is really little different than Tesla's claims in the twenties or Herman Hesse's Strange News From Another Star.

The Celestine Prophecy.
and
the Tenth Insight. James Redfield.

After the writing of "the Prophet" by Kahlil Gibran, a series of works claiming to be divinely inspired followed from a variety of sources, foremost amongst which was the offering by James Redfield.

Revelations of the Metatron. Anonymous.

Whoever wrote this was wise to keep their name off it. A retelling of the war in heaven and the fall of the rebel angels with a hollisitc, sophistic spin in high English, immitating biblical writing, poorly.

Maitreya's Mission (vol.2). Benjamin Creme.
   Comes shrink wrapped and I haven't opened it yet. The person claiming to be the Maitreya, or last Buddha, also claims to have predicted the tech bubble burst, and also made a speech at the United Nations.

Scientology: the Fundamentals of Thought. L. Ron Hubbard.
   L. Ron Hubbard doesn't really claim to be channeling, but he does claim divine origins for his religious front the Church of Scientology. His writing style is desperately cultish.

Many Lives, Many Masters. Brian L. Weiss.
   The prmeire western offering on the eastern concept of reincarnation.

Ancient Cosmologies. Baker and Loewe.
   A good text chronicling gensis myths from around the world.

Nostradamus: the Complete Prophecies. John Hogue.

Saint Germaine on Alchemy. Elizabeth Claire Prophet.
   Written in the voice of the elusive comte himself.

The Hermetic Tradition (Alchemy). Evola.
   One of the better works on the subject.

Story of Jesus. Edgar Cayce.
   The sleeping prophet's readings on the lifetime of the Christian savior.

The Lost Years of Jesus. Elizabeth Claire Prophet.
   Supposedly, according to this researched work, Jesus spent his missing time in Issa, Tibet.

The Mystical Christ. Manly P. Hall.
   A thirty third degree freemason talks about Christ and reincarnation.

Mystic Christianity. Yogi Ramacharaka.
   Eastern philosopher discusses western religious elements shared in common.

The Protestant Mystics. Fremantle.
   Very short, and a little old.

The Teachings of the Mystics. Walter T. Stace.
   A short journey among past masters.

The Nature of Faith. Gerhard Ebeling.

A Protestant mystic comments on the substance of essence.

From Whom God Hid Nothing. Meister Eckhart.
A mideval German Dominican Catholic mystic.

Life is Real Only Then, When 'I AM.' G.I. Gurdjieff.
The reknowned Russian mystic and author of Beelzebub's tales to His Grandchildren.

JPS Hebrew-English Tanakh. Jewish Publication Society.
The Old Testament in Hebrew and English.

Cathehism of the Catholic Church. Libreria Editice Vaticana.
The official rules for the Catholic faith.

The Oxford Concise Concordance. Revised Standard Addition.
The Catholic prayer book.

The Koran. Attributed to Muhammed.
The Penguin Classics edition.

The Book of Mormon. The Church of Jesus Christ of Latter Day Saints.
Given to me on my own doorstep by some friendly strangers, or maybe they were just strangely befrienders.

The Satanic Bible.
and
the Satanic Rituals,
and
Satan Speaks!
and
The Devil's Notebook. Anton Szandor LaVey.
You can't keep a good man down.

Apocrypha
These are works of significant religious impact and antiquity that have been excluded from the canonized vulgate and from the Torah.

the Book of Jasher. trans. Flaccus Albinus Alcuinus.
Contains an alternative description of Genesis and a first hand account of Moses at Mt. Horeb.

The Lost Book of Enki. Sitchin.
composited together from a cosmogony, an Epic of Creation called the Eridu Genesis, and the Atra Hasis.

Inanna. Wolkstein and Kramer.
Inanna and the God of Wisdom. Inanna and Dumuzi. The Descent of Inanna. Seven Hymns to Inanna.

The Epic of Gilgamesh.
The classic poem thought for a while to be the oldest writing in existence.

Necronomicon. attributed to Abdul Alhazred.
The Seven Gates of the Zones. The Fifty Names of Power of the Elders.

Forbidden Mysteries of Enoch. Prophet.
The Book of Enoch and the Book of the Secrets of Enoch.

Kebra Nagast. Brooks.
Describing the thaking of the ark of the covenant to Axum by Menelik the son of Solomon the King with the Queen of Sheeba.

6th and 7th Books of Moses. Tice.
Containing the magical charms and spells used by Moses and Aaron.

the Nag Hammadi Library. Robinson.
Prayer of Paul. Apocryphon of James. Gospel of Truth. The Treatise on the Resurrection. The Tripartite Tractate. Apocryphon of John. Gospel of Thomas. Gospel of Philip. Hypostasis of the Archons. On the Origin of the World. Exegesis on the Soul. Thomas the Contender. Gospel of the Egyptians. Eugnostos the Blessed. Dialogue of the Savior. The Apocalypse of Paul. 1st Apocalypse of James. 2nd Apocalypse of James. Apocalypse of Adam. The Acts of Peter and the Apostles. The Thunder, Perfect Mind. Authoritative Teaching. The Concept of Our Great Power. Plato's Republic. The Discourse on the Eight and the Ninth. The Prayer of Thanksgiving. Aesclepius. The Paraphrase of Shem. Second Treatise of Seth. Apocalypse of Peter. Teachings of Silvanus. Three Steles of Seth. Zostrianos. Letter of Peter to Philip. Melchizedek. Thoughts of Norea. Testimony of Truth. Marsanes. Interpretation of Knowledge. Valentinian Exposition. Allogenes. Hypsiphrone. Sentences of Sextus. Fragments. Trimorphic Protennoia. Gospel of Mary. Act of Peter.

The Lost Books of the Bible. Testament Books.
Mary. Protevangelion. I. Infancy. II. Infancy. Christ and Abgarus. Nicodemus. Apostles' Creed. Laodiceans. Paul and Seneca. Paul and Thecla. I. Clement. II Clement. Barnabus. Ephesians. Magnesians. Trallians. Romans. Philadelphians. Smyrnaeans. Polycarp. Philippians. I. Hermas — Visions. II. Hermas — Commands. III. Hermas — Similitudes. Letters of Herod to Pilate. Last Gospel According to Peter.

the Other Gospels. Cameron.
Gospel of Thomas. Dialogue of the Savior. Gospel of the Egyptians. Papyrus Oxyrhynchus 840. Apocryphon of James. Secret Gospel of Mark. Papyrus Egerton 2. Gospel of Peter. Gospel of the Hebrews. Acts of John 87 - 105. Gospel of the Nazoreneans. Gospel of the Ebionites. Protevangelium of James. Infancy Gospel of Thomas. Epistula Apostolorum. Acts of Pilate.

The Gospel of Thomas. Dart and Riegert.
One hundred fourteen sayings of Jesus.

The Apocrypha. Edgar J, Goodspeed.
I. Esdras. II. Esdras. Tobit. Judith. Additions to Esther. Wisdom of Solomon. Baruch. Susanna. Song of Three Children. Bel and the Dragon. Prayer of Mnasseh. I. Maccabees. II. Maccabees.

the Dead Sea Scrolls Uncovered. Eisenman and Wise.
Fifty new fragments with transliteration. Messianic and Visionary Recitals. Prophets and Pseudo-Prophets. Biblical Interpretation. Calendrical Texts and Priestly Courses. Testaments and Admonitions. Works Reckoned as Righteousness — Legal Texts. Hymns and Mysteries. Divination, Magic and Misc.

The Dead Sea Scrolls Translated. Garcia Martinez.
Rules 1. Halakhic Texts. Literature with Eschatological Content (War Scroll and New Jerusalem). Exegetical Literature. Para-Biblical Literature. (Book of Giants, Pseudo-Moses). Poetic Texts. Liturgical Texts. Astronomical Texts, Calendars and Horroscopes. The Copper Scroll.

The Essene Gospel of Peace (parts 1 - 4). trans. Edmond Bordeaux Szekely
First published from the secret Vatican archive in 1928.

the Other Bible. Willis Barnstone.
Creation Myths.
Secrets of Enoch. Book of Jubilees. Haggadah. Manichean Gnosis. Secret Book of John. Hypostasis of the Archons. Apocalypse of Adam. Gospel of Philip. Paraphrase of Shem. Second Treatise of Seth. Mandaean Gnosticism. Kabbalah.
History and Narratives
Martyrdom of Isiah. IV. Maccabees. Perpetua and Felicity. Ahikar. Genesis Apocryphon, Manual of Discipline, Damascus Document, War Scrolls (Dead Sea Scrolls). Letter of Aristeas.
Wisdom Literature and Poetry
Psalm 16 of Solomon. Dead Sea Scroll Psalms. Odes of Solomon. Gospel of Truth and Valentinian Speculation. Gospel of Thomas. Hymn of the Pearl. Manichaean Hymn-Cycles. Coptic Pasalm Book.
Gospels
Gospel of the Hebrews. Gospel of the Ebionites. Secret Gospel of Mark. Apocryphon of James. Gospel of Bartholomew. Gospel of Nicodemus.

Infancy Gospels
of James. of Pseudo Mathew. of Thomas. Latin Infancy Gospel. Arabic Infancy Gospel.

Acts
Acts of the Apostles and John. Acts of John. Acts of Peter. Acts of Paul. Acts of Andrew. Acts of Thomas

Apocalypses
1 Enoch. 2 Enoch. Sibylline Oracles. Apocalypse of Baruch. Apocalypse of Ezra. Ascension of Isaiah. Apocalypse of Peter. Apocalypse of Paul. Apocalypse of Thomas. Christian Sibyllines. Hermes: Poimandres. Hermes: Asclepius. Hermes: Hermaphroditism of God. Thomas the Cntender. Trimorphic Protennoia. The Thunder, Perfect Mind.

Diverse Gnostic Texts
Simon Magus. Valentinus and the Valentinian System of Ptolemaeus. Ptolemaeus' Letter to Flora. Basilides. The Naassene Psalm. Baruch by Justin. Marcion. Carpocrates. The Cainites. The Sethians. The Sethian-Ophites. Ophite Diagrams.

Manichaeans and Mandaean Gnostic Texts
Mani and Manichaeism. Faust Concerning Good and Evil. Augustine's Letters Against the Manichaeans. Evodius Against the Manichaeans. The Kephalaia of the Teacher. Diverse Manichaean Documents. Mandaean Salvation and Ethics.

Mystical Documents
The Divine Throne Chariot. The Zohar. The Mystical Theology of Pseudo-Dionysius.

Hebrew Book of the Dead. Zhenya Senyak.
A guideline for the soul moving through the underworld as it leaves the body either to return or enter the afterlife as described by the wanderings of the Hebrew people in the wilderness of Sinai; in forty four bazaks.

The Seven Golden Chapters of Hermes Trismegistus.
How to make the philosopher's stone.

The Emerald Tablet. Hermes Trismegistus.
As Above, So Below.

The Chemical Wedding of Christian Rosencreutz. trans. Joscelyn Godwin.
The probable autobiography of Chretian DeTroye.

Mayan/Taoist/Egyptian
These are grouped together for no better reason than that I have so few of them. The Mayan and the Taoist sections kind of run together, and the Egyptian section is almost entirely only one author.

Popul Vuh. Dennis Tedlock. trans.

The Mayan book of the dawn of life and the glories of Gods and kings.

Secrets of the Ancient Incas. Langevin.
　　The natives of Peru, where the Nazca lines and Ica skulls were found.

The Cherokee Sacred Calendar. Raven Hail.
　　Includes ephemeris.

Earth Ascending.
and
Time and the Technosphere.
and
The Mayan Factor. Jose Arguelles.
　　A truly unique, original and innovative concept: the Mayan Loom, similar to the galactic circuit of the Enochian Communications System. It is based on a number system, but integrates with the evolutionary stages of the brain, as given in the MPDR as Oc, Chiccan and Men.

The Mayan Prophecies. Adrian G. Gilbert and Maurice M. Cotterell.
　　How the Mayan calendar, by measuring the sidereal rotations of Venus relative to the phases of the moon, also measures the eleven year sunspot cycle.

An Introduction to the Study of Mayan Hieroglyphics. Morley.
　　The Mayan days of the month and the Mayan calendrical numbering system.

the Mayan Calendar. Calleman.
　　Unravelling novelty as the factors of evolution relative to galactic position.

Secrets of Mayan Science/Religion. Hunbatz Men.
　　Descriptions of shamanic rites and rituals by a descendent of the tribe.

the Maya. Edgar Cayce.
　　Predictions about archaeological and anthropological finds by the sleeping prophet.

Invisible Landscape. Terrence and Dennis McKenna.
　　A psilocybin trip in Mexico reveals a hidden numerical system in the I Ching lunar calendar synchronized to the Mayan sunspot cycle calendar, counting down what the gonzo McKenna boys call "novelty" — the first cause for all trivia.

I Ching. Sam Reifler.
　　The sixty four hexagrams with commentary by the three Chinese sages Ariha, Kama and Moksha.

Psychadelic Prayers and other meditations. Timothy Leary.

The Tao Te Ching as modernized by Dr. Timothy Leary to refer to psychoanalysis.

Tao Te Ching. Aleister Crowley.
The Tao Te Ching as modernized somewhat earlier in the century by Crowley.

Chronicle of the Pharaohs. Clayton.
Contains the names and dates of all the dynastic era rulers of ancient Egypt.

The Secret History of Ancient Egypt. Herbie Brennan.

The Great Pyramid. Tom Valentine.
This is that book that started the religious inches = years to the second coming movement among new age Christian eschatologists. Short and thin.

The Great Pyramid. Piazzi Smythe.
A detailed book containing intricate measures of every detail. Long and thick.

The Ancient Kingdoms of the Nile. Fairservis.
A Mentor book from 1962. Shows how rapidly the fringe speculations change and how slow and dodgey the establishment is to make corrections.

The Truth About Egyptian Magick. Gerald and Betty Schueler.
Egyptian and occult equivalency diagrams.

The Seven Souls in the Mysteries of the Ancients. Gerald Massey.
A pamphlet concerning the seven baw of Re.

Egyptian Religion.
and
Egyptian Magic.
and
the Egyptian Book of the Dead.
and
the Egyptian Heaven and Hell.
and
the Gods of the Egyptians.
and
Legends of the Egyptian Gods.
and
Osiris. Budge.
Sir E. A. Wallis Budge is, and has been for almost the past fifty years, the foremost western authority on the antiquities of the Egyptians. Most of these books include reprints from texts in their original hieroglyphic with the phonetic as well as literary translations running along underneath like clef notes.

Magic

    The Great Work of the magician is simple: One must learn first to harness and to calm the lesser aspects of the self, to bring the base instinctual into line with a hgiher purpose. Following this the sense of self itself must be abolished, leaving only the purpose. In this crucible the Gold of the Heart will be transubstantiated into the Philosopher's Stone. When this is imbibed as the Elixer of Life from the Grail Chalice, the magician will have accomplished the goal of the Great Work, an eternal life of egolessness.

Egyptian Mysteries. Lucie Lamy.

    From which I learned the correpsondence between the ba and ruach and akh and nefesh.

Books of the Dead. Stanislav Grof.

    A post-modern analysis of ancient books read to the dying as guidebooks through hallucinogenous psychotropic states of altered consciousness. A good book to have around in case thereof.

Astrology. Warren Kenton.

    Reviews of the psychological traits commonly attributed to each sign. Does not go very far in depth regarding the rising sign as opposed to the birth sign, the roles of the planets ascending or descending, the decanates, or precession.

Sacred Geometry. Robert Lawlor.

    Geometry of sacred religious art and architecture from Egypt and India to Rome and England, as well as patterns in nature and certain wave forms, with full color diagrams.

Alchemy. Stanislas Klossowski de Rola.

    Large glossy prints of many of the symbolism rich paintings and plates done on this mercurial science in the middle ages. Tends to take the more psychological approach than that of hardline mettalurgy.

Freemasonry. W. Kirk MacNulty.

    Beautiful tyling boards that reveal none of the finer mysteries of the craft.

Beyond Death. Stanislav and Christina Grof.

    A book about historical concepts of the afterlife, including eastern and western heaven and hell and reincarnation.

Kabbalah. Zev Ben Shimon Halevi.

    Lots of diagrams, very little explanation.

the Magus. William Barrett.

An essential work for studying the shemhamforash. Not without its inaccuracies.

Isis Unveiled (in two volumes).
and
an Abridgement of the Secret Doctrine. Madame Helena P. Blavatsky.

The primary works of Theosophy. Madame Blavatsky claimed to have studied under a school she called the "Secret Chiefs" while travelling in the far east, and brought back their teachings on yoga, tantra and kundalini, as well as reincarnation to mix together with existing western mysticism, creating new ideas such as "the Akashic Records." If the Golden Dawn had not fallen all over themselves to claim contact with these same "Secret Chiefs," it is unlikely either that they would have adopted the unsubstantiable new notion of the "Akashic records" either.

Knowledge of the Higher Worlds and its Attainment. Rudolf Steiner.

Steiner was one of those who contacted the Golden Dawn after Madame Sprengel crossed the veil of the abyss forever, and he became actively involved in what he called "anthropism" or the scientific study of humanity. While this seems to have made little impact on the philsophies of occult orders of the day, his participation in the push to create a global occult bund is still making waves to this day, at least among the enlightened and hopeful.

The Book of Secrets. Osho.

A book of eastern yogic practices.

The Woman's Encyclopedia of Myths and Secrets. Walker.

An invaluable tool for referencing almost any mythological character or symbol known to wo/man.

Standard Dictionary of Folklore, Mythology and Legend. Funk and Wagnels.

A solid reference tool.

The Magician's Companion. Whitcomb.

The modern day equivalent of William Barrett's the Magus, this work is essential for teaching the tools of the craft as well as answering some of the more difficult preliminary questions of speculative philosophy. I found it helpful with the shemhamforash.

Necronomicon Spellbook. ed. Simon.

These entities were designed for use in combat, but can also answer many questions regarding the nature of the universe.

Goetic Evokation. Steve Savedow.

A truly indespnsible book. The only source I've found for the tree of death and order of qliphotic princes besides 777, and the only source for the demonic dekans of the zodiac.

The Goetia. The Lesser Key of Solomon the King. Mathers/Crowley.
The illustrations notwithstanding, this work is a cornerstone of understanding the precession of the equinoxes.

The Key of Solomon the King. Mathers.
The spirits ruling over the planets and their signs and sigils.

The Book of the Sacred Magic of Abramelin the Mage. Mathers.
Magic letter squares whose mystery is unlocked by gematria.

The Grimoire of Armadel. Mathers.
A much less known work, there is much enlightenment herein awaiting.

Summoning Spirits. Konstantinos.
A helpful book for making talismans, beforming the aforesaid art, with pictures rendering the appearances these conjurations might take on.

A Dictionary of Angels. Davidson.
A very thourough reference book listing almost every angel ever known to man by name and describing their history in brief. Particularly of interest is a list of fallen angels that does not not tally either with the 200 angels of Enoch nor the 1/3 of the heavenly host accorded by the Catholic Church in their interpretation of Revelation.

The Archangel Michael. Rudolf Steiner.
An expostualtion on the coming to reign of the seventh regent and the end of the kali yuga.

On the Akashic Records. Edgar Cayce.
The Sleeping Prophet on from where his own visions come.

The Enochian Magick of Dr. John Dee. James.
The four watchtowers, thirty ayres, seven princes, et al. of the Enochian system devised by magician John Dee.

An Advanced Guide to Enochian Magick. Schueler.
The magickal ceremonies associated with working the Enochian system.

Enochian Sex Magick. Crowley, DuQuette, Hyatt.
Copulation while in God form; the core point of producing a moon child.

The Projection of the Astral Body. Muldoon and Carrington.

The full and conscious activation of the atman and its smudging out of the body.

The Astrology of Fate. Greene.
Teaches more about rising signs and houses.

On the Akashic Records. Edgar Cayce.
The Sleeping Prophet on from whence his visions derived.

Remote Viewing Secrets. McMoneagle.
Most interesting for a correlation between accurate remote viewing "hits" and a certain time of day (i.e. position of the earth).

Skrying for Beginners. Tyson.
Where I learned about Nephotes.

The Pictoral Key to the Tarot. A.E. Waite.
Diagrams and descriptions. Not the standard revised Rider-Waite pack though, but the original Golden Dawn deck. The differences are subtle, especially in black and white.

Magical Alphabets. Pennick.
Runes, Futhark, Linear A and Linear B, Masonic Cipher, etc.

The Complete Magic Primer. David Conway.
Instructs in the right proper creation of ritual space, mood, atmosphere and finally practise of consecration.

Zolar's Encyclopedia of Ancient and Forbidden Knowledge.
A crash course in the occult and mystical.

The Dictionary of the Esoteric. Neville Drury.
Very modern and recent. Good for looking up subjects that have only lately come to light, such as characters from the apocrypha, etc.

Amulets and Talismans. Budge.
The complete book of hexes, charms, fascinations and how to ward them off.

A Beginner's Guide to Constructing the Universe. Scheider.
Mathematically based number systems from one through ten.

The Black Arts. Richard Cavendish.

The Book of Black Magic. A.E. Waite.

A collection of sigils by which to summon "countless" numbers of underworldly sevrnats to do as to which they are bidden.

The Magic Circle. Rev. Yaj Nomolos.
A small pamphlet about organizing a group of magicians for the purpose of performing ritual ceremonies.

The Magician: His Training and Work. W. E. Butler.
A short course in the essentials of ceremonial practise.

The Great Secret. Maurice Meterlinck.
On the occult.

The Great Secret. Eliphas Levi.
On chaining the Great Beast. Contemporary magical philosophy to Freudian Id/Ego/Superego dialectics.

The Complete Book of Spells, Ceremonies and Magic. Gonzalez-Wippler.
A very helpful book with thin pages and small print, describing the bridge in ceremony between the pagan, green witchcraft, wicca and the black/white dualism of traditional sorcery. Not to be engaged in by the inexperienced aspirant.

Modern Ritual Magic. Francis King.
A very helpful, first hand history of the occult trends throughout the various different orders and splinter organizations of the last century.

A History of Secret Societies. Arkon Daraul.
The introductory work for most young initiates seeking entry into the higher mysteries shrouded in the veil of arcane mystery schools.

Keepers of the Secrets. Siblerud.
An updated version of Daraul's Secret Societies that focuses more on the positive, public personae of the organizations it discovers.

Initiation. Elizabeth Haich.
A treatise on the sacred art from a feminist's point of view.

Freemasonry of the Ancient Egyptians.
and
the Lost Keys of Freemasonry. Manly Palmer Hall.
Two excellent treatises on the philosophical origins of the high craft by a thirty third degree member.

Duncan's Ritual and Monitor.
The Yorke Rite through the Royal Arch.

A Dictionary of Freemasonry. Macoy.
  A comprehensive and affordable Masonic encyclopedia.

Freemasonry and Its Ancient Mystic Rites. Leadbeater.
  A sophisticated examination of the different schools of Masonry and to what each traces the Order's origins.

Mystic Masonry. J.D. Buck.
  A different working ritual than the Yorke Rite with which I am unfamiliar.

Mahabone. Anonymous.

The Golden Dawn. Regardie.
  The Outer Order Rituals as they were when Regardie was a member.

Ritual Magic of the Golden Dawn. Francis King.
  Lots of flying roles. I don't remember what else.

The Secret Inner Order Rituals of the Golden Dawn. Regardie, Zalewski.
  The inner Order rituals which were not changed due to Regardie's having printed the outer Order rituals.

Self-Initiation Into the Golden Dawn Tradition. Cicero.
  The modern Golden Dawn's enlistment brouchure.

The Equinox Volume III. Number 10. Crowley, et al.
  In which the right proper organisation and structure of the OTO is described.

The True and Invisible Rosicrucian Order. Paul Foster Case.
  Primarily one of the all time classic works on Tarot, this new and expanded edition also includes a history section on the grave of our beloved and departed frater C.R.

The Confessions. Aleister Crowley.
  His autohagiogrphy.

Aleister Crowley and the Hidden God. Grant.
  A rather unflattering portrayal of Crowley in his Great Beast persona.

The Eye in the Triangle. Regardie.
  A biography of Crowley by his secretary and friend.

The Illustrated Beast: the Aleister Crowley scrapbook. Robertson.
  Some amusing excerpts from the yellow journalism and hot button trials of the day.

The Magick of Thelema. Don Luis DuQuette.
　　The simple rituals of the pentagram, invoking evoking, etc.

Magick in Theory and Practise.
and
Magick Without Tears.
and
Book Four.
and
The Book of Thoth.
and
Little Essays Toward Truth.
and
Liber Aleph vel CXI.
and
The Holy Books of Thelema.
and
The Equinox of the Gods.
and
The Heart of the Master.
and
Eight Lectures on Yoga.
and
The Book of Lies.
and
The Book of the Law. Aleister Crowley.

the QBLH

　　The running and returning that typified QBLHistic experience is the oneness with the flow of a higher dimensional energy. This energy is active and passive; when it is active, we must learn to be passive relative to it; when it is passive we must learn to be active relative to it. In the west the serpent of the teli is associated with tsimtsum, contraction; the descent from the heights.

The Kybalion. Three Initiates.
　　This is by far one of the best works on the phsyics of QBLH ever conceived, dealing with the seven hermetic principles.

The Hermetica. Freke and Gandy.
　　the interperated saying of Hermes the Thrice Greatest in 20 chapters.

Ancient Future. Wayne B. Chandler.
　　The Hermetic laws as black nationalism.

The Greek Qabalah. Barry.

Gematria of the Greek alphabet.

Jesus Christ: the Number of His Name. Gaunt.
 Gematria with Greek phrases from the new testament.

777 and other Qabalistic Writings. Aleister Crowley.
 An indespensible work of comparative theology, cosmology, mythology and numerology.

The Kabbalah Unveiled. Mathers.
 Translated, with tediously redundant commentary, by Mathers. An essential.

The Tree of Life. Regardie.
 The book that got Israel Regardie into the Golden Dawn.

The Secret Doctrine of the Kabbalah. Leet.
 The tree of life as a physics lattice.

The Dimensions of Paradise. John Michell.
 "the porportions and symbolic numbers of ancient cosmology."

The Sacred Magic of the Qabbalah. Manly P. Hall.
 A short pamphlet overflowing enlightenment.

The Chicken Qabalah. DuQuette.
 By drawing associations between the left, right or middle pillars, the first, second, third or fourth world planes, etc. different lessons can be learned from the sephirot on the tree of life diagram.

Qabalah, Tarot, and the Western Mystery Tradition. Bias.
 Essentially a work of Tarot, this book spills over stylistically to touch on many topics of QBLH and mysticism.

On the Kabbalah and Its Symbolism.
and
Kabbalah. Gershom Scholem.
 Two reknowned works on the QBLH by one of the twentieth century's premiere Jewish mystics.

The Early Kabbalah.
 Traces the history of QBLH through saffhardic Spain.

The Essential Kabbalah. Matt.
 Parables for meditation.

Judaic Mysticism. Davis and Mascetti.
A thourough book about the different applications of QBLH throughout history.

The Truth About Cabala. David Godwin.
A short, but instructive, pocket book.

The Work of the Kabbalist. Z'ev ben Shimon Halevi.
Straightforward talk to those practising the mystical arts. Honestly refreshing.

The Guide for the Perplexed. Moses Maimonides.
A sixteenth century biblical scholar and Jewish mystic instructs.

The Bahir. Aryeh Kaplan. ed.
A first century school of rabbis instructs in the round. Illuminating!

Zohar: annotated and explained. Matt.
Some myths and parables.

Zohar: the book of Splendour. Gershom Scholem.
Some myths and parables.

The Essential Zohar. Rav P. S. Berg.
Some myths and parables.

Ten Luminous Emanations Vol I. Rabbi Yehuda Ashlag.
The actual Zohar. A must read.

Sefer Yetzirah. Aryeh Kaplan. ed.
The traditional QBLHist's Bible, metaphorically speaking.

Buddhist/Hindu
The highest state of attainment is enlightenment, as Leary defined it, the nervous system devoid of all mental conceptual activity. It has also been said that you shall know the truth when you are calm, at rest. In the east the serpent is associated with Kundalini, the rising power that passes through our chakras.

The Rig Veda.

Bhagavad-Gita.

The Upanishads.

The Rubayyat of Omar Khayyam.

The Tibetan Book of the Dead.

Buddha: His Life and His Teaching. Walter Henry Nelson.

The Dhammapada. ed. Ananda Maitreya.

Buddhist scriptures.

The Middle Length and Long Discourses of the Buddha.

Teachings of the Buddha. ed. Jack Kornfield.
    Shambala Pocket Classics.

Teachings of the Compassionate Buddha. ed. E.A. Burtt.

The Threefold Lotus Sutra.

A Popular Dictionary of Buddhism. ed. Christmas Humphreys.

The Four Noble Truths. His Holiness the Dalai Lama.

Metaphysical Meditations. Pranayama Yogananda.

Be As You Are: the Teachings of Sri Ramana Maharshi. ed. David Godman.

Sri Isopnisad. Swami Prabhupada.

Zen and the Birds of Appetite. Thomas Merton.

An Introduction to Zen Buddhism. D.T. Suzuki.
    Foreword by Carl Jung.

Zen Flesh, Zen Bones. Paul Reps. ed.

The Way of Life. Lao Tzu.

Zen Comments on the Mumonkan. Zenkei Shibayama.

Raja Yoga.
and
Karma Yoga and Bhakti Yoga. Swami Vivekananda.

Kundalini: the Arousal of the Inner Energy. Ajit Mookerjee.

Occult Tibet. J.H. Brennan.

Appendix 1:

Adam Kadmon and the kabbalistic worlds:
ayin: "no thing;" the God Time; ylem; Chikhai Bardo; beyond the one singularity
ayin soph: "without end;" the multiverse of singularities

ayin soph aur: limitless light; tachyonic projection; illumination; a singularity

zimzum: "contraction;" the abyss between the ideal supernal and the real aethyreal
(AK)   Jechidah: the true will; the magic theatre of the superego
1)   atziluth: pure will; radiant aspects; conception; divine calling
(AK)   chiah: self-preservation; the will to power; the audience of the id
2)   beriah: intellect; gaseous aspects; manifestation; divine creation
(AK)   nefesh (akh): the spirit; associated with immortality; the ego characters
3)   yetzirah: emotion; liquid aspects; formation
(AK)   ruach (ba): the soul; seven channels (baw); the mediated forms
4)   assiah: action; solidity; divine making
(AK)   neschamah (ka): the double; potential energy, fourteen channels (kaw)
         qliphoth (kha): quantum information vortices

Appendix 2:

the THIRTY AYRES: each 91 Parts of the Earth Imposed by God has its
(ANGELIC KING)
1   LIL: Occodon (ZARZILG), Pascomb (ZINGGEN), Valgars (ALPUDUS)
2   ARN: Doagnis (ZARNAAH), Pacasna (ZIRACAH), Dialioai (ZIRACAH)
3   ZOM: Samapha (ZARZILG), Virooli (ALPUDUS), Andispi (LAVAVOTH)
4   PAZ: Thotanp (LAVAVOTH), Axziarg (LAVAVOTH), Pothnir (ARFAOLG)
5   LIT: Lazdixi (OLPAGED), Nocamal (ALPUDUS), Tiarpax (ZINGGEN)
6   MAZ: Saxtomp (GEBABAL), Vavaamp (ARFAOLG), Zirzird (GEBABAL)
7   DEO: Opmacas (ZARNAAH), Genadol (HONONOL), Aspiaon (ZINGGEN)
8   ZID: Zamfres (GEBABAL), Todnaon (OLPAGED), Pristac (ZARZILG)
9   ZIP: Oddiorg (HONONOL), Cralpir (LAVAVOTH), Doanzin (ZARZILG)

10 ZAX: Lexarph (ZINGGEN), Comanan (ALPUDUS), Tabitom (ZARZILG)
11 ICH: Molpand (LAVAVOTH), Usnarda (ZURCHOL), Ponodol (HONONOL)
12 LOE: Tapamal (ZURCHOL), Gedoons (CADAAMP), Ambriol (ZIRACAH)
13 ZIM: Gecaond (LAVAVOTH), Laparin (OLPAGED), Docepax (ALPUDUS)
14 UTA: Tedoond (GEBABAL), Vivipos (ALPUDUS), Ooanamb (ARFAOLG)
15 OXO: Tahamdo (ZARZILG), Nociabi (LAVAVOTH), Tastoxo (ARFAOLG)
16 LEA: Cucarpt (ZIRACAH), Lauacon (HONONOL), Sochial (ARFAOLG)
17 TAN: Sigmorf (ZIRACAH), Avdropt (OLPAGED), Tocarzi (ZARZILG)
18 ZEN: Nabaomi (GEBABAL), Zafasai (ALPUDUS), Yalpamb (ARFAOLG)
19 POP: Torzoxi (ARFAOLG), Abriond (CADAAMP), Omagrap (ZINGGEN)

20 CHR: Zildron (GEBABAL), Parziba (HONONOL), Totocan (ALPUDUS)
21 ASP: Cirzpa (ARFAOLG), Toantom (CADAAMP), Vixpalg (ZURCHOL)
22 LIN: Ozidaia (ARFAOLG), Paraoan (OLPAGED), Calzirg (ARFAOLG)

Jon Gee

23 TOR: Ronoomb (ZARNAAH), Onizimp (LAVAVOTH), Zaxanin (ZINGGEN)
24 NIA: Orcanir (ZARNAAH), Chialps (LAVAVOTH), Soageel (ZINGGEN)
25 UTI: Mirzind (ZARNAAH), Obvaors (ZIRACAH), Ranglam (ARFAOLG)
26 DES: Pophand (ARFAOLG), Nigrana (CADAAMP), Bazchim (ARFAOLG)
27 ZAA: Saziami (ZIRACAH), Mathula (ZARNAAH), Orpanib (GEBABAL)
28 BAG: Labnixp (LAVAVOTH), Pocisni (ZARZILG), Oxlopar (ZURCHOL)
29 RII: Vastrim (HONONOL), Odraxti (ZARNAAH), Gomziam (ARFAOLG)
30 TEX: Taoagla (ARFAOLG), Gemnimb (ZARNAAH),
            Advorpt (HONONOL), Doxinal (ZURCHOL)

ZARZILG (7), ZINGGEN (7), ALPUDUS (8), ZARNAAH (8), ZIRACAH (4),
LAVAVOTH (10), ARFAOLG (15), OLPAGED (5), GEBABAL (7), HONONOL
(7), ZURCHOL (5), CADAAMP (4)

Element:	Great Secret Holy Names of God	Great Kings	Six Seniors
Water:	MPH ARSL GAIOL	RAAGIDEL	LSRAHPM/SAIINOU/LAUAXRP SLGAIOL/SONAZNT/LIGDISA
Air:	ORO IBAH AOZPI	BATAIVAH	HABIORO/AAOXAIF/HTMORDA AHAOZPI/AVTOTAR/HIPOTGA
Fire:	OIP TEAA PDOCE	EDLPRNAA	AAETPIO/ADREOET/ALNCVOD AAPDOCE/ANODOIN/ARINNAO
Earth:	MOR DIAL HCTGA	ICZHIHAL	LAIDROM/ACZINOR/LZINOPO ALHCTGA/AHMILCV/LIIANSA

Glossary:

active consciousness:
that part of consciousness which is aware of itself as internal; as opposed to passive consciousness, that is that part of consciousness which is aware of the external realm of others and the environment.

antipode:
diametrically opposed; antithesis of conjoined or nested

archetypes:
higher dimensional anthropomorphications
(see also: metaforms)

astral travel:
the releasing of the light body, or atman, in the aethyreal realm

Atlantis:
the western hemisphere mythical Lost Continent; in recent times archaeology has increasingly come to accept that large amounts of known land masses were uncovered by water or ice during the last ice age, and that it is possible more land was known to ancient peoples of that era than has previously been thought

atman (or light body):
the internal electrochemical functioning of the body

aura:
the unique electromagnetic field surrounding an individual

body jumping:
sensing through somebody else's body, or gross vessel

body surfing:
moving the sense perception through a series of gross vessel hosts

Cabala:
the maintenance of tradition, often through secrecy, in sects and denominations
(see other derivations)

closed set:
a geometrically or algebraically finite continuum; the antithesis of open set

coincidence:
an unexpected similarity between two or more probabilistic events
(see also: synchronicity)

dimensionality:
traditionally the measure of quantifiable qualities of mass, such as length, bredth, depth; under the relativity of duration and distance expanded to include the measure of time as a fourth dimension, perpendicular to the all of the preceeding other three; in its fourth form it has spatial as well as temporal; theoretically expandable under such modern theories of physics as superstring and M-brane theory to include an infinite number of dimensions, though it is usually thought that factors would act to conserve the number of dimensions present to the minimum possible number.
(see also: nth dimensionality)

event:
herein quantified geometrically as a cube, such as the mome of chemistry, the basic unit of time; measured as anything from the instantaneous temporal singularity to a duration of any given span; pertainant to all scales of physical size.

fields of potential:
a generally idealized concept describing such configurations of potential (as opposed to kinetic) energy as the electron cloud (as opposed to the electron singularity).

formal system of metaphysics:
a hexagon surrounded by the six fundamental questions of reasoning in the order, clockwise from top: How, When, Where, What, Who, Why.

fractal:
self-embedding, self-replicating pattern
(see also: gnomon)

galactic filaments, walls and voids:
due to the conservation of dimensionality, according to superstring theory and M-brane theory, the additional, "extra" dimensions more than the minimum needed to sustain the universalized laws of physics are coiled up very tightly into gravitational strings, which I also identify with the microwavelength sum over histories of tachyons, and these pull all the stellar matter in the universe together towards them, thus creating filaments, or strands, clusters, or walls, of galaxies, and voids, or empty spaces, in between them.

gnomon:
a self-repeating pattern, such as the part of a parallelogram remaining after a similar, smaller parallelogram has been taken from one of its corners

gravitational singularity:
a singularity with a strong gravitational pull, such as those in the center of black holes

Gondwanaland:
One of the two major precambrian land masses, which would later break apart to become become North and South America, Europe and greater Asia.

holognomon:
a hologram of a gnomon; a hologram is a three dimensional holograph, or single image taken stereoscopically (or with multiple lenses from multiple angles); a gnomon is a self-repeating pattern, such as the part of a parallelogram remaining after a similar, smaller parallelogram has been taken from one of its corners

hypercross:
a fifth dimensional extrapolation of the hypercube, cube, hexagon, point fractal propagation in which six cubes surround all sides of a central seventh

hypercube:
nested, a cube within the center of a cube

hypercubic hypercross:
a hypercross, or cross of six cubes surrounding a central seventh, whose several cubes, themselves, contain a cube nested within each of their centers

hypercubic hypercross of hypercubic hypercrosses:
a hypercross each of whose six cubes as well as its central seventh are apportioned into twenty seven (three cubed) smaller cubes, the center cubes of each face of which as well as the central cube of each is its own hypercross, and the central cube of each a nested hypercube within the greater cubes of the larger hypercross

hyperdimension:
refers to multiple dimensionality greater than the minimized or conserved amount but less than infinite

hypereality:
reality which is all inclusive of the nth dimensional tachyonic multiverse on the other side of timespace and the hyperdimensional metaforms which reside there
hypersphere:
nested, a sphere within the center of a sphere

impossible feedback loops:
self-referential systems that are self-embedded
(see also: closed sets)

instantaneous manifestation:
immediate thought and reality correspondence

irrational numbers:
numbers that cannot fully be calculated
inversion:
the opposite of any given thing immediately juxtposed to it

Kabballah:
the body of God
(see also alternate spellings)

Laurasia:
the precambrian supercontinent that would break apart due to plate tectonics to become Africa, India, Australia and eastern Asia

Lemuria (or Mu):
the eastern hemisphere mythical Lost Continent; counterpart to Atlantis; in recent times Lemuria has come to be associated with the ice age land bridge that joined much of modern day Oceania, Australia and New Zealand to southeast Asia

magick:
the science and art of causing change to occur in conformity with will

manifestation:
the causing of a symbolic synchronicity or direct formal apparition by projection

memory castle:
a theoretical construct using creative visualization often used by professors to help them remember the curricula of their lectures

mental projective space:
the conceptual realm inside which exists the sense perceptual realm

metaform:
fourth spatial dimensional (or higher) geometrical shapes, in particular the fractal and gnomonic series of the Platonic regular solids and Archimedian semi-regular solids

metaphysics:
metaphysics is like looking into a fish pond and seeing the fish that live there, the algae on the water's surface, and the reflection of the sky all at the same time

microwave:
a wavelength so small that it cannot be measured using modern apparati

nodes:

herein referring to the points at the corners of the formal system of metaphysics where the six fundamental questions of reasoning are positioned as upon a lattice

nth dimensionality:
a term applied in physics to denote dimensionality that is greater than infinite

open set:
any geometrically or arithmetically non-finite set; the antithesis of a closed set

phi/pi:
a transcendental, transfinite ratio whose components may be found relative to one another ubiquitously throughout nature on all dimensional levels

possibilities:
less specific than probabilities, an generally idealized form of the fact that, within a closed system such as the known laws of physics, certain events are preconceived as being more likely, while others are thought of as altogether impossible

potential:
usually applied to potential energy (as opposed to kinetic energy), herein used as a generally idealized description of virtual particles and their characteristic ambiguity as defined by the quantum uncertainty principle

precession:
the movement of the earth's pole around its own orbital path, determined by the earth's 23.5 degree angle of inclination from perfect perpendicularity to the gravitational plane of the sun; the cause of the cyling seasons of the Great Ages, such as ice ages alternating with global warming, as well as the apparent movement of the signs of the zodiac over the millennia opposite their apparent movement throughout the year

prediction:
calculation of hypothetical future events using a rigorous system

primary clear light:
seen at the moment of Ego Loss — he who "has the power to die consciously," who can immediately grasp the concept of the empty mind and, at the supreme moment of quitting the ego, can recognize and become one with the consequential ecstasy, "is awakenned into reality." Should the participant be experienced he will reach Chikhai Bardo quickly and seek to maintain or prolong it.

probabilities:
a term most commonly applied in quantum mechanics to refer to the pure chance, given the conditions, of an event occurring

projection:
the forcing of thought from inside the brain outside into mental conceptual space

prophecy:
calculation of hypothetical future events using abstraction

psychic driving:
the repetition of a subliminal message

psylodon:
a reptile mammal hypbrid species who had legs that grew from the sides of its body and a tail that grew straight out from its hips, the ancestor of all mammals

Qabala:
the numerological and grammatical aspects of Jewish esoterica
(see also: alternate spellings)

QBLH:
the mathematical and geometrical aspects of Jewish esoterica
(see also: alternate spellings)

quantum foam:
the sub-atomic continuum populated exclusively by particles and waves which obey the calculations for the uncertainty principle

quantum tunneling:
the apparently anomolous movement of a quantum from ine side of a solid barrier, such as a square well, to the other, seemingly uninterupted

quantum uncertainty:
the formula, calculated by Erwin Schrodinger, stating that it is impossible to measure with conventional instruments the velocity and spin of a quantum particle at the same time; according to the calculations this is because it would require interrupting either one or the other with a photon, at which point it would then be possible to measure whichever was not effected; one interesting side effect of this theory has appeared to be quantum tunneling such as occurs for photons in the double slit experiment, where a single photon acts holographically, splitting to go through both of two slits only one micron wide

remote viewing:
distance projection either through a gross vessel or astrally

self-embedded:

such as a fractal or a gnomon, something that is comprised of a hologram of its whole

self-propagating:
something that is auto-correlated, or self-replicating, as a fractal or a gnomon

self-referential:
referring back to its own contents

simulacrum and simulacra:
simulation and simulations

singularity: the smallest possible point in physics; herein I have differentiated between gravitational singularities (with infinite density and zero volume), and temporal singularities (with infinite dimension and zero mass)

sleeper agents:
brain washed assassins whose memories of training are unconscious

Socratic method:
in short, the method of rhettoric that proposes only answering an interlocuter's questions with questions until the solution is discovered

soul:
the spark of life inherent to all living things; the archetypal aura

spacetime:
the side of the fabric of the continuum on which gravitational distortions occur

sphere of the mind:
the mental projection, or creative visualization, of the mental projective space itself

spirit:
the phi/pi pattern of the torus shaped aura; the archetypal soul

subspace:
the regular concept of space as according to dimensional conservation
(see also: dimensionality)

synchronicity:
a meaningful coincidence implying a meaningful cross-connection
(see also: coincidence)

telechanneling:
the use of artificial technology to effect the psyche, inducing a slight altered state

telesurfing:
the use of artificial technology to channel messages from a series of multiple source

temporal singularity:
a singularity of time, such as a wormhole connecting two distant points in spacetime such that information can flow between them simultaneously

tesseract:
a fourth dimensional hypercube at antipode viewed at a forty five degree angle from above one of its edges

Thoth:
the ancient Egyptian god of Time; also called Hermes by the Greeks

timespace:
the side of the fabric of the continuum on which dimensional distortions occur

torus:
a fourth dimensional nested hypersphere viewed at antipode

transcendental numbers:
numbers whose decimal place numbers continue on infinitely

the ureaus (or **ajna**):
the third eye, associated with the pineal gland and the thalami

wave of potential:
herein used as a generally idealized description of the wave nature of potential energy

wells of probability:
herein used as a generally idealized description of the relative predictability of real particles
(see antithesis: fields of potential)

ylem:
the mixture of primordial matter prior to the big bang

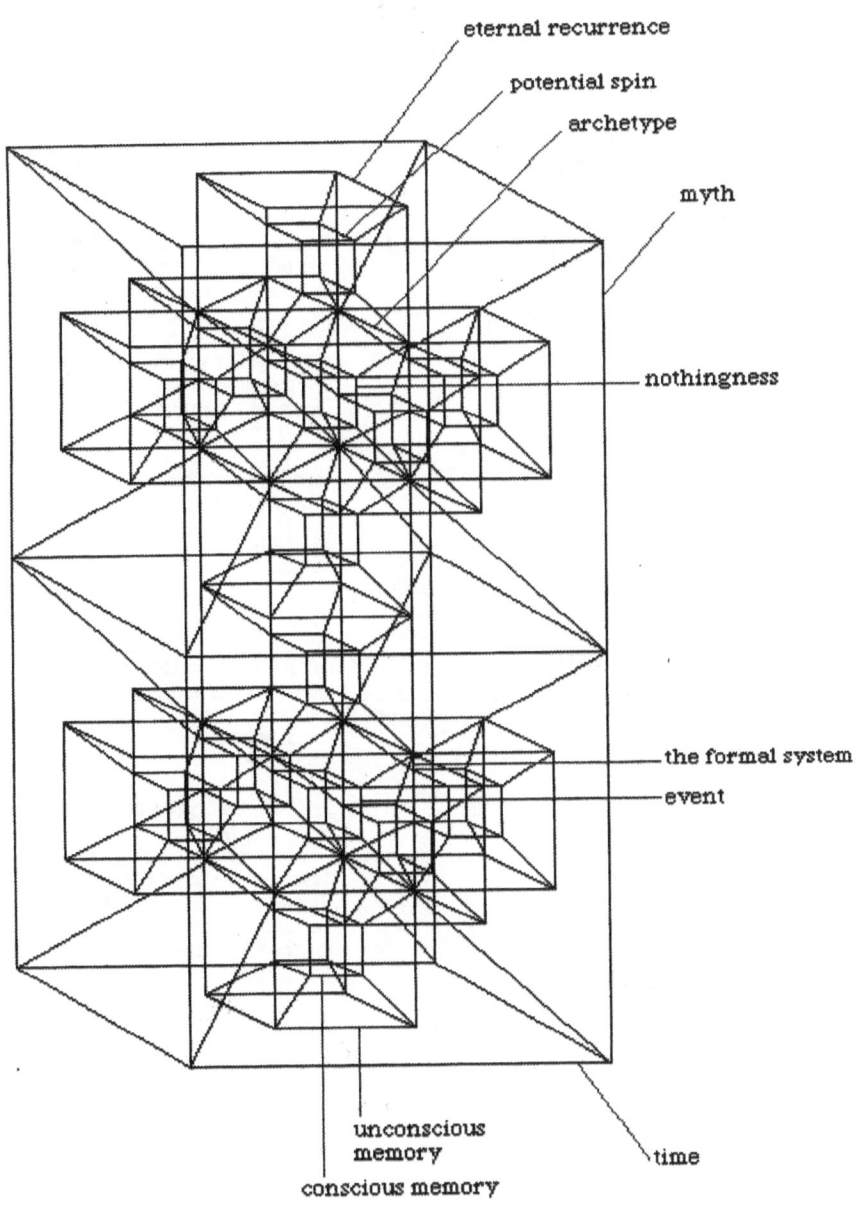

eternal recurrence

potential spin

archetype

myth

nothingness

the formal system

event

unconscious
memory

time

conscious memory

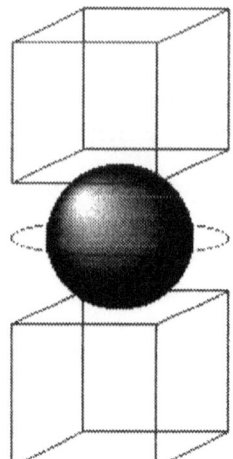

the hypercross and spin

The Hypercross is created by starting with a regular three dimensional cube.

First imagine that the three dimensional cube is in front of you. Then, as you turn it to one side or the other, you see it move into parogee (two cubes of the same size sharing one common wall), until finally, looking at it from the opposite side you were before, you see it at apogee (a cube within a cube) and now you are holding a hypercube.

The Hypercube is a fourth dimensional cube, or, in other words, a geometric expression of the force of time.

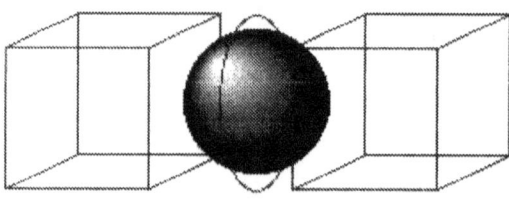

Now turn this hypercube around, so that you see each of the six possible cubes that can be thus expanded from each face of the first cube, until you see all six at parogee.

This is the hypercross. Next take the hypercross apart into three pairs of cubes. Then, instead of a three dimensional cube over time in the center, substitute a sphere, to represent particles.

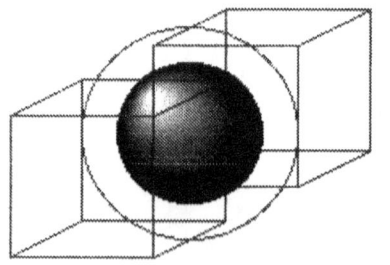

Lastly observe how each of the cubes turns its own way, and these spin the particle.

the wave-mechanical hyper-sphere at parogee

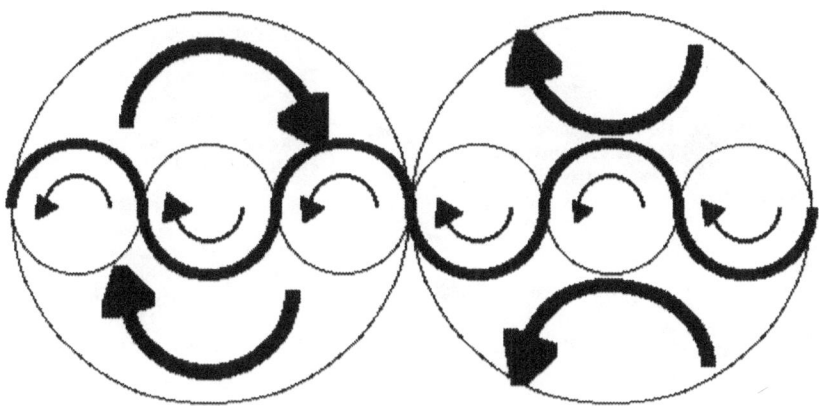

Key:

the big band depicts a wavelength.

the gap between the two large spheres represents the super-position of potential
directional fronts upon the wavelength.

the relationship between these is such that it causes the wavelength to
become a standing wave-form, that is, the measure of a static field.

the smaller spheres represent the lesser versions of the interference patterns
that cause wavelengths to form.

the medium arrows show the motion of particles in a static field.

the large arrows indicate the way peaks and troughs form on the wavelength. These are only two directions however, accounting for two
types of vector, longitudinal and lattitudinal; therefore the third
dimensional vector must come from the shape of a torus.

the apparently three dimensional shapes of the large spheres are really fourth
dimensional probability wells.

Ten Dimensional Model of Active Consciousness

Second Dimension:

Thought. Actually a fractal infinitude surrounding the sphere of true consciousness.

Third Dimension:

The Sphere of the mind or sphere of true consciousness. Comprised of infinite thoughts categorized by focus into conscious, subconscious, and unconscious.

First Dimension:

the stelloctahedron of fact/fiction. simple probability in the form of clear possibility

Fourth Dimension:

the hypersphere of awareness of time. This has scale correspondence to the sphere of the mind.

the cross section of a torus is a sphere. Thus, at its *antipode* point, the torus will appear to be two identical spheres.

Fifth Dimension:

the torus of access to unconscious memory

Ninth Dimension:

Implosion of concentration. Links to the first dimension.

Sixth Dimension:

the externally projected hypersphere of active consciousness. Interacts with probabilities in event to create new possibilities.

Eighth Dimension:

Concentration. Pertains to the Central Nervous System.

Seventh Dimension:

the sphere of mental projective space. the Five Senses. Pertains to the measurement of time in field form.

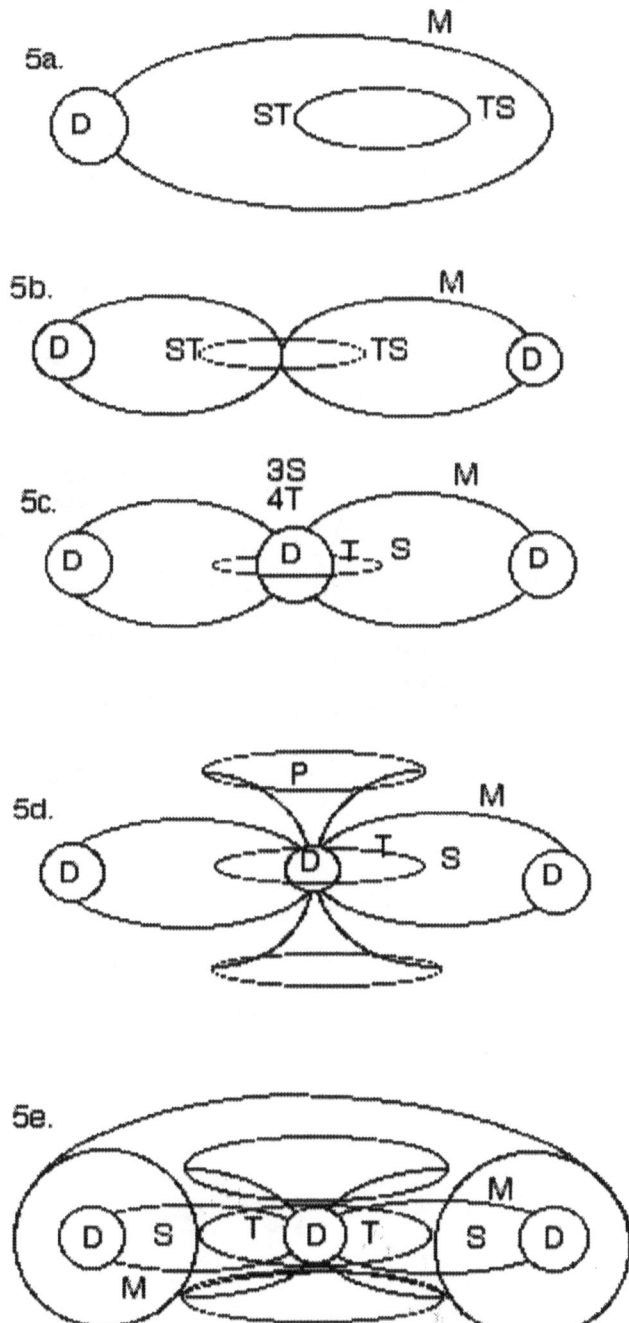

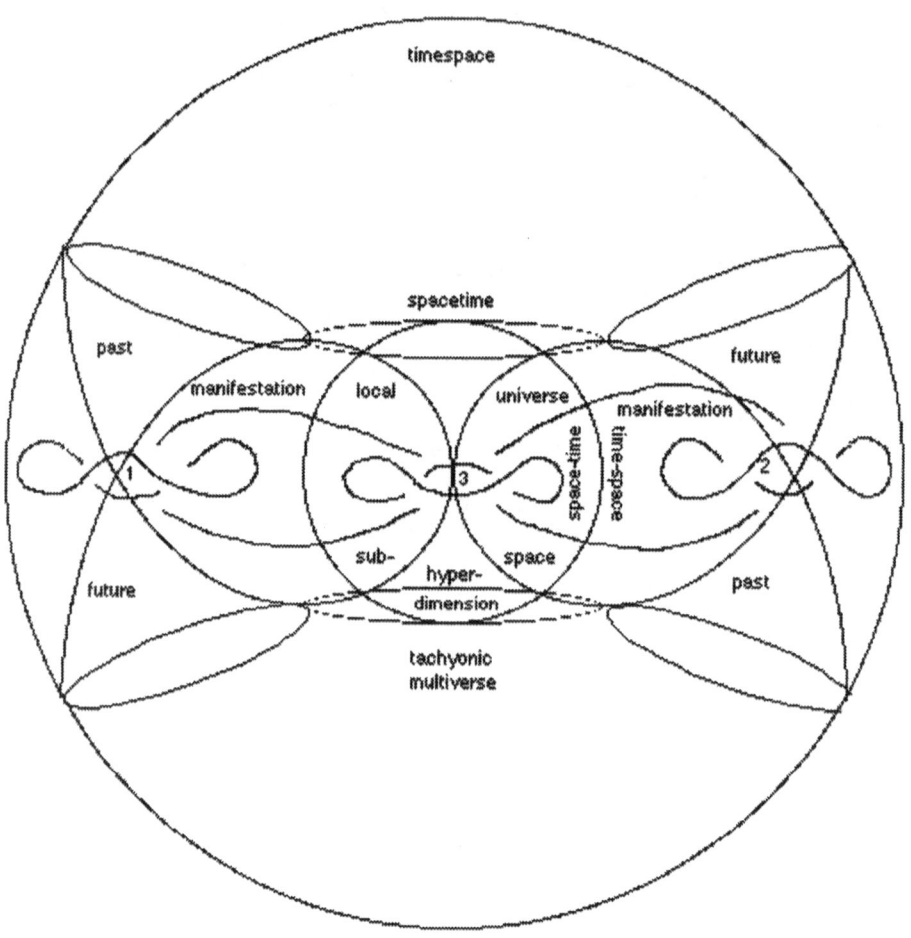